Oats: wild and cultivated

A monograph of the genus *Avena* L. (Poaceae)

Bernard R. Baum

Biosystematics Research Institute,
Canada Department of Agriculture, Research Branch,
Ottawa, Ontario, Canada
Monograph No. 14
1977

©Minister of Supply and Services Canada 1977

Available by mail from

Printing and Publishing
Supply and Services Canada
Ottawa, Canada K1A 0S9

or through your bookseller.

Catalogue No. A54-3/14 Canada: $17.50
ISBN 0-660-00513-1 Other countries: $21.00

Price subject to change without notice.

Thorn Press Limited

Contract No. 10KT.01A05-6-38627

When the problem of form is examined closely one sees that the reason it is so difficult to solve is that it requires a different line of attack from the one used in most other scientific problems. In approaching it the scientist is on relatively unfamiliar ground. The powerful tool he has learned to use so well and with which he feels at home is that of analysis, the breaking down of something into its parts. A study of form, however, requires the use of synthesis, an ability to put these parts together into a formed whole. To analyze an organism into a hierarchy of progressively smaller units, and ultimately chemical ones, does indeed make possible the discovery of a very great deal about it, not only as to its construction but particularly as to the activities that are carried on within it. This in itself, however, does not provide an understanding of what it is that draws these parts and particles into a formed and organized system. Synthesis at any level is much more difficult than analysis.

E. W. Sinnot

The problem of organic form.
Yale University Press, New Haven. 1963. p. 200.

Preface

This monograph is based mainly on the results of the dissection of many *Avena* specimens for the study of micromorphological characters. This tedious and painstaking effort was necessary in order to discover many aspects of the organization and variation in *Avena*. This method also resulted in the detection of discontinuities in that variation.

Recognition of taxa need not be based on conspicuous characters merely because the plants themselves are conspicuous. In bryophytes and fungi, accurate results may only be obtained by the use of micromorphological characters. In angiosperms, the degree of accuracy in identification and classification can be increased by using these kinds of characters. In some cases, this is the only way to reveal the "true" state of nature.

Although the level of organization of angiosperms is different from that in other classes of plants, some parts of vascular plants are analogous with those of bryophytes and fungi, and are somewhat similar in the degree of magnification. For example, for taxonomic purposes, the size and shape of "leaf" cells in bryophytes are useful, whereas the size and shape of guard cells or epidermic cells in angiosperms are also useful. Sculpture, size, and shape of spores in fungi are important, whereas the same traits in pollen grains of angiosperms are equally important.

Micromorphological characters formed the basis of the delimitation of *Avena* species and of important identification keys to species and hybrids. These characters are rather difficult to use, but for *Avena* they are useful and accurate. In order to master them, you must dissect properly and observe with patience.

Acknowledgments

A number of people besides myself carried out dissections of *Avena* plants; among these, I wish to express my gratitude to Mrs. J. N. Findlay (née Horton) and Mrs. V. E. Hartman (née Hadland) for their assistance in this and other aspects of the work. I received continuing encouragement from Dr. T. Rajhathy, Head of the Cytogenetics Section, Ottawa Research Station, of this Department, while he was working on *Avena* cytogenetics; we went on collecting trips and shared a common interest in *Avena* — from different aspects perhaps but for the same goal — that fostered generous cooperation. The exchange of material and ideas influenced the course of my work for the better. I thank my colleague Dr. Rajhathy. I also thank Mr. E. Brandon of Data Processing Division, Financial and Administration Branch, in this Department, for help in computer mapping, and Mr. Thomas Wray of the Department of Energy, Mines and Resources for providing the various sources of formulas for computer mapping.

I express my gratitude to Dr. J. W. Morrison, Research Coordinator (Cereal Crops), Research Branch, of this Department, for his interest, encouragement, and fruitful discussions during the research phase; and to the Research Program Service for editorial work.

The following material was reproduced by permission of the National Research Council of Canada, from articles written by me:

Photographs from Plates I and II from the article entitled, "The role of the lodicule and epiblast in determining natural hybrids of *Avena sativa* × *fatua* in cultivated oats." Can. J. Bot. 47:85-91. 1969.

Figures 1 and 2 (Plate I) and Figure 9 (Plate III) from the article entitled, "The use of lodicule type in assessing the origin of *Avena* fatuoids." Can. J. Bot. 47:931-944. 1969.

Figures 1 and 2 (Plate I) and Figure 7 (Plate II) from the article entitled, "*Avena occidentalis,* a hitherto overlooked species of oats." Can. J. Bot. 49:1055-1057. 1971.

Figures 6 and 7 (Plate I) from the article entitled, "Taxonomic studies in *Avena abyssinica* and *A. vaviloviana,* and some related species." Can. J. Bot. 49:2227-2232. 1971.

Figures 9 and 10 (Plate II) from the article entitled, *"Avena septentrionalis,* and the semispecies concept." Can. J. Bot. 50:2063-2066. 1972.

Table 1 (page 2243), Table 3 (page 2246), Table 4 (page 2247), Table 5 (page 2251), Table 7 (pages 2252-2253), Table 9 (page 2255), and Table 10 (page 2258); Figure 1 (page 2249), Figure 2 (page 2250), Figure 3 (page 2256), Figures 4 and 5 (page 2257) and Figures 6 and 7 (page 2258) from the article entitled, "Classification of the oat species (*Avena,* Poaceae) using various taximetric methods and an information-theoretic model." Can. J. Bot. 52:2241-2262. 1974.

Table 1 (page 2117), Figures 3 to 7 (page 2124), and Figures 9 to 11 (page 2125) from the article entitled, "Cladistic analysis of the diploid and hexaploid oats (*Avena,* Poaceae) using numerical techniques." Can. J. Bot. 53:2115-2127. 1975.

Contents

Part 1. General

1. Introduction ... 1
2. Background ... 2
   - Etymology ... 2
   - Archeology ... 4
   - Anatomy and histology ... 6
   - Palynology ... 8
   - Cytogenetics ... 9
   - Polymorphism and population biology ... 10
   - Literature survey ... 10
3. Materials and methods ... 11
4. Species concept with respect to taxonomy of cultivated plants, with emphasis on *Avena* taxonomy ... 20
5. The generic limits of the genus *Avena* ... 23
6. Comments on previous systems of classification of *Avena* ... 26
7. The classification of *Avena* species in this work: an account of how the classes were obtained and a description of the system ... 28
8. Speciation and evolutionary inferences ... 56
9. The epicuticular waxes on the glumes of *Avena* examined under the scanning electron microscope ... 76
10. Cultivars ... 83
11. Fatuoids ... 86
12. Characters for discrimination and identification of oat species and for their classification into groupings within the genus ... 88
13. Identification keys ... 96
    - (1) Unweighted key, i.e., all characters given equal weight ... 108
    - (2) Weighted key with the most important diagnostic characters only ... 111
    - (3) Weighted key excluding some characters and excluding chromosome numbers ... 114
    - (4) Weighted key with characters 10–25 only ... 116
    - (5) Weighted key according to results from "CHARANAL" ... 119
    - (6) Weighted key for cytologists and breeders ... 122
    - (7) Weighted key with a few obvious characters for quick identification ... 123

Part 2. Systematic treatment of species

14. Introduction ... 127
15. Description of genus ... 129
16. Section *Avenotrichon* ... 132
    - (1) *A. macrostachya* ... 132
17. Section *Ventricosa* ... 137
    - (2) *A. clauda* ... 137
    - (3) *A. eriantha* ... 145
    - (4) *A. ventricosa* ... 153
    - (5) *A. clauda* × *eriantha* F_1 hybrid ... 159
18. Section *Agraria* ... 161
    - (6) *A. brevis* ... 163
    - (7) *A. hispanica* ... 170
    - (8) *A. nuda* ... 178
    - (9) *A. strigosa* ... 185
19. Section *Tenuicarpa* ... 194
    - (10) *A. barbata* ... 195

	(11) *A. canariensis*	205
	(12) *A. damascena*	210
	(13) *A. hirtula*	215
	(14) *A. longiglumis*	221
	(15) *A. lusitanica*	227
	(16) *A. matritensis*	233
	(17) *A. wiestii*	239
	(18) *A. lusitanica* × *longiglumis* F$_1$ hybrid	245
20.	Section *Ethiopica*	246
	(19) *A. abyssinica*	248
	(20) *A. vaviloviana*	255
	(21) *A. abyssinica* × *vaviloviana* F$_1$ hybrid	261
21.	Section *Pachycarpa*	265
	(22) *A. maroccana*	266
	(23) *A. murphyi*	271
22.	Section *Avena*	274
	(24) *A. atherantha*	276
	(25) *A. fatua*	280
	(26) *A. hybrida*	291
	(27) *A. occidentalis*	300
	(28) *A. sativa*	306
	(29) *A. sterilis*	334
	(30) *A. trichophylla*	350
	(31) *A. sativa* fatuoid	356
	(32) *A. sativa* heterozygous fatuoid	362
	(33) *A. sativa* × *fatua* F$_1$ hybrid	364
	(34) *A. sativa* × *sterilis* F$_1$ hybrid	368
23.	Non Satis Notae	374
24.	Names wrongly cited	376

Part 3. Digest

Digest	379
Literature cited	456

List of figures

Figs. 1–3.	Overgrazing	15, 16
Figs. 4–5.	Admissible phenograms	40, 41
Fig. 6.	Relative position of the 21 admissible classifications on the coordinates of the first 3 principal axes	49
Fig. 7.	Plot of the 28 OTU's on the first and second canonical axes	51
Fig. 8.	Plot of the 28 OTU's on the first and third canonical axes	51
Fig. 9.	Overlapping clusters obtained from characters 9, 13, 15, 16, 18, 20, 24–29	52
Fig. 10.	Overlapping clusters obtained from all 29 characters	53
Fig. 11.	Output of procladogram of diploid *Avena* species	67
Fig. 12.	Output of final cladogram of the diploid *Avena* species	68
Figs. 13–17.	Representative trees of diploid *Avena* species	69
Fig. 18.	Minimum spanning tree computed from the similarity matrix of the diploid OTU's and suggested ancestors combined	70
Figs. 19–21.	Representative trees of hexaploid *Avena* species	71
Fig. 22.	Minimum spanning tree computed from the similarity matrix of the hexaploid OTU's and the HTU's selected as ancestors, all combined	71
Fig. 23.	Evolutionary scheme of speciation in *Avena*	73
Figs. 24–29.	Patterns of epicuticular wax crystals	78
Fig. 30.	Cytological processes in the generation of fatuoids	87
Fig. 31.	Two kinds of juvenile growth in *Avena*	89
Figs. 32–35.	Two basic shapes of the ligules in *Avena*	90

Figs. 36–43.	Configurations of the sativa-type lodicules	95
Fig. 44.	World distribution of the genus *Avena*	131
Figs. 45–50.	Morphological and micromorphological diagnostic details of *A. macrostachya*	133
Fig. 51.	Distribution map of *A. macrostachya*	135
Figs. 52–53.	Two sites of *A. macrostachya*	136
Fig. 54.	Holotype of *A. macrostachya*	138
Fig. 55.	Distribution map of section *Ventricosa*	139
Figs. 56–62.	Morphological and micromorphological details of *A. clauda*	141
Fig. 63.	Distribution map of *A. clauda*	142
Figs. 64–66.	Three sites of *A. clauda*	143, 144
Fig. 67.	Lectotype of *A. clauda*	146
Figs. 68–72.	Morphological and micromorphological diagnostic details of *A. eriantha*	148
Fig. 73.	Distribution map of *A. eriantha*	149
Fig. 74.	A site of *A. eriantha*	150
Fig. 75.	Lectotype of *A. eriantha*	152
Figs. 76–80.	Morphological and micromorphological diagnostic details of *A. ventricosa*	154
Fig. 81.	Distribution map of *A. ventricosa*	155
Fig. 82.	A site of *A. ventricosa*	156
Fig. 83.	Isolectotype of *A. ventricosa*	158
Fig. 84.	Morphological diagnostic details of *A. clauda* × *eriantha* F_1 hybrid	159
Fig. 85.	Distribution map of *A. clauda* × *eriantha* F_1 hybrids	160
Fig. 86.	Distribution map of section *Agraria*	162
Figs. 87–96.	Morphological and micromorphological diagnostic details of *A. brevis*	165, 166
Fig. 97.	Distribution map of *A. brevis*	166
Fig. 98.	Isotype of *A. brevis*	169
Figs. 99–107.	Morphological and micromorphological diagnostic details of *A. hispanica*	172, 173
Fig. 108.	Distribution map of *A. hispanica*	175
Figs. 109–116.	Morphological and micromorphological diagnostic details of *A. nuda*	180
Fig. 117.	Distribution map of *A. nuda*	181
Fig. 118.	Authentic specimen of *A. nuda*	183
Figs. 119–129.	Morphological and micromorphological diagnostic details of *A. strigosa*	188
Fig. 130.	Distribution map of *A. strigosa*	190
Fig. 131.	Holotype of *A. strigosa*	192
Fig. 132.	Distribution map of section *Tenuicarpa*	195
Figs. 133–139.	Morphological and micromorphological diagnostic detail of *A. barbata*	198
Fig. 140.	Distribution map of *A. barbata*	199
Fig. 141.	Lectotype of *A. barbata*	202
Figs. 142–146.	Morphological and micromorphological diagnostic details of *A. canariensis*	206
Fig. 147.	Distribution map of *A. canariensis*	207
Fig. 148.	A site of *A. canariensis*, *A. occidentalis*, and *A. barbata*	208
Fig. 149.	Holotype of *A. canariensis*	209
Figs. 150–154.	Morphological and micromorphological diagnostic details of *A. damascena*	211
Fig. 155.	Distribution map of *A. damascena*	212
Fig. 156.	Holotype of *A. damascena*	214
Figs. 157–163.	Morphological and micromorphological diagnostic details of *A. hirtula*	216
Fig. 164.	Distribution map of *A. hirtula*	217
Fig. 165.	Lectotype of *A. hirtula*	220
Figs. 166–172.	Morphological and micromorphological diagnostic details of *A. longiglumis*	222
Fig. 173.	Distribution map of *A. longiglumis*	223
Fig. 174.	A site of *A. longiglumis*	224
Fig. 175.	Lectotype of *A. longiglumis*	226

Figs. 176–181.	Some morphological and micromorphological diagnostic details of	
	A. lusitanica	229
Fig. 182.	Distribution map of *A. lusitanica*	230
Fig. 183.	Lectotype of *A. lusitanica*	232
Figs. 184–189.	Morphological and micromorphological diagnostic details of	
	A. matritensis	234
Fig. 190.	Distribution map of *A. matritensis*	235
Fig. 191.	Holotype of *A. matritensis*	238
Figs. 192–200.	Morphological and micromorphological diagnostic details of	
	A. wiestii	240
Fig. 201.	Distribution map of *A. wiestii*	242
Figs. 202–203.	Two sites of *A. wiestii*	243
Fig. 204.	Holotype of *A. wiestii*	245
Fig. 205.	Morphological details of *A. lusitanica* × *longiglumis* F_1 hybrid	246
Fig. 206.	Distribution map of section *Ethiopica*	247
Figs. 207–211.	Morphological and micromorphological diagnostic details of	
	A. abyssinica	249
Fig. 212.	Distribution map of *A. abyssinica*	251
Fig. 213.	A site of *A. abyssinica*	252
Fig. 214.	A site of *A. sterilis*, *A. abyssinica*, and *A. vaviloviana*	252
Fig. 215.	Isotype of *A. abyssinica*	254
Figs. 216–222.	Morphological and micromorphological diagnostic details of	
	A. vaviloviana	256
Fig. 223.	Distribution map of *A. vaviloviana*	258
Figs. 224–225.	Two sites of *A. vaviloviana*	259
Fig. 226.	Lectotype of *A. vaviloviana*	260
Figs. 227–230.	Morphological and micromorphological diagnostic details of	
	A. abyssinica × *vaviloviana* F_1 hybrids	262
Fig. 231.	Distribution map of *A. abyssinica* × *vaviloviana* F_1 hybrids	264
Fig. 232.	Distribution map of section *Pachycarpa*	265
Figs. 233–238.	Morphological and micromorphological diagnostic details of	
	A. maroccana	267
Fig. 239.	Distribution map of *A. maroccana*	268
Fig. 240.	A site of *A. maroccana*	269
Fig. 241.	Holotype of *A. maroccana*	270
Figs. 242–247.	Morphological and micromorphological diagnostic details of	
	A. murphyi	272
Fig. 248.	Distribution map of *A. murphyi*	273
Fig. 249.	Distribution map of section *Avena*	275
Figs. 250–254.	Morphological and micromorphological diagnostic details of	
	A. atherantha	277
Fig. 255.	Distribution map of *A. atherantha*	278
Figs. 256–261.	Morphological and micromorphological diagnostic details of	
	A. fatua	283
Fig. 262.	Distribution map of *A. fatua*	284
Fig. 263.	Lectotype of *A. fatua*	288
Figs. 264–273.	Morphological and micromorphological diagnostic details of	
	A. hybrida	294, 295
Fig. 274.	Distribution map of *A. hybrida*	296
Fig. 275.	Lectotype of *A. hybrida*	298
Figs. 276–281.	Morphological and micromorphological diagnostic details of	
	A. occidentalis	302
Fig. 282.	Distribution map of *A. occidentalis*	303
Fig. 283.	Lectotype of *A. occidentalis*	305
Figs. 284–292.	Morphological and micromorphological diagnostic details of	
	A. sativa	316
Fig. 293.	Distribution map of *A. sativa*	317
Fig. 294.	Lectotype of *A. sativa*	321
Figs. 295–301.	Morphological and micromorphological diagnostic details of	
	A. sterilis	338
Fig. 302.	Distribution map of *A. sterilis*	339
Figs. 303–308.	Six sites of *A. sterilis*	340–342
Fig. 309.	Lectotype of *A. sterilis*	345

Figs. 310–315.	Morphological and micromorphological diagnostic details of *A. trichophylla*	352
Fig. 316.	Distribution map of *A. trichophylla*	353
Fig. 317.	Neotype of *A. trichophylla*	355
Figs. 318–325.	Morphological and micromorphological diagnostic details of *A. sativa* fatuoid	358
Fig. 326.	Distribution map of *A. sativa* fatuoids	359
Fig. 327.	Spikelet of *A. sativa* heterozygous fatuoid	363
Fig. 328.	Spikelet of *A. sativa* × *fatua* F_1 hybrid	365
Fig. 329.	Distribution map of *A. sativa* × *fatua* F_1 hybrids	367
Fig. 330.	Spikelets of *A. sativa* × *sterilis* F_1 hybrid	370
Fig. 331.	Distribution map of *A. sativa* × *sterilis* F_1 hybrids	371

List of tables

Table 1.	World cereal production	1
Table 2.	Characters used in the classificatory study	31–35
Table 3.	Basic data matrix used in the classificatory study	36
Table 4.	Basic data matrix used in the character analysis	37
Table 5.	Correlation analysis (characters and classifications)	46, 47
Table 6.	Classification by classification matrix of the distances (*D* values) between pairs of admissible classifications	48
Table 7.	Characters used in the cladistic analysis	58–62
Table 8.	OTU's for cladistic analysis	63
Table 9.	Basic data matrix of diploid *Avena* species for computing Camin and Sokal cladograms	64
Table 10.	Basic data matrix of diploid *Avena* species for computing Wagner networks	65
Table 11.	Basic data matrix of hexaploid *Avena* species for computing Wagner networks	65
Table 12.	Distribution of diploid species	74, 75
Table 13.	Epicuticular wax patterns	79–81
Table 14.	Characters used in identification keys	98
Table 15.	Character-states in the identification keys	99, 100
Table 16.	Data matrix of all the objects used as input into the keys generating program	101–106
Table 17.	Summary of new genera	380

Part 1. General

1. Introduction

Oats occupy a significant place among the eight most important cereals in the world. Like other cereals, oats has actually increased in total world production since the early 1960's, although its relative position among the eight has remained the same. In total world production, oats consistently ranked fifth and barley fourth, while maize fluctuated from third to second and rice from second to third; wheat consistently ranked first. This opinion is based on tabular data published in *Agricultural Economics and Statistics, Vol. 23, No. 2, February 1974, FAO, Rome*.

In Table 1, I indicate the ranking of the major eight cereals.

TABLE 1. WORLD CEREAL PRODUCTION.

Area, yield, and grain production of the eight major cereals in 1972, ranked in order of importance*

Crop	Area (1000 ha)	Rank	Average yield (kg/ha)	Rank	Total production (1000 tonnes)	Rank
Wheat	213 494	1	1628	6	347 603	1
Rice	131 230	2	2251	2	295 377	3
Barley	84 915	4	1793	3	152 238	4
Maize	108 208	3	2785	1	301 392	2
Rye	17 026	8	1654	4	28 170	8
Oats	31 282	7	1640	5	51 293	5
Millet	65 089	5	660	8	42 956	7
Sorghum	39 929	6	1170	7	46 709	6

*Source: Production Yearbook 1972, Vol. 26, FAO, Rome, 1973.

Because of the importance of oats on a worldwide basis and because of its relatively greater importance in Canadian agriculture (inasmuch as Canada grows no rice and only a restricted quantity of maize), we recognize the need for extensive research in *Avena* to improve the oat crop.

Initially, researchers directly involved in oat improvement felt that they needed more information about the genetics and cytogenetics of *Avena* so that they could work on a wider spectrum. As a result of the work of Rajhathy and co-workers in Ottawa, and others, new information on this subject gradually became available. Some of this information was significant enough to make oat workers realize that the last comprehensive monograph of *Avena* (Malzew 1930) had outlived its usefulness; a reassessment of the species of the genus *Avena* was imperative if the necessary wide scope research on the genus was to progress. So, in 1967 I started to work on the taxonomy of the genus *Avena* with the aim of preparing this account.

Some aspects are not covered by this monograph because excellent accounts of these have been already published. *Oats and Oat Improvement* (Coffman 1961) gives an excellent agronomic treatment. Abridged accounts, similar to the above, have been published by various authors, for example, Plarre (1970), Coffman and MacKey (1959), and Brouwer (1971). Most important are the accounts on oat cytogenetics by Rajhathy and Thomas (1974) and on histology by Bonnett (1961).

The purpose of this monograph is therefore primarily taxonomic, that is, classification, nomenclature, identification, cladistics, distribution of taxa, and other fields not covered by the treatments mentioned.

2. Background

ETYMOLOGY

Ancient written records about the genus *Avena* are scarce. According to De Candolle, the ancient Greeks (400 BC to 200 AD) knew oats very well and called it *Vromos,* or *Bromos* (Theophrastos, Galenos, Dioskorides), which is not to be confused with *Bromus* of the present-day meaning. The Romans (250 BC to 50 AD) called it *avena* (Plinius, Columella, and others).

Oats were apparently cultivated in China as early as 386-534 AD, or perhaps even earlier (Nakao 1950), according to a book of that period by "Kia Sz'niu" entitled "Tsi min yao shu" (Bretschneider 1881). The title of this book was translated by Bretschneider as: Important Rules for the People to gain their living in peace. One of the plants referred to in the book was identified by Bretschneider as *Avena nuda,* which I judge to be the *naked form* of *A. sativa.* This crop is still cultivated as a staple food by the Chinese (Nakao 1950) in North China and Mongolia (personal communication from a Chinese delegate who visited Ottawa in 1973).

As far as the contemporary scientific name of the genus is concerned, *Avena* was taken up by Bauhin (1671) from Plinius, and subsequently by Tournefort (1700). This name was then designated by Linnaeus (1753, 1754) as the generic name.

It seems that pre-Linnaean botanists, for example, Bauhin, conceived for *Avena* the species *A. sativa, A. strigosa,* and possibly also *A. brevis* and *A. hispanica* of the present work, as can be inferred from Thellung (1912).

Relatively little is known about oats (wild and cultivated) among the ancient folk taxonomies. Thellung (1912), Haussknecht (1885), De Candolle (1882), Malzew (1930), and Coffman (1961) summarize the origin and history of oats with particular reference to cultivated oats.

A review of the names existing for oats (that is, the names for the genus *Avena*) might help assess the importance of that genus to different groups of people. "In most folk taxonomies, taxa that are members of the category *generic* are more numerous than life form taxa, but are nonetheless finite in number, usually about 500" (Raven et al. 1971). Similarly, consider some of the present and past etymological equivalents of the word "oats":

English	Oat
Anglo-Saxon	Ata, Ate
French	Avoine
Latin	Avena
Spanish	Avena
Italian	Vena
Portuguese	Avea, Veia
Breton	Kerch
Irish	Coirce, Cuirce, Corca
Welsh	Kerch, Cerch
Arabic (Egypt)	Khafoor
Finnish	Kaura
Estonian	Kaer
Georgian	Kari
Hungarian	Zab
Croat	Zob
Danish	Havre
Flemish	Haver
Norwegian	Havre
Swedish	Hafre
Old Nordic	Hafri
Old Saxon	Hafero
Old high german	Habero
Frankish	Havero, Haparo
Polish	Owies
Romanian	Ovesul, Ovaz
Russian	Ovies
Serbian	Ovas
Ancient Slavic	Oviser, Ovesu, Ovsa
Lithuanian	Awiza
Lettonian	Ausas
Japanese	Enbaku
Turc	Ioulaf or Yulaf
Basque	Olba, Oloa
Tartar	Sulu
Berber	Zekhoum
Greek (recent)	Vromi
Indian (recent)	Jai, Javi
Hebrew (recent)	Shibolet shual

Clearly, there are many roots to the word denoting "oats." It might be interesting to make philological inferences associating these root variations with past migrations of groups of people.

3

ARCHEOLOGY

Cultivated oats have been known in Europe since the Early Bronze Period, about 2000 BC onward, although they have become much more common since the Early Iron Age. Archeological findings are chiefly from Germany, Poland, England, Sweden, Denmark, Czechoslovakia, and Switzerland.

The findings by Villaret-von-Rochow (1971) of *A. ludoviciana,* that is, *A. sterilis,* do not support her theory about the origin of oats; these findings just indicate that *A. sterilis* was probably a weed among cereals as it is today.

Werth (1944) claimed that the Anglo-Saxons used oats as their main cereal to make bread and related foods.

In most archeological findings, it is difficult to make conclusive identifications at the species level; therefore, interpretations of their findings by various authors must be read with caution. Jessen and Helbaek (1944) comment on this:

> If the carbonised oat flowers had been preserved intact it would not have been difficult to decide whether they represent *A. sativa* or forms of the *A. strigosa* group, for instance by means of the tip of the flowering glume. However, the ultimate part of the flowering glume is lacking from all the carbonised flowers, and the awn itself is always broken off, so that at most its base is preserved. For identification it has therefore been necessary to use the shape, particularly that of the base of the flowers in conjunction with the presence or absence of awns ...
> In many cases, however, the condition of the flowers was such as to make definite identification impossible.

Thus, *A. sativa* is difficult to distinguish from *A. fatua,* and *A. strigosa* from *A. barbata,* and so on.

Specific details are documented in the following list of archeological contributions. Keep in mind that the interpretation of carbonized grains is difficult. Probably much more must be discovered to obtain a valid picture. We know that oats have been cultivated since the fourth century in China, and probably long before that, but archeological evidence is still incomplete on this question.

Berggren, G. 1958. Vaxtmaterial Fran Traskboplatsen I Dagsmosse. Sven. Bot. Tidskr. 50(1):97-112.
Bertsch, K., and Bertsch, F. 1949. Geschichte Unserer Kulturpflanzen. Stuttgart, Wissenschafth. Verlag.
Buschan, G. 1890. Die Heimath und Das Alter der Europaischenkulturpflanzen. Corresp. der Dtsch. Anthropol. No. 10, p. 127-134.
Buschan, G. 1895. Vorgeschichtliche Botanik der Cultur und Nutzpflanzen der alten Welt auf Grund prähistorischer Funde. Breslau.
Deininger, J. 1891. Adatok Kulturnovenyeink Tortenetehez. Jahrb, D. Kgl. Ung. Landwirthschaftl. Lehranstalt zu Keszthelyfur. 1891. p. 21-49.

Hatt, G. 1937. Landbrug i Danmarks Oldtid. Köbenhavn.
Heer, O. 1865. Die Pflanzen der Pfahlbauten. Neujahrsbl. d. Naturf. Ges. Zürich 68, auf das Jahr 1866. Zürich.
Helbaek, H. 1966. Vendeltime farming products at Eketorp on Oland, Sweden. Acta Archaeol. Acad. Sci. Hung. 37:216-221.
Hjelmqvist, H. 1953. Ett Bidrag till Kulturvaxternas Aldstahistoria I sydsverige. Bot. Not. 4:420-430.
Hjelmqvist, H. 1955. Die Alteste Geschichte Der Kulturpflanzen In Schweden. Opera Bot. 1(3):1-186.
Holmboe, J. 1906. Studier Over Norske Planters Historie, III. En Samling Kulturplanter Og Ugraes Fra Vikingetiden. NYT Mag. F. Naturv. 44:61-71.
Jäger, K. D. 1966. Die pflanzlichen Grossreste aus der Burgwallgrabung Tornow Kr. Calau. *In* Herrmann, J., Tornow und Vorberg. Ein Beitrag zur Frühgeschichte der Lausitz. Schr. d. Section für Vor- u. Frühgeschichte d. Dt. Akad. Wiss. Berlin 21:164-189, Taf. 41-44 (Beilage 12-13).
Jessen, K. 1954. Plantefund Fra Vikingetiden I Danmark. Bot. Tidsskr. 50(2):125-139.
Jessen, K., and Helbaek, H. 1944. Cereals in Great Britain and Ireland in Prehistoric and Early Historic Times. Kgl. Dan. Vidensk. Selsk. Biol. Skr. 3(2):1-68.
Kir'janova, N. A. 1972. O nachodkach zeren sel'skochozjajstvennych Kul'tur pri raskopkach gorodišča Druck. Sov. Archeol. (3)355-357.
Klichowska, M. 1966. Szczątki roślinne z wykopalisk archeologicznych w Kruszwicy, pow. Inowroclaw, z lat 1959-1964. Spraw. Archeol. 18:387-390.
Klichowska, M. 1966. Wyniki badań próbki zboża z Zamku w Tykocinie. Spraw. Archeol. 18:396.
Klichowska, M. 1967. Możliwości konsumpcyjne zbóż i motylkowych w północno-zachodniej Polsce od neolitu do konca XII.w. Stud. z Dziejów Gospod. Wiejsk. 9:31-47.
Klichowska, M. 1967. Szczatki róslinne ze stanowiska archeologicznego Wolin-Mlynówka. Mater. Zachodniopomorskie 11: 577-579.
Klichowska, M. 1969. Odciski róslin na polepie ze stan. 1 w. Kotlinie, pow. Jarocin, z lat 1963–1964. Spraw. Archeol. 20:419-421.
Klichowska, M. 1969. Badania nad polepa z stan. 1 w Golańczy Pomorskiej, pow. Gryfice, z roku 1961. Spraw. Archeol. 20:449-452.
Klichowska, M. 1969. Wyniki badań polepy z Bard, pow. Kilobrzeg, z 1964 roku. Spraw. Archeol. 20:453-454.
Klichowska, M. 1969. Rósliny uprawne i dziko rosnące z grodsiska z VII-IX wieku w Bruszczewie, pow. Kóscian. Spraw. Archeol. 20:
Klichowska, M. 1971. Z dalszych badań paleoetnobotanicznych. Spraw. Archeol. 23:239-257.
Klichowska, M. 1972. Vascular plants in archaeological excavations of northwestern Poland from the neolithic to the early middle ages. Poznan. Tow. Przyj. Nauk Wydz. Mat.-Przyr. Pr. Kom. Biol. 35(2):3-74.
Knörzer, K.-H. 1970. Römerzeitliche Pflanzenfunde aus Neuss. Novaesium IV, Limesforschungen, Stud. zur Organ. der Römischen Reichsgrenze an Rhein und Donau. Band 10.
Opravil, E. 1965. Pflanzenfunde der im Jahre 1962 durchgeführten archäologischen, Erforschung der mittelalterlichen Stadt Opava. Acta Mus. Silesiae. Ser. A. Sci. Nat. 14:77-83.

Schlatter, T. 1894. Die einfuhrung der Kulturpflanzen in den cantonen St. Gallen und Appenzell. Jahresber. St. Gallischen Naturwiss. Ges. 1893-94. 40 pp.

Schultze-Motel, J. 1971. Literatur über archäologische Kulturpflanzenreste (1968). Jahresschr. für mitteldtsch. Vorgeschichte 55:55-63.

Schulz, A. 1913. Die Geschichte der Kultivierten Getreide I. Halle, A. S. 134 pp.

Schulz, A. 1914. Über Kulturpflanzen und Unkrauter Deutschlands In Prahistorischer Zeit. I. Z. Naturwiss.-Med. Grundlagenforsch 85:329-341.

Schulz, A. 1915. Über Einen Neuen Fund Von Hallstattzeitlichenkulturflanzen-und Unkrauterresten in Mitteldeutschland. Ber. Dtsch. Bot. Ges. 33(1):11-19.

Schulz, A. 1915. Über Neue Funde Von Getreideresten Ausprahistorischer Zeit in den Thuringish-Sachsischen Landern. Naturwiss. Wochenschr. 14:266-270.

Tempír, Z. 1966. Results of paleoethnobotanical studies on the cultivation of agricultural plants in the CSSR. Ved. Pr. Cesk. Zemed. Muz.:27-144.

Villaret-Von-Rochow, M. 1971. *Avena Ludoviciana* Dur. in the Late Neolithic of Switzerland, a Contribution to the origin of the oat (*Avena Sativa* L.), Ber. Dtsch. Bot. Ges. 84(5):243-248.

Werth, E. 1944. Der Hafer, Eine urnordische Getreideart. (Zur Geographie und Geschichte der Kulturpflanzen und Haustiere. XXV). Z. Pflanzenzuecht. 26(1/2):92-102.

Wieserowa, A. 1967. Early mediaeval remains of cereals and weeds from Przemyśl, SE Poland. Folia Quat. 28:1-16.

Willerding, U. 1970. Vor- und frühgeschichtliche Kulturpflanzenfunde in Mitteleuropa. Neue Ausgrabungen u. Forsch. in Niedersachsen 5:287-375.

Wimmer, J. 1905. Geschichte des Deutschen Bodens Mit Seinem Pflanzen- und Tierleben von der Keltisch-Romischen Urzeit Bis zurgegenwart. Historisch-Geographische Darstellungen, Halle A.S. 475 pp.

Wittmack, L., and Buchwald, J. 1902. Pflanzenreste aus der Hunenburg Bei Rinteln A.D. Weser und Eine Verbesserte Methode Zurherstellung Von Schnitten Durch Verkohlte Holzer. Ber. Dtsch. Bot. Ges. 20:21-81.

Zeist, W. van. 1970. Prehistoric and early historic food plants in the Netherlands. Palaeohistoria 14:41-173.

Zhukovski, P. M. 1924. An Investigation of the Seed Material of the Peasants of Eastern Georgia. Bull. Sect. Sci. Appl. Bot., Gard. Tiflis. Fasc. 3:127-223.

ANATOMY AND HISTOLOGY

The reader interested in the anatomy and histology of *Avena sativa* should consult the excellent treatment in *The Oat Plant: Its Histology and Development* (Bonnett 1961). This book gives an account on the vegetative parts, the reproductive parts, the reproductive processes, and the seed parts. It is a comprehensive description of *A. sativa*.

For *A. fatua*, Cannon (1900) described the development of the spikelet and the floret, the fertilization process, and the early embryogeny.

I was unable to find any publication concerned with the anatomy and organogenesis of the other species of *Avena*. The following titles, published after 1961, might provide additional details on various aspects not covered by Bonnett.

Arrigoni, O., and Rossi, G. 1962. Ricerche sull infrastruttura del coleottile di Avena. I. Il fascio vascolare in corso di differenziaione. G. Bot. Ital. 69:112-118.

Arrigoni, O., and Rossi, G. 1964. Ricerche sull infrastruttura del coleottile di avena. 3. I Tubi Cribrosi. G. Bot. Ital. 71:96-112.

Baitulin, I. O. 1973. Regularities of the formation of seminal roots in annual cereals. Izv. Akad. Nauk. Kaz. SSR. Ser. Biol. Nauk. 3:28-37.

Chonan, N. 1970. Studies on the photosynthetic tissues in the leaves of cereal crops: V. Comparison of the mesophyll structure among seedling leaves of cereal crops. Proc. Crop. Sci. Soc. Jap. 39:418-425.

Colvin, J. R. 1965. The absence of elastic deformation in dried bent cellulose microfibrils in plant cell walls. Can. J. Bot. 43:339-343.

Cronshaw, J. 1965. Cytoplasmic fine structure and cell wall development in differentiating zylem elements *(Avena sativa, Acer rubrum)*. *In* Proceedings of the advanced science seminar on cellular ultrastructure of woody plants, 20-26 September, 1964, Upper Saranac Lake, N.Y., Syracuse University Press, Syracuse, N.Y. p. 99-124.

Cronshaw, J., and Bouck, G. B. 1965. The fine structure of differentiating xylem elements. J. Cell. Biol. 24:415-431.

Fulcher, R. G., O'Brien, T. P., and Simmonds, D. H. 1972. Localization of arginine-rich proteins in mature seeds of some members of the gramineae. Aust. J. Biol. Sci. 25:487-497.

Gunning, B. E. S., and Jagoe, M. P. 1967. The prolamellar body (of Avena sativa, structure, organization, and response to illumination and darkness.) *In* Proceedings of a NATO advanced study institute on biochemistry of chloroplasts, August, 1965, Aberystwyth, Wales, UK., Vol. 2. Academic Press, London and New York. p. 655-676.

Jellum, M. D. 1962. Relationships between lodging resistance and certain culm characters in oats. Crop. Sci. 2:263-267.

Kaufman, P. B. 1965. The effects of growth substances on intercalary growth and cellular differentiation in developing internodes of *Avena sativa*. II. The effects of gibberellic acid. Physiol. Plant. 18:703-724.

Kaufman, P. B., Cassell, S. J., and Adams, P. A. 1965. On nature of intercalary growth and cellular differentiation in internodes of *Avena sativa*. Bot. Gaz. 126:1-13.

Laurema, S., Markkula, M., and Raatikainen, M. 1966. The effect of virus diseases transmitted by the leaf hopper *Javesella pellucida* (F.) on the concentration of free amino acids in oats and on the reproduction of aphids. Ann. Agric. Fenn. 5:94-99.

Luke, H. H., Warmke, H. F., and Hanchey, P. 1966. Effects of the pathotoxin victorin (from *Helminthosporium victoriae*) on ultra-structure of root and leaf tissue of *Avena* species. Phytopathology 56:1178-1183.

Lunden, A. O., and Wallace, A. T. 1961. Some effects of ultraviolet light on barley and oat embryos. Crop. Sci. 1:212-215.

O'Brien, T. P. 1967. Observations on the fine structure of the oat coleoptile: I. The epidermal cells of the extreme apex. Protoplasma 63:385-416.

O'Brien, T. P., and Thimann, K. V. 1967. Observations on the fine structure of the oat coleoptile. III. Correlated light and electron microscopy of the vascular tissues. Protoplasma 63:443-478.

O'Brien, T. P., and Thimann, K. V. 1967. Observations on the fine structure of the oat coleoptile: II. The parenchyma cells of the apex. Protoplasma 63:417-442.

Paunero, E. 1968. Notas sobre gramineas: IV. Contribucion a la anatomia foliar de algunas aveneas. Collect. Bot. (Barcinone) 7:917-937.

Rodionova, N. A. 1964. Nekotorye morfologo-anatomicheskieosobennosti razlichnykh po ustoichivosti k poleganiyu obraztsovovsa. Tr. Prikl. Bot. Genet. Sel. 36:185-194.

Tataeoka, Tuguo. 1969. Notes on some grasses. XX. Systematic significance of the vascular bundle system in the mesocotyl. Bot. Mag. (Tokyo) 82:387-391. (in Jap.).

Tayal, M. S., Kaushik, M. P., and Nanda, K. K. 1961. Growth and phasic development of oat. I. A developmental study of the shoot apex in two varieties of *Avena sativa* under differential photoperiodic treatments. Indian J. Plant Physiol. 4:132-149.

Tayal, M. S., and Nanda, K. K. 1962. Growth and phasic development of oat. II. Development of the inflorescence and floral organs in two varieties under two photoperiods. Phyton. Rev. Int. Bot. Exp. 19:141-153.

Thimann, K. V., and O'Brien, T. P. 1965. Histological studies on the coleoptile. II. Comparative vascular anatomy of coleoptiles of *Avena* and *Triticum*. Am. J. Bot. 52:918-923.

Wardrop, A. B., and Foster, R. C. 1964. A cytological study of the oat coleoptile. Aust. J. Bot. 12:135-141.

Webster, J. M. 1966. Production of oat callus and its susceptibility to a plant (Chrysanthemum) parasitic nematode *(Aphelenchoides ritzemabosi)*. Nature 212:1472.

Wilkins, R. J. 1972. The potential digestibility of cellulose in grasses and its relationship with chemical and anatomical parameters. J. Agric. Sci. 78:457-464.

PALYNOLOGY

A few authors have studied the pollen of *Avena* to define variations in pollen grain structure in the Gramineae in general or to identify differential characteristics between various cereals and related wild species. I do not know of an in-depth study in the palynology of the species of *Avena* (although I have gathered the material for such a study).

Faegri and Iversen (1964) stated that grass (Gramineae) pollen, because of its homogeneity, is difficult to observe under the light microscope. However, their key distinguishes between *A. sativa* and *A. fatua*: *A. sativa* is distinctly areolate whereas *A. fatua* is verrucate-rugulate. Thus their key essentially discriminates different types of pollen for the Gramineae, but not particular species nor particular genera.

Faegri and Iversen also point out that the microstructure and microsculpturing of the exine appear to be important in discriminating genera and species in the Gramineae. In this respect, the scanning electron microscope appears promising as shown in the paper of Andersen and Bertelsen (1972). Of the *Avena* species, only *A. sativa* (and only one sample) has been studied by these authors. Andersen and Bertelsen investigated only a few samples from cereal species, using the scanning electron microscope; their purpose was not taxonomic but to clarify disagreement among previous findings by different authors. Romanov (1970) also studied microsporogenesis in *A. sativa*.

The following references summarize palynological research relating to *Avena*:

Andersen, T. S., and Bertelsen, F. 1972. Scanning electron microscope studies of pollen of cereals and other grasses. Grana 12:79-86.
Beng, J.-J. 1961. Leitfaden der Pollenbestimmung für Mitteleuropa und agrenzende Gebiete. Lief. 1. Gustav Fischer Verlag, Stuttgart.
Erdtman, G. 1944. Pollen morphology of the cereals, with notes on the pollen morphology in Triticale. Sven. Bot. Tidskr. 38:73-80.
Erdtman, G. 1956. Current trends in palynological research work. Grana Palynol. 1:127-139.
Erdtman, G., and Praglowski, J. R. 1959. Six notes on pollen morphology and pollen-morphological techniques. Bot. Not. 112:175-184.
Faegri, K., and Iversen, J. 1964. Textbook of pollen analysis. 2nd ed. Blackwell, Oxford, 237 pp.
Grohne, V. 1957. Die Bedeutung des Phasenkontrastverfahrens für die Pollenanalyse, dargelegt am Beispiel der Gramineenpollen vom Getreidetyp. Photogr. Forsch. 7:237-248.
Ludwig, F. 1880. Biologische Mittheilungen II. Heteranthieanemophiler Pflanzen. Bot. Centralbl. 1880:861-862.
Romanov, I. D. 1966. Spetsificheskie osobennosti razvitiyapyl tsy zlakov. Dokl. Akad. Nauk SSSR Ser. Biol. 169(2):456-459.
Romanov, I. D. 1970. Features of Poaceae pollen development and their importance for some genetical studies. Genetika 6:11-24.
Rowley, J. R. 1960. The exine structure of "Cereal" and "wild" type grass pollen. Grana Palynol. 2:9-15.

CYTOGENETICS

An authoritative and complete account of the karyotypes, reproductive biology, chromosome pairing within species and among species hybrids, and genome analysis has been recently summarized in *Cytogenetics of Oats (Avena L.)* (Rajhathy and Thomas 1974). In this work, the authors corroborate their findings with data on the biochemical relationships of the genomes and provide chapters on aneuploidy and handling of oat chromosomes.

Briefly, the taxa recognized by Rajhathy and Thomas fall into six groups: (i) diploids with subterminal chromosomes — these are labeled as the species with the C-genome; (ii) diploids with a single pair of subterminals, that is, the A-genome species; (iii) diploids without subterminals — also A-genome species; (iv) tetraploids that are nearly autotetraploids; these are labeled AB genome; (v) a tetraploid with the AC genome, and another species related to it; (vi) hexaploids with the ACD-genome. No D-genome diploid species has been found to date.

On the combined basis of genome, karyotype, and reproductive isolation, Rajhathy and Thomas recognize eleven biological species, that is, seven diploids, three tetraploids, and a single hexaploid. Obviously the number of biological species differs from the number of taxonomic species recognized in this work. See, however, the chapter on species concept in this work.

POLYMORPHISM AND POPULATION BIOLOGY

Genetic polymorphism is a well-known phenomenon in oats. Extensive work has been carried out by Allard and associates, especially Jain, on Californian material of *A. fatua* and *A. barbata*. The extent of polymorphism has been investigated in terms of the number of factors that determine a certain trait, such as hairiness and color of lemma, and the distribution of those traits within different populations. The findings showed that *A. barbata* is largely monomorphic, whereas *A. fatua* is highly polymorphic in Californian populations, and furthermore that the larger nongenetic component in the former species accounts for its greater phenotypic variation. On that basis Jain and Marshall (1967) postulate that *A. barbata* relies less on genetic diversity and more on phenetic plasticity than *A. fatua* in adapting to heterogeneous environments. At least in *A. fatua,* random drift appeared important in changes of populations in achieving local monomorphism of different heterogeneous populations.

Studies in lemma color, lemma pubescence, leaf sheath hairiness, node hairiness, and three isoenzymatic marker loci show that the patterns of variation of Californian *A. barbata* are similar to those found in the Mediterranean area (Singh and Jain 1971). Furthermore, various studies (Marshall and Allard 1970; Murray et al. 1970; Craig et al. 1972, 1974; Jain and Singh, in press) in *Avena* species fail to reveal a one-to-one relationship between the different allozyme variation patterns and the genomic constitutions of the species. This situation suggests the existence of latent polymorphisms to a generally great degree in the species, but varying among them.

A study of isozymes at the cultivar level of *Avena sativa* (Singh et al. 1973) has shown that 7 out of 10 cultivars have no intravarietal variation, and that each cultivar has a characteristic isozyme pattern.

LITERATURE SURVEY

I gathered about 9000 references on *Avena,* which I initially intended to use for a literature survey. These publications, abstracts, and titles cover the period 1870–1973.

Most of these publications are concerned with *A. sativa* (i.e., cultivated oats) and many deal with research on, or application of, the *Avena coleoptile*. The most common topics are physiology, ecology, germination, phenology, and such agronomic problems as weeds, herbicides, fungicides, insecticides, and fertilizers. Less frequent topics are genetics, fertilization, pollination, competition, cytology, microstructure of cells and organelles, biochemistry and chemical compounds, and food and forage. A few publications deal with plant and animal life that live on or infect cultivated oats, such as insects, other fauna, and various fungi, including micorrhiza and viruses.

Some titles, at first glance, seem to be unrelated to the subject; in content, these can be classified under miscellaneous uses. A few publications are concerned with hybrids, others with *Helictotrichon* and related genera, which have species that were once considered as *Avena*. A very few publications deal with mutagens and teratology.

3. Materials and Methods

MATERIALS

This study is based on herbarium material; on live plants grown in Ottawa obtained as seeds from various institutes and agencies; on material collected in the Near East, Mediterranean, and related areas during various expeditions organized and sponsored by the Canada Department of Agriculture (Baum et al. 1972; Baum et al. 1975); and on experimental material and additional material.

Herbarium material

I have obtained on loan about 9000 herbarium sheets of *Avena* from the institutes listed below, and have studied 3000 herbarium sheets in a visit to major European herbaria during the summer of 1973. Of these 12,000 sheets, 4200 have been dissected, and lodicule and epiblast preparations have been made. In the other 7800 specimens, all characters listed in Table 2 have been examined except for the lodicules and epiblasts.

In the following lists of herbaria, the acronyms are according to Lanjouw and Stafleu (1964).

ACAD	Acadia University, Wolfville, Nova Scotia, Canada.
ALTA	University of Alberta, Edmonton, Canada.
B	Botanisches Museum, Berlin, Germany.
BAA	Herbario de la Facultad de Agronomia y Veterinaria de la Universidad de Buenos Aires, Buenos Aires, Argentina.
BAG	National Herbarium of Iraq, Baghdad, Iraq.
BAK	Herbarium of the Systematic Division, Botanical Institute of Azerbaijan Academy of Sciences, Baku, USSR.
BAS	Botanisches Institut der Universität Basel, Basel, Switzerland.
BG	Universitetets Botaniske Museum, Bergen, Norway.
BM	British Museum (Natural History) London, Great Britain.

Materials

BP	Museum of Natural History, Department of Botany, Budapest, Hungary.
BR	Jardin Botanique de l'Etat, Bruxelles, Belgium.
BRA	Slovenské národné múzeum, Bratislava, Czechoslovakia.
BRI	Botanic Museum and Herbarium, Brisbane, Australia.
C	Botanical Museum and Herbarium, Copenhagen, Denmark.
CAL	Central National Herbarium, Calcutta, India.
CAN	National Museum of Canada, Natural History Branch, Ottawa, Canada.
CGE	Botany School, University of Cambridge, Cambridge, Great Britain.
COI	Botanical Institute of the University of Coimbra, Coimbra, Portugal.
DAO	Phanerogamic Herbarium, Biosystematics Research Institute, Research Branch, Department of Agriculture, Ottawa, Canada.
DAS	Experimental Farm, Canada Department of Agriculture, Regina, Saskatchewan, Canada.
ELVE	Estaçao de Melhoramento de Plantas, Elvas, Portugal.
FI	Herbarium Universitatis Florentinae, Instituto Botanico, Firenze, Italy.
G	Conservatoires et Jardin botaniques, Geneve, Switzerland.
GB	Herbarium, Institute of Systematic Botany, University of Göteborg, Göteborg, Sweden.
HAL	Institut für Systematische Botanik und Pflanzengeographie der Martin-Luther-Universität, Halle, Germany.
JE	Institut für Spezielle Botanik und Herbarium Haussknecht, Jena, Germany.
K	The Herbarium and Library, Kew, Great Britain.
KW	Botanical Institute of the Academy of Sciences of the Ukrainian SSR.
L	Rijksherbarium, Leiden, Netherlands.
LAU	Musée botanique cantonal, Lausanne, Switzerland.
LD	Botanical Muséum, Lund, Sweden.
LE	Herbarium of the Komarov Botanical Institute of the Academy of Sciences of the USSR.
LINN	The Linnean Society of London, London, Great Britain.
LY	Herbiers de la Faculté des Sciences de Lyon, Lyon, France.
M	Botanische Staatssamlung, München, Germany.
MA	Instituto "Antonio José Cavanilles", Jardin Botánico, Madrid, Spain.
MPU	Institut de Botanique, Université de Montpellier, Montpellier, France.
MT	Herbier Marie-Victorin, Institut Botanique, Université de Montréal, Canada.
O	Botanisk Museum, Oslo, Norway.
OAC	Ontario Agricultural College, Guelph, Ontario, Canada.
P	Muséum National d'Histoire Naturelle, Laboratoire de Phanérogamie, Paris, France.
PRC	Universitatis Carolinae facultatis biologicae scientiae ćathedra, Praha, Czechoslovakia.
QFA	Faculty of Agriculture, Quebec, Canada.
S	Botanical Department, Naturhistoriska Riksmuseum, Stockholm, Sweden.

SASK	The W. P. Fraser Memorial Herbarium, Department of Plant Ecology, University of Saskatchewan, Saskatoon, Saskatchewan, Canada.
TI	Botanical Institute, Tokyo, Japan.
TRT	Department of Botany, University of Toronto, Toronto, Ontario, Canada.
TUB	Institut für spezielle Botanik und Pharmakognosie der Universität, Tübingen, Germany.
UBC	The University of British Columbia, Vancouver, British Columbia, Canada.
UPS	Institute of Systematic Botany, University of Uppsala, Uppsala, Sweden.
US	U.S. National Museum (Department of Botany), Washington, D.C., USA.
V	Provincial Museum, Victoria, British Columbia, Canada.
W	Naturhistorisches Museum, Wien, Austria.
WIN	Herbarium of the Department of Botany, University of Manitoba, Winnipeg, Manitoba, Canada.
WIR	Herbarium of Cultivated Plants, Institute of Plant Industry, Leningrad, USSR.
WU	Botanisches Institut und Botanischer Garten der Universität Wien, Wien, Austria.
Z	Botanischer Garten und Institut für Systematische Botanik der Universität Zürich, Zürich, Switzerland.

Live plants

I have obtained 5000 packets of seeds of *Avena* from the World Collection maintained by the Crops Research Division, Agricultural Research Service, USDA, Beltsville, Maryland 20705, USA. These accessions consisted chiefly of cultivars and strains of *A. sativa* from various countries covering practically the whole distribution of that species. All the accessions were grown in Ottawa in 1967; specimens were collected and examined; their floral parts were dissected and studied; and seeds (diaspores) were harvested, dissected, and examined. This material is now conserved in DAO together with seed packets affixed to the herbarium sheets, and the 5000 microscopic slides of lodicules and epiblasts were prepared and examined. Duplicate sets were distributed to WIR.

Material collected

About 5000 accessions of wild species of *Avena* have been collected, and are presently maintained by the Canada Department of Agriculture (Baum et al. 1975). Seeds (diaspores) from each of these accessions have been dissected, observations have been made on floral characters, and many specimens grown from these seeds have been examined. Details about this collection of wild oats, including the origin of each accession, appear in Baum et al. (1975).

Materials

The following table summarizes the number of samples from each area or country; at least one specimen from each of these samples has been studied.

Country	Number of samples
Afghanistan	25
Algeria	622
Canary Islands	91
Corsica	13
Crete	50
Cyprus	1
Ethiopia	530
Gibraltar	5
Greece	66
Iran	562
Iraq	199
Israel	603
Italy	46
Kenya	31
Lebanon	108
Libya	83
Morocco	570
Portugal	2
Sardinia	32
Sicily	77
Spain	13
Syria	68
Tunisia	538
Turkey	945
USA	1
USSR	2
Uruguay	1
Total samples	5264

Certain species were not always easy to collect. Some species, such as *A. ventricosa, A. clauda,* and *A. eriantha,* are difficult to find, not only because they are often found in small populations, but mainly because they are disappearing as a result of overgrazing (Figs. 1, 2). Indeed overgrazing must be at least partly responsible for the small size of some populations because quite large or very large populations of the same species are found (Fig. 3) in places protected from grazing animals or inaccessible to them. Weedy species are not uncommon because they are protected from grazing in cultivated fields (e.g., *A. sterilis,* Fig. 30). Although other factors also affect the abundance of every species, overgrazing is now seriously affecting populations in some areas that are important for gene diversity and concentration of germplasm.

Oats: Wild and Cultivated
BIBLIOGRAPHY

B. R. Baum
Biosystematics Research Institute
Research Branch, Agriculture Canada
Ottawa, Ont. K1A 0C5

A list of about 9000 references on the genus *Avena* has been compiled by the author. Topics covered include anatomy, genetics, cytology, physiology, ecology, agronomy, pests, and other subjects. The publications cover the period 1751–1973.

A copy of the *Bibliography* is available from the author free on request.

Figs. 1–2. Overgrazing can deplete germ plasm and make material difficult to find. 1. Cattle grazing in *Quercus brantii* park forest, Zagros Mountains, 41 km S of Shanabad-Gharb, E. Iran. 2. Overgrazed area on hills near Malayer, Iran.

Fig. 3. Ungrazed area on hills in eastern Anatolia, Turkey.

Experimental and additional material

Drs. Rajhathy and Sampson, of the Ottawa Research Station, Canada Department of Agriculture, artificially produced many interspecific hybrids for cytogenetic and genetic studies. These hybrids were examined by myself and studied morphologically. The monosomics, nullisomics, and various artificially-induced chromosome aberrations produced by Dr. Rajhathy were particularly useful in taxonomic studies of fatuoids in this work (and see also Baum 1969b for more details). In addition to the World Collection of USDA mentioned above, seed material (diaspores) obtained from various institutions was dissected and examined. Also examined were seeds from pedigreed seed growers in Canada and from various commercial oat growers in Canada. These seeds were particularly useful in the taxonomic elucidation of A. fatua $\times$ sativa F_1 hybrids, their parents and various off-types.

METHODS

Taxonomical aspects

Attributes of various organs were checked in a preliminary study on a wide-ranging sample. Useful attributes were retained for scoring, others were eliminated. The usefulness of each attribute was assessed according to the following criteria: (i) *invariability of the attribute state within a specimen,* or alternatively, states were defined in a manner to fulfill this criterion; for example, there is great variability in lemma tips, as can be seen from a study in *A. sativa* (Baum 1971*b*), but the attribute states of the lemma tips, as defined in Table 2, are consistent within specimens; (ii) *clear-cut, or well-defined states,* that is, states with practically no intermediates.

In scoring the various attributes and attribute states on the data sheets (see below how these were prepared), these criteria were constantly reviewed while examining each specimen. Among the attributes that were scored (see Table 2), some could easily be seen with the naked eye, some had to be examined under a dissecting microscope, and others could only be studied under a compound microscope (for details on characters, see Chapter 12).

Fresh material can be dissected easily. In most cases, however, mature diaspores and herbarium specimens had to be dissected. To soften this material, the dry parts or the diaspores were dipped into simmering water for a few minutes; this method proved very successful. The various organs were then carefully dissected, and the attribute states were scored. The lodicules and epiblasts were mounted on microscopic slides in Amman's lactophenol (Sass 1958).

The micromorphology of the lodicules and epiblasts was effectively examined under a Carl Zeiss research microscope RA 38 with Nomarski interference contrast attachment.

Morphological features of scars, lemmas, and paleas were examined on a Carl Zeiss Stereomicroscope IV with combination of epi-illumination and transmitted light. This equipment was also used for epiphotomicrography of scars, tips of lemmas, and backs and keels of paleas.

Data sheets were used for every specimen. Each data sheet included the collector's name, place and time of collection, herbarium code (i.e., the acronyms taken from Lanjouw and Stafleu 1964), the latitude and longitude coordinates that were found for mapping purposes, the attribute states scored for the specimen, and drawing and comments on additional attributes or features. After the species were defined, the specimens were identified accordingly and each identification was also incorporated into the appropriate data sheet.

Definition of the species was almost a by-product of the process of scoring. Some combinations of attribute states emerged recurrently and in association with general habit of the specimens. During this process I became so familiar with the material that I could make predictions, on the basis of only a few attributes, about the rest of the attributes on a particular specimen. In other words, species were formulated this way, and because some attributes were obviously better predictors than others, I gradually gave more weight to these. This intuitive approach and process of defining the species was satisfactory to me. To define the hybrids, however, I had to use experimental material because of a greater need to substantiate and because of the different purpose. Furthermore, for classificatory purpose above the species level and for identification purposes (i.e., for generation of keys and for cladistic inferences), I had to use taximetric techniques (see Chapters 7, 8, and 13 for details).

The data sheets were subsequently transcribed onto punched cards in code form for various purposes: (i) assessing the variability in each species hitherto defined by summarizing the states within each attribute for each species, and thus also verifying identifications. This assessment led to the formulation of the species circumscriptions and to elaborations of the detailed descriptions; (ii) computer mapping; (iii) generating listings, such as specimens collected for every species, specimen inventory for each herbarium, collector's inventory, and data on habitat and distribution of each species or entity thus defined.

Computer mapping

Two kinds of maps were produced, mercator projections and azimuthal equiareal projections. Individual species maps were done by mercator projection, whereas maps of sections were done with the azimuthal equiareal projection centered on the north pole. The raw data, consisting of the latitude and longitude coordinates on punched cards described above, was input to the plot program, which is essentially a conversion program using the following formulae taken from Maling (1973).

The mercator projection is: $X = A \log \tan \left(\frac{\pi}{4} + \frac{\Phi}{2} \right)$
$Y = A\omega$

The azimuthal equiareal projection centered on the north pole is:

$$\mu = 2A \sin \left(\frac{\pi}{4} - \frac{\Phi}{2} \right)$$
$$\Theta = \pi - \omega$$

where,
Φ = latitude (positive northward)
ω = longitude (positive eastward)
X = northing on map
Y = easting on map
μ = distance from center of map
Θ = azimuth from center of map
A = radius of earth × nominal scale map

The output was written on a plot tape that was input to a Gerber flatbed plotter model 600 (PL/1).

Infrageneric classification

Only taximetric methods were used. The data sheets and the punched data cards served to assess the states of each of the characters. The characters and their states were coded as in Table 2. For full details on the taximetric methods, see Chapter 7.

Evolutionary relationships

Cladograms of the diploid and hexaploid species were elaborated by using numerical cladistic methods. The data input are based on assumptions of organophyletic relationships among the various states within each of the characters or on no assumptions of this kind. For further details see Chapter 8.

Morphology of epicuticular waxes of glumes

The glumes were studied under a scanning electron microscope. Further details on the preparation of the glumes and on the microscope are given in the method section of Chapter 9.

Nomenclature

Typification for most of the names has been effected by following the common practice of studying the protolog for each name; learning the methodology of each authority of a name; locating type collections; and identifying the status of types according to the guidelines given in the *Code of Nomenclature*. As well as many individual bibliographical accounts on authors and collectors, the following sources were useful in detecting where type collections are conserved: De Candolle's Phytographie, Lasègue's Herbier Delessert, the "Index herbariorum: Collectors" of Lanjouw and Stafleu, and Stafleu's Taxonomic Literature. Of course, final authentication of types was confirmed with the aid of autographs conserved in major herbaria.

The next step was to identify each of the types according to the taxonomic concepts outlined in this work. Once this was achieved, a list of names for each of the species and hybrids recognized in this work was established. In these lists, the names were arranged according to the date of publication of the protologs. Dates were verified with the aid of bibliographic accounts about authors and in many cases with the aid of Stafleu's "Taxonomic Literature."

Subsequently, the correct name for the species was chosen from each list of names and the synonymy established according to the principles and rules of the *International Code of Botanical Nomenclature* (Stafleu et al. 1972). Each name in the lists of synonyms is, therefore, referred to an explanatory note about its nomenclature.

Some specific cases:

Dr. F. W. Meyer (personal communications, 1972), JE, wrote me that Haussknecht did not write the epithets of his new taxa on all his labels. Those labels, however, have descriptive notes that helped determine the status of types — with the aid of the protolog, obviously.

Thellung (1929) published new names and new combinations indicating two alternative ranks for those, "var. vel f.," that is, variety or form. Article 35 of the Code stipulates that "for such names published before 1 Jan. 1953 the choice made by the first author who assigned a definite rank must be followed." Unless otherwise indicated in the various nomenclatural notes of the different species in this work, I remain responsible for making the decision of the definite rank. The same applies for a few other names published in a similar manner without definite rank by some authors.

According to Ascherson and Graebner (1899), *A. diffusa* should not be conceived as species but as subspecies. They state "Zerfällt in zahlreiche Unterarten, Rassen und Abarten . . . ," explicitly indicating that *A. diffusa* is a subspecies. The citation of that taxon should therefore be *A. sativa* ssp. *diffusa* (Neilr.) Ascherson and Graebner. I have listed *A. diffusa,* with many of the varieties attributed to it, in the Digest because it is often quoted as a specific epithet; see for instance Note 88 of *A. sativa.*

4. Species concept with respect to taxonomy of cultivated plants, with emphasis on *Avena* taxonomy

Taxonomy of cultivated plants was neglected until the late fifties. Edgar Anderson (1952) in *Plants, Man and Life* has lucidly made the point and given the reasons for that neglect. A number of workers, mainly because of their participation in germ-plasm exploration, have recently become interested in the taxonomy of cultivated plants.

Since Vavilov's time, biosystematic work in the cultivated plants and their wild allies has been increasingly recognized and needed.

The taxonomy of cultivated plants and their wild allies is particularly difficult, not only because of the neglect already mentioned, but because man is growing them in various habitats and geographical areas, he is giving them different treatments, and he is manipulating them genetically. These difficulties are compounded by the interactions and intimate association between cultivated plants and their weedy and wild counterparts. In addition, an inherent taxonomic problem in the study of cultivated plants demands the classification and identification of cultivars, a task that is practically irrelevant to strictly wild plants.

Briefly, there is a taxonomic problem at the species level and a different problem of infraspecific nature.

I approach taxonomy this way. New methods and research tools are continually being devised and new data accumulated; taxonomists should use as many tools and as much data as possible, everything from classical to modern. Various combinations of data and methods will yield different representations of relationships among the OTU's (Operational Taxonomic

Units), and taxonomists must interpret these representations, then choose the best taxonomy to propose for the test of time.

It is also important to recognize that classification, identification, and cladistics are separate procedures. Each has its own purpose and methods. This viewpoint is contrary to the general belief, especially of classical taxonomists, that only one procedure is sufficient and necessary to achieve the three purposes simultaneously. In this work, therefore, the delimitation of species was done separately and differently from the classification of these species; another method was used to elaborate the different keys and yet another method to assess the cladistic relationships among the species.

It is well-known that lexical diversification will be highest in those areas of major cultural concern. In most folk taxonomies, polytypic taxa increase directly with the cultural significance of the plants to which they refer (Berlin, Breedlove, and Raven 1974). In *Avena,* the species and the cultivar taxa are of major concern. Of secondary concern are the grouping of species into classes and the grouping of cultivars into classes. Of minor concern, at this point, is the infraspecific classification of each of the different wild *Avena* species.

Species: I refrain from categorizing the concept that I followed. The species were assessed and delimited primarily upon micromorphologic attributes. Using experimental material, F_1 hybrids were assessed and delimited in the same manner and on the same attributes. As a result, (i) some species do not have well-marked sterility barriers, whereas (ii) other species are reproductively isolated.

Examples of the first situation arise when two or more species may be called conspecific according to the biological species concept; I used the term 'semispecies' to describe these "micromorphological" species (Baum 1972). So, for instance, all the species of section *Avena* are semispecies to each other; moreover, on several occasions interspecific hybrids occur and, with respect to the attributes here investigated, only the F_1 hybrids may be identified. Furthermore, the evidence strongly suggests that the species thus defined and recognized in this work retain their identity in nature.

Examples of the second situation occur when the species are reproductively isolated and also recognized by me on a micromorphological basis, for example, *A. ventricosa* and *A. eriantha,* or *A. longiglumis* and *A. wiestii;* the biological species is in complete accord with the taxonomic species.

Although most *Avena* species are very similar to each other in their general habit and gross morphology, the micromorphologic criteria appear to be the most definitive and immediate (time-consuming experiments are not needed, although they may be useful) for recognizing the specific taxa of *Avena.* Hawkes and Hjerting (1969) claimed that the biological species concept is by and large a very satisfactory one for delimitation of species in *Solanum,* but in the same work they claim that hybrids between species may be obtained with ease. Furthermore, they claim that most of these hybrids are fertile, but only a few were detected by them in nature. Consequently, Hawkes and Hjerting infer that because of eco-geographical isolation, gene flow is in fact restricted. This seems to be precisely the situation in *Avena* except that isolation is also effected by differential phenology and is enhanced because the species is predominantly self-pollinated. In view of what has already been mentioned, I cannot concur with Hawkes and Hjerting in the applicability of the biological species concept; on the contrary, the taxonomic species, which is primarily based on morphology or broadly speaking on phenetics, seems the operational and applicable concept for potatoes and certainly for *Avena*. The point I want to make is that although cytogenetic data (in the broad sense to include breeding behavior etc . . .) are very important, they should be regarded only as another kind of phenetic data, that is data of primary importance, but not the sole criterion for species delimitation.

Cultivars: Here too, I prefer not to adhere to a particular concept. Cultivars are man-made utilitarian entities. Although the breeding and agronomic procedures used to generate the varieties are controlled, one can safely say that many other uncontrolled biological processes occur at the same time resulting in natural variability. Just as variability may lie unsuspected within apparently uniform populations, so may hidden selective forces operate in agronomic system (Walker 1969). Thus cultivars can be regarded as minor infraspecific variants, which are more or less closely similar to each other, and in addition the variability within each cultivar is such that in the majority of cases cultivars intergrade with one another. At this level and with the constraints under which cultivars are determined, recognized, and used by man, one need not be concerned with discovering them; instead, one need only discriminate one from another and find an adequate definition or circumscription of each. This can effectively be done by using numerical methods coupled with such phenetic characters as those of morphology, anatomy, chemistry, physiology, serology, and amino acid sequences of proteins. (See also Chapter 10.)

GROUPING OF TAXA INTO CLASSES

Species: The classification of species into groupings, such as sections within a genus, can serve many purposes; the more purposes achieved, the better the classification. There should, therefore, be a constant search for the best classification, which I define as the classification that conveys maximum information about the characters while the characters convey maximum information about the classification. Information can be measured in various terms, such as Shanon's information. The constant search for the best classification also involves the search for new characters. By constantly refining a classification of species, I believe, we might attain the goal of helping the plant breeder to understand the structure of a genus so that he can plan and utilize the material available to him for a more efficacious improvement of the crop.

Cultivars: The grouping of cultivars of one species into classes may have some merit and some uses. If these groupings convey genetic kinship and if the cladistic relationships are unknown, the groupings may be of great help to the breeder and geneticist. But, at best, these groupings can provide only an approximation of cladistic relationships, and many times there may be disparity between the phenetic relationships and the true cladogram, as has been suggested by Baum and Lefkovitch (1973). When the cladistic relationships among the cultivars are known, for example from the pedigree charts of the oat cultivars (Baum 1973*a*), the classification of cultivars may then be useful for identification purposes. To identify cultivars, however, I would prefer a scheme independent of groupings; see, for instance, Baum and Brach (1975) and Baum and Thompson (1976).

INFRASPECIFIC CLASSIFICATION OF WILD SPECIES

This is of minor concern perhaps because very few of the infraspecific taxa produced by taxonomists, at least in *Avena,* have been used by others at all. Perhaps these infraspecific taxa have never well enough reflected biological reality. Superimposed on this is the inherent difficulty of assessing infraspecific units. Not only do we lack objective criteria, as has been so well described by E. O. Wilson (1974), but we have no clear methodology for finding infraspecific taxa. In many cases this classification activity would be very laborious or economically unsound, especially when studying a group on a worldwide scale.

When the Canada Department of Agriculture parties went on collecting trips to gather germ plasm for oat improvement (Baum et al. 1975), there was no time to concentrate on sampling to find infraspecific taxa. Instead, they tried to collect within large areas to obtain a wide geographic variation that would provide an adequate basis for planning future trips that could place greater emphasis on local genetic variation.

At least in oats, attempts aimed at infraspecific classification are not of major concern now, although their potential contribution to the planning and effective utilization of gene pools is promising. Such attempts should clearly be undertaken in the future in connection with gene banks.

5. The generic limits of the genus *Avena*

Linnaeus described 16 species of *Avena* in his various publications, as can be drawn from Richter (1840). After new species were discovered, many of his species were removed from *Avena* because various authorities then realized that new genera should be established (e.g., *Trisetum, Arrhenatherum, Gaudinia, Danthonia, Cheataria,* and *Helictotrichon,* among others). A glance at the index at the end of this work reveals the many transfers of names from *Avena* to other genera. Some names have been transferred back and forth, such as *A. macrostachya, Danthoniastrum compactum (A. compacta),* and *Duthiea oligostachya (A. oligostachya),* and reclassified many times (see Baum 1973*b*).

The genera most closely allied to *Avena* are *Helictotrichon, Avenochloa,* and *Arrhenatherum,* but until a decade ago many authors of floras used to classify the species of these three genera under *Avena* with a broad generic circumscription. Obviously, there are various circumscriptions of these genera, and different authors adhere to different schemes. Short historical reviews of *Avena* and its allied genera were published earlier (Baum 1968*a*; Holub 1962) and this topic will not be treated here except for main items related to *Avena* as delimited in this work.

Not only the delimitation of *Avena* has long been in dispute but also the question of rank attributed to each taxon (Baum 1968*a*). The annual species were for a long time recognized as a group or a class. Many authors used to call this group *Aveneae genuineae,* that is, the genuine oats, which is equivalent to what the ancient Greeks called "Vromos" (see Chapter 2), or equivalent to what the Romans called *"Avena,"* or similar names of ancient folk taxonomies.

Besser (1826-1827, in a letter to Schultes and Schultes), among others, proposed to regard *Avena* as a genus comprising only the annual species, and proposed the generic name *Helictotrichon* for the perennial species. This taxonomy remained obscure, perhaps because it was published in note form and was consequently overlooked by many taxonomists until Hubbard (1936, 1937). Most taxonomists, including the monographers of the annual species of *Avena,* kept this group at the sectional level or at some other rank below genus. Even Malzew (1929, 1930), the most prominent monographer of this group of species, conceived it as Section *Eu-Avena* following Grisebach (1853) who actually followed Reichenbach's (1830) concepts.

A few authors of the latter part of the eighteen hundreds, such as Stapf (1899), circumscribed *Avena* at the generic rank in a manner that included the annual species only. The Russian botanists Rozhevits (1934) and Nevski (1934) also recognized *Avena* as a genus separate from *Avenastrum* (i.e., *Helictotrichon*). Since the nineteen thirties, this circumscription of *Avena* has been increasingly adopted by taxonomists, either directly or indirectly by recognizing *Helictotrichon* (Schweickderdt 1937; Henrard 1940; Pilger 1954; Paunero 1959; Holub 1958; and Bor 1960; to name a few). Some botanists, however, followed Malzew until recently, for example, Maire (1953).

Although Potztal (1951) recognized *Avena* in its narrow sense, that is, the annual species only, she transferred (using only the histotaxic method for leaf blades) all the perennial species with the "Avena-type" and "Stipavena-type" of blade to the genus *Arrhenatherum* and classified the other perennial species, that is, those with the "Avenastrum-type" of blades, in the genus *Helictotrichon.* (A glance in the index will reveal all the transfers made by her.) In my opinion, Potztal put too much weight on one character. Moreover, Henrard (1940) and Metcalfe (1960) have, among others, already expressed the view that histotaxic investigations alone are not a good method for discrimination of taxa and are not sufficient to provide criteria for determination of genera.

The problem of delimitation of the genus will remain a matter of dispute because in various aspects *A. macrostachya* occupies an intermediate position. For this reason St.-Yves (1931) decided to keep *Avena* as a genus in the broad sense, that is, together with all the perennial species. Four approaches to classification appear possible: (i) to accept one large genus as was done traditionally, (ii) to regard *Helictotrichon, Avena,* and *A. macrostachya* as three separate genera, (iii) to include *A. macrostachya* with *Helictotrichon,* and (iv) to include *A. macrostachya* with *Avena.*

I have earlier (Baum 1968*a*) taken the fourth alternative, and have given reasons for it. At that time, I did not fully understand that discrimination is a separate exercise from classification and that the two have different purposes and different principles, and so I was concerned with a mixture of both. It now appears to me that such evidence is not sufficient to justify a broad-based conclusion. For instance, we have not yet crossed *A. macrostachya* with all the species of *Avena* and with all the species of *Helictotrichon* to study breeding behavior and chromosome behavior of artificial hybrids. Such work, but only with the *Avena* species, is now under way. We are lacking a great deal of biosystematic data, even comparative morphology with *Helictotrichon* species, not to speak of protein relationships and other data. When placed in *Avena, A. macrostachya* appears to be isolated in one section (see Chapter 7); and a similar exercise ought to be done for *Helictotrichon* and *Avena* together. Most genera, it can be stipulated, have not yet been investigated thoroughly to assess their circumscription. What has been done in general is discrimination, and this has been used for classification also. I do not think that it is germane here to discuss the conceptual problem of classification and discrimination, but it is noteworthy that I have shown (see Chapter 7) that characters that are useful for discrimination do not carry the same weight for classification purposes.

I would be safe in stating with present knowledge that *A. macrostachya* should be placed in *Avena,* based on overall similarity assessed intuitively on gross morphological habit, on shape and structure of the glumes, on the anatomy of the awns, on the histotaxis of the blades, and on the nature of the endosperm. It has been shown, however, that *A. macrostachya* occupies an isolated place in *Avena* (see Chapter 7). The loose panicles, the relatively large florets and spikelets, and the mode of insertion of the awns in *A. macrostachya* are more similar to those of *Avena* than *Helictotrichon.* The configuration and structure of the scars, and the perennial habit are similar to *Helictotrichon.* A few more details are given in Baum (1968*a*) and it appears that the smooth, rounded back of the glumes is an excellent charracter for discrimination of *Avena* (in my sense, i.e., including *A. macrostachya*) from its allies *Helictotrichon, Avenochloa,* and *Arrhenatherum.* The lodicules are more complex than those I described earlier (Baum 1968*a*), as can be seen from my later publications and from the data in this work. The lodicules should be more thoroughly investigated in *Helictotrichon* for generic delimitation and other work needs to be done (see above remarks).

The inclusion of *A. macrostachya* in the genus *Avena* has been supported by Gervais (1973) on the basis of the structure of the cross section of the roots and on the basis of gross morphology.

The circumscription of the genus *Avena* (including *A. macrostachya*) is given in the generic description (see Chapter 15).

6. Comments on previous systems of classification of *Avena*

(1) Cosson (1854) recognized 12 species. His groupings are monothetic and based on only one character, that is, the mode of disarticulation of the florets. Thus subsection *Sativae* includes all the species with non-disarticulating florets, that is, all the cultivated species, whereas subsection *Agrestes* includes all the wild species with at least the lowermost floret disarticulating at maturity. Subsection *Agrestes* is further subdivided into two groupings, also on the basis of the mode of disarticulation. Those species with only the lowermost floret disarticulating belong to series *Biformes,* whereas the species with all the florets disarticulating belong to series *Conformes.*

Cosson's classification remained in use for almost 80 years and was used by prominent investigators of *Avena,* such as Thellung. It has outlived its usefulness because it has been superseded by different classifications, which are based on additional characters, especially cytological ones. Both in the Malzewian context and in view of new findings, this classification should be abandoned because it does not convey much about the species and the characters. It is, however, the first classification of the genus based on a character with a strong discriminative weight.

(2) Husnot (1897) treated only 6 species, the ones confined to the geographical area that he dealt with. His classification is essentially the same as that of Cosson. It differs in that the two subsections of Cosson are relegated to the category of series. The two series have the same circumscriptions and the name of one is changed from "Agrestes Cosson" to "Fragiles Husnot." Furthermore, Husnot does not recognize, or at least does not mention, the two groupings of Cosson's *Agrestes*. In other words, Series *Fragiles* is not subdivided into groupings.

(3) Malzew (1929, 1930) recognized 7 species and a number of subspecies. His classification is monothetic and based chiefly on two characters, that is, the lemma tips and the chromosome number, for the two major groupings, subsection *Aristulatae* and subsection *Denticulatae*. Three groupings are recognized in subsection *Aristulatae* and these are based mainly on the relative size of the two glumes and the shape of the scar. This classification is still followed by many oat workers.

(4) The classification of Nevski (1934), in contrast to the preceding ones, appears to be polythetic and treats 26 species. Although it is presented as a key to species like the previous systems, many groups obviously overlap in their character-states. Nevski's system differs in that 7 groupings are recognized. These are at the level of series and some are subdivided into *stirpes*. In Nevski's classification, although some characters are adopted from Malzew, a number of new ones are used, such as: number of florets in the spikelets; and range of sizes of spikelets, of glumes, and of lemmas.

For further information, see Chapter 12, "Comments on characters for discrimination of oat species." A resumé of various classificatory schemes follows.

SYNOPSIS OF VARIOUS SYSTEMS OF CLASSIFICATION OF *AVENA*

(1) Cosson, M. E. (1854) Bull. Soc. Bot. France 1:13-18
    Subsection *Sativae* Coss. et Dur. in Coss. p. 13
        Lectotype species: *A. sativa* L.
    Subsection *Agrestes* Coss. et Dur. in Coss. p. 14
        Lectotype species: *A. sterilis* L.
        Series *Biformes* Coss. et Dur. in Coss. p. 14
            Lectotype species: *A. sterilis* L.
        Series *Conformes* Coss. et Dur. in Coss. p. 14
            Lectotype species: *A. fatua* L.
(2) Cosson and Durieu (1855) Exploration Scientifique de l'Algérie, Botanique 2:105, same system as (1).
(3) Husnot, T. Graminées. 1896–1899
    Series *Sativae* Husnot p. 38 (1897)
        Lectotype species: *A. sativa* L.
    Series *Fragiles* Husnot p. 39 (1897)
        Lectotype species: *A. fatua* L.
(4) Malzew, A. I. (1930) Suppl. 38th of Bull. Appl. Bot. Genet. Pl. Breed.
    Subsection *Aristulatae* Malz., p. 225
    Lectotype species: *A. strigosa* Schreb.
        Series *Inaequaliglumes* Malz., p. 225
        Lectotype species: *A. clauda* Dur.
        Series *Stipitatae* Malz., p. 225
        Lectotype species: *A. longiglumis* Dur.
        Series *Eubarbatae* Malz., p. 225
        Lectotype species: *A. strigosa* Schreb.
    Subsection *Denticulatae* Malz., p. 226
    Lectotype species: *A. fatua* L.
(5) Nevski, S. A. (1934) Acta Univ. As. Med. Ser. 8b, Bot. Fasc. 17 (1934)
    Series *Claudae* Nevski
    Type species: *A. clauda* Dur.
    Series *Erianthae* Nevski
    Type species: *A. eriantha* Dur.
    Series *Longiglumes* Nevski
    Type species: *A. longiglumis* Dur.
    Series *Ventricosae* Nevski
    Lectotype species *A. ventricosa* Bal. ex Coss.
    Series *Barbatae* Nevski
    Lectotype species *A. barbata* Pott ex Link
        Stirps *Strigosae* Nevski
        Lectotype species: *A. strigosa* Schreb.
    Series *Fatuae* Nevski
    Lectotype: *A. fatua* L.
        Stirps *Volgenses* Nevski
        Type species: *A. volgensis* Nevski
        Stirps *Macranthae* Nevski
        Lectotype species: *A. macrantha* (Hack.) Nevski
        Stirps *Sativae* Nevski
        Lectotype species: *A. sativa* L.
    Series *Steriles* Nevski

Lectotype species: *A. sterilis* L.
> Stirps *Pseudosativae* Nevski
> Type species: *A. thellungii* Nevski
> Stirps *Byzantinae* Nevski
> Lectotype species: *A. byzantina* Koch

7. The classification of *Avena* species in this work: an account of how the classes were obtained and a description of the system

The present classificatory exercise is based on 27 species recognized and described in detail in this work, and uses some of the new techniques and approaches of classification now available, that is, numerical taxonomy in its broad sense and computer. This chapter is an excerpt of my earlier publication (Baum 1975) and has been slightly modified.

PURPOSE:

- to determine possible classifications (groupings) using a number of clustering procedures based on a set of characters or subsets of these;
- to evaluate these classifications with respect to their information content;
- to study the relationships between the groupings of the best classification;
- to devise a means of discrimination among these groupings;
- to document and describe the classification system thus obtained.

A note on information and diagnostic value:

(1) Information: The best classification, among the classifications attempted, is one that tells us the maximum about the greatest number of characters, that is, it is one with maximum information content. In this paper the characters and the character-states make up the measures of information on the objects, that is, the species. The information-theoretic measure of information used here is based on Shannon's formula of entropy:

$$H = - \sum_{i=1}^{n} P_i \log_2 P_i$$

where H is the amount of entropy of an information system; P_i is the probability of an object being in state i of a given character.

(2) Diagnostic value: It is not necessarily true that characters with high diagnostic value for objects possess the same value for distinguishing between the groupings of the best classification. This will be demonstrated below. Briefly, the *interdependence of characters* (based on entropy values) tells us something about their diagnostic value for purposes of identification of the objects, whereas *interdependence of characters and classifications* (also based on entropy values) tells us about the various amounts of information in each of the characters that the different classifications have preserved.

MATERIALS AND METHODS

Although 27 species are recognized in this monograph, this classificatory study includes 28 OTU's; OTU No. 28 is conspecific with *A. sativa*. This was done purposely so that any classification in which OTU's 28 and 20 fell into separate groupings was automatically discarded as inadmissible. The following table lists the 28 OTU's with their appropriate codes used in the classificatory study.

Code	Name of species	Code	Name of species
1	A. abyssinica	15	A. maroccana
2	A. hispanica	16	A. atherantha
3	A. barbata	17	A. murphyi
4	A. brevis	18	A. nuda
5	A. canariensis	19	A. occidentalis
6	A. clauda	20	A. sativa
7	A. damascena	21	A. hybrida
8	A. eriantha	22	A. sterilis
9	A. fatua	23	A. strigosa
10	A. matritensis	24	A. trichophylla
11	A. lusitanica	25	A. vaviloviana
12	A. hirtula	26	A. ventricosa
13	A. longiglumis	27	A. wiestii
14	A. macrostachya	28	A. sativa fatuoid

The character-states have been subjectively assessed and defined (see Chapter 3, "Methods" and Chapter 12), and they are listed in Table 2 with their assigned codes. The numeric codes were used for all the exercises except for input into CHARANAL, which used the alphabetic coding. The actual values of the quantitative characters were used as input for all exercises except for input into CHARANAL for which the equivalent values are given in the last two columns to the right together with their alphabetical codes.

The 28 OTU's were subsequently coded according to Table 2, and according to their descriptions or circumscriptions. At this stage, it should be kept in mind that the circumscriptions were intuitively assessed. The Basic Data Matrix thus obtained is given in Tables 3 and 4. The latter served for input into character analysis.

Because the data are *mixed* (i.e., they consist of dichotomic, alternative, multistate unordered, multistate ordered, and quantitative characters), Gower's coefficient of resemblance, S_G (Gower 1971), was used to compute the similarities between all pairs of characters. S_G has another advantage because it can handle a hierarchical system of characters (i.e., weighting of primary, secondary, . . . characters) in the sense of Kendrick and Proctor (1964), such as characters 11–15.

Dissimilarity is defined as $(1 - S_{ij}^2)^{1/2}$. The OTU by OTU matrix of dissimilarities thus obtained was subsequently used as input for cluster analysis. A number of other dissimilarity matrices were computed for various subsets of characters; namely, all characters except 29; characters 10–19, 22, and 24–29; characters 10–19, 22, and 24–27; and characters 9, 13, 16, 18, 20, 24, and 25–29. These subsets of characters, except the last, were intuitively chosen *a priori*.

The rationale of the choice of various subsets of characters is as follows. In the subset "all characters except character 29", I wanted to see how effective the data was without the information on the genome. The subset "characters 10–19, 22, 24–29" contains those characters that I consider, on intuitive grounds, to be most important for taxonomic purposes. The subset "characters 10–19, 22, 24–27" contains the same characters as the preceding subset except the characters on the genome and chromosome number. Character 28 was also omitted because different species have, at least partly, different genes on their chromosomes; therefore species with the same chromosome number are not necessarily similar. The last subset, however, was obtained from a character analysis exercise (see below, Flowchart and "Results of clustering and ordination analysis").

The following clustering methods were used in combination with the above-mentioned character sets: nearest neighbor, farthest neighbor, unweighted centroid, weighted centroid, average linkage, incremental sum of squares, flexible sort (using various α values, 0.1–0.9), and single linkage (a different algorithm than the one used for the nearest neighbor method).

To investigate the interrelationship among the characters, the information theoretic model of Estabrook (1967) and his program CHARANAL (Legendre and Rogers 1972) were used. The last of the above-mentioned subsets of characters was selected by this method. The CHARANAL program was also used, after clustering of the OTU's, to find the association between the various characters and the various classifications. Similarly, from the measures of entropy for each classification, a classification matrix was obtained by computing

$$D(I,J) = \frac{H(J/I) + H(I/J)}{H(I.J)},$$

where $H(J/I)$ = information held exclusively by the Ith classification
$H(I/J)$ = information held exclusively by the Jth classification
$H(I.J)$ = information possessed by the union of I and J.

This classification by classification matrix was defined as a distance matrix. To study the relationships among classifications, the distance matrix was studied by principal coordinate analysis (Gower 1966,1967) and also by finding the minimum spanning tree (Prim 1957; Gower and Ross 1969) connecting the classifications.

A non-hierarchical clustering was also performed using similarities based on the set of all characters and also on the subset of characters obtained from CHARANAL described above. Overlapping clusters were found by the method of Jardine and Sibson (1971, Appendix 5:238–239) for computing maximal complete subgraphs using an algorithm by Poushinsky and Sheldrake (unpublished); the same dissimilarity matrices mentioned above were used.

Finally, the groupings for the classification considered to be the best were studied by a canonical analysis (Seal 1964).

All computer programs were written in FORTRAN V or adapted for the UNIVAC 1108 by the Statistical Research Service, Agriculture Canada.

The following flowchart summarizes the steps taken in the classificatory exercise.

TABLE 2. CHARACTERS USED IN THE CLASSIFICATORY STUDY.

List of characters and character-states scored for the *Avena* OTU's. The organs from which the characters were taken and the appropriate codes of states are given. By karyotype-genome is meant the genome formula without going into genomic variability; for instance A genomes are known (Rajhathy and Thomas 1974) but these are not distinguished.

Organs or part of plant	Character	State		Codes of state	
General habit	1) Longevity	(1) Perennial		0	A
		(2) Annual		1	B
	2) Juvenile growth	(3) Erect		10	A
		(4) Prostrate to erect		15	B
		(5) Prostrate		20	C
	3) Color appearance	(6) Glaucous		0	A
		(7) Green		1	B
	4) Culm	(8) Geniculate or prostrate		0	A
		(9) Erect		1	B
		(10) Prostrate or erect		2	C
	5) Height	(11) Low plants, not exceeding 95 cm (15–95 cm)		0	A
		(12) Tall plants, exceeding 100 cm (30–180 cm)		1	B
Ligules	6) Shape at anthesis or younger	(13) Obtuse		0	A
		(14) Acute		1	B
Panicle	7) Shape	(15) Equilateral		0	A
		(16) Slightly flagged, equilateral or flagged, equilateral and sometimes flagged		1	B
Spikelets	8) Length without awns	(17) Short; not exceeding 25 mm in upper limit		10	A
		(18) Indeterminate; lower limit less than 20 mm, upper limit exceeding 25 mm		15	B
		(19) Long; lower limit not less than 20 mm, upper limit exceeding 25 mm		20	C

Classification

TABLE 2. (Cont'd)

Organs or part of plant	Character		State		Codes of state
	9) Number of florets	(20)	1–2 or 2 only	0	A
		(21)	2–3	1	B
		(22)	2–4	2	C
		(23)	2–5	3	D
		(24)	2–6	4	E
		(25)	3–4	5	F
		(26)	3–6	6	G
		(27)	1–7	7	H
Glumes	10) Relative length	(28)	Equal	0	A
		(29)	Unequal	1	B
Dispersal unit	11) Mode of disarticulation	(30)	Non-disarticulating	0	A
		(31)	Disarticulating	1	B
Florets	12) Mode of disarticulation	(32)	Only lowermost floret disarticulating	0	A
		(33)	All florets disarticulating	1	B
Scars	13) Shape of 1st floret scar	(34)	Oval to round	0	A
		(35)	Heart-shaped or slightly so	1	B
		(36)	Elliptic	2	C
		(37)	Narrow elliptic	3	D
		(38)	Linear	4	E
	14) Shape of 3rd floret scar	(39)	Heart-shaped	0	A
		(40)	Not heart-shaped	1	B
	15) Periphery ring	(41)	All around scar	0	A
		(42)	Confined to ½ scar	1	B
		(43)	Confined to ⅓–½ scar	2	C
		(44)	Confined to ⅓ scar	3	D
		(45)	Confined to ¼ scar	4	E
		(46)	Confined to ⅛ scar	5	F
		(47)	Confined to ⅛–¹⁄₁₆ scar	6	G
Lemma	16) Place of awn insertion	(48)	Absent or rudimentary, or inserted about middle of lemma to below tip	0	A

Classification

	(50)	Inserted at about ⅓	2	C
	(51)	Inserted between lower ⅓ and ½	3	D
	(52)	Inserted at about middle	4	E
	(53)	Inserted at about upper ⅓	5	F
	(54)	Inserted between upper ⅓ and ¼	6	G
	(55)	Inserted upper ¼ below tip	7	H
17) Structure	(56)	Tough	0	A
	(57)	Resembling glumes	1	B
	(58)	Tough or resembling glumes	2	C
18) Shape of tip	(59)	Bilobed	0	A
	(60)	Bidenticulate	1	B
	(61)	Bisubulate	2	C
	(62)	Biaristulate	3	D
	(63)	Bisetulate-biaristulate	4	E
	(64)	Shortly bisubulate or mucronate	5	F
	(65)	Bisubulate or biaristulate	6	G
	(66)	Bidenticulate to bisubulate	7	H
	(67)	Bisetulate or bisetulate-biaristulate	8	I
19) Vestiture below awn insertion	(68)	Beset densely with macrohairs	0	A
	(69)	Without macrohairs or a few around awn insertion	1	B
	(70)	Both of the above present	2	C
20) Rows of cilia along edges of keels	(71)	1	0	A
	(72)	1–2	1	B
	(73)	1–3	2	C
	(74)	2	3	D
21) Vestiture of back	(75)	Glabrous	0	A
	(76)	Prickles	1	B
	(77)	Macrohairs	2	C
	(78)	Prickles or macrohairs	3	D
22) Type	(79)	Sativa	0	A
	(80)	Fatua	1	B
	(81)	Strigosa	2	C
	(82)	Fatua and sativa	3	D
	(83)	Fatua and strigosa	4	E

Palea

Lodicule

TABLE 2. (Cont'd)

Organs or part of plant	Character	State		Codes of state
	23) Prickles	(84) Always absent	0	A
		(85) Rarely present	1	B
		(86) Often present	2	C
Epiblasts	24) Type	(87) Brevis	0	A
		(88) Fatua	1	B
		(89) Septentrionalis	2	C
		(90) Sativa	3	D
		(91) Sativa and fatua	4	E
	25) Lower limit of width	In actual mm, or	0.2	A
			0.25	B
			0.3	C
			0.35	D
			0.4	E
			0.45	F
			0.5	G
			0.55	H
			0.6	I
			0.8	J
	26) Upper limit of width	In actual mm, or	0.2	A
			0.25	B
			0.3	C
			0.35	D
			0.4	E
			0.45	F
			0.5	G
			0.6	H
			0.65	I
			0.7	J
			1.0	K
	27) Median of range	In actual mm, or	0.2	A

			0.3		B
			0.325		C
			0.35		D
			0.4		E
			0.425		F
			0.5		G
			0.525		H
			0.55		I
			0.6		J
			0.65		K
			0.9		L
					M
Chromosomes	28) Number	$2n = 14$	0	A
		$2n = 28$	1	B
		$2n = 42$	2	C
		$2n = 40, 41$ or 42	3	D
		$2n = 14$ and 48	4	E
	29) Karyotype-genome	CC	0	A
		AA	1	B
		AABB	2	C
		AACC	3	D
		AACCDD	4	E

Classification

TABLE 3. BASIC DATA MATRIX USED IN THE CLASSIFICATORY STUDY.

The 28 OTU's and 29 characters are used as input in various computational procedures. Missing value code is −1.

Characters OTU's	1	2	3	4	5	6	7	8	9	10	11	12	13	14	15	16	17	18	19	20	21	22	23	24	25	26	27	28	29
1	1	10	1	1	0	0	0	10	1	0	0	−1	−1	−1	−1	4	0	2	2	1	1	0	0	0	0.5	0.5	0.5	1	2
2	1	15	0	1	1	1	0	10	0	0	0	−1	−1	−1	−1	6	0	4	1	0	1	1	1	0	0.4	0.4	0.4	0	1
3	1	20	1	0	0	0	0	20	0	0	1	−1	2	−1	2	4	0	6	0	0	1	1	0	0	0.3	0.4	0.35	1	2
4	1	10	0	1	0	0	1	10	1	0	0	0	−1	−1	−1	7	0	5	0	0	1	1	0	0	0.3	0.4	0.35	0	1
5	1	15	1	1	1	0	0	10	0	1	1	−1	−1	−1	−1	5	0	−1	1	1	0	1	1	3	0.3	0.4	0.35	0	0
6	1	20	1	1	0	0	1	20	4	0	1	1	4	−1	4	3	0	4	0	2	0	0	0	2	0.3	0.3	0.3	0	0
7	1	15	1	1	0	1	0	20	1	1	1	1	2	−1	5	2	0	4	0	2	3	1	0	3	0.25	0.25	0.25	0	−1
8	1	20	0	1	0	0	0	10	1	1	1	0	4	−1	4	4	0	4	1	0	1	0	0	2	0.3	0.3	0.3	0	1
9	1	15	1	1	−1	0	0	10	1	0	1	1	0	−1	5	2	0	2	2	2	3	1	0	1	0.45	0.6	0.525	2	4
10	1	−1	0	1	0	1	0	20	1	1	1	0	3	−1	2	4	0	7	0	0	−1	1	0	3	0.3	0.35	0.325	−1	−1
11	1	20	1	1	−1	0	1	15	1	0	1	−1	3	−1	4	2	0	3	0	0	−1	0	0	0	0.3	0.3	0.3	0	1
12	1	20	1	1	0	0	0	10	1	0	1	−1	2	−1	4	2	0	4	0	2	3	3	0	3	0.3	0.3	0.3	0	−1
13	1	20	0	2	−1	0	0	20	1	0	1	−1	4	−1	4	3	0	2	0	0	−1	3	1	2	0.2	0.2	0.2	0	1
14	0	−1	−1	1	−1	1	0	15	6	0	1	1	3	0	6	4	0	8	0	0	−1	1	0	−1	0.3	0.35	0.325	−1	−1
15	1	20	1	1	1	1	1	20	1	1	0	0	0	−1	4	4	1	2	1	−1	−1	−1	0	−1	−1	−1	−1	−1	3
16	1	15	1	1	1	0	1	15	1	1	1	−1	2	−1	2	4	0	1	0	3	2	0	0	1	0.8	1.0	0.9	1	4
17	1	10	1	1	0	1	0	10	3	0	1	−1	1	0	2	0	2	7	1	2	2	0	1	1	0.5	0.6	0.55	2	4
18	1	10	0	1	0	0	1	20	2	0	1	0	0	−1	2	4	0	6	2	2	2	0	0	1	0.6	0.7	0.65	2	3
19	1	15	1	1	0	1	1	20	6	0	0	−1	−1	−1	−1	1	1	7	0	1	1	4	0	0	−1	−1	−1	0	1
20	1	15	1	1	1	1	1	20	5	0	1	1	2	−1	2	4	0	2	2	2	1	1	0	0	0.6	0.6	0.6	2	4
21	1	−1	−1	1	1	0	0	15	7	0	0	1	1	−1	4	0	2	1	1	2	1	0	0	2	0.45	0.65	0.55	2	4
22	1	15	1	1	0	1	0	10	2	0	1	0	0	0	2	4	0	7	2	2	1	0	0	4	0.35	0.35	0.35	2	4
23	1	10	0	1	1	0	1	15	3	0	1	−1	0	−1	2	3	0	7	0	2	1	0	0	2	0.55	0.65	0.6	2	4
24	1	−1	1	1	1	0	0	20	1	0	0	0	−1	0	−1	2	0	4	2	1	−1	2	2	0	0.3	0.35	0.325	2	4
25	1	10	0	1	0	1	0	10	1	0	1	0	1	−1	3	2	0	2	0	1	1	2	3	2	0.35	0.35	0.35	1	4
26	1	10	0	1	0	0	1	20	5	0	1	1	4	0	5	4	0	2	2	1	−1	1	0	0	0.4	0.45	0.425	1	2
27	1	10	1	1	1	1	1	15	0	0	1	1	2	−1	2	4	0	8	0	0	−1	0	0	2	0.35	0.35	0.35	0	0
28	1	15	−1	1	1	1	1	10	1	0	1	1	1	1	2	4	0	7	2	2	0	0	0	4	0.5	0.6	0.55	3	4

TABLE 4. BASIC DATA MATRIX USED IN THE CHARACTER ANALYSIS.

The 28 OTU's and 29 characters were recoded for input into CHARANAL.

Characters	1	2	3	4	5	6	7	8	9	10	11	12	13	14	15	16	17	18	19	20	21	22	23	24	25	26	27	28	29
OTU's																													
1	B	A	B	B	A	B	A	A	B	A	A	A	O	O	O	E	A	C	B	B	B	A	A	A	G	G	G	B	C
2	B	B	A	A	B	B	A	A	B	A	A	O	O	O	O	E	A	E	A	A	B	B	B	A	E	E	F	A	B
3	B	C	B	B	B	A	B	C	B	A	B	B	O	B	C	E	A	G	B	B	B	B	A	A	E	E	D	B	B
4	B	A	A	B	A	A	A	A	A	A	A	O	B	B	O	H	A	F	B	A	B	B	A	A	C	E	D	A	B
5	B	B	B	B	A	B	B	A	B	B	A	A	B	B	O	E	A	B	A	B	A	B	B	A	C	E	D	A	B
6	B	B	B	B	A	A	A	C	A	A	B	B	B	B	E	O	A	E	B	B	A	B	B	D	D	C	D	A	A
7	B	A	B	B	A	A	B	A	B	B	B	B	E	B	F	C	A	E	A	B	A	A	A	D	B	B	C	A	B
8	B	C	A	B	A	A	A	A	B	A	B	A	A	B	E	E	A	C	B	C	B	A	A	B	C	C	B	A	A
9	B	B	B	B	B	B	B	B	B	B	B	B	D	O	C	E	A	H	B	A	B	A	B	D	F	H	D	C	E
10	B	B	A	B	O	B	A	C	B	A	B	B	D	B	E	E	A	D	B	A	B	D	B	D	C	D	D	A	O
11	B	O	B	B	A	A	B	A	B	B	B	B	D	B	E	C	A	E	A	A	B	A	A	A	C	C	C	A	B
12	B	O	O	A	A	A	A	A	G	B	B	B	C	B	G	D	A	C	A	A	O	B	B	A	A	A	A	B	B
13	B	C	A	B	A	A	B	B	D	A	B	A	A	B	E	E	A	E	A	A	B	A	B	B	A	O	D	B	O
14	A	C	C	A	B	A	A	C	C	A	A	A	A	B	E	E	B	C	A	A	C	D	A	A	O	H	O	O	B
15	B	O	A	B	B	B	B	C	F	B	B	A	O	O	O	B	A	G	B	D	B	B	B	C	J	J	K	C	O
16	B	B	B	B	B	A	B	C	B	B	B	B	A	O	C	A	A	H	B	C	O	B	A	B	I	H	H	C	D
17	B	B	A	B	A	A	A	C	B	A	B	A	O	O	O	E	B	G	C	C	B	B	A	A	O	J	J	O	B
18	B	B	A	B	B	B	B	B	B	A	B	B	A	O	O	E	A	C	B	C	B	A	A	B	I	O	L	C	E
19	B	B	B	B	A	B	B	C	B	A	B	B	B	B	O	C	A	B	A	O	B	B	A	A	O	H	O	C	E
20	B	B	O	B	O	B	B	B	B	A	A	O	B	O	O	E	A	H	A	C	O	E	B	C	I	I	K	C	E
21	B	O	A	B	B	B	A	A	F	B	B	A	B	O	D	E	A	E	A	C	D	A	B	B	F	O	J	C	E
22	B	C	B	B	B	A	B	A	B	A	B	A	O	B	O	E	A	C	D	O	D	B	B	A	D	H	K	C	B
23	A	A	B	B	A	B	B	A	B	A	B	B	B	B	F	E	A	E	A	B	O	O	C	C	H	I	D	O	E
24	B	O	A	B	A	A	A	A	B	A	A	A	E	B	A	E	A	A	C	B	B	A	A	A	E	D	E	C	E
25	B	A	A	B	B	B	B	A	B	A	B	A	B	B	F	E	A	C	C	C	B	B	B	D	D	E	G	B	B
26	B	B	A	B	B	A	A	A	B	A	B	A	E	B	C	E	A	C	B	O	O	A	A	O	O	F	E	C	C
27	B	A	B	A	A	B	B	A	F	A	B	B	C	B	C	E	A	—	A	B	B	A	A	A	G	D	B	A	A
28	B	B	O	B	B	A	B	A	B	A	B	B	B	B	C	E	A	H	C	O	B	A	A	E	H	H	J	D	E

37

RESULTS OF CLUSTERING AND ORDINATION ANALYSES

80 phenograms were obtained by the eight clustering methods using similarities based on the various subsets of characters; 16 of these were chosen as admissible based on the following criteria:
(1) Presence of obvious clusters — acceptable; absence of clusters (i.e., chaining or great unevenness) — rejected.
(2) Too many clusters — rejected. The phenon, which determines the clusters, was determined according to the parsimonious compromise (Baum and Lefkovitch 1972a), that is, one group is clearly too few and more than a certain number is too many.
(3) If the former two criteria are met, but too many OTU's form isolated groups containing only one OTU — rejected.
(4) Because OTU 28 is the same species as OTU 20, this was used as another criterion for choice of phenograms. A phenogram that met the first three criteria was rejected if OTU 20 and 28 were separated into two different groupings.

These 16 phenograms are shown in Figures 4 and 5, with the phenon lines and corresponding groups thus formed. In five phenograms, two acceptable phenon lines were drawn using the same criteria, resulting in 21 different classifications of the 28 OTU's. The 21 admissible classifications are listed in the following table with details complementary to the phenograms in Figures 4 and 5.

These admissible classifications resulted from various combinations of different character sets and different clustering methods, and these are summarized in the table below:

SUMMARY OF THE ADMISSIBLE CLASSIFICATIONS GIVING DETAILS ON CHARACTER INPUT, CLUSTERING METHOD, NUMBER OF CLUSTERS, AND PHENON LEVEL.

The numbers in the right-hand column indicate levels of dissimilarity except for method 6 and 7. In method 7 the alpha value is specified. Strategies: 1 = nearest neighbor; 2 = furthest neighbor; 3 = unweighted centroid; 4 = weighted centroid; 5 = average linkage; 6 = incremental sums of squares; 7 = flexible sort (alpha 0.1–0.9); 8 = single linkage (different algorithm than method 1).

Classification code	Characters (codes)	Clustering method	Number of clusters	Phenon level
C 30	9,13,15,16,18,20,24–29	1,8	4	0.73
C 31	9,13,15,16,18,20,24–29	5	6	0.78
C 32	9,18,15,16,18,20,24–29	6	5	1.1
C 33	9,13,15,16,18,20,24–29	6	6	0.97
C 34	9,13,15,16,18,20,24–29	$7(\alpha = 0.6)$	3	1.18
C 35	9,13,15,16,18,20,24–29	$7(\alpha = 0.6)$	6	0.94
C 36	10–19,22,24–27	2	2	0.98
C 37	10–19,22,24–27	2	3	0.95
C 38	10–19,22,24–27	6	4	1.20
C 39	10–19,22,24–29	2	6	0.89
C 40	10–19,22,24–29	6	2	1.50
C 41	10–19,22,24–29	$7(\alpha = 0.6)$	5	0.98
C 42	10–19,22,24–29	$7(\alpha = 0.6)$	2	1.16
C 43	all characters except 29	2	5	0.90
C 44	all characters except 29	6	2	1.35
C 45	all characters except 29	6	4	1.03
C 46	all characters except 29	$7(\alpha = 0.6)$	6	0.89
C 47	all characters	1,8	4	0.72145
C 48	all characters	2	2	0.945
C 49	all characters	6	6	0.98
C 50	all characters	$7(\alpha = 0.6)$	7	0.89

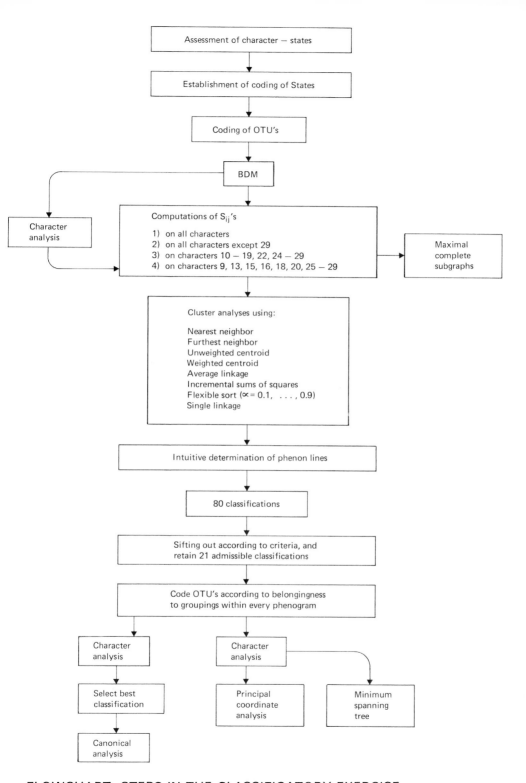

FLOWCHART: STEPS IN THE CLASSIFICATORY EXERCISE.

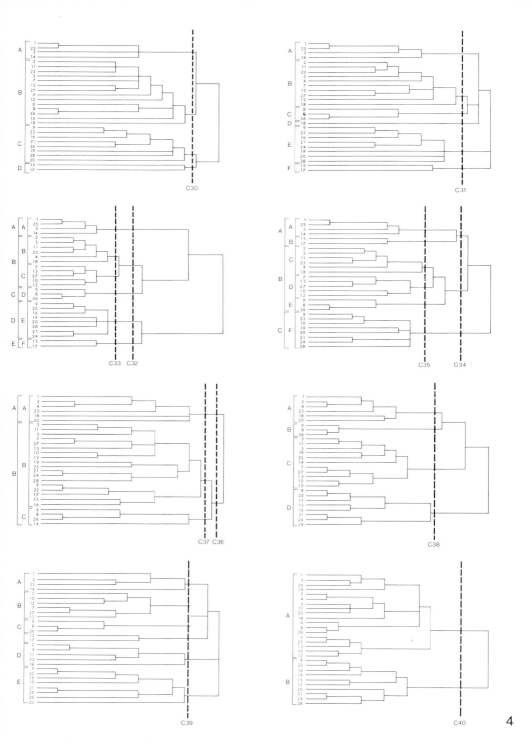

Figs. 4–5. Admissible phenograms. Phenon lines are shown, and the groupings thus formed are traced above the OTU codings. The groupings are coded A, B, C ... for reference in the text and for input in CHARANAL. See table on p. 38 for further details about these 21 admissible classifications that are coded to the lower end of each phenon line.

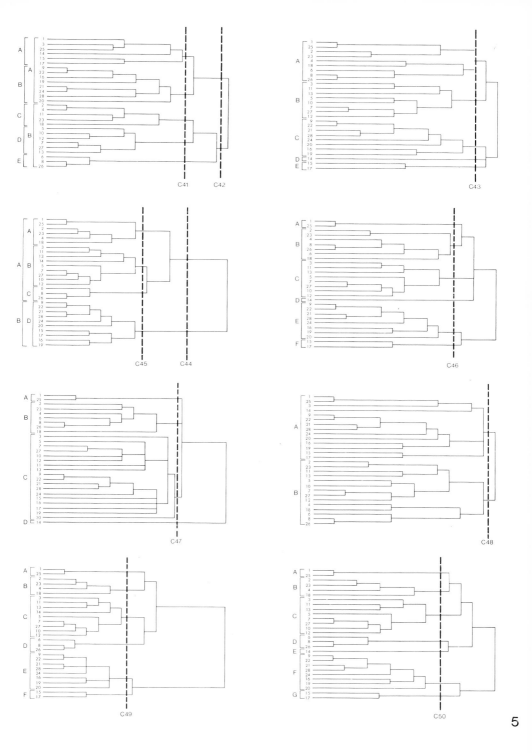

The subset "characters 9, 13, 15, 16, 18, 20, and 24–29" was determined from the character analysis by selecting the 12 characters with the highest SUMRAT and SAMRAT values obtained from CHARANAL. These values are listed below.

LIST OF SUMRAT AND SAMRAT VALUES OF THE 29 CHARACTERS ARRANGED IN ORDER OF INCREASING SUMRAT VALUES. THE LAST 12 CHARACTERS WERE SELECTED FOR INPUT IN THE CLUSTER ANALYSES.

Character	SUMRAT	SAMRAT
1	1.04824	6.04187
17	1.98184	8.37718
4	2.12901	6.15392
11	2.42975	5.06223
10	2.45763	8.17946
12	2.55913	4.66937
7	2.59752	4.31072
5	3.00719	5.18233
23	3.11733	4.85378
3	3.22577	5.59299
14	3.50989	7.71632
6	3.99882	6.73025
8	4.22949	4.70912
21	4.76347	6.80315
2	5.11606	5.48219
19	5.54241	6.01092
22	6.34724	5.72610
28	7.38538	7.87506
20	8.70585	8.24493
24	9.58462	7.43692
16	10.13459	7.14003
13	10.22528	7.27382
9	10.25774	7.21837
29	10.92658	8.93648
15	11.82685	7.98560
18	12.05978	6.88310
25	13.26459	7.43658
26	14.49523	7.55644
27	16.65287	7.99090

SUMRAT (I) is the sum of fractions representing the amount of information that character I has in common with each of the other characters divided by the amount of information of the character with which I is compared in that ratio, namely

$$\text{SAMRAT (I)} = \sum_{j=1}^{n} \left(\frac{H[I.J_j]}{H(J_j)} \right)$$

SAMRAT (I) is also the sum of ratios, but differs from SUMRAT in that its denominator is always the amount of information in I instead of the amount of information in the various J_j, that is,

$$\text{SUMRAT (I)} = \sum_{j=1}^{n} \left(\frac{H[I.J_j]}{H(I)} \right)$$

It is interesting that the subset of characters selected from the character analysis resembles the subset that I selected, on intuitive grounds, as most important for taxonomic purposes. Because the "intuitive" subset comprises 17 characters and the "CHARANAL" subset only 12, if we were to add to the "CHARANAL" subset the next characters with highest SUMRAT and SAMRAT values from the table above (i.e., character 19 and 22), the resemblance between the two subsets would be remarkable.

Some comments on the admissible classifications (Figs. 4 and 5) can now be made together with a description of each.

C30: In groups A and D are the tetraploid species; in B are the diploids; and in C are the hexaploids.

C31: In C are the C-genome species; in F are the sterilis-like tetraploids; and in E are all the hexaploids; the rest I am unable to interpret intuitively.

C32: In A and E are the tetraploids; in C are the C-genome species; in B are all the diploids; and in D are the hexaploids.

C33: This classification is very similar to C32 except that the diploid species are grouped into two different groupings, and I am unable to comment upon this, heuristically.

C34: The groupings here are simple: in A, the tetraploids; in B, the diploids; and in C, the hexaploids.

C35: This classification resembles C33. In A and B are the tetraploids; in C and D are the diploids, on which I am unable to comment; in E are the C-genome species; and in F are the hexaploids.

C36: In A are all the cultivated species and in B are all the wild species; this classification is identical with that of Cosson (1854).

C37: This classification is similar to C36 except that in C are an additional group containing the 3 species having the C-genome together with *A. macrostachya* — the only perennial oat species and with a hitherto unknown genome. These four species all have more or less unequal glumes.

C38: In A are all the cultivated species; in B are the 3 C-genome species; in C are the wild diploids and tetraploids; and in D, are the wild hexaploids together with *A. sativa* fatuoid.

C39: In A and D are the tetraploids; in C are the C-genome species; in F are all the hexaploids; and I am unable to interpret heuristically the groups of diploids in B and E.

C40: In A are all the diploids and tetraploids and in B are all hexaploids.

C41: In A are all the tetraploids; in B are all the hexaploids; in E are the three C-genome species; and in C and D are other diploids in two groups that I am unable to comment upon.

C42: This classification is simple. In A are the tetraploids and hexaploids; and in B are all the diploids.

C43: In A is a heterogeneous group; the group in B, I cannot comment upon because part of the diploids are in A; in C are all the hexaploids; in D is *A. macrostachya,* the only perennial oat species; in E are the two tetraploid oats with the AC genome.

C44: This classification is similar to C40.

C45: All the groups are heterogeneous, heuristically speaking, except group C, which contains the 3 species with the C-genome.

C46: In A are the tetraploid oats growing in Ethiopia; in B is a heterogeneous grouping consisting of the four cultivated diploids and the three C-genome species; in C are the wild diploids and the tetraploid *A. barbata*; in D is the only perennial oat species, *A. macrostachya*; in E are all the hexaploids; and in F are the two tetraploid, sterilis-like oats having the AC genome.

C47: A, B, and D are the same as in C46, but C is a group upon which I am unable to comment.

C48: This classification is similar to C42 in membership of groupings and the classification is simple.

C49: In A are the two Ethiopian tetraploid oats; in B are the 4 diploid cultivated oats; in C is a heterogeneous group containing the diploid wild species of oats together with the tetraploids, *A. barbata* and *A. macrostachya*; in D are the three C-genome species; in E are all the hexaploids; and in F are the tetraploid, sterilis-like species.

C50: This classification is similar to C49 except that *A. macrostachya* is isolated (group E). It is also very similar to C46.

From the foregoing comments it is difficult to decide which classification to adopt. To decide, I looked first at the information content of each classification, and next determined which classification contained the maximum number of characters having the greatest correlation with their grouping. The criterion for the best classification will be that in which the greatest number of characters tell us something about the classification and vice versa.

From the technical point of view, one can get a feeling about which classification is good by looking at the conditional entropies of the characters of a particular classification. In order to determine which classification meets the criterion proposed above, it is supposed that the $D(I,J)$'s between the I characters and the J classifications measure the degree of independence.

These values are summarized in Table 5.

Clearly the lower the $D(I,J)$, the greater the correlation of character I with classification J. $D(11, C36)$ is 0, which indicates complete correlation between character 11 and classification C36. But C36 is the worst classification, according to the previously discussed criterion, because all other characters are almost unrelated to that classification. Hawksworth et al. (1968) used the following evaluation of D's:

$D \leq 0.5$, very highly correlated;
$0.5 < D \leq 0.7$, highly correlated;
$0.7 < D \leq 1.0$, correlated;
D is very close to 1.0, unrelated.

This criteria can be taken into account in evaluating Table 5 above. But since the D's depend on the entropy values, so do the SUMRAT and SAMRAT values and therefore an additional way to locate which classification has maximum information content is to look at the SUMRAT and SAMRAT values of the classifications given in the next table.

This investigation suggests that C50 is the best classification; Table 5 with the D values shows that C50 has more characters with low D's than any other classification; the last two tables further suggest that some other classifications are almost equally good, viz. C49 and C46. In C49 there are 11 characters with D's < 0.7, whereas C50 has only 10 such characters; however, C50 has more characters with low D values than C49. In C46 there are 10 characters with D's < 0.7, as in C50, but their overall value is slightly higher than those in C50.

SUMMARY OF THE SUM OF RATIOS (SUMRAT AND SAMRAT) OF THE VARIOUS ADMISSIBLE CLASSIFICATIONS. THE CLASSIFICATIONS ARE ARRANGED IN ORDER OF INCREASING SUMRAT VALUES.

Classification	SUMRAT	SAMRAT
C36	6.31735	13.35289
C42	13.08290	23.22342
C48	13.08290	23.22342
C40	13.64220	25.99746
C44	13.64220	25.99746
C37	13.68709	17.51623
C47	16.75734	20.91023
C34	19.28047	22.95389
C38	21.49217	19.46567
C30	22.76824	23.60281
C45	24.76859	22.60681
C43	26.31074	22.36436
C32	27.42500	23.10099
C41	27.97619	21.50574
C31	29.05768	22.49925
C46	30.33523	22.82007
C33	30.81931	21.75005
C35	30.81931	21.75005
C49	31.12313	22.98137
C39	31.46396	22.07856
C50	33.77878	22.84083

TABLE 5. CORRELATION ANALYSIS (CHARACTERS AND CLASSIFICATIONS).

Summary of the distances ($D(I,J)$) between pairs of characters (I) and admiss[ible] classifications (J). N.I. = no information. The lower the $D(I,J)$ value, the greate[r] character I correlated with classification J. Note the perfect correlation between C[..] and character 11, and between C32 and character 29. Note also the overall low [val]ues of C50.

Characters	1	2	3	4	5	6	7	8	9	10	11	12	13	
Classifications														
C30	94107	86484	86142	95100	91792	82381	95821	95312	77935	96216	98042	90598	77795	86
C31	95478	84576	87444	94613	88867	84234	90035	94141	70820	88701	94203	88523	71884	88
C32	95120	85804	88215	94700	89628	83679	91636	95220	75781	87845	96051	88523	71884	88
C33	95861	86140	86348	92760	90737	85768	91699	91887	76766	89623	90364	85916	66522	89
C34	94914	85632	89106	95845	93886	83724	96493	98546	79708	97089	98944	99557	84782	86
C35	95861	86140	86348	96942	90737	85768	91699	91887	76766	89623	90364	85916	66522	89
C36	98678	90519	88784	95932	99907	99202	99942	97060	93144	96692	00000	N.I.	N.I.	N
C37	92417	88845	88833	95972	97357	91454	90835	96308	85394	73246	41759	99226	83675	95
C38	97567	90533	88770	93075	85228	89474	94944	96352	81435	89703	59859	88007	61838	97
C39	95861	85801	80595	92760	90737	83002	90651	93202	76766	89623	87758	88840	71884	89
C40	97955	93692	97168	94963	89774	78891	99894	98737	79077	94598	97939	95673	79604	91
C41	96538	85090	82578	93167	92300	84058	90892	95780	78220	89840	87839	96310	78559	89
C42	96907	98403	88631	97634	93018	80505	99814	98525	85672	99340	98715	99451	82190	82
C43	88901	83830	88598	90794	89119	86204	93868	87408	75166	89077	88867	90688	69689	84
C44	97955	93692	97168	94963	89774	78891	99894	98737	79077	94598	97939	95673	79604	91
C45	97033	84424	92375	91370	88669	84471	94602	92947	79760	89194	79735	91237	66777	74
C46	90114	83205	82353	91657	88669	76884	90701	87326	75166	88821	89838	88972	65018	78
C47	83576	88591	80470	96076	89186	80917	92655	86962	85164	82446	87548	95002	73868	86
C48	96907	98403	88631	97634	93018	80505	99814	98525	85672	99340	98715	99451	82190	82
C49	97574	79306	80977	92809	87544	78222	90617	90534	77665	91088	79573	88451	67462	74
C50	91090	79306	81729	92374	86897	78860	88893	88164	73355	84850	80787	88972	65018	78

In order to study the relationships between the various admissible classifications, the correlation between each pair of classifications was computed in terms of D values. This classification by classification matrix of the D values can be defined as a dissimilarity matrix, and is given in Table 6.

The minimum spanning tree and principal coordinates for the classifications were computed from this dissimilarity matrix. The classifications, plotted on the three principal axes obtained from the principal coordinate analysis, are represented in Fig. 6 together with the minimum spanning tree superimposed on the points. This gives us a four-dimensional representation of the relationships among the various admissible classifications and also a measure of the relative distance of these against the best classification, C50. Clearly C49 and C46 are very close to C50. It is also apparent that C33 and C35 are identical in the content of their groupings, as are C40 with C44, and C42 with C48. It is interesting that these identical classifications were obtained from different sets of characters, or different methods of clustering, or both. For the other classifications, it can be construed that different clustering

	16	17	18	19	20	21	22	23	24	25	26	27	28	29
00	85056	94989	73461	91801	38468	90842	93256	89897	69077	67508	61327	60581	44280	18944
16	77454	87988	66613	85476	36769	83649	83710	89419	62110	65923	61239	54018	55619	08062
16	82484	95205	71164	87346	37272	82668	88881	90059	61446	65923	61239	54018	52564	00000
36	79572	94614	64995	83040	39267	83352	81751	90124	48088	63367	57886	51125	58480	15365
50	87078	94451	78978	92263	47487	93189	94741	97556	75253	72889	66753	65796	40423	28021
36	79572	94614	64995	83040	39267	83352	81751	90124	48088	63367	57886	51125	58480	15365
.	83191	82860	92204	85961	96300	94965	91092	94161	86026	94733	90659	91625	95069	95092
57	78810	88783	86680	77171	83903	86765	87999	95321	74031	91523	86480	85311	88674	77741
54	82399	91831	72578	65999	70302	81643	88058	94570	61051	75947	77362	70394	78828	61763
39	76258	94215	64995	80676	36766	83779	77185	89211	54997	64302	61922	51414	58635	15752
53	94647	94171	80186	94440	48765	91542	95940	98792	71324	76149	77402	75068	60695	53270
20	77800	93767	70028	80492	44253	85450	77853	95277	60658	69029	66580	56235	56788	23400
36	93026	94791	81981	94324	70063	93401	96930	97124	83984	80439	80722	79220	55791	51386
23	80885	94586	72783	67964	34556	89159	83227	93190	67513	65679	68739	61837	66695	46068
53	94647	94171	80186	94440	48765	91542	95940	98792	71324	76149	77402	75068	60695	53270
17	86893	93782	77052	68418	52699	82307	85229	95471	57802	68079	70944	61410	71176	44483
70	79140	94535	71160	69475	39959	89256	80685	91249	67996	61658	62131	55178	57364	35639
56	82456	94369	83422	79433	75033	95118	80824	93264	80622	85731	82876	80891	79926	71782
36	93026	94791	81981	94324	70063	93401	96930	97124	83984	80439	80722	79220	55791	51386
95	78148	93416	69904	68936	45306	83862	81787	90555	60102	61185	58895	49021	59170	25181
70	73567	93801	68198	70544	43453	83862	79538	90274	60102	61185	58895	49021	61012	25181

strategies, or different sets of characters, or both may yield fairly close results depending on the combinations.

According to the distance measure used, C46 is clearly very close to the best classification, viz. C50. This is very significant because it appears that, at least for the *Avena* data, the information on the genome is not necessary for forming a classification that fairly closely approximates the best classification (see also table on p. 36 for further details of input). The only difference between C46 and C50 is that group B of the former is split into group B and D of the latter (see Fig. 5).

It is noteworthy that it was the *flexible sort* method (Lance and Williams 1967), with all characters, which gave the best classification. Second best was the incremental sum of squares strategy also with all characters. Third best was C46; the table on p. 38 and Fig. 6 give more details about the other classifications. The subset of characters selected from the data obtained from CHARANAL did not yield good classifications with the various clustering

TABLE 6. CLASSIFICATION BY CLASSIFICATION MATRIX OF THE DISTANCES (D VALUES) BETWEEN PAIRS OF ADMISSIBLE CLASSIFICATIONS.

These were obtained from CHARANAL after coding the classifications as characters for each of the 28 OTU's.

	30	31	32	33	34	35	36	37	38	39	40	41	42	43	44	45	46	47	48	49
31	.22474																			
32	.18157	.07719																		
33	.31177	.19138	.15909																	
34	.11648	.33272	.27690	.39194																
35	.31177	.19138	.15909	.00000	.39194															
36	.98042	.94203	.96051	.90364	.98944	.90364														
37	.93899	.79367	.80538	.76145	.95597	.76145	.41759													
38	.75410	.61843	.62282	.59242	.82303	.59242	.59859	.45375												
39	.31177	.18412	.15909	.10739	.39194	.10739	.87758	.73299	.56061											
40	.44342	.57964	.54448	.61694	.55990	.61694	.97939	.92358	.71169	.61694										
41	.39194	.26028	.23925	.18325	.33894	.18325	.87839	.73860	.62614	.08016	.69711									
42	.40807	.55294	.51555	.59262	.33003	.59262	.98715	.96956	.82276	.59262	.64464	.55712								
43	.46397	.46718	.46530	.44459	.54583	.44459	.88867	.73103	.58775	.40686	.53050	.47720	.75077							
44	.44342	.57964	.54448	.61694	.55990	.61694	.97939	.92358	.71169	.61694	.00000	.69711	.64464	.53050						
45	.61640	.47273	.47104	.44936	.69271	.44936	.79735	.65123	.40394	.41076	.49940	.48294	.75974	.31105	.49940					
46	.36177	.37838	.37261	.35888	.44364	.35888	.89838	.70089	.56559	.31907	.58179	.38940	.64858	.10925	.58179	.34581				
47	.73410	.70194	.71033	.67117	.76543	.67117	.87548	.58809	.69906	.63894	.84770	.66222	.80730	.50730	.84770	.55523	.39804			
48	.40807	.55294	.51555	.59262	.33003	.59262	.98715	.96956	.82276	.59262	.64464	.55712	.00000	.75077	.64464	.75974	.64858	.80730		
49	.44009	.30401	.29474	.28667	.51640	.28667	.79573	.66982	.44867	.24403	.59695	.31620	.70745	.26204	.59695	.19487	.16358	.52232	.70745	
50	.42110	.28932	.27966	.27293	.49535	.27293	.80787	.61038	.47975	.23104	.62308	.30137	.68124	.19720	.62308	.24707	.09873	.45748	.68124	.06484

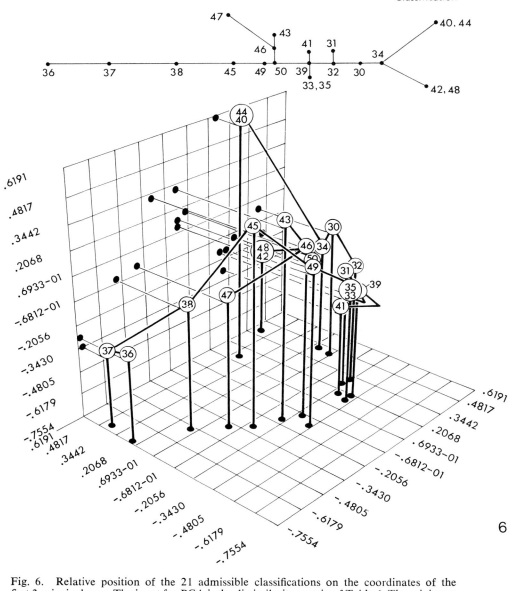

Fig. 6. Relative position of the 21 admissible classifications on the coordinates of the first 3 principal axes. The input for PCA is the dissimilarity matrix of Table 6. The minimum spanning tree is superimposed between the points and is therefore greatly distorted, and is for this reason reproduced on top of the diagram. See table on p. 29 for the OTU codings.

Classification

methods used here. Furthermore, one of the worst classifications among the admissible ones was C36, which was based on the subset of characters considered on intuitive grounds to be the most important subset for classificatory purposes and for discrimination among species in this genus. Another classification, C32, has perfect correlation with the genome ($D[29,32]=0$), but it cannot be rated highly on the basis of the criteria (see Table 5 and table on p. 45). Some differences between the groupings of C32 and those of the best classification, C50 can be grasped heuristically with the aid of Figs. 4 and 5.

The 7 groupings of C50 were subjected to canonical analyses using only characters 4, 6–11, 16–22, and 23 to avoid missing values (including no comparisons). The analyses were performed because initially these multivariate populations were not readily separable by an ordinary and conventional key, that is by yes or no characters. This became apparent after reorganizing the BDM according to the 7 populations followed by the computations of the mean and variance of each of the 29 variates. For canonical analysis the variates should be normally distributed, but here quite a few characters are discontinuous. From Rao (1952) and Gilbert (1968), it appears that canonical analysis need not necessarily be disqualified in this case.

The 28 OTU's were plotted on the first 3 canonical axes (Figs. 7 and 8) from the 6 axes obtained.

Clearly the plot of axis 1 on 2 (Fig. 7) separates populations A, B, E, F, and G on axis 1, whereas C and D are separated on axis 2. Furthermore, the plot of axis 1 on 3 (Fig. 8) separates populations C, D, and F more clearly on axis 3, whereas axis 1 separates A, B, E, and G. The canonical loadings are given instead of an ordinary key to groupings because of the same rationale for which the canonical analyses were performed. In other words, there are no discrete character-states that can be used for constructing a traditional key. These are the canonical loadings:

AXES		1	2	3	4	5	6
VARIATES							
(1)	4	1.17545	0.78304	0.61316	1.00421	−0.10015	−0.02087
(2)	6	1.15481	1.24420	0.18478	−2.82934	0.39621	−1.04644
(3)	7	1.14969	−0.49508	−0.66917	1.49742	−0.92604	0.64784
(4)	8	0.07505	0.33984	−0.09430	0.24829	−0.10644	−0.15112
(5)	9	−0.92789	−0.67387	0.61049	−0.83974	−0.03312	0.68006
(6)	10	0.35038	−4.19521	0.66405	1.65852	−0.14924	−2.43116
(7)	11	4.52909	−5.20661	−3.19306	−1.43955	−1.06617	0.39073
(8)	16	−0.42008	−0.11284	0.32500	0.01659	0.02072	0.18022
(9)	17	3.26704	−1.71066	−2.27461	0.97652	0.24713	0.42711
(10)	18	0.28498	−0.24723	−0.14684	−0.32129	0.26549	0.40422
(11)	19	−2.09372	1.89323	0.88092	1.26482	−1.88927	−0.35369
(12)	20	3.77440	−0.49227	2.43206	0.77441	0.18892	0.22968
(13)	22	−1.10919	0.53358	0.72230	−0.16429	0.69020	−0.66175
(14)	23	1.18435	−1.80459	−0.28285	0.98628	0.88184	0.88790

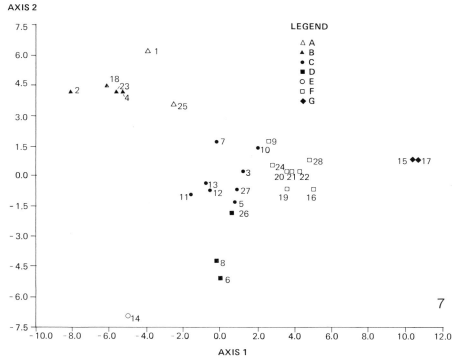

Fig. 7. Plot of the 28 OTU's according to the first and second canonical axes. The letters refer to the grouping code of C50 (see Fig. 5), and the numbers refer to the codes of the OTU's (see pp. 36–37).

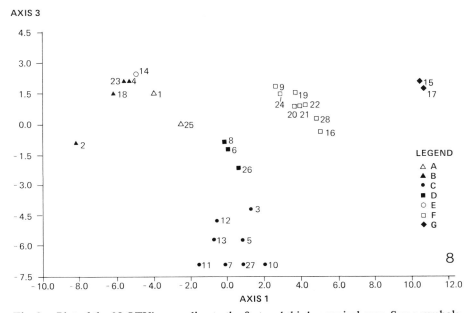

Fig. 8. Plot of the 28 OTU's according to the first and third canonical axes. Same symbols as in Fig. 7.

A sequential F-test (Dixon 1970), listing the characters according to their apparent importance, was also performed. The character with the highest F ratio is regarded as the most important and so on in descending order. The most important characters so identified are 20, 10, and 11 (cf. Table 2), and they can be used as major "key characters" for identification of the groupings with the aid of the canonical loadings.

Some relationships between the groupings and of species within the groupings can be construed from the positions of the OTU's on their canonical axes (Figs. 7 and 8). Such plots, however, because of their inherent limitations (i.e., the limited number of characters used and the distortion resulting from using few axes) give us only one point of view on the relationships.

In addition, I decided to look at the relationships between the species by using overlapping clusters. The method of Jardine and Sibson of maximal complete subgraphs was used, and the level chosen was the lowest in which most subgraphs contained more than one OTU. Two overlapping clusters were thus obtained (Figs. 9 and 10), among several others, using the subset of characters 9, 13, 15, 16, 20, 24–29 (Fig. 9) and the set of all the characters (Fig. 10).

Fig. 9. Overlapping clusters obtained from characters 9, 13, 15, 16, 18, 20, 24–29. They are used as input into the Poushinsky and Sheldrake program of Jardine and Sibson's algorithm of maximum complete subgraphs. The threshold level is 0.715. The link between the OTU's is the minimum spanning tree.

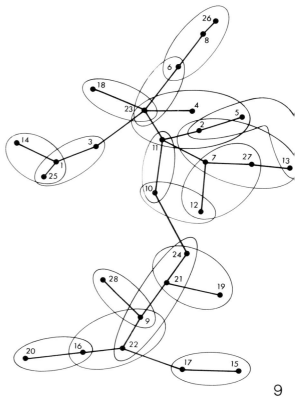

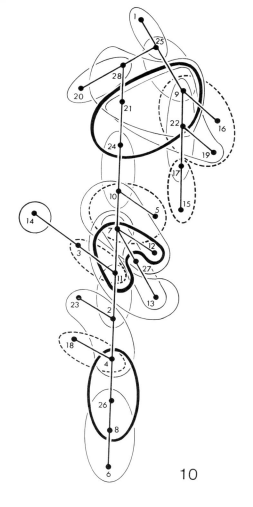

Fig. 10. Overlapping clusters obtained from all 29 characters; threshold level 0.772. The same method was used as in Fig. 9.

Other subsets of characters yielded overlapping clusters of complicated structure whose graphical representation is confused or topologically difficult to reproduce and comprehend. The overlapping clusters based on the input of the "CHARNAL" characters (Fig. 9) not surprisingly yielded groupings identical with those of C30 (Fig. 4) because C30 is based on single linkage (see table, p. 38). Similarly, the overlapping clusters based on input of all the characters (Fig. 10) resembles C47 (Fig. 5), which is also based on single linkage (see table, p. 38).

The relationships among the species in the overlapping clusters formed from all characters (Fig. 10) are close to those implied in C50 (Fig. 5). Thus, OTU's 1 and 25 form a group comparable to A; OTU's 9, 16, 19, 20, 21, 22, 24, and 28 are comparable to F; OTU's 15 and 17 are a group as in G; OTU's 3, 5, 7, 10, 11, 12, 13, and 27 form a group as in C; OTU's 2, 4, 18, and 23 are comparable to B; OTU's 6, 8, and 26 are related as in D; and OTU 14 or E is isolated. Furthermore, it is significant that the gross, morphologically similar OTU 17 (*A. murphyi*) and OTU 22 (*A. sterilis*) are connected, and that the conspecific OTU 28 (*A. sativa* fatuoid) and OTU 20 (*A. sativa*) are also connected. The significance of the group 4, 26, and 8 is obscure to me.

THE NEW CLASSIFICATION: DESCRIPTION OF THE NEW SYSTEM OF *AVENA*

I define and designate the groupings of C50, the best classification, as taxa of the category of sections. I shall list the sections in the sequence adopted in this book and refer to C50 using square brackets, the letters indicating the groupings and the numbers indicating the OTU's, but see Fig. 5 for reference.

Section 1. *Avenotrichon* (Holub) Baum [E]
Avena macrostachya [14]
Section 2. *Ventricosa* Baum [D]
Avena clauda [6], *A. eriantha* [8], *A. ventricosa* [26]
Section 3. *Agraria* Baum [B]
Avena brevis [4], *A. hispanica* [2], *A. nuda* [18], *A. stigosa* [23]
Section 4. *Tenuicarpa* Baum [C]
Avena barbata [3], *A. canariensis* [5], *A. damascena* [7], *A. hirtula* [12], *A. longiglumis* [13], *A. lusitanica* [11], *A. matritensis* [10], *A. wiestii* [27]
Section 5. *Ethiopica* Baum [A]
Avena abyssinica [1], *A. vaviloviana* [25]
Section 6. *Pachycarpa* Baum [G]
Avena maroccana [15], *A. murphyi* [17]
Section 7. *Avena* [F]
Avena atherantha [16], *A. fatua* [9], *A. hybrida* [21], *A. occidentalis* [19], *A. sativa* [20], *A. sterilis* [22], *A. trichophylla* [24]

Because the different sections are separated by the various canonical axes, the description of these are the canonical loadings mentioned above.

Unfortunately there are no provisions in the Code (Stafleu et al. 1972) that stipulate on the validity of numerical functions to serve as diagnoses. Article 36 is neither in favor nor against usage of these in the protolog. I have made a proposal to the International Botanical Congress, 1975, to change Article 36 of the Code to accommodate this problem (Baum 1974).

COMMENTS ON THE RESULTS OF THE CLASSIFICATORY EXERCISE

The eight sequential, agglomerative, hierarchical, non-overlapping (SAHN), clustering strategies used in this study are among the most frequently employed methods (Sneath and Sokal 1973).

The "best" classification resulted from the flexible *clustering strategy,* and C46, which is very close to it, also resulted from the same method. According to Lance and Williams (1967) when the four constraints $\alpha_j + \alpha_k + \beta = 1$, $\alpha_j = \alpha_k$, $\beta < 1$, and $\gamma = 0$ are applied to their combinatorical formula this becomes the *flexible clustering* strategy. The best clusterings including C50 were obtained when $\alpha = 0.6$, that is when $\beta = -0.2$. Lance and Williams suggest $\beta = -0.25$ for general use.

I have not tried all the possible methods of clustering, and certainly not tried all the possible subsets of characters. Furthermore, it is to be expected that some of the various combinations of character sets with different methods will give similar results. Nevertheless, it is apparent that many reduced character sets yield classifications that do not come close to the best classification, when the criterion for assessment is the one used in this exercise. Previous classifications of *Avena* were monothetic and were based only on very few and gross morphological characters.

Another important phenomenon became apparent from this exercise. Characters thought to be most important for discrimination between OTU's (on intuitive grounds or after a character-analysis exercise) are not necessarily most important for discrimination between groupings of OTU's, that is, the sections in this instance. The "good" characters by which the species can be keyed out do not possess at all the same discriminatory capacity for the sections. A key to species, which is obviously based on characters with high diagnostic power, can be regarded as a hierarchical classification. This kind of classification, however important, serves only for the restricted purpose of identification. The aim in this exercise was to produce a classification with maximum information content according to some assumptions and criteria. Clearly, gradually sifting out characters from the basic data matrix on the basis of a measure of correlation for classificatory purposes (e.g., Legendre and Rogers 1972) results in producing a subset of characters that are only useful for a diagnostic key or perhaps for cladistic classification. Both diagnostic keys and cladistics may be very remote from or inconsistent with phenetic relationships (Baum and Lefkovitch 1973).

From the foregoing the message is clear; different studies and exercises have to be carried for different purposes in systematics. If we are interested in classification of species we have to undertake certain steps; if we want to find a means of identifying these species we need to perform different studies. For cladistics other methods need to be used, and so forth. This pluralistic approach in systematics and of course the recognition of different purposes in taxonomy has been enunciated and recognized by a number of writers, such as Sneath and Sokal (1973, and authors referenced therein), and with emphasis on oats by Baum (1970) and Baum and Lefkovitch (1973).

CONCLUDING REMARKS ON THE CLASSIFICATION

The new classification of *Avena* possesses more information content (in the sense of Shannon information) than the other 79 classifications that were investigated and consequently it appears to be the best one to propose for the test of time. Furthermore, it seems to reflect biological reality. The species within the sections seem to be related to each other as semispecies (Baum 1972): thus one may naturally encounter interspecific hybrids within sections but not between sections. In this monograph these hybrids are described within each section after the species.

So, in the new classification in section *Ethiopica*: *A. abyssinica* and *A. vaviloviana* are more or less endemic to Ethiopia and have more genes in common than any other species. Section *Agraria* comprises all the diploid species with a cultivated base of their spikelets. Section *Ventricosa* contains the three diploid species with the C-genome, and with unequal glumes. Section *Tenuicarpa* contains the rest of the diploid species together with "nearly autotetraploid" *A. barbata* (Holden 1966; Sadasivaiah and Rajhathy 1968; and Rajhathy and Thomas 1974). Section *Avenotrichon* contains the only perennial oat species *A. macrostachya*. Section *Sativa* contains all the hexaploid species with the AACCDD genome. Section *Pachycarpa* contains the two AACC-genome species with the sterilis-like spikelets.

A SHORT COMMENT ON SOME RELATED CLASSIFICATIONS

Classification C46 (Fig. 5) is very similar to the new classification (C50, Fig. 5), the difference being that Section *Ventricosa* is merged together with Section *Agraria*. It is based on the same set of characters except for the genome character (character no. 29), and is based on the same clustering method. This indicates that the glume characters and disarticulation characters, which differentiate Section *Ventricosa* from Section *Agraria*, do not have enough weight; but it might indicate, on the other hand, because of the strong similarity between the two classifications with respect to information content, that enough information is included in the 28 characters (excluding no. 29) to closely reflect upon the genome.

Classification C32 (Fig. 4) is completely correlated with the genome, but the classification itself clearly does not possess maximum information, whereas in the new classification (C50) the genome is adequately represented.

8. Speciation and evolutionary inferences

A full account on speciation was outlined by Rajhathy and Thomas (1974) with emphasis on cytology and reproductive biology. The opinion of these authors is that two main evolutionary mechanisms have been responsible for speciation in *Avena:* (i) structural differentiation of chromosomes, and (ii) hybridization and chromosome doubling. The first mechanism resulted in production of two distinct genomic groups, namely the AA and the CC species, and in production of a range of modified karyotypes within each group. The second mechanism brought about the creation of the tetraploid and hexaploid oats. The tetraploid oats can be divided into three groups: (i) *A. macrostachya* — not yet fully understood; (ii) the group with the AABB genome, which is nearly autoploid; and (iii) the *A. maroccana–murphyi* complex having the AACC genome. The hexaploid oats were designated the genomic formula AACCDD, in which the origin of the DD genome and its identity is still unknown.

The classification of the species in this work more or less reflects the genome situation. Section *Avenotrichon* contains the not yet fully understood tetraploid *A. macrostachya*; Section *Agraria* contains the AA-genome species with a cultivated base; Section *Tenuicarpa* has the AA-genome species and one AABB species, viz., *A. barbata*; Section *Ventricosa* contains the CC-genome species exclusively; Section *Ethiopica* has two tetraploid species that belong to the AABB group; Section *Pachycarpa* includes two tetraploid species, one of these with the AACC-genome; and Section *Avena* contains all the hexaploid species having the AACCDD genome.

Cytology and cytogenetics is a prerequisite for elucidating evolutionary relationships of species within genera. But, these disciplines have their limitations when used alone. For instance, to establish the sequence of the phyletic relationships within the group of the diploid species, and certainly within the hexaploid species, one needs to do it through phenetics, and the characters can be taken from a wide selection, ranging from morphology to amino-acid sequences.

Cladistic analysis, that is, the finding of cladistic relationships from neontological data on the basis of morphology, has recently been formulated by Hennig (1950, 1966) as a methodology. More recently, with the use of computers, cladistic analysis has developed into a scientific method; Estabrook (1972) made a comprehensive review.

I have applied a few numerical cladistic methods to the morphological data (see Table 2) of the diploid and hexaploid species. This chapter deals with the results that I have obtained.

MATERIALS AND METHODS

Some recoding of the characters and their states was necessary (Table 7). The OTU's studied are listed in Table 8; 14 diploids and 8 hexaploids are considered. For each group slightly different characters were used (see Table 7). Some characters were eliminated altogether (from the original data, Table 2) because they were invariant within the subset of OTU's. Some OTU's had previously missing values, primarily because of the nature of the data (i.e., inapplicable characters) and thus had to be recoded appropriately in some instances. Quantitative characters had to be recoded into simple multistate characters.

Two numerical methods were used for generating cladograms, namely the method of Camin and Sokal (1965) and the method of Farris (1970).

The two methods differ in some basic assumptions, and certainly in their algorithms. In the method of Camin and Sokal the characters must be coded in discrete states; the states within each character must be coded in some sequence according to some prior knowledge or assumption of whether they are pleisomorphous or apomorphous; and evolutionary steps are irreversible. Every character is fitted to a pattern cladogram from which a compatibility matrix is computed; "good" and "bad" characters are assessed on the basis of the compatibilities. A procladogram is then created through a procedure that, among others, takes into consideration the number of primitive states within OTU's (that, with most such states, is branching off first). Finally the procladogram is subjected to an iterative procedure by changing location of steps together with moving branches so as to achieve a good approximation to a maximally parsimonious tree.

Evolution

TABLE 7. CHARACTERS USED IN THE CLADISTIC ANALYSIS.

These are recorded from Table 2. The double-bracketed numbers in the three right-hand columns indicate the new character number. As can be seen, less characters were used in the cladistic analysis than in the classificatory study (Table 2). The reason for the different coding of the character-states in the second column on the right is implicit, but because these character-states are based on my assumptions of pleisomorphism and apomorphism, they are, of course, debatable.

Organs or part of plant	Character	State	Codes in Table 2	Coding for Wagner trees, diploids only	Coding for the method of Camin and Sokal, diploids only	Coding for Wagner trees, hexaploids only
General habit	1) Longevity	(1) Perennial	0	—	—	—
		(2) Annual	1	—	—	—
	2) Juvenile growth	(3) Erect	10	—	—	—
		(4) Prostrate to erect	15	—	—	—
		(5) Prostrate	20	—	—	—
	3) Color appearance	(6) Glaucous	0	—	—	—
		(7) Green	1	—	—	—
	4) Culm	(8) Geniculate or prostrate	0	(1) 0	(1) 0	—
		(9) Erect	1	1	1	—
		(10) Prostrate or erect	2	2	2	—
	5) Height	(11) Low plants, not exceeding 95 cm (15–95 cm)	0	—	—	—
		(12) Tall plants, exceeding 100 cm (30–180 cm)	1	—	—	—
Ligules	6) Shape at anthesis	(13) Obtuse	0	(2) 0	(2) 0	(1) 0
		(14) Acute	1	1	1	1
Panicle	7) Shape	(15) Equilateral	0	(3) 0	(3) 0	(2) 0
		(16) Slightly flagged, equilateral or flagged, equilateral and sometimes flagged	1	1	1	1

				upper limit more than 25 mm (19) Long, lower limit not less than 20 mm, upper limit exceeding 25 mm	20	2	2	2	2
	9) Number of florets	(20)	1–2 or 2 only		0	(5) 0	(5) 0	(5) 0	(4) –
		(21)	2–3		1	1	1	1	0
		(22)	2–4		2	–	–	–	1
		(23)	2–5		3	2	2	2	2
		(24)	2–6		4	–	–	–	–
		(25)	3–4		5	3	3	3	3
		(26)	3–6		6	–	–	–	–
		(27)	1–7		7				4
Glumes	10) Relative length	(28)	Equal		0	(6) 1	(6) 0	(6) 0	–
		(29)	Unequal		1	0	1	1	–
Dispersal unit	11) Mode of disarticulation	(30)	Non-disarticulating		0	(7) 1	(7) 0	(7) 0	(5) 0
		(31)	Disarticulating		1	0	1	1	1
Florets	12) Mode of disarticulation	(32)	Not disarticulating		–	(8) 2	(8) 0	(8) 0	(6) 0
		(32)	Only lowermost floret disarticulating		0	1	1	1	1
		(33)	All florets disarticulating		1	0	2	2	2
Scars	13) Shape of 1st floret scar	----	No scar		–	(9) 4	(9) 0	(9) 0	(7) 0
		(34)	Oval to round		0	–	–	1	1
		(35)	Heart-shaped or slightly so		1	3	1	2	2
		(36)	Elliptic		2	2	2	3	3
		(37)	Narrow elliptic		3	1	3	4	4
		(38)	Linear		4	0	4	–	–
	14) Shape of 3rd floret scar	----	No scar		–	–	–	–	(8) 0
		(39)	Heart-shaped		0	–	–	–	1
		(40)	Not heart-shaped		1	–	–	–	2
	15) Periphery ring	----	No periphery ring		–	(10) 4	(10) 0	(10) 4	(9) 0
		(41)	All around scar		0	–	–	–	–
		(42)	Confined to ½ scar		1	–	–	–	–
		(43)	Confined to ⅓–½ scar		2	3	1	3	1
		(44)	Confined to ⅓ scar		3	–	–	–	2

TABLE 7. (Cont'd)

Organs or part of plant	Character	State		Codes in Baum 1975	Coding for Wagner trees, diploids only	Coding for the method of Camin and Sokal, diploids only	Coding for Wagner trees, hexaploids only
Lemma	16) Place of awn insertion	(45)	Confined to ¼ scar	4	2	2	3
		(46)	Confined to ⅛ scar	5	3	1	–
		(47)	Confined to ⅛–¹⁄₁₆ scar	6	4	0	–
		(48)	Absent or rudimentary, or inserted about middle of lemma to below tip	0	(11) 0	(11) 0	(10) 0
		(49)	Inserted at about lower ¼	1	–	–	–
		(50)	Inserted at about ⅓	2	1	1	1
		(51)	Inserted between lower ⅓ and ½	3	2	2	–
		(52)	Inserted at about middle	4	3	3	2
		(53)	Inserted at about upper ⅓	5	–	–	–
		(54)	Inserted between upper ⅓ and ¼	6	4	4	–
		(55)	Inserted upper ¼ to below tip	7	5	5	–
	17) Structure	(56)	Tough	0	(12) 0	(12) 0	(11) 0
		(57)	Resembling glumes	1	1	1	–
		(58)	Tough or resembling glumes	2	–	–	1
	18) Shape of tip	(59)	Bilobed	0	(13) –	(13) –	(12) –
		(60)	Bidenticulate	1	0	0	0
		(61)	Bisubulate	2	1	1	1
		(62)	Biaristulate	3	2	2	–
		(63)	Bisetulate-biaristulate	4	3	4	–
		(64)	Shortly bisubulate or mucronate	5	4	6	2
		(65)	Bisubulate or biaristulate	6	5	3	3
		(66)	Bidenticulate to bisubulate	7	–	–	–
		(67)	Bisetulate or bisetulate-biaristulate	8	6	5	–
	19) Vestiture below awn insertion	(68)	Beset densely with macrohairs	0	(14) 0	(14) 0	(13) 0

Evolution

Palea	20) Rows of cilia along edges of keels	(71) ?	0	—	—	—	—	(13) 0	(14) 1
		(72) 1–2	1	—	—	—	1	1	—
		(73) 1–3	2	—	—	—	—	—	—
		(74) 2	3	—	—	—	—	—	—
	21) Vestiture of back	(75) Glabrous	0	—	—	—	—	(16) 2	(14) 0
		(76) Prickles	1	—	—	—	—	4	1
		(77) Macrohairs	2	—	—	—	—	0	—
		(78) Prickles or macrohairs	3	—	—	—	—	3	—
								1	
Lodicule	22) Type	(79) Sativa	0	—	—	—	(16) 0	(17) 2	(15) 0
		(80) Fatua	1	—	—	—	1	1	1
		(81) Strigosa	2	—	—	—	2	0	—
		(82) Fatua and sativa	3	—	—	—	3		—
		(83) Fatua and strigosa	4	—	—	—	4		
	23) Prickles	(84) Always absent	0	—	—	—	(17) 0	(18) 0	(16) —
		(85) Rarely present	1	—	—	—	1	1	0
		(86) Often present	2	—	—	—	2	2	1
									2
Epiblasts	24) Type	(87) Brevis	0	—	—	—	(18) 0	(19) 0	(17) —
		(88) Fatua	1	—	—	—	1	1	0
		(89) Septentionalis	2	—	—	—	2	2	1
		(90) Sativa	3	—	—	—	—	3	1
		(91) Sativa and fatua	4	—	—	—	—	4	2
									3
									4
	25) Lower limit of width	In actual mm, or	.2	—	—	—	(19) 0	(20) 0	(18) —
			.25	—	—	—	1	1	—
			.3	—	—	—	2	2	—
			.35	—	—	—	3	3	0
			.4	—	—	—	4		
			.45	—	—	—			
			.5	—	—	—			
			.55	—	—	—			
			.6	—	—	—			
			.8	—	—	—			
	26) Upper limit of width	In actual mm, or	.2	—	—	—	(20) 0		
			.25	—	—	—	1		
			.3	—	—	—	2		
			.35	—	—	—	3		

TABLE 7. (Cont'd)

Organs or part of plant	Character	State	Codes in Baum 1975	Coding for Wagner trees, diploids only	Coding for the method of Camin and Sokal, diploids only	Coding for Wagner trees, hexaploids only
	27) Median of range	In actual mm, or	.4	4	4	—
			.45	—	—	—
			.5	—	—	1
			.6	—	—	2
			.65	—	—	—
			.7	—	—	—
			1.0	—	—	—
			.2	(21) 0	(21) 0	(19) —
			.25	1	1	—
			.3	2	2	—
			.325	3	3	—
			.35	4	4	0
			.4	5	5	—
			.425	—	—	—
			.5	—	—	1
			.525	—	—	2
			.55	—	—	3
			.6	—	—	—
			.65	—	—	—
			.9	—	—	—
Chromosomes	28) Number	$2n = 14$	0	—	—	—
		$2n = 28$	1	—	—	—
		$2n = 42$	2	—	—	—
		$2n = 40, 41$ or 42	3	—	—	—
		$2n = 14$ and 28	4	—	—	—
	29) Karyotype-genome	CC	0	(22) 0	(22) 1	—
		AA	1	1	0	—

TABLE 8. OTU'S FOR CLADISTIC ANALYSIS.

A list of their names and codes.

Diploids

2	*A. hispanica*
4	*A. brevis*
5	*A. canariensis*
6	*A. clauda*
7	*A. damascena*
8	*A. eriantha*
10	*A. matritensis*
11	*A. lusitanica*
12	*A. hirtula*
13	*A. longiglumis*
18	*A. nuda*
23	*A. strigosa*
26	*A. ventricosa*
27	*A. wiestii*

Hexaploids

9	*A. fatua*
16	*A. atherantha*
19	*A. occidentalis*
20	*A. sativa*
21	*A. hybrida*
22	*A. sterilis*
24	*A. trichophylla*
28	*A. sativa fatuoid*

In the computation of Wagner networks by the method of Farris the characters need not be coded in discrete states. (Quantitative characters are admissible, but in my experience mixed data leads to faulty results and therefore the data has to be coded in a uniform manner.) There is no need for assuming organophyletic trends, that is, to determine which state is pleisomorphous and which are apomorphous and to what degree. Furthermore, the assumption of irreversibility is not made. The network is produced by stepwise addition of OTU's to the tree through production of HTU's (Hypothetical Taxonomic Units) and parsimony is achieved through an optimizing procedure on the HTU's.

The BDM (Table 9) of the diploid OTU's served as input into Bartcher's (1966) package of programs for computing trees by the method of Camin and Sokal.

TABLE 9. BASIC DATA MATRIX OF DIPLOID *AVENA* SPECIES FOR COMPUTING CAMIN AND SOKAL CLADOGRAMS.

The OTU names appear on Table 8, and the characters are figured on Table 7 with details on their states.

OTU character	2	4	5	6	7	8	10	11	12	13	18	23	26	27
1	1	1	1	1	1	1	1	0	2	1	1	1	1	1
2	0	0	1	0	0	0	1	0	0	0	0	0	0	0
3	0	1	0	1	0	1	0	1	0	1	1	0	1	0
4	0	0	0	2	2	0	2	1	0	2	2	0	0	1
5	0	1	0	2	1	1	1	1	1	1	3	1	1	0
6	1	1	1	0	1	0	1	1	1	1	1	1	1	1
7	1	1	0	0	0	0	0	0	0	0	1	1	0	0
8	2	2	1	0	0	1	0	0	0	0	2	2	1	0
9	4	4	3	0	2	0	1	1	2	0	4	4	0	2
10	4	4	2	1	2	1	2	2	2	0	4	4	1	3
11	4	5	3	2	1	3	1	1	2	3	0	1	3	3
12	0	0	0	0	0	0	0	0	0	0	1	0	0	0
13	4	6	0	4	4	1	2	4	1	5	3	4	1	5
14	1	1	0	1	0	1	0	0	0	0	1	2	1	0
15	0	1	0	1	0	1	0	0	0	0	1	1	1	0
16	4	4	2	4	2	4	2	4	2	3	1	0	4	2
17	1	2	1	1	2	2	1	2	2	1	2	0	1	2
18	0	0	0	2	1	2	2	0	2	1	0	0	2	1
19	4	2	2	2	1	2	2	2	0	2	3	2	3	0
20	4	4	4	2	1	2	3	2	0	3	3	3	3	2
21	5	1	4	2	1	2	3	2	0	3	4	3	4	1
22	0	0	0	1	0	1	0	0	0	0	0	0	1	0

The BDM (Tables 10 and 11) of the diploid and of the hexaploid OTU's served as input into the CLAD/OS package of programs provided by Farris (personal communication) for computing Wagner networks together with related computations. Every run was done twice: the first run included the karyotype-genome character whereas the second run omitted it.

In the various networks thus obtained, at least one HTU from each was chosen as a probable ancestor of the cladogram. The vectors of these root HTU's, including that of the Camin and Sokal tree after recoding, were added each one separately to the BDM (Table 10) in order to compute similarities between each pair of OTU's and the appropriate HTU. In addition the same was done with all three root HTU's combined. Subsequently, from the similarity matrices thus obtained, minimum spanning trees (Prim 1957; Gower and Ross 1969) were generated connecting the OTU's and the HTU, or in the combined exercise it included all the different root HTU's, for assessing which OTU or OTU's resemble most the hypothetical ancestor. The same computations were done for the hexaploid OTU's.

Unit character consistency C_i (Farris 1969) was computed for each character i on each cladogram that was generated after the parsimonious exercise had been achieved. $C_i = \dfrac{r_i}{l_i}$ where: $r_i =$ range of character i; $l_i =$ the patristic unit character length of character i, i.e., the amount of change implied by the tree for that character.

CLAD/OS was run on an IBM/370-168 computer; Bartcher's program CLADON I, II, and III, and other runs were done on a UNIVAC 1108 computer. Farris recently obtained a much faster and exact solution for the Camin and Sokal model (talk given at the 1975 meeting of the Classification Society, Iowa City, Iowa, USA).

TABLE 10. BASIC DATA MATRIX OF DIPLOID *AVENA* SPECIES FOR COMPUTING WAGNER NETWORKS.

The OTU names appear on Table 8, and the characters on Table 7.

OTU character	1	2	3	4	5	6	7	8	9	10	11	12	13	14	15	16	17	18	19	20	21	22
2	1	0	0	0	0	0	0	0	0	0	4	0	3	1	0	1	1	0	4	4	5	1
4	1	0	1	0	1	0	0	0	0	0	5	0	4	1	1	1	0	0	2	4	4	1
5	1	1	0	0	0	0	1	1	1	2	3	0	0	0	0	0	1	0	2	4	4	1
6	1	0	1	2	2	1	1	2	4	3	2	0	3	1	1	1	1	2	2	2	2	0
7	1	0	0	2	1	0	1	2	2	2	1	0	3	0	0	0	0	1	1	1	1	1
8	1	0	1	0	1	1	1	1	4	3	3	0	1	1	1	1	0	2	2	2	2	0
10	1	1	0	2	1	0	1	2	3	2	1	0	2	0	0	0	1	2	2	3	3	1
11	0	0	1	1	1	0	1	2	3	2	1	0	3	0	0	1	0	0	2	2	2	1
12	2	0	0	0	1	0	1	2	2	2	2	0	1	0	0	0	0	2	0	0	0	1
13	1	0	1	2	1	0	1	2	4	4	3	0	6	0	0	3	1	1	2	3	3	1
18	1	0	1	2	3	0	0	0	0	0	0	1	5	1	1	4	0	0	3	3	4	1
23	1	0	0	0	1	0	0	0	0	0	1	0	3	2	1	2	2	0	2	3	3	1
26	1	0	1	0	1	0	1	1	4	3	3	0	1	1	1	1	1	2	3	3	4	0
27	1	0	0	1	0	0	1	2	2	1	3	0	6	0	0	0	0	1	0	2	1	1

TABLE 11. BASIC DATA MATRIX OF HEXAPLOID *AVENA* SPECIES FOR COMPUTING WAGNER NETWORKS.

The OTU names appear on Table 8, and the characters on Table 7.

OTU character	1	2	3	4	5	6	7	8	9	10	11	12	13	14	15	16	17	18	19
9	1	0	0	0	1	2	1	2	1	2	0	3	2	1	0	0	1	1	1
16	0	0	2	2	1	1	1	0	3	1	0	2	0	1	0	0	2	1	2
19	1	1	2	3	1	2	3	1	1	2	0	1	0	1	0	1	4	1	3
20	1	1	1	4	0	0	0	0	0	0	1	0	1	0	0	2	1	2	2
21	1	1	0	1	1	2	1	1	1	2	0	3	2	0	0	1	0	0	0
22	1	1	1	2	1	1	1	0	2	1	0	3	2	1	1	0	3	2	3
24	1	0	2	3	1	1	4	0	2	1	0	3	2	0	1	1	0	0	0
28	1	1	0	0	1	2	2	2	1	2	0	3	2	0	0	2	2	1	2

Evolution

RESULTS OF THE CLADISTIC ANALYSES

Diploids

The method of Camin and Sokal gave similar results with and without character 29. The procladogram (Fig. 11) contains 235 steps with the karyotype-genome character, and 233 without. After the parsimonious exercise, the final cladogram (Fig. 12) has been reduced to 161 and 160 steps respectively. The ancestor possesses by definition the following vector of character states:
{0,0,0,0,0,0,0,0,0,0,0,0,0,0,0,0,0,0,0,0,0,0}

The method of Farris gave various results, partly because a different OTU was taken as starting point for elaborating the network. As a result different groups of more or less similar networks were obtained. The following cladograms represent these groups (Figs. 13–17). The total number of steps (i.e., 114–117) in these groups is considerably less than those obtained through the Camin and Sokal procedure, and with more characters. The designated roots have the following vectors of character-states:

50 {1,0,1,2,1,0,1,2,4,3,3,0,3,0,0,1,1,1,2,3,3,1}
51 {1,0,1,1,1,0,1,3,4,3,2,0,3,0,0,1,1,1,2,2,2,1}
52 {1,0,0,2,1,0,1,2,3,2,3,0,3,0,0,1,1,1,2,3,3,1}
55 {1,0,1,0,1,0,1,1,4,3,3,0,1,0,0,1,1,1,2,3,3,1}
56 {1,0,1,2,1,0,1,2,4,3,3,0,3,0,0,1,1,1,2,3,3,1}
61 {1,0,1,2,1,0,1,2,4,3,2,0,3,0,0,1,1,2,2,2,2,1}

When similarities are computed for each pair of OTU's and the appropriate root, it appears that the most similar OTU to any one of these putative ancestors is OTU 13 or OTU 11 or both, except for the Camin and Sokal ancestor (CS), which is closest to OTU 12.

The minimum spanning tree (Fig. 18) drawn from the matrix of similarities with all the root HTU's shows this situation. Except for CS all the root HTU's are phenetically close to each other and they resemble OTU's 6, 13, 11, 26, and 7 more so than the other OTU's. A plausible interpretation would be to regard OTU 13 as the extant species, which retained a set of characters close to those that are assumed to have existed in the putative ancestor. On the restricted basis of the characters here considered, the difference between *A. longiglumis* (OTU 13) and the putative ancestor is that the latter possesses a narrow elliptic scar, bisetulate-biaristulate lemma tip and a fatua-type lodicule instead. If one compares the two, the only real difference lies in the shape of the scar because *A. longiglumis* by definition possesses two states (65) and (82) of characters 13 and 16 respectively (of Table 7).

Unit character consistencies were computed for each character within each network, and it was found that there is general agreement for those characters that indicate homoplasy. The culm (1), the relative length of glumes (6), the mode of disarticulation of the dispersal unit (7), and the structure of the lemma (12) are all characters with no homoplasy in the networks

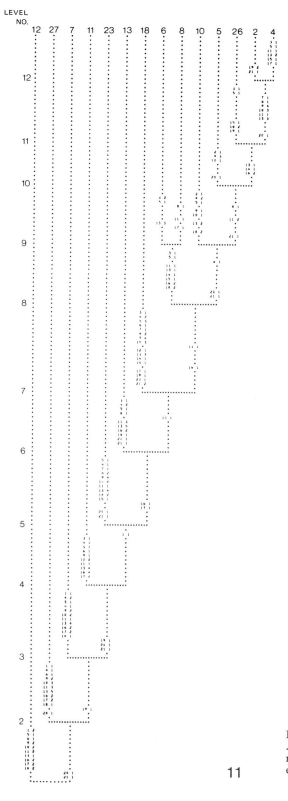

Fig. 11. Output of procladogram of diploid *Avena* species. Obtained from using the method of Camin and Sokal. See further explanations in the legend of Fig. 12.

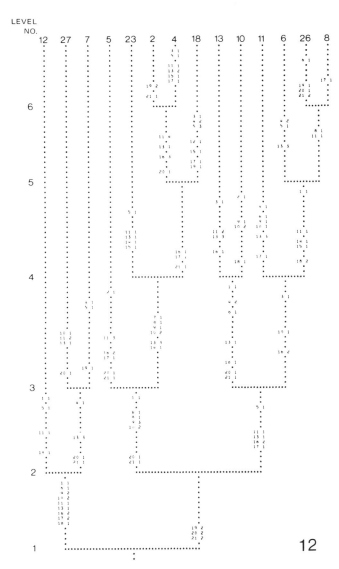

Fig. 12. Output of final cladogram of the diploid *Avena* species. Obtained as a result of the parsimonious exercise on the procladogram (Fig. 11). The numbers to the left of the lines indicate the character codes, and the corresponding numbers to the right of the lines indicate the number of evolutionary steps.

Figs. 13–17. Representative trees of diploid *Avena* species. Obtained from Wagner networks generated by the method of Farris. The numbers to the sides of each node indicate the patristic differences between each OTU and its corresponding HTU. Fig. 13. Suggested root: 61; total number of steps 114; tree generated without the karyotype-genome character. Fig. 14. Suggested roots: 50 and 52; total number of steps 114 with and 115 without the karyotype-genome characters. Fig. 15. Suggested roots: 55 and 56; total steps 116, without the karyotype-genome character. Fig. 16. Suggested roots: 55 and 56; total steps 117, with the karyotype-genome character. Fig. 17. Suggested root: 51; total steps 116, with the karyotype-genome character.

Evolution

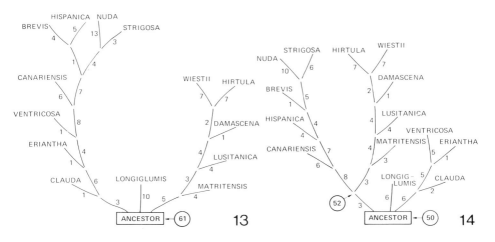

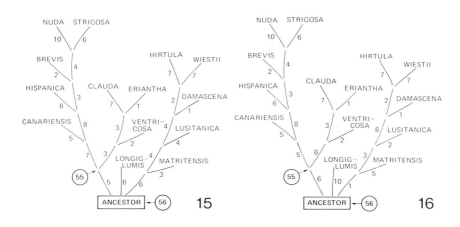

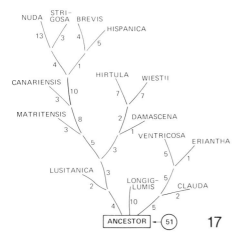

Evolution

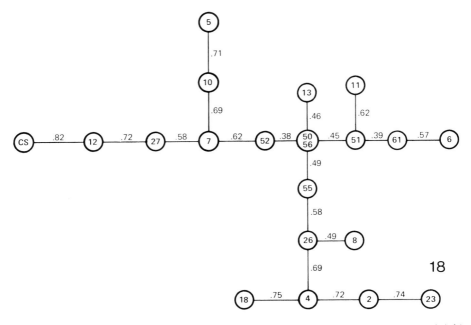

Fig. 18. Minimum spanning tree computed from the similarity matrix of the diploid OTU's and suggested ancestors combined. Similarities, converted to distances, are indicated on the nodes. The numbers in circles denote the OTU's (see Table 8) and the HTU's chosen as ancestors (see Figs. 13–17 and text for details).

thus obtained. In addition an overall low degree of homoplasy was demonstrated in the shape of the first floret scar (9), and in the degree of confinement of the periphery ring around the scar (10). However, a very high degree of homoplasy was demonstrated in the shape of the panicle (3), the spikelet length (4), and in the shape of the lemma tip (13).

Hexaploids

Using Farris's method, I generated a number of Wagner networks. Some were done with OTU 28 and some without. The representative cladograms are given in Figs. 19–21. The designated roots have the following vectors of character states:

HTU 1 {1,0,0,1,1,2,1,1,1,2,0,3,2,1,0,0,1,1,1}
HTU 4 {1,1,0,1,1,2,1,1,1,2,0,3,2,0,0,1,0,0,0}
HTU 5 {1,1,1,2,1,1,1,0,1,1,0,3,2,1,0,1,2,1,2}

When similarities are computed for each pair of OTU's and one the HTU's above, the pairs with highest similarities involve a different OTU. The minimum spanning tree based on the matrix of similarities between the OTU's and all three HTU's demonstrates the situation (Fig. 22). So HTU 1 (Fig. 20) resembles closely *A. fatua*; HTU 4 (Fig. 19) has zero patristic and phenetic distance to *A. hybrida*, but resembles to a certain degree *A. fatua*; and HTU 5 (Fig. 21) resembles more closely *A. sterilis* than any other species.

Evolution

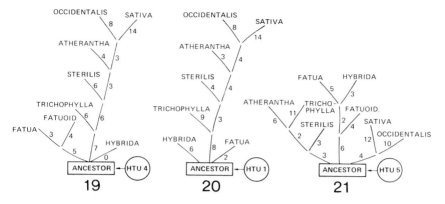

Figs. 19–21. Representative trees of hexaploid *Avena* species. Obtained from Wagner networks generated by the method of Farris. The numbers to the sides of each node indicate the patristic differences between each OTU and its corresponding HTU. Fig. 19. Total numbers of steps: 69. Fig. 20. Total number of steps: 65, excluding the Fatuoid clade. Fig. 21. Total number of steps: 71.

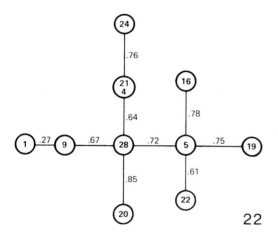

Fig. 22. Minimum spanning tree computed from the similarity matrix of the hexaploid OTU's and the HTU's selected as ancestors, all combined. Similarities were converted to distances. The numbers in circles indicate OTU's (see Table 8) and the appropriate roots HTU's (see Figs. 19–21 and text for details).

Unit character consistencies were also computed for each character within each network thus obtained. This analysis revealed that the characters lacking homoplasy are: the shape of the ligules (1); the mode of disarticulation of the dispersal unit (5); the lemma structure (11); and except for the cladogram of Fig. 11, the shape of the tip of the lemma (12). In Fig. 21 the confinement of the periphery ring around the scar (9) lacks homoplasy as well. Furthermore, a very high degree of homoplasy is found in the panicle shape (2) and to some extent in the lodicule type (14). The remaining characters show a fairly low degree of homoplasy.

Discussion on the results of the cladistic analyses

Cladistic analysis was performed separately on the diploid and hexaploid species only because the methods used in these exercises cannot handle OTU's of hybrid origin and because I assume that each group (i.e., the diploids and the hexaploids) has arisen from a common ancestor and then has undergone the evolutionary processes of adaptive radiation. The tetraploid species consist of groups too small to be analyzed by computer or to be analyzed at all with these methods; and certainly they cannot be lumped together to be investigated as one group. Section *Pachycarpa* has a different origin from section *Ethiopica;* the former has the AACC genome and the latter the AABB genome. Section *Avenotrichon* contains only *A. macrostachya,* which is still not fully understood cytogenetically, although all indications suggest that it does not affiliate with the other two sections of tetraploid species.

The cladograms reported here should be regarded as hypotheses of evolutionary relationships among the species within each of the two groups concerned. Among the different hypotheses for the diploid species, I regard as superior the cladogram with HTU 50 as ancestor (Fig. 14), because it is the most parsimonious tree and because the root is the HTU from which the lineages of the CC-genome species and the AA-genome species start. Furthermore HTU 50 is phenetically closest to *A. longiglumis* (Fig. 18), which is one of three species in this genus with isobrachial chromosomes (Rajhathy and Thomas 1974), a condition being regarded as primitive.

Of the different hypotheses presented here for the hexaploid species, I am undecided as to which one to regard as superior. The tree with HTU 4 as ancestor (Fig. 19) is equally as parsimonious as the tree with HTU 1 as ancestor (Fig. 20), but the former is the only tree obtained with a recent species, *A. hybrida,* with zero patristic distance from the root of the tree. I have designated previously (Baum 1973) the very same species as a probable survivor of the ancestor of the predomesticated oat.

If we treat the fatuoids (see Chapter 11) as an OTU (OTU 28), which we really should not for that purpose, the result is that in all the different networks the patristic distance between *A. sativa* and the former is great. One would expect in such a cladistic exercise to have the OTU Fatuoid come on the same clade with *A. sativa,* because the former is a mutant of the latter.

Perhaps, the best compromise is to regard the cladogram with HTU 5 as ancestor (Fig. 21) as the most plausible alternative of the three, because the patristic distance between *A. sativa* and its mutant OTU Fatuoid is the shortest. The root of that cladogram is phenetically similar (Fig. 22) to a sterilis-type oat. This is in agreement with Coffman's (1946) hypothesis on the origin of cultivated oats, but in disaccord with the general belief that the fatua-lineage should precede the sterilis-lineage.

It might be useful at this stage to study the amino-acid sequences of some proteins in *Avena* and use also this data for elucidating the evolutionary relationships among the diploids and among the hexaploids.

FURTHER EVOLUTIONARY SPECULATIONS BASED ON THE DISTRIBUTION OF THE SPECIES

A summary of the distribution of the diploid species is given in Table 12. From this summary and from the details given about every species in this work (see habitat and distributional notes under appropriate species), it becomes evident that at the diploid level there are species that are: endemic and locally rare (e.g., *A. damascena*); endemic and locally common (e.g., *A. canariensis*); restricted and locally common (e.g., *A. hirtula*); scattered and locally rare (e.g., *A. ventricosa*); scattered and locally common (e.g., *A. eriantha*); and relatively common (e.g., *A. lusitanica*). Also among the diploid species, we have cultivated (e.g., *A. strigosa*) and weedy species (e.g., *A. clauda*).

At the tetraploid and hexaploid levels one encounters the same different categories, which are summarized below.

This situation can be interpreted into the following evolutionary scheme (Fig. 23). At a relatively early stage of speciation of the diploids, tetraploidy (namely allotetraploidy and segmental allotetraploidy) occurred among the diploid species common at that time. Some newly formed species were probably very successful until changing environmental conditions brought about their restricted distribution, such as *A. maroccana* and *A. abyssinica,* or complete disappearance and extinction. This does not rule out the possibility of constant attempts to reproduce similar tetraploids from almost similar diploids or from the survivors of those early diploid species. The tetraploid *A. murphyi* might be such a case in relation to *A. maroccana,* or vice versa in relation to time. It is at this point impossible to know which species is a neotetraploid and which is a paleotetraploid.

The same scheme is applicable to the hexaploid species and their tetraploid and diploid ancestors.

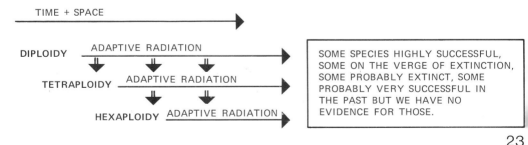

Fig. 23. Evolutionary scheme of speciation in *Avena*. Double arrows symbolize hybridization among species followed by doubling of chromosome number leading to amphiploid new species.

TABLE 12. DISTRIBUTION OF DIPLOID SPECIES.

Their presence in various countries or in well-defined geographic entities that may belong politically to some countries already enumerated.

	Afghanistan	Algeria	Austria	Azores	Belgium	Bulgaria	Canary Islands	Corsica	Crete	Cyprus	Czechoslovakia	Denmark	Egypt	Finland	France	Germany	Greece	Hungary	Iran	Iraq
A. clauda		x				x			x		x						x		x	x
A. eriantha		x				x											x		x	x
A. ventricosa		x								x										x
A. brevis			x	x	x		x				x				x	x		x		
A. hispanica														x						
A. nuda			x	x							x	x			x	x				
A. strigosa			x	x				x			x	x		x	x	x		x		
A. canariensis							x													
A. damascena																				
A. hirtula		x																		
A. longiglumis		x											x							
A. lusitanica		x					x	x	x						x		x			
A. matritensis		x							x											
A. wiestii	x	x				x							x				x		x	x

Adaptive radiation is probably the most significant evolutionary process in oat species, whereas polyploidy seems just to increase the scope and divergence of the former process.

Tetraploids

A. macrostachya	Endemic, locally rare
A. abyssinica	Endemic, locally common, cultivated
A. vaviloviana	Endemic, locally common, weedy
A. barbata	Very common, weedy
A. murphyi	Endemic, locally rare
A. maroccana	Endemic, locally common

	Jordan	Lebanon	Lithuania	Luxemburg	Lybia	Madeira	Morocco	Netherlands	Norway	Poland	Portugal	Sardinia	Saudi Arabia	Sicily	Spain	Sweden	Switzerland	Syria	Turkey	United Kingdom	USSR	Yugoslavia	No. of countries present	Percent of presence	Ranking
	x			x														x	x		x		13	29.5	9
	x			x														x	x		x		11	25.0	8
			x											x							x		6	13.6	4
		x	x	x		x	x							x	x					x	x	x	19	43.1	11
								x					x										3	6.8	3
																				x			7	15.9	5
	x	x					x	x					x	x	x					x	x		18	40.9	12
				x																			2	4.5	2
																x							1	2.2	1
			x											x									3	6.8	3
x		x	x									x x		x									9	20.4	6
	x	x	x	x		x	x	x					x	x		x	x					x	19	43.1	11
		x									x	x	x	x				x			x		10	22.7	7
	x													x	x				x	x			14	31.8	10

Hexaploids

A. occidentalis	Scattered, locally common
A. atherantha	Scattered, locally rare
A. trichophylla	Scattered, locally rare
A. hybrida	Common, sometimes weedy
A. fatua	Very common, weedy
A. sterilis	Very common, weedy
A. sativa	Very common, cultivated

The existence and distribution of vicarious species in *Avena* (e.g., *A. damascena* in Syria and *A. hirtula* in Algeria, Morocco, and Spain) certainly supports the views mentioned above; moreover, the distribution of similar character-states in different species of *Avena,* for example, the heart-shaped scar in *A. canariensis, A. occidentalis, A. abyssinica,* and *A. hybrida,* also give support to this contention. Of course, one cannot deny the possibility of homoplasy in a number of characters, but even in the scar character a very low degree of homoplasy was found among the results of the cladistic analysis. The phenomenon of vicarism itself certainly supports the proposed evolutionary scheme of the *Avena* species; the numerous examples of vicarism known in the plant kingdom point very clearly to changing areals of species in time and space. A recent and well documented example is that of *Pinus canariensis* and *P. roxburghii* (Page 1974).

Given the assumption of the high probability of changes occurring in distributions of species in time and space, and the cytogenetic information that we have on *Avena* (Rajhathy and Thomas 1974), together with the present cladistic analysis, the evolutionary scheme here proposed must await the test of time.

9. The epicuticular waxes on the glumes of *Avena* examined under the scanning electron microscope

The epicuticular waxes in a number of taxa were studied by various workers for different purposes. A review on the subject appeared in Martin and Juniper (1970*a*). Among the more recent studies, some are concerned with chemical variability (e.g., Tribe et al. 1972; Tulloch 1973), others with morphological variability (e.g., Baker and Parsons 1971); and most revealing are those studies aimed at the understanding of the factors contributing to that variability and the causes associated with that variability (e.g., Wettstein-Knowles 1974).

The different patterns of epicuticular waxes depend on various factors, one of the most important factors being chemical composition (Wettstein-Knowles 1974 and references therein). Other factors are the number of pores from which the waxes are exuded, the closeness of the pores and their arrangement (Hall 1967), and influence of the environment. In addition, Wettstein-Knowles (1974) proposed a hypothesis that the rate of exudation might be another factor, and stated that most of the factors are controlled by genes. The factors cited by Hall, indeed, seem to be controlled by genes. Certainly, genes do control at least the chemical composition of the waxes; because the wax patterns are somehow dependent on that composition, it might be worthwhile to survey the genus *Avena* in order to assess the usefulness of wax patterns as a character for discrimination and identification of species and to explore the possibility of this character's role in classification and evolutionary relationship.

Similar studies for taxonomic purposes were done for *Eucalyptus* (Hallam and Chambers 1970) and for apple cultivars (Faust and Shear 1972).

MATERIALS AND METHODS

Glumes, at various stages from postanthesis to maturity, were collected from 357 samples grown in the greenhouse of the Ottawa Research Station and from herbarium material (see Chapter 3, "Materials"). A summary of the number of samples for every species and their country of origin is shown in Table 13.

Small rectangles (but big enough to cover most of the width of the glume) were excised from about the middle of the glumes, and affixed to the specimen stubs with silver paint; fresh material was first frozen in Freon 12 at $-150°C$ and dried at $-80°C$ in an Edwards "Speedivac-Peace" freeze dryer and attached to specimen stubs as above. Subsequently, the specimen stubs were placed in a vacuum chamber, rotated, and coated with gold 20–30 nm (200–300 Å) thick and deposited at angles of 60° and 30°. Examination and ultramicrographic photography of the epicuticular waxes were performed on a Cambridge Stereoscan Mark IIa electron microscope.

RESULTS

Three basic patterns of epicuticular wax crystals are found on the glumes: filament-shaped (Figs. 24 and 27), platelet-shaped (Figs. 25 and 28), and knob-shaped (Figs. 26 and 29). These patterns we shall call states. As can be seen in Table 13, some samples have only one state over the surface of their glumes, whereas in other samples two or all three states are found. In those cases where more than one state was present, sometimes the states were segregated in different areas and in other instances the states were intermixed in the same area.

The same basic pattern may vary. For example, the filament-shaped waxes may be long and dense or short and appressed; the platelet-shaped waxes may have fringed or smooth margins; the knob-shaped waxes are often arranged in circles or may be randomly dispersed (for more details, see Baum and Hadland 1975).

The observations are summarized in Table 13 where the arrangement of species follows the classification and sequence adopted in this work.

> Section *Avenotrichon*. In *A. macrostachya* all the specimens possess knobs and filaments on the same glumes.
> Section *Ventricosa*. Plates prevail; most specimens of *A. clauda* and *A. eriantha* have exclusively platelets and knobs.
> Section *Agraria*. Most species have at least two patterns intermixed in the same area of the glume; *A. brevis* has mostly knobs and filaments or knobs only; *A. hispanica* and *A. strigosa* have knobs and plates or exclusively filaments; *A. nuda* is poorly understood because there is a high proportion of unstructured wax.
> Section *Tenuicarpa*. *A. canariensis, A. longiglumis, A. lusitanica*, and *A. wiestii* have mostly knobs and platelets or only knobs, and occasionally a few specimens of *A. wiestii* of Algerian origin have

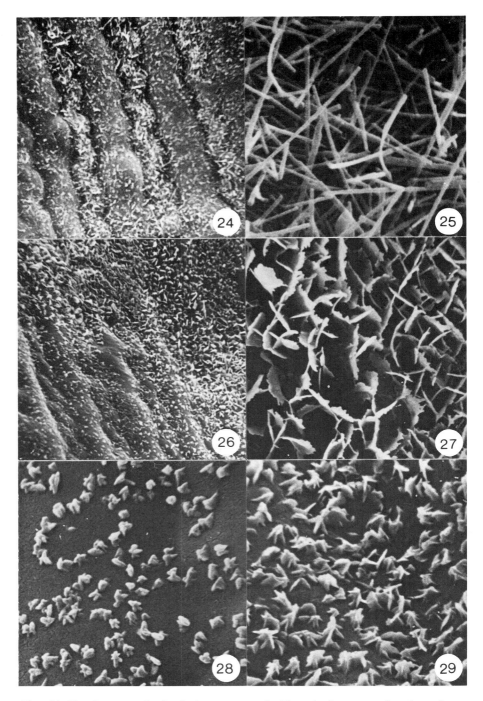

Figs. 24–29. Patterns of epicuticular wax crystals. Three basic patterns found on glumes of various *Avena* species. Figs. 24 and 27, filaments in *A. sterilis*, × 1280 and 7400 respectively; Figs. 25 and 28, plates in *A. murphyi*, × 1350 and 6350; Figs. 26 and 29, knobs in *A. lusitanica* × 6700 and in *A. vaviloviana* × 6000.

TABLE 13. EPICUTICULAR WAX PATTERNS.

Summary of observations on 357 samples. The species are arranged according to sections in this work. For each species the country source is given. The pattern scores are given in the appropriate columns together with the total scores for every species.

Species	Country	K	KP	P	KF	PF	F	KPF	U	Total samples
Section *Avenotrichon*										
A. macrostachya	Algeria			3						
Total of section				3						3
Section *Ventricosa*										
A. clauda	Algeria		2						1	
	Iran			2						
	Iraq			2						
	Turkey			1						
Total			2	5					1	8
A. eriantha	Algeria			1		1				
	Iran			1						
	USSR	1	1							
Total		1	1	2		1				5
A. ventricosa	Algeria			2						
	Libya		4							
	USSR		2							
Total			6	2						8
Total of section		1	9	9		1			1	21
Section *Agraria*										
A. brevis	Belgium	3								
	France	2	3		1		2	1		
	Germany	1	2				1	1		
	Spain	2			1		1			
Total		8	5		2		4	2		21
A. hispanica	Portugal	2	1				2			5
A. nuda	W. Europe	3					1			4
A. strigosa	England	1		1			1			
	Germany		1							
	Greenhouse		1							
	Spain					1	2			
	Wales		3							
Total		1	5	1		1	3			11
Total of section		14	11	1	2	1	10	2		41
Section *Tenuicarpa*										
A. barbata	Algeria		3							
	Canary Is.	2								
	Greenhouse		1							
	Iraq	1	1	1						
	Italy						2			
	Morocco		2			1				
	Spain					1				
	USSR			1						
Total		3	7	2	2		2			16

TABLE 13. (Cont'd)

Species	Country	K	KP	P	KF	PF	F	KPF	U	Total samples
A. canariensis	Canary Is.		6	6			2			
	Cape Verde Is.		2							
	Morocco		2	1						
Total			10	7			2			19
A. damascena	Syria				1		1			2
A. hirtula	Spain	1					11			12
A. longiglumis	Algeria		5							
	Israel	1	1							
	Morocco						3			
	Portugal	1		1						
	Spain	1	2							
Total		3	8	1			3			15
A. lusitanica	Algeria		5				2			
	Corsica		1							
	Crete						1			
	Morocco		1							
	Sicily				1					
	Turkey						1			
Total			7		1		4			12
A. matritensis	Algeria						9			
	France	3	5					1		
	Sardinia		1					1		
Total		3	6				9	2		20
A. wiestii	Algeria				1		2			
	Egypt	1			1					
	Iran		3							
	Israel		3	2						
	Libya		3	1						
	Sinai		3							
Total		1	12	3	2		2			20
Total of section		11	50	13	6		34	2		116
Section *Ethiopica*										
A. abyssinica	Algeria			1						
	Egypt		1							
	Ethiopia	4	5	1	1			2		
Total		4	6	2	1			2		15
A. viloviana	Ethiopia	2	9	1						12
Total of section		6	15	3	1			2		27
Section *Pachycarpa*										
A. maroccana	Morocco	1	4	8			5			18
A. murphyi	Spain			4						4
Total of section		1	4	12			5			22
Section *Avena*										
A. atherantha	France						3			
	Kenya		1							
	Sicily							1		
	USSR						1			
Total			1				4	1		6

TABLE 13. (Cont'd)

Species	Country	K	KP	P	KF	PF	F	KPF	U	Total samples
A. fatua	Austria				1			3		
	Canada		1				4		1	
	Germany		2			2				
	Iran						2			
	Iraq						3			
	Japan	2								
	Spain	1								
	Switzerland						1	1		
	Turkey						6			
	USA						3			
Total		3	3		1	2	19	4	1	33
A. hybrida	Afghanistan						1			
	Germany		1				1			
	Japan						2			
	USSR						3			
Total			1				7			8
A. occidentalis	Algeria						3			
	Canary Is.	1			1		8	2		
Total		1			1		11	2		15
A. sativa	Brazil								1	
	Canada						3			
	Ethiopia			1		1	3	1		
	Germany						1			
	Iran						2			
	Portugal						2			
	Spain						1			
	Turkey						9			
	USSR						2			
Total				1		1	23	1	1	27
A. sterilis	Algeria						3			
	Italy						2			
	Lebanon						3			
	Tunisia						3			
	Turkey					1	5			
Total						1	16			17
A. trichophylla	Afghanistan						2			
	Malta	2								
	Portugal	4	2		1				3	
	Spain		1							
	USSR				1		4			
Total		6	3		2		6	1	3	21
Total of section		10	8	1	4	4	86	9	5	127
Total of genus		43	97	39	16	6	135	15	6	357

exclusively filaments; *A. hirtula* and *A. damascena* have filaments almost exclusively; *A. barbata* contains at least two states on each specimen; and *A. matritensis* has specimens with knobs and platelets, or in Algeria filaments only.

Section *Ethiopica*. Here knobs and plates prevail, even when a species is grown outside its natural habitat as was *A. abyssinica* in Algeria (this was taken from a specimen from Trabut). Sometimes only knobs occur as in *A. vaviloviana* or platelets and filaments as in *A. abyssinica*.

Section *Pachycarpa*. In *A. murphyi* there are exclusively platelets but this is based on a small sample, the only one available. In *A. maroccana* there are exclusively platelets or platelets and knobs and a few samples have exclusively filaments.

Section *Avena*. In this section filaments prevail. In *A. fatua, A. hybrida, A. occidentalis, A. sativa,* and *A. sterilis* the majority of specimens possess exclusively filaments. In *A. fatua,* two specimens from Germany and two from Japan have exclusively knobs and platelets; and a specimen from Austria and another from Switzerland have all three states and also unstructured wax. All the *A. atherantha* samples have filaments except one from Kenya, which has knobs and plates instead, and except another, but poor, specimen from Sicily, which has all three states. *A. trichophylla* also has filaments except for the majority of specimens from Portugal and Malta, which have knobs exclusively.

Little variability is apparent within the species of section *Avena* as far as the type of exudation of wax on the glumes is concerned; most specimens show exclusively filaments. Furthermore, in other sections some species are much more variable and some have similar narrow variability.

Species cannot be identified based solely on epicuticular wax morphology of the glume, except in a few cases and then only when the alternatives are restricted. Obviously, the morphology of the epicuticular waxes on the glumes is not species-specific, but when a few alternatives are given these may be very useful in identification.

DISCUSSION

Tulloch and Hoffman (1973), investigating the chemical content of leaf waxes on *Avena sativa* 'Kelsey', reported a low beta-diketone (5.5%) proportion in the waxes and stated that the beta-diketone is hentriacontane-14, 16 dione. Wettstein-Knowles (1974) claimed that thin wax tubes (i.e., filaments) are present only when the wax contains beta-diketones, and that on surfaces bearing essentially filaments, beta-diketones amount to no less than 27% of the wax. Therefore, the glumes of the hexaploid oat species (section *Avena*), which have almost exclusively filaments, are likely to have not less than 27% beta-diketones in the wax. The other species have different wax components, and, accordingly, have less than 27% beta-diketones. Chemotaxonomical investigations might be in place here in order to define the proportions of the wax components for every species and to test their usefulness for identification purposes. (As we have already concluded, morphology of the epicuticular waxes has limited value for identification.)

Different organs on the plant or different parts of the plant may have different proportions of wax components (Tulloch 1973), which are certainly expressed in, or associated with, the different morphological patterns found in different parts of the same plant (Baum and Hadland 1975 and unpublished data). Information on the various proportions of wax components on different parts may not only assist in identification to species but can contribute to classification. Furthermore, data on waxes from different parts of the plant can contribute to assessment of physiological traits and to susceptibility to fungal pathogens (Martin and Juniper 1970b and references therein; Takahoshi et al. 1972 for oats, among others).

Fungi affect wax morphology. When fungal hyphae were present, Baum and Hadland (1975) observed unusual, long, plate-shaped waxes, as though two or many plates were aligned and fused side by side.

Why do the hexaploid oats have almost exclusively filaments? One possibility is that genes that control beta-diketone synthesis (Wettstein-Knowles 1972) are present in double or triple ratio (because the hexaploid oats have the AACCDD genome); another possibility is that the hexaploids have evolved from biotypes of those diploid progenitors with filaments only. Alternatively, since *A. maroccana* contains the AACC genome (Rajhathy and Sadasivaiah 1969) it is quite probable that only those biotypes with filamentous waxes on glumes are progenitors of the hexaploid oats.

10. Cultivars

Oats occupy an important place among the main coarse grains: (in order of importance) corn, barley, oats, sorghum, and rye. The relative importance of oats was maintained by the breeding effort of various agencies in different countries. Most efforts concentrate on improving the hexaploid cultivated oats, *A. sativa*. Section *Agraria* consists of four species that have been cultivated in the past to a much greater extent than they now are. One species of section *Ethiopica*, namely *A. abyssinica*, has been cultivated in the highlands of Ethiopia.

An excellent account on oat breeding is given by Coffman, Murphy, and Chapman (Coffman 1961). From this account and from my pedigree charts (Baum 1973a), it can be seen that oat improvement until recently was based on gene exchange within the species *A. sativa*. More recently wild species of Section *Avena*, chiefly *A. sterilis*, have been taken as sources of genes.

With the invention and the introduction of new breeding techniques, even distant species in the genus become important sources of genes for oat improvement. Even material from related genera is increasingly becoming useful in genetic engineering. The interest in wild species led to the establishment and maintenance of gene banks in various countries, for example, USA, Canada, USSR, Japan, Australia, and Sweden, to name only a few.

NOMENCLATURE

So far, at least 4000 names have been given to various oat cultivars in different parts of the world. These names have been put together in a registry (Baum 1973a), which includes commercial synonyms, translations, transliterations, and cross-references. Most names in this registry apply to *A. sativa* but a few belong to species of section *Agraria*.

PHYLOGENETIC RELATIONSHIPS

Pedigree charts for many cultivars of oats have been elaborated with computer techniques using a procedure given in detail by Baum and Thompson (1970).

The pedigree charts describe the phylogenetic relationships among the cultivars. These were published in a separate volume with the registry mentioned above (Baum 1973a) and can be used for various cladistic aspects, for example, finding the coefficients of inbreeding and common parentage.

It was already realized (Baum 1970) that no classificatory system was congruent with the cladistic relationships — nor could any system be — at least as far as oats are concerned (Baum and Lefkovitch 1973). Therefore in *Avena* cultivars, very likely in cereal cultivars, and possibly in cultivars in general, finding phylogenetic relationships must be done separately from classification into cultivar groupings, using different sources of data and certainly different principles and procedures.

PHENETIC RELATIONSHIPS AND TAXONOMY

Classification and identification of cultivars in general and of oat cultivars in particular poses many difficulties. In the introductory section of the "Material for an International Oat Register" (Baum 1973a), I summarized the publications of authors from many countries since 1752 (the starting date for the nomenclature of cultivated plants) until recently, together with references to their systems, or identification schemes and characters they used, or both.

I have attempted recently to establish cultivar groupings using numerical techniques (Baum and Lefkovitch 1972a, 1972b). Fourteen groupings were thus established and an identification procedure for them, based on a probabilistic model, was provided. The 14 cultivar groupings can form the basis of a worldwide classification to be used for future placement or determination of belongingness of a particular cultivar to one of these groupings.

The determination of belongingness of a cultivar to a particular grouping may be useful, depending on the need or the purpose. Most useful, however, is the identification of the different cultivars themselves. In other words, given a sample of plants or seeds, what is the identity of that sample in terms of belongingness to a cultivar? This is a critical problem, and one which appears pressing in view of the legislation in various aspects of seed trade. A farmer, an inspector, or a merchant will not be primarily interested in whether 'Garry' belongs to cultivar grouping No. 1, 5, or 11. Instead, such people will be interested in knowing if the cultivar in question is indeed 'Garry', or perhaps 'Rodney'? Furthermore, the breeder will be interested in knowing if his new cultivar can be distinguished from the other cultivars. This distinction should not be confused with the related issue of whether the same cultivar is a cultivar different from other cultivars. Belongingness of a particular cultivar to a cultivar grouping will give the farmer or the merchant only part of the answer.

I have attempted for a number of years to devise an identification scheme for cultivars, a scheme that will be accurate, fast, objective, reliable, and flexible. We need to minimize the probability of misclassification; we normally require an answer as fast as possible; we need criteria, devices, and processes that yield reliable and consistent identification; and we need a system flexible enough to accommodate the changes that may occur in different cultivars from year to year, or from place to place, because of the interactions between the genotype and the environment. Those interactions express themselves more greatly at the cultivar level than at the species level. This, of course, depends on our degree of sensitivity in measuring at the cultivar level the fluctuations of characters, which are greater than those at the species level (which are often ignored).

To achieve this aim I recognized the need of devising an automatic identification scheme. The first breakthrough in this direction was achieved by using inflorescence spectrographic techniques (Baum and Brach 1975; Brach and Baum 1975). Work is now under way to elaborate on that beginning, so as to make the identification process fully automated.

Here is the rationale behind using spectrographic techniques. Cultivars that belong to the same species, although closely related genetically, are still different from one another with respect to one, or a few, or many traits. Among the many different traits, the differential chemical content in cultivars seems to be universal. There is a growing body of evidence for this contention, for example, with grain proteins (Ellis 1971; Singh et al. 1973; and Cabezas et al. 1972), with monoterpenes (Rottink and Hanover 1972), with phenols (Bose 1972), with peroxidases (Chu et al. 1972), and with many others; McKee (1973) cites many references prior to 1972. All these references document cultivar specificity in patterns or ranges, albeit with various degrees of overlap.

From the foregoing, McKee (1973) has concluded that "if varieties are distinct, there should be corresponding chemical differences." On this basis we are interested in finding the different patterns that might emerge as a result of radiant energy exciting the atoms in these chemicals. In other words we are concerned with establishing the differential patterns for identification purposes, and not the chemical causes.

Still aiming at making identification fully automatic, I tackled the problem from a morphological point of view. Because many oat breeders recognize different cultivars by the shape of their seeds, I sought statistical evidence to determine if indeed seeds of different varieties differ in shape. So, in the summer of 1972 I collected samples of seeds of various varieties grown in Canada, from different provinces and at different stages of propagation, viz., breeders seeds, select, foundation, registered, and certified. During most of 1973 we measured the size-shape of the seeds in the samples. A series of analyses of variance indicated that in general there were no significant differences between samples of the same cultivar, between the different propagation stages of the same cultivar and between the different localities. Most important however, there were significant differences between different cultivars. Moreover, although the size-shape measurements were obtained by people using identical guidelines and criteria, the results differed from worker to worker. Subsequent canonical analyses indicated that many cultivars belong to different or separate populations on the basis of the size-shape measurements. The results seem to justify the development of an automatic scanning device for recording the size-shape of oat cultivars. The technical details will be found in forthcoming publications on this subject (first is Baum and Thompson, In press).

11. Fatuoids

Fatuoids are essentially fatua-like plants arising in populations of cultivated oats. A good review on the subject is given by Huskins (1946) and O'Mara (1961). The classical work of Nilsson-Ehle (1907, 1911, 1914, 1921) on this subject demonstrated that fatuoids arise by spontaneous mutations or chromosome aberrations in cultivated oats. This evidence, however, does not preclude the other alternatives of fatuoid origin: that fatuoids could arise (i) through hybridization, that is, segregation of *A. sativa* × *fatua* hybrids, as is believed by many authors, or (ii) by genetic mutations (Jones 1930).

Malzew (1930) did not find any difference between *A. fatua* and fatuoids and therefore concluded that Nilsson-Ehle's findings are thus supported morphologically, that is, the lack of differences obviously proves that these are mutations. Following Malzew, many researchers in this subject contended that it must be difficult, if not impossible, to discriminate morphologically between fatuoids and *A. fatua* unless the parents are well known; nonetheless the attempt was often made.

Earlier I found (Baum 1969*a*, 1969*b*, 1971, and related papers) reliable markers that discriminate between *A. sativa, A. fatua, A. sativa* × *fatua* F_1 hybrids and F_1-like phenotypes, and *A. sativa* fatuoids of various kinds and their hybrids with *A. fatua*. These markers are documented here and appear in the descriptions of the various entities.

According to my findings, fatuoids belong to the species *A. sativa* essentially, and not to *A. fatua*. The micromorphological evidence that I found

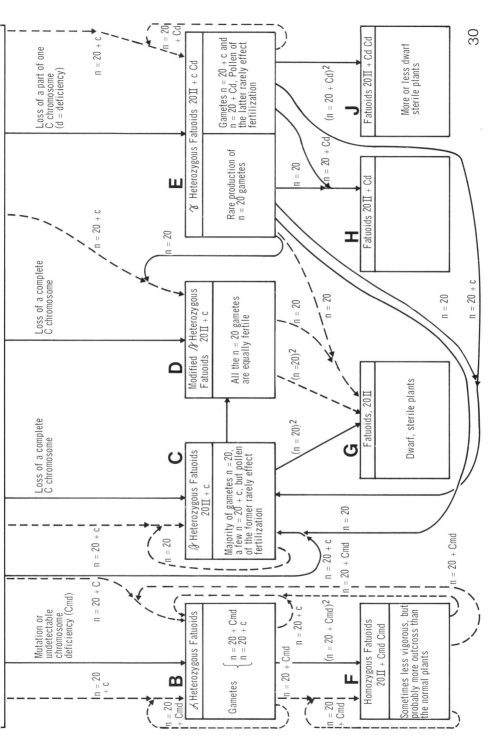

Fig. 30. Cytological processes involved in the generation of fatuoids. A schematic representation of these processes and the relationships between fatuoids and normal plants, based on Huskins' (1946) review paper.

supports the mutational and chromosome-aberration theories of origin of fatuoids. Furthermore, the hybrids have been found to be different from fatuoids and obviously not of the same nature. A brief description on the nature of fatuoids is germane, and is summarized and adapted from Huskins (1946). A schematic representation (Fig. 30) will help me to describe the process of the origin of fatuoids in crops. The C chromosome carries the factor that suppresses the fatuoid configuration; this is a recessive or partly recessive factor. A mutant gamete from a normal plant together with a normal gamete yields the alpha heterozygous fatuoids (Fig. 30, box B). Loss of a complete C chromosome in one gamete only leads to a beta heterozygous fatuoid (Fig. 30, box C). When the gametes produced by such beta heterozygous fatuoids are all $n=20$, modified beta heterozygous fatuoids are produced (Fig. 30, box D), and these may also arise directly from normal plants. Gamma heterozygous fatuoids (Fig. 30, box E) are caused by a C-chromosome deficiency, where the deficient part is the one that carried the fatuoid depressant factor. Homozygous fatuoids (Fig. 30, boxes F, G, H, and J) originate from each of the three kinds (alpha, beta, and gamma) of heterozygous fatuoids. More ways of origin of the various kinds, and more relationships are shown in Fig. 30 by arrows (broken arrows indicate normal gametes unless otherwise noted).

In the past, the alpha homozygous fatuoids and the various heterozygous fatuoids have been hard to distinguish from *A. fatua* on a morphological basis. The lodicules (Figs. 320–323) alone and certainly in combination with other traits provide good markers. The F_1 hybrids do not have lodicules of the sativa type, but instead, have ones of the fatua type (see Baum 1968*b*, 1969*a*, 1969*b*, for further details and proof based on experimental material).

It is noteworthy that the incidence of fatuoids among various cultivars is associated with the parentage of these cultivars (Baum 1969*b*).

12. Characters for discrimination and identification of oat species and for their classification into groupings within the genus

Most characters and their states are listed in Table 2 with their codes for classificatory purposes. This chapter provides additional information on each character and is arranged in the same sequence as in Table 2. The character numbers are given in single brackets and the character-state numbers in double brackets, using the same numbers as in Table 2.

1) Longevity (1) Perennial (2) Annual. This character is fairly simple to recognize. Only one species is perennial and its distribution is very restricted. The perennial species is readily distinguishable from the annual; it is densely tufted with many remnants of last years growth and the underground parts are thick and rhizomatous.

2) Juvenile growth (3) Erect (4) Prostrate to erect (5) Prostrate. Etheridge (1916) and Stanton (1955), among a few others, recognized three kinds of

growth habits characteristic of cultivars at the early stages of development, namely: prostrate, semiprostrate, and erect. I found it very difficult to recognize the semiprostrate kind at the species level; moreover I found that some species may have two kinds of early growth habits. This situation is implicit in *A. sativa* according to data of both Etheridge and Stanton, because different cultivars within *A. sativa* have different early growth habits. For these reasons I recognize erect (Fig. 31, right), prostrate (Fig. 31, left), and prostrate to erect. The last state includes species that have both kinds of early growth habits and species with an intermediate habit similar to Stanton's semiprostrate.

3) Color appearance (6) Glaucous (7) Green. The color of the leaves and, consequently, the general appearance of the plant, especially before or at anthesis, is green or dark grayish green. In some species the whole range is found. Although a wide degree of variability exists within species, most of them can be classified into one of the two states of foliage color. The states are used in identification and for classificatory purposes, and the whole range is described for every species.

4) Culm (8) Geniculate or prostrate (9) Erect (10) Prostrate or Erect. Some species are prostrate for a relatively long period during vegetative growth, although at anthesis their panicles are upright. This growth pattern is achieved through a gradual bending of the stems at the internodes, thus giving

Fig. 31. Two kinds of juvenile growth in *Avena*: left, prostate; right, erect. In some species the two kinds exist.

the plant the geniculate habit of its culms. Many species have erect culms, and moreover some species, such as *A. hirtula,* have both prostrate and erect states.

5)   Height (11) Low plants (12) Tall plants.   The actual range observed is reported in the descriptions. For purposes of discrimination, description, and classification, however, two overlapping states have been distinguished. The two states with their definitions are given in Table 2.

6)   Shape of ligule (13) Obtuse (14) Acute.   These character-states are not always easy to distinguish because the ligule matures at different times on different leaves on the same plant. Before or at anthesis seems to be the appropriate time to observe the shape of the ligule for use as a morphological character. Obtuse ligules (Figs. 32, 33) seem to be associated with low ploidy levels, whereas acute ligules (Figs. 34, 35) are found on species of high ploidy levels. Because ligules occur in a wide variety of shapes, use Figs. 32–35 as guidelines only.

7)   Shape of panicle (15) Equilateral (16) Slightly flagged, equilateral or flagged, equilateral and sometimes flagged.   In considering the shape of the

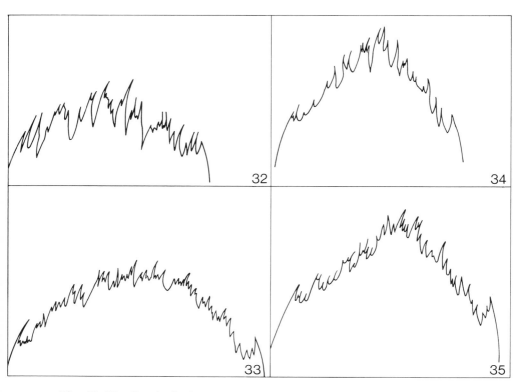

Figs. 32–35.   Two basic shapes of the ligules in *Avena*. Figs. 32, 33 obtuse ligules; Figs. 34, 35 acute ligules.

panicle as an attribute, you can distinguish three states: equilateral, intermediate, and flagged; and these states are used for identification. Because of the variability within species, however, only the two states mentioned above are recognized as characters and these states are used for classificatory purposes.

8) Length of spikelet (17) Short (18) Indeterminate (19) Long. The actual range is given in the descriptions. The upper limits and lower limits of the range determine the states and these are itemized in Table 2. The length is measured from the extreme base to the lemma tips of the uppermost floret without taking the awns into consideration. If the lemma tips are aristulate, the aristulae of the uppermost floret are taken into consideration for measurements. These states are used in identification and classification.

9) Number of florets in the spikelets. The range varies for each species and is given in the descriptions. The various ranges were classified into eight states for convenience (Table 2).

10) Relative length of glumes (28) Equal (29) Unequal. Various authors use more states, such as subequal, subinequal. In the descriptions I did not use rigid states but described the situation; for example, I used words such as "almost equal." For classificatory purposes I found it practical to distinguish two character-states. For example, *A. sterilis* has equal glumes (Fig. 295), and this state is common to many species, whereas *A. clauda* (Fig. 56) has unequal glumes.

In addition, the range of the number of nerves on the glumes, and the range of length of the glumes are given in the descriptions of the species.

11) Mode of disarticulation of the dispersal unit (30) Non-disarticulating (31) Disarticulating. In the non-disarticulating mode, the lowermost floret remains attached. Consequently the whole spikelet remains in its place, does not shed at maturity, and is said to possess a cultivated base. In the disarticulating mode, the lowermost floret develops a scar, sometimes called suckermouth, and sheds at maturity.

12) Mode of disarticulation of the florets (32) Only the lowermost disarticulating (33) All the florets disarticulating. As a rule, if the dispersal unit is non-disarticulating, then all the florets are non-disarticulating; if the dispersal unit is disarticulating, then there are two possibilities, (i) the lowermost floret is disarticulating and the remaining florets in the spikelet do not develop scars, and (ii) all the florets develop a scar complex at their base and will shed at maturity. For identification purposes, characters 11 and 12 are merged so that the combined attribute states then are: non-disarticulating, lowermost disarticulating, and all florets disarticulating.

13) Shape of scar of 1st floret (34) Oval to round (35) Heart-shaped (36) Elliptic (37) Narrow elliptic (38) Linear. The states are distinguished for classificatory purposes, but more details are given in the descriptions. The following figures may be useful guidelines for recognizing each state: Fig. 297, oval to round scar; Fig. 266, heart-shaped scar; Fig. 152, elliptic scar; Fig. 178, narrow elliptic scar; and Fig. 59, linear scar. These states are also used for identification.

14) Shape of scar of 3rd floret (39) Heart-shaped (40) Not heart-shaped. Usually all the scars, when present, are similar in shape. In some species, however, the scar of the third floret has a different shape and may be heart-shaped (Fig. 278), or all the scars may be heart-shaped. Sometimes the second floret may have a heart-shaped scar. In fatuoids, the scar of the 1st floret is often slightly heart-shaped (Fig. 319). The descriptions provide the whole range of shapes within a species, but for classification purposes and for discrimination the two states above, despite being crude, are very useful.

15) Relative size of periphery ring. The term periphery ring is taken from Coffman (1964) who stated that the scar is a cavity surrounded by a periphery ring. The periphery ring is very smooth and shiny (Fig. 244). In some species it is found at the distal point of the scar, in some it is confined to ¼ of the scar, and in other species it extends up to the entire periphery of the scar (e.g., *A. vaviloviana*). The descriptions provide the range of confinement of the periphery ring and appropriate figures accompany these descriptions for each species. A number of states have been distinguished for classificatory purposes (see Table 2) as well as for identification.

16) Place of insertion of the awn. In order to observe this carefully, the whole length of the lemma must be taken into consideration, especially in species with bisetulate lemma tips. The range is given in the descriptions, and these were grouped into eight states (see Table 2) for classification and identification.

17) Lemma structure (56) Tough (57) Resembling glumes (58) Tough or resembling glumes. In considering the states of lemma structure as attribute-states, it is more appropriate to distinguish between the first two states only. At anthesis, and especially after that, the lemma becomes hard, thick, and very firmly encloses the caryopsis. In some cases the lemma resembles the glumes in structure and also in morphology; moreover the caryopsis as a result remains loose and sheds at maturity. The first two states are used for identification, and the third is used together with these two, as character-states, for classificatory purposes.

18) Shape of the lemma tip. Lemma tips can take various forms from bilobed or bimucronate to bisetulate and bisetulate-biaristulate. This range has been divided into various states for classificatory purposes and identification. The range for each species is given in the descriptions, and the 9 formulated states are reported in Table 2.

It is important to carefully distinguish between a real setula and a membranous tooth. The former has a protruding nerve whereas the latter does not possess a nerve at all. The aristula is, in a practical sense, a protruding nerve from the attenuate tip of the lemma. When the biaristulate lemma is also bisetulate, then each of the setulae emerge at the sides adjacent to the base of the aristulae (see Fig. 177).

For identification purposes only six states are used: bilobed, bidenticulate, bisubulate, biaristulate, bisetulate-biaristulate, and shortly bisubulate or mucronate.

19) Vestiture of lemma below the insertion of the awn (68) Covered densely with macrohairs (69) Without macrohairs or a few around the insertion of the awn. These attribute-states are used for identification. For use as characters, a third state is necessary, that is, (70) Both kinds present in the same species, and this state is used together with the first two states for classificatory purposes. In the descriptions of species full details are given.

20) Number of rows of cilia along the edges of the keels of the palea. The palea of *Avena* has two keels. Each of the keels is beset with cilia along the length or on part of it only (cf., *A. brevis,* for example). These cilia occur in rows, which sometimes cannot be completely distinguished; therefore the four states: (71) 1 row (72) 1–2 rows (73) 1–3 rows (74) 2 rows. These four states are used for classificatory purposes and for identification, but the descriptions also contain information on whether the keels are entirely or partly covered. If the keels are only partly covered, the cilia occur towards the tip of the palea.

21) Vestiture of the back of the palea (75) Glabrous (76) With prickles (77) With macrohairs. This character ought to be checked towards the tip of the palea, especially to ascertain hairiness. The macrohairs, as a rule, are longer than the cilia of the keels, and certainly much longer than the prickles, see Fig. 237. Because some species have both prickles and macrohairs, an additional character-state is needed for classificatory purposes (category 78 of Table 2); for identification purposes, only the first three states are used.

22) Lodicule type (79) Sativa type (80) Fatua type (81) Strigosa type. The lodicule is a transparent body triangular in outline. Close to its place of attachment, it is cylindrical, and about ¼ or ⅛ way up it is open instead of ending as a closed cone. I call this opening a troughing. At the level of troughing some lodicules have an appendage attached like a rudder, partly to the cylindrical part and partly to the lower part of the open conical body. These rudder-like appendages I call side lobes. When side lobes are present, then the configuration of the lodicule is sativa type (Fig. 286). Extreme care is needed to detect the presence of this side lobe because it varies in size and shape and quite often it may be concealed due to the layout on the microscopic slide. The next 8 figures demonstrate the structures (Figs. 36–43).

One should always check for sativa-type lodicules. If no side lobe is present, then the configuration of the lodicule is fatua type (Fig. 259).

In the strigosa-type lodicule (Fig. 123) the side lobe is larger and is fused to the conical part of the lodicule above the level of troughing, sometimes near the very tip of the lodicule.

In addition to these various parts, every lodicule possesses a wing-like appendage attached to the cylindrical part of the base. This appendage can never be confused with the side lobes in the sativa-type lodicules because it is membranous, auricle-shaped, and often has a wavy margin; whereas the side lobes have a thicker structure, elongated cells arranged more or less in parallel and thus appearing as in rows, and of course they have a different shape. This wing-like appendage sometimes has characteristic configurations, which are species-specific as in *A. abyssinica* and *A. vaviloviana*.

The character-states are used in the keys. These and other details are given in the descriptions, but for classificatory purposes two additional states were recognized, (82) and (83) (see Table 2).

23) Presence or absence of prickles on lodicules. Unicellular prickles, when present, occur mostly at the tip of the lodicules (Fig. 61), but do occur also on side lobes. The three states (84) Always absent, (85) Rarely present, and (86) Often present, should be assessed on samples, not on individuals. The usefulness of this character is mainly for classificatory purposes.

24) Epiblast type (87) Brevis type (Figs. 95–96), (88) Fatua type (Fig. 260), (89) Septentrionalis type (Figs. 271–273), (90) Sativa type (Fig. 290), (91) Sativa and Fatua types. The configuration of the upper part of the epiblast determines its states. In the brevis and septentrionalis types the margin is always convex; in the fatua and sativa types it is concave. As a rule, the brevis type and septentrionalis type of epiblasts are smaller, that is, narrower and also shorter, than the fatua and sativa types.

25) 26) 27) Width range of epiblast. This is given in the descriptions. For identification purposes three states are established (see list of values on Table 2) and for classificatory purposes the actual lower limits, upper limits, and median of range are used.

28) Chromosome number. The number is known for most species, but it has been reassessed using seeds from the CAV collection (Baum et al. 1975). Dr. Rajhathy of the Ottawa Research Station counted the chromosomes for the entire collection.

29) Genome. The information on the genome, drawn from publications by Dr. Rajhathy, has recently been summarized (Rajhathy and Thomas 1974). I have simplified this information in a manner useful for classificatory purposes. Thus five states are recognized: AA, CC, AABB, AACC, and AACCDD. These, however, are not used in the keys to identification.

COMMENTS ON CHARACTERS FOR DISCRIMINATION OF OAT SPECIES

The attributes which I considered most important for the formulation of the species while scoring individual plants were: the lodicule type (22); epiblast type and size (24–27); mode of disarticulation (11–12); relative

---

Figs. 36–43. Configurations of the sativa-type lodicules when they have been mounted on microscopic slides. These slides demonstrate the extreme care needed to evaluate the presence of the sativa-type state. In Fig. 36 the side lobe is very small; in Fig. 37 it is large relative to the small body; in Fig. 38 it appears as a piece torn from the main body; in Fig. 39 it is folded underneath the main body; in Fig. 40 the side lobe is very close to the main body, slightly touching it and therefore difficult to discern; in Fig. 41 the side lobe is very small and folded inwards to the body; Figs. 42 and 43 are similar to Fig. 41, but in Fig. 42 the side lobe is easy to perceive. (All × 90.)

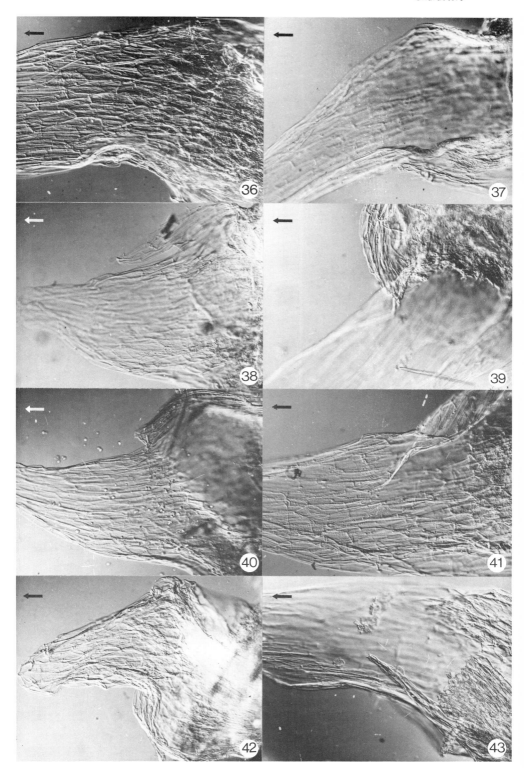

length of glumes (10); shape of scars, all of them and third ones (13, 14); degree of confinement of periphery ring to the scar (15); configuration of lemma tips (18); and vestiture of back of palea (21). After the species were assessed and circumscribed, all the above characters were subjected to character analysis using Estabrook's method. Input of these into CHARANAL (Hawksworth et al. 1968) revealed a similar, but not identical list of important characters (see table on p. 42 for further details).

Many characters used in this work are new, that is, they were never treated before by other *Avena* workers, or else were defined differently. Among these characters are the lodicule type, epiblast type and size, periphery ring, number of rows of cilia on the keels of the palea, vestiture of back of palea, and ligule shape. Malzew (1930) used quite a few characters in his descriptions but he used only a few for classificatory purposes and for identification. Furthermore, one of his most important characters (the shape of the lemma tip) has been here reinterpreted, that is, although Malzew and I both used the same name to designate that character, I have a new definition of the components of the lemma tip, the aristula and setula. In other words Malzew relied on the external shape of the lemma tip, whereas I rely on the structure as to nervation in order to decide whether an aristula or a setula or both are present. The shape of the scar was used restrictively by Malzew, and the heart-shaped configuration was discovered and described by Durieu for *A. occidentalis*. Durieu used the scar configuration of the third floret for his *A. occidentalis* only. The scar itself is here interpreted on a morphological basis, whereas it has been interpreted on a functional basis by previous authors.

Other authors, for example, Cosson (1854) and Thellung (1912), used very few characters. Even Nevski (1934) used few characters, although he was the first to use the number of florets in spikelets, spikelet length, glume length, and lemma length in his key to species. Stanton (in Coffman 1961) used chiefly lemma characters and base of floret characters, such as scars, which he interpreted on the basis of function and morphology. (See also Chapter 6 for additional remarks about earlier classification systems of *Avena*.)

13. Identification keys and automatic identification of species and hybrids

INTRODUCTION

A number of keys are given here to facilitate identification. Select for use the keys appropriate to the characters available on specimens to be identified, and also consult the descriptions, figures, and "Identification hints." The matrix of input and the character codes are provided for incorporation into an automatic identification scheme or any form of data bank.

METHOD OF PRODUCING KEYS

The various keys were obtained by using the algorithm developed by Pankhurst (1970). The program was also written by the same author, but has been modified by Mr. D. L. Paulhus of the Statistical Research Service, Canada Department of Agriculture for use on the UNIVAC 1108 computer.

The input for the construction of keys is a table of character-states for the taxa, here called objects. When the taxa have several states, each state is included; this method results in different objects bearing the same taxon name. The main computation finds characters that separate the taxa into equal groups with preference given to dichotomies over polychotomies. The basic separation function is

$$F = F_1 + F_2,$$
where $F_1 = (k-2)^2$
$$F_2 = \sum_{i=1}^{k} |1 - n_i k/N|$$

where at a given node with N objects there are k subgroups each containing n_i objects. The separations on different possible characters are evaluated, and the one with the minimum F is preferred, subject to the condition that characters with higher weight are considered first. The different weights are given *a priori* according to various criteria, which may be intuitive, or the result of character analysis, or selected to achieve a particular purpose.

Data input

Twenty-five characters are used. The characters have been edited for the special purpose of input into the program, and are itemized in Tables 14-15. They are essentially the same as those in Table 2, except that (i) character 11, the mode of disarticulation of the florets, incorporates characters 11 and 12 of Table 2, (ii) character 24, the epiblast width, incorporates the whole range listed in Table 2 as characters 25, 26, and 27; and (iii) character 29 of Table 2 is omitted, because the genome obviously is impractical at this point for identification purposes.

Table 15, which shows the possible values, that is, states, that each character can be assigned, should be used in conjunction with the table of characters (Table 14 above).

The matrix of the character-state values for the taxa (i.e., the objects by values matrix) given in Table 16 can be used for automatic identification or for producing different keys. Missing values appear as blanks. Sometimes a blank was left on purpose when I thought it uneconomical to represent a taxon many times (i.e., as many different objects) because of the presence of many states, or for that matter because it seemed to me purposeless to re-edit those states into appropriate values.

Identification

TABLE 14. CHARACTERS USED IN IDENTIFICATION KEYS.

They are used in coding the objects (i.e., species, hybrids or other units) and the range of possible values of the various objects.

Character number	Character name	Possible character values
1	Plant	1, 2
2	Juvenile growth	3, 4, 5
3	Color appearance	6, 7
4	Culm	8, 9
5	Relatively (height)	10, 11
6	Ligule (shape at anthesis or younger)	12, 13
7	Panicle (shape)	14, 15, 16
8	Spikelet (length range without awns)	17, 18, 19
9	Spikelet (number of florets range)	20, 21, 22, 23, 24, 25, 26, 27, 28, 29
10	Glumes (relative length)	30, 31
11	Mode of disarticulation	32, 33, 34
12	Scars (general shape, usually of at least 1st floret)	35, 36, 37, 38, 39, 40
13	Scar of third floret (shape in contrast to first, or same in 2nd or 4th)	40, 41
14	Periphery ring (length in relation to scar)	42, 43, 44, 45, 46, 47, 48
15	Awn (point of attachment or insertion)	49, 50, 51, 52, 53, 54, 55, 56
16	Lemma structure	57, 58
17	Lemma tip (shape)	60, 61, 62, 63, 64
18	Lemma vestiture below awn insertion	65, 66
19	Palea (range in number of rows of cilia found along edges of keels)	67, 68, 69, 70
20	Palea (vestiture of back)	71, 72, 73
21	Lodicule (type)	74, 75, 76
22	Lodicule (presence of prickles)	77, 78, 79
23	Epiblast (type)	74, 75, 80, 81, 90
24	Epiblast (width)	82, 83, 84, 89
25	Chromosome number (2n number)	85, 86, 87, 88

TABLE 15. CHARACTER-STATES IN THE IDENTIFICATION KEYS.

These are the values that may obtain the various characters (i.e., all the states within characters). Note that the code and names of a particular value may serve a number of characters simultaneously. For instance, the name of the value *fatua type* may serve the lodicule and epiblast characters simultaneously.

Value number	Value name
1	Annual
2	Perennial
3	Erect
4	Prostrate
5	Prostrate to erect
6	Green
7	Glaucous
8	Erect
9	Geniculate or prostrate
10	Low plants not exceeding 95 cm (15–95)
11	Tall plants exceeding 100 cm (30–180)
12	Acute
13	Obtuse
14	Equilateral
15	Flagged (often dense and contracted branches)
16	Slightly flagged (often contracted branches)
17	Short (not exceeding 25 mm in upper limit)
18	Long (lower limit not less than 20 mm, upper limit exceeding 25 mm)
20	With 1 floret
21	With 2 florets
22	With 2–3 florets
23	With 2–4 florets
24	With 2–5 florets
25	With 2–6 florets
26	With 3–4 florets
27	With 3–6 florets
28	With 1–7 florets
29	With 3 florets
30	Equal (or nearly so, the upper sometimes up to one-quarter longer than the lower)
31	Unequal (the lowest being half to two-thirds as long as the upper)
32	Non-disarticulating (no scars present)
33	Lowermost floret only disarticulating (no scars in the other florets)
34	All florets disarticulating (scars present in all florets)
35	Oval to round
36	Heart-shaped or slightly so
37	Elliptic
38	Narrow elliptic
39	Linear
40	Heart-shaped
41	Not heart-shaped (similar to other ones)
42	All around scar
43	Only confined to one-eighth (of scar, or less)
44	Only confined to one-third to one-half (of scar)
45	Only confined to one-quarter (of scar, or less)
46	Only confined to one-eighth to one-sixteenth (of scar)
47	Only confined to one-half (of scar)
48	Only confined to one-third (of scar)
49	Inserted at about middle of lemma
50	Absent or rudimentary

TABLE 15. (Cont'd)

Value number	Value name
51	Inserted at about lower fourth of lemma
52	Inserted at about lower third of lemma
53	Inserted between lower third to middle of lemma
54	Inserted at about upper third of lemma
55	Inserted between upper third to upper fourth of lemma
56	Inserted from upper fourth to below tip of lemma
57	Tough (of much harder and different structure)
58	Resembling glumes
59	Bilobed (with two nerves ending in each lobe)
60	Bidenticulate
61	Bisubulate (can be very long and resembling next value but not in structure)
62	Biaristulate
63	Bisetulate-biaristulate
64	Shortly bisubulate (or mucronate)
65	Beset densely with macrohairs
66	Without macrohairs or only a few often found around awn insertion
67	With 1 row of cilia along edges of keels
68	With 1–2 rows of cilia along edges of keels
69	With 1–3 rows of cilia along edges of keels
70	With 2 rows of cilia along edges of keels
71	Back beset with hairs
72	Back beset with prickles
73	Back glabrous (no hairs and no prickles)
74	Sativa type
75	Fatua type
76	Strigosa type
77	Never bearing prickles
78	Often with prickles
79	Rarely with a few prickles
80	Septentrionalis type
81	Brevis type
82	0.3 mm wide or less
83	More than 0.3 up to 0.4 mm wide
84	0.45–0.7 mm wide
85	$2n = 14$
86	$2n = 28$
87	$2n = 42$
88	$2n = 41, 40,$ or 42
89	0.75–1.0 mm wide
90	Sativa type and fatua type

Identification

TABLE 16. DATA MATRIX OF ALL THE OBJECTS USED AS INPUT INTO THE KEYS GENERATING PROGRAM (PANKHURST'S).

It can be used for identification by computer.

A. macrostachya/Balansa ex Cosson et Durieu																			
2	6	8	11	13	16	19	27	31	34	38	41	45	54	57	61	66	68	74	77	86
A. clauda/Durieu																			
1	4	6	8	10	13	16	18	25	31	34	39	41	43	53	57	63	66	68	75	79	74	83	85
A. eriantha/Durieu																			
1	4	7	8	10	13	16	17	22	31	33	39		43	49	57	61	66	68	75	77	74	83	85
A. ventricosa/Balansa ex Cosson																			
1	3	7	8	10	13	16	17	21	30	33	39		43	54	57	61	66	68	75	79	74	83	85
A. clauda × *eriantha* (as in *eriantha* but spikelets narrower and longer)																			
1	7	8	10	13	16	17	22	31	33	39		43	49	57	63	66	68	75	77	74	83	85
A. brevis/Roth																			
1	3	7	8	10	13	16	17	22	30	32			56	57	64	66	68	72	75	77	81	83	85
A. hispanica/Ard. ex Saggi																			
1	5	7	8	11	13	14	17	21	30	32			55	57	63	66	67	72	75	79	81	83	85
A. nuda/Linneus																			
1	3	7	8	10	13	16	18	27	30	32			49	58	61	66	68	72	75	77	81	83	85
A. nuda/																			
1	3	7	8	10	13	16	18	27	30	32			49	58	61	66	68	72	76	77	81	83	85
A. nuda/																			
1	3	7	8	10	13	16	18	27	30	32			50	58	61	66	68	72	75	77	81	83	85

TABLE 16. (Cont'd)

A. nuda/																				
1	3	7	8	10	13	16	18	27	30	32						76	77	81	83	85
A. strigosa/Schreber																				
1	3	7	8	11	13	14	17	22	30	32			50	58	61	66	68	72	76	77	81	83	85
A. strigosa/																				
1	3	7	8	11	13	14	17	22	30	32			52	57	63	65	68	72	76	78	81	83	85
A. barbata/Pott ex Link																				
1	4	6	9	11	13	14	18	22	30	34	37			52	57	63	66	68	72	76	78	81	83	85
A. barbata/																				
1	4	6	9	11	13	14	18	22	30	34	37	41	44	49	57	61	65	68	72	75	77	81	83	86
A. canariensis/Baum, Rajhathy et Sampson																				
1	5	6	8	10	12	14	17	21	30	33	36		45	49	57	60	65	67	72	74	79	81	83	85
A. damascena/Rajhathy et Baum																				
1	5	6	8	10	13	14	18	22	30	34	37	41	45	52	57	63	65	67	72	74	77	80	82	85
A. hirtula/Lagasca																				
1	4		8	10	13	14	17	22	30	34	37		45	53	57	61	65	67	72	74	77	74	82	85
A. longiglumis/Durieu																				
1	4	7	8	11	13	16	18	22	30	34	39		46	49	57	62	65	67	72	75	79	80	83	85
A. longiglumis/																				
1		7	8	11	13	16	18	22	30	34	39		46	49	57	63	65	67	72	75	79	80	83	85
A. longiglumis/																				
1		7	8	11	13	16	18	22	30	34	39		46	49	57	62	65	67	72	74	79	80	83	85

Identification

					13	14	18	22	30	34	38	41	45	52	57	63	65	67	72	74	77	81	82	63		
A. matritensis/Baum																										
1																										
6	8				12	14	18	22	30	34	38	41	45	52	57	62	65	67	72	74	79	81	83	85		
A. wiestii/Steudel																										
1	3	6	8	11	13	14	19	21	30	34	37	41	44	49	57	63	65	67		74	77	80	82	85		
A. wiestii/																										
1	3	6	8	11	13	14	19	21	30	34	37	41	44	49	57	62	65	67		74	77	80	82	86		
A. wiestii/																										
1	3	6	8	11	13	14	19	21	30	34	37	41	44	49	57	63	65	67		74	77	80	82	86		
A. wiestii/																										
1	3	6	8	11	13	14	19	21	30	34	37	41	44	49	57	62	65	67		74	77	80	82	86		
A. abyssinica/Hochstetter																										
1	3	6	8	10	12	14	17	22	30	32				49	57	61	66	68	72	74	77	81	84	86		
A. vaviloviana/(Malzew) Mordvinkina																										
1	3	6	8	10	12	14	17	22	30	34	36	40	42	49	57	61	65	68	72	75	77	81	84	86		
A. vaviloviana/																										
1	3	6	8	10	12	14	17	22	30	34	36	40	42	49	57	61	66	68	72	75	77	81	84	86		
A. abyssinica × *vaviloviana*/																										
1		8		10	12	14	17	22	30	34	36	40	42	49	57	61	66	68	72	74	77	81	84	86		
A. maroccana/Gandoger																										
1	4	7	8	11	12	14	18	22	30	33	35		45	49	57	59	65	70	71	75	79	75	89	86		
A. murphyi/Ladizinsky																										
1	3	6	8	11	12	14	18	23	30	33	35		47	51	57	60	66	70	72	75	79	75	84	86		
A. murphyi/																										
1	3	6	8	11	12	14	18	23	30	33	35		47	51	57	61	66	70	72	75	79	75	84	86		

TABLE 16. (Cont'd)

A. atherantha/Presl																			
1		8		13	14	18	22	30	33	35		45	52	57	61	65	69	71	75	77	75	84	87
5	6																						
A. fatua/Linneus																			
1		8	11	12	14	17	22	30	34	35	41	44	49	57	60	65	69	72	75	77	75	84	87
5	6																						
A. fatua/																			
1		8	11	12	14	17	22	30	34	35	41	44	49	57	61	65	69	72	75	77	75	84	87
5	6																						
A. fatua/																			
1		8	11	12	14	17	22	30	34	35	41	44	49	57	60	66	69	72	75	77	75	84	87
5	6																						
A. fatua/																			
1		8	11	12	14	17	22	30	34	35	41	44	49	57	61	66	69	72	75	77	75	84	87
5	6																						
A. hybrida/Petermann																			
1		8		12	14	17	23	30	34	35	40	44	49	57	60	65	69	72	74	77	80	83	87
A. hybrida/																			
1		8		12	14	17	23	30	34	35	40	44	49	57	60	66	69	72	74	77	80	83	87
A. hybrida/																			
1		8		12	14	17	23	30	34	35	40	44	49	57	61	65	69	72	74	77	80	83	87
A. hybrida/																			
1		8		12	14	17	23	30	34	35	40	44	49	57	61	66	69	72	74	77	80	83	87
A. occidentalis/Durieu																			
1		8	10	12	16	18	26	30	34	37	40	44	49	57	61	65	69	72	75	77	80	84	87
5	6																						
A. sativa/Linneus																			
1		8	11	12		19	28	30	32				49	57	60	66	69		74	77	90	84	87
5																							
A. sativa/																			

104

Identification

Taxon																								
(row cut off)	1	5		8	11	12		19	28	30	32			50	57	60	66	69		74	77	90	84	87
A. sativa /	1	5		8	11	12	14	19	28	30	32					58	60	66	69		74	77	90	84	87
A. sterilis/Linneus	1	5	6	8	11	12	14	19	24	30	33	35		48	52	57	60	65	69	72	75	79	75	84	87
A. sterilis /	1	5	6	8	11	12	14	19	24	30	33	35		48	52	57	61	65	69	72	75	79	75	84	87
A. sterilis /	1	5	6	8	11	12	14	19	24	30	33	35		48	52	57	60	66	69	72	75	79	75	84	87
A. sterilis /	1	5	6	8	11	12	14	19	24	30	33	35		48	52	57	61	66	69	71	75	79	75	84	87
A. trichophylla/C. Koch	1	6		8	11	12	14	18	26	30	33	37		48	52	57	61	65	69	72	74	79	80	83	87
A. trichophylla /	1	6		8	11	12	14	18	26	30	33	37		48	52	57	61	65	69	71	74	79	80	83	87
A. trichophylla /	1	6		8	11	12	14	18	26	30	33	37		48	52	57	61	66	69	72	74	79	80	83	87
A. trichophylla /	1	6		8	11	12	14	18	26	30	33	37		48	52	57	61	66	69	72	74	79	80	83	87
A. sativa fatuoid /	1	5		8	11	12	14	17	22	30	34	36	41	44	49	57	61	65	69	72	74	77		84	88
A. sativa fatuoid /	1	5		8	11	12	14	17	22	30	34	36	41	44	49	57	60	65	69	72	74	77		84	88
A. sativa fatuoid /	1	5		8	11	12	14	17	22	30	34	36	41	44	49	57	61	66	69	72	74	77		84	88

TABLE 16. (Cont'd)

A. sativa fatuoid /																													
1	5	8	11	12	14	17		22		30		34	36	41	44	49		57		60	66	69	72	74		77	84		88
A. sativa × *fatua* or *sativa* × *sterilis* (each resembling more parent) /																													
1	5	8	11	12			19		28	30	32					49		57		60	66	69			75	77	84	87	
A. sativa × *fatua* or *sativa* × *sterilis* (each resembling more parent) /																													
1	5	8	11	12			19		28	30	32					49			58	60	66	69			75	77	84	87	
A. sativa × *fatua* or *sativa* × *sterilis* (each resembling more parent) /																													
1	5	8	11	12			19		28	30	32						50	57		60	66	69			75	77	84	87	
A. sativa × *fatua* or *sativa* × *sterilis* (each resembling more parent) /																													
1	5	8	11	12			19		28	30	32						50		58	60	66	69			75	77	84	87	

The assignment of weight was effected by using integers 0, 1, 2, 3; 0 denotes the lowest weight and 3 the highest. This assignment was done differently for each character, and separately for each key.

THE KEYS

Seven keys are provided; each is titled and lists the characters on which the key is based with the appropriate given weights.

(i) The first key uses all characters unweighted (i.e., all characters are given equal weight) and will not prove very useful as a general key. It is however useful for comparative purposes, that is, for checking the identity of a specimen in doubt, especially when the specimen to be identified may belong to one of two very similar taxa. You will notice that this key is relatively longer than the other keys because in most cases the remaining distinctive characters are furnished for each object that is keyed out.

(ii) The second key uses the characters that I consider to be most important for diagnostic purposes; with the appropriate weight, it is one of the most useful keys. In this key the chromosome number has been purposely omitted.

(iii) The third key is a variation of the second key; it differs from the former in a few characters and in the weight given to the characters. It is also a most useful key.

(iv) The fourth key uses only the reproductive characters (i.e., 10–25). It puts low weight on the chromosome number because adequate material for chromosome counts may not be available on the specimens.

(v) The fifth key includes all the characters, but their weight is based on the SUMRAT values of the character analysis exercise (Table on p. 42). Those characters with high SUMRAT values were given high weight. Cutoff points were arbitrarily selected in the list of these values and the code-weights were assigned accordingly.

(vi) The sixth key, entitled for cytologists and breeders, is based chiefly on vegetative characters and on the chromosome numbers. Because many characters were omitted, some species key out together and cannot be identified conclusively. It may prove useful, however, when the origin of the specimen is known.

(vii) The seventh and last key given here is designed for quick identification, but by itself it cannot conclusively identify some species that key out as groups.

The user may select a key according to his preference. Normally, when a species or a hybrid keys out, all its remaining distinctive characters are furnished by the program to assist in identification; this is not always true of the keys given here, especially the last mentioned ones, because the options PARTIAL and COMPRESS have been used. As a result, when different objects with the same taxon name key out as a group and are *compressed* under one name, not all the distinctive characters of the objects can be furnished. They cannot be furnished because the objects appear together under one name and thus do not possess distinctive characters.

1. UNWEIGHTED KEY (i.e., ALL CHARACTERS GIVEN EQUAL WEIGHT)

1 Lemma vestiture below awn insertion: beset densely with macrohairs 2
  2 Ligule acute 3
    3 Lodicule fatua type 4
      4 Mode of disarticulation: lowermost floret only disarticulating 5
        5 Juvenile growth prostrate; color appearance glaucous; spikelet long; spikelet with 2–3 florets; periphery ring only confined to one-quarter; awn inserted at about middle of lemma; lemma tip bilobed; palea with 2 rows of cilia along edges of keels; palea back beset with hairs; epiblast 0.75–1.0 mm wide; chromosome number $2n = 28$ .... *A. maroccana*
        5 Juvenile growth prostrate to erect .... *A. sterilis*
      4 Mode of disarticulation: all florets disarticulating 6
        6 Relatively low plants not exceeding 95 cm (15–95) 7
          7 Juvenile growth prostrate to erect; panicle slightly flagged; spikelet long; spikelet with 3–4 florets; scars elliptic; periphery ring only confined to one-third to one-half; palea with 1–3 rows of cilia along edges of keels; epiblast septentrionalis type; chromosome number $2n = 42$ ...... *A. occidentalis*
          7 Juvenile growth erect; panicle equilateral; spikelet short; spikelet with 2–3 florets; scars heart-shaped or slightly so; periphery ring all around scar; palea with 1–2 rows of cilia along edges of keels; epiblast brevis type; chromosome number $2n = 28$ .... *A. vaviloviana*
        6 Relatively tall plants exceeding 100 cm (30–180) .... *A. fatua*
    3 Lodicule sativa type 8
      8 Lodicule never bearing prickles 9
        9 Spikelet with 2–3 florets .... *A. sativa fatuoid*
        9 Spikelet with 2–4 florets .... *A. hybrida*
      8 Lodicule rarely with a few prickles 10
      10 Periphery ring only confined to one-third .... *A. trichophylla*
      10 Periphery ring only confined to one-quarter 11
        11 Spikelet short; spikelet with 2 florets; mode of disarticulation: lowermost floret only disarticulating; scars heart-shaped or slightly so; awn inserted at about middle of lemma; lemma tip bidenticulate . *A. canariensis*
        11 Spikelet long; spikelet with 2–3 florets; mode of disarticulation: all florets disarticulating; scars narrow elliptic; awn inserted at about lower third of lemma; lemma tip biaristulate .... *A. matritensis*
  2 Ligule obtuse 12
  12 Panicle slightly flagged 13
    13 Lemma tip biaristulate .... *A. longiglumis*
    13 Lemma tip bisetulate-biaristulate 14
      14 Culm geniculate or prostrate; spikelet indeterminate; scars narrow elliptic; periphery ring only confined to one-quarter; awn inserted at about lower third of lemma; lodicule never bearing prickles; epiblast brevis type; epiblast 0.3 mm wide or less .... *A. lusitanica*

	14	Culm erect .... *A. longiglumis*
	12	Panicle equilateral 15
15		Spikelet with 2 florets .... *A. wiestii*
15		Spikelet with 2–3 florets 16
	16	Culm geniculate or prostrate .... *A. barbata*
	16	Culm erect 17
		17 Spikelet long 18

        18 Mode of disarticulation: lowermost floret only disarticulating; scars oval to round; lemma tip bisubulate; palea with 1–3 rows of cilia along edges of keels; palea back beset with hairs; lodicule fatua type; epiblast fatua type; epiblast 0.45–0.7 mm wide; chromosome number $2n = 42$. *A. atherantha*

        18 Mode of disarticulation: all florets disarticulating; scars elliptic; lemma tip bisetulate-biaristulate; palea with 1 row of cilia along edges of keels; palea back beset with prickles; lodicule sativa type; epiblast septentrionalis type; epiblast 0.3 mm wide or less; chromosome number $2n = 14$ .... *A. damascena*

    17 Spikelet short 19

        19 Juvenile growth prostrate; relatively low plants not exceeding 95 cm (15–19); mode of disarticulation: all florets disarticulating; awn inserted between lower third to middle of lemma; lemma tip bisubulate; palea with 1 row of cilia along edges of keels; lodicule sativa type; lodicule never bearing prickles; epiblast sativa type; epiblast 0.3 mm wide or less .. *A. hirtula*

        19 Juvenile growth erect; relatively tall plants exceeding 100 cm (30–180); mode of disarticulation: non-disarticulating; awn inserted at about lower third of lemma; lemma tip bisetulate-biaristulate; palea with 1–2 rows of cilia along edges of keels; lodicule strigosa type; lodicule often with prickles; epiblast brevis type; epiblast more than 0.3 up to 0.4 mm wide ......
                                                                                              A. strigosa

1		Lemma vestiture below awn insertion: without macrohairs or only a few often found around awn insertion 20
20		Ligule obtuse 21
	21	Glumes unequal 22
		22 Color appearance glaucous 23
		23 Lemma tip bisubulate .... *A. eriantha*
		23 Lemma tip bisetulate-biaristulate .... *A. clauda* × *eriantha*

            (as in *eriantha* but spikelets narrower and longer)

      22 Color appearance green 24

        24 Plant annual; relatively low plants not exceeding 95 cm (15–95); spikelet long; spikelet with 2–6 florets; scars linear; periphery ring only confined to one-eighth; awn inserted between lower third to middle of lemma; lemma tip bisetulate-biaristulate; lodicule fatua type; lodicule rarely with a few prickles; chromosome number $2n = 14$ .... *A. clauda*

        24 Plant perennial; relatively tall plants exceeding 100 cm (30–180); spikelet indeterminate; spikelet with 3–6 florets; scars narrow elliptic; periphery ring only confined to one-quarter; awn inserted at about upper third of lemma; lemma tip bisubulate; lodicule sativa type; lodicule never bearing prickles; chromosome number $2n = 28$ .... *A. macrostachya*

    21 Glumes equal 25

        25 Spikelet long .... *A. nuda*

- 25 Spikelet short 26
  - 26 Relatively low plants not exceeding 95 cm (15–95) 27
    - 27 Spikelet with 2–3 florets; mode of disarticulation: non-disarticulating; awn inserted from upper fourth to below tip of lemma; lemma tip shortly bisubulate; lodicule never bearing prickles; epiblast brevis type .... *A. brevis*
    - 27 Spikelet with 2 florets; mode of disarticulation: lowermost floret only disarticulating; awn inserted at about upper third of lemma; lemma tip bisubulate; lodicule rarely with a few prickles; epiblast sativa type .... *A. ventricosa*
  - 26 Relatively tall plants exceeding 100 cm (30–180) 28
    - 28 Juvenile growth prostrate to erect; spikelet with 2 florets; awn inserted between upper third to upper fourth of lemma; palea with 1 row of cilia along edges of keels; lodicule fatua type; lodicule rarely with a few prickles .... *A. hispanica*
    - 28 Juvenile growth erect; spikelet with 2–3 florets; awn inserted at about lower third of lemma; palea with 1–2 rows of cilia along edges of keels; lodicule strigosa type; lodicule often with prickles .... *A. strigosa*
- 20 Ligule acute 29
  - 29 Lodicule fatua type 30
    - 30 Lemma tip bisubulate 31
      - 31 Juvenile growth erect 32
        - 32 Relatively tall plants exceeding 100 cm (30–180); spikelet long; spikelet with 2–4 florets; mode of disarticulation: lowermost floret only disarticulating; scars oval to round; periphery ring only confined to one-half; awn inserted at about lower fourth of lemma; palea with 2 rows of cilia along edges of keels; lodicule rarely with a few prickles; epiblast fatua type .... *A. murphyi*
        - 32 Relatively low plants not exceeding 95 cm (15–95); spikelet short; spikelet with 2–3 florets; mode of disarticulation: all florets disarticulating; scars heart-shaped or slightly so; periphery ring all around scar; awn inserted at about middle of lemma; palea with 1–2 rows of cilia along edges of keels; lodicule never bearing prickles; epiblast brevis type .... *A. vaviloviana*
      - 31 Juvenile growth prostrate to erect 33
        - 33 Spikelet short; spikelet with 2–3 florets; mode of disarticulation: all florets disarticulating; periphery ring only confined to one-third to one-half; awn inserted at about middle of lemma; lodicule never bearing prickles .... *A. fatua*
        - 33 Spikelet indeterminate; spikelet with 2–5 florets; mode of disarticulation: lowermost floret only disarticulating; periphery ring only confined to one-third; awn inserted at about lower third of lemma; lodicule rarely with a few prickles .... *A. sterilis*
    - 30 Lemma tip bidenticulate 34
      - 34 Lemma structure resembling glumes ... *A. sativa* × *fatua* or *sativa* × *sterilis* (each resembling more parent)
      - 34 Lemma structure tough 35
        - 35 Lodicule rarely with a few prickles 36
          - 36 Juvenile growth erect; spikelet long; spikelet with 2–4 florets; periphery ring only confined to one-half; awn inserted at about lower

		fourth of lemma; palea with 2 rows of cilia along edges of keels; chromosome number $2n=28$ .... ***A. murphyi***

  36 Juvenile growth prostrate to erect; spikelet indeterminate; spikelet with 2–5 florets; periphery ring only confined to one-third; awn inserted at about lower third of lemma; palea with 1–3 rows of cilia along edges of keels; chromosome number $2n=42$ .... ***A. sterilis***

 35 Lodicule never bearing prickles 37

  37 Spikelet short; spikelet with 2–3 florets; mode of disarticulation: all florets disarticulating .... ***A. fatua***

  37 Spikelet indeterminate .... ***A. sativa*** × ***fatua*** or ***sativa*** × ***sterilis***
  (each resembling more parent)

29 Lodicule sativa type 38

 38 Lemma tip bidenticulate 39

  39 Spikelet short 40

   40 Spikelet with 2–4 florets; scars oval to round; scar of third floret heart-shaped; epiblast more than 0.3 up to 0.4 mm wide; chromosome number $2n=42$ .... ***A. hybrida***

   40 Spikelet with 2–3 florets; scars heart-shaped or slightly so; scar of third floret not heart-shaped; epiblast 0.45–0.7 mm wide; chromosome number $2n=41$, 40 or 42 .... ***A. sativa fatuoid***

  39 Spikelet indeterminate .... ***A. sativa***

 38 Lemma tip bisubulate 41

  41 Epiblast more than 0.3 up to 0.5 mm wide 42

   42 Spikelet short; spikelet with 2–4 florets; mode of disarticulation: all florets disarticulating; scars oval to round; periphery ring only confined to one-third to one-half; awn inserted at about middle of lemma; lodicule never bearing prickles .... ***A. hybrida***

   42 Spikelet long .... ***A. trichophylla***

  41 Epiblast 0.45–0.7 mm wide 43

   43 Relatively tall plants exceeding 100 cm (30–180); palea with 1–3 rows of cilia along edges of keels; chromosome number $2n=41$, 40, or 42 ....
   A. sativa fatuoid

   43 Relatively low plants not exceeding 95 cm (15–95) 44

    44 Mode of disarticulation: non-disarticulating .... ***A. abyssinica***

    44 Mode of disarticulation: all florets disarticulating ....
    A. abyssinica × ***vaviloviana***

2. WEIGHTED KEY WITH THE MOST IMPORTANT DIAGNOSTIC CHARACTERS ONLY (Characters 1, 10–13, 15–18, 20, 21, 23). WEIGHT (1, 12, 15, 18, 23 = 0; 10, 11, 13, 16, 17, 20, 21 = 1)

1 Lemma structure resembling glumes 2

 2 Lemma tip bidenticulate 3

  3 Lodicule fatua type .... ***A. sativa*** × ***fatua*** or ***sativa*** × ***sterilis***
  (each resembling more parent)

		3	Lodicule sativa type .... ***A. sativa***
	2	Lemma tip bisubulate .... ***A. nuda***
1	Lemma structure tough 4
	4	Glumes unequal 5
		5	Mode of disarticulation: lowermost floret only disarticulating 6
			6	Lemma tip bisubulate .... ***A. eriantha***
			6	Lemma tip bisetulate-biaristulate .... ***A. clauda*** × ***eriantha*** (as in *eriantha* but spikelets narrower and longer)
		5	Mode of disarticulation: all florets disarticulating 7
			7	Lemma tip bisetulate-biaristulate; plant annual scars linear, awn inserted between lower third to middle of lemma; lodicule fatua type .... ***A. clauda***
			7	Lemma tip bisubulate; plant perennial; scars narrow elliptic; awn inserted at about upper third of lemma; lodicule sativa type .... ***A. macrostachya***
	4	Glumes equal 8
		8	Mode of disarticulation: non-disarticulating 9
			9	Lodicule strigosa type .... ***A. strigosa***
			9	Lodicule sativa type 10
				10	Lemma tip bisubulate; epiblast brevis type .... ***A. abyssinica***
				10	Lemma tip bidenticulate; epiblast sativa type or fatua type .... ***A. sativa***
			9	Lodicule fatua type 11
				11	Lemma tip bisetulate-biaristulate; awn inserted between upper third to upper fourth of lemma .... ***A. hispanica***
				11	Lemma tip shortly bisubulate; awn inserted from upper fourth to below tip of lemma .... ***A. brevis***
				11	Lemma tip bidenticulate .... ***A. sativa*** × ***fatua*** or ***sativa*** × ***sterilis*** (each resembling more parent)
		8	Mode of disarticulation: lowermost floret only disarticulating 12
			12	Lodicule sativa type 13
				13	Palea back beset with hairs .... ***A. trichophylla***
				13	Palea back beset with prickles 14
					14	Lemma tip bidenticulate; scars heart-shaped or slightly so; awn inserted at about middle of lemma; epiblast brevis type ... ***A. canariensis***
					14	Lemma tip bisubulate .... ***A. trichophylla***
			12	Lodicule fatua type 15
				15	Lemma tip bilobed; awn inserted at about middle of lemma ...***A. maroccana***
				15	Lemma tip bidenticulate 16
					16	Awn inserted at about lower fourth of lemma .... ***A. murphyi***
					16	Awn inserted at about lower third of lemma .... ***A. sterilis***
				15	Lemma tip bisubulate 17
					17	Lemma vestiture below awn insertion: beset densely with macrohairs ..18
						18	Palea back beset with hairs .... ***A. atherantha***

		18	Palea back beset with prickles .... *A. sterilis*

        17   Lemma vestiture below awn insertion: without macrohairs or only a few often found around awn insertion 19

            19   Scars linear; awn inserted at about upper third of lemma; epiblast sativa type .... *A. ventricosa*

            19   Scars oval to round 20

                20   Awn inserted at about lower fourth of lemma .... *A. murphyi*

                20   Awn inserted at about lower third of lemma .... *A. sterilis*

8   Mode of disarticulation: all florets disarticulating 21

21   Lodicule fatua type 22

        22   Lemma tip biaristulate 23

            23   Palea back glabrous; scars elliptic; epiblast brevis type .... *A. barbata*

            23   Palea back beset with prickles; scars linear; epiblast septentrionalis type .... *A. longiglumis*

        22   Lemma tip bidenticulate .... *A. fatua*

        22   Lemma tip bisetulate-biaristulate 24

            24   Scars narrow elliptic; awn inserted at about lower third of lemma; epiblast brevis type .... *A. lusitanica*

            24   Scars linear; awn inserted at about middle of lemma; epiblast septentrionalis type .... *A. longiglumis*

        22   Lemma tip bisubulate 25

            25   Scar of third floret heart-shaped 26

                26   Scars elliptic; epiblast septentrionalis type .... *A. occidentalis*

                26   Scars heart-shaped or slightly so .... *A. vaviloviana*

            25   Scar of third floret not heart-shaped 27

                27   Scars elliptic; epiblast brevis type .... *A. barbata*

                27   Scars oval to round .... *A. fatua*

21   Lodicule sativa type 28

        28   Lemma tip bidenticulate 29

            29   Scar of third floret not heart-shaped .... *A. sativa fatuoid*

            29   Scar of third floret heart-shaped .... *A. hybrida*

        28   Lemma tip bisetulate-biaristulate 30

            30   Scars linear .... *A. longiglumis*

            30   Scars elliptic 31

                31   Awn inserted at about lower third of lemma .... *A. damascena*

                31   Awn inserted at about middle of lemma .... *A. wiestii*

        28   Lemma tip biaristulate 32

            32   Awn inserted at about lower third of lemma; scars narrow elliptic; epiblast brevis type .... *A. matritensis*

            32   Awn inserted at about middle of lemma 33

                33   Scars linear .... *A. longiglumis*

		33	Scars elliptic .... *A. wiestii*

    28 Lemma tip bisubulate 34
        34 Lemma vestiture below awn insertion: without macrohairs or only a
            few often found around awn insertion 35
            35 Scar of third floret not heart-shaped .... *A. sativa fatuoid*
            35 Scar of third floret heart-shaped 36
                36 Scars oval to round; epiblast septentrionalis type .... *A. hybrida*
                36 Scars heart-shaped or slightly so; epiblast brevis type ....
                                                            A. abyssinica × *vaviloviana*
        34 Lemma vestiture below awn insertion: beset densely with macrohairs 37
            37 Awn inserted between lower third to middle of lemma; scars elliptic .....
                                                                           A. hirtula
            37 Awn inserted at about middle of lemma 38
                38 Scar of third floret heart-shaped; scars oval to round ...... *A. hybrida*
                38 Scar of third floret not heart-shaped; scars heart-shaped or slightly
                    so .... *A. sativa fatuoid*

3. **WEIGHTED KEY EXCLUDING SOME CHARACTERS AND EX-CLUDING CHROMOSOME NUMBERS** (Characters: 1, 10–17, 20, 21, 23, 24)
   WEIGHT (1, 12, 16, 17, 23 = 0; 14, 15, 20 = 1; 13, 24 = 2; 10, 11, 21 = 3)

1 Glumes unequal 2
    2 Mode of disarticulation: lowermost floret only disarticulating 3
        3 Lemma tip bisubulate .... *A. eriantha*
        3 Lemma tip bisetulate-biaristulate .... *A. clauda* × *eriantha*
                                (as in *eriantha* but spikelets narrower and longer)
    2 Mode of disarticulation: all florets disarticulating 4
        4 Lodicule fatua type; plant annual; scars linear; periphery ring only confined to
          one-eighth; awn inserted between lower third to middle of lemma; lemma tip
          bisetulate-biaristulate .... *A. clauda*
        4 Lodicule sativa type; plant perennial; scars narrow elliptic; periphery ring only
          confined to one-quarter; awn inserted at about upper third of lemma; lemma
          tip bisubulate .... *A. macrostachya*
1 Glumes equal 5
    5 Mode of disarticulation: lowermost floret only disarticulating 6
        6 Lodicule sativa type 7
            7 Palea back beset with hairs .... *A. trichophylla*
            7 Palea back beset with prickles 8
                8 Periphery ring only confined to one-quarter; scars heart-shaped or slightly
                  so; awn inserted at about middle of lemma; lemma tip bidenticulate; epi-
                  blast brevis type .... *A. canariensis*

		8 Periphery ring only confined to one-third .... *A. trichophylla*
	6	Lodicule fatua type 9
		9 Epiblast 0.75–1.0 mm wide; awn inserted at about middle of lemma; lemma tip bilobed .... *A. maroccana*
		9 Epiblast more than 0.3 up to 0.4 mm wide; scars linear; periphery ring only confined to one-eighth; awn inserted at about upper third of lemma; epiblast sativa type .... *A. ventricosa*
		9 Epiblast 0.45–0.7 mm wide 10
		10 Awn inserted at about lower fourth of lemma .... *A. murphyi*
		10 Awn inserted at about lower third of lemma 11
		11 Periphery ring only confined to one-quarter; palea back beset with hairs .... *A. atherantha*
		11 Periphery ring only confined to one-third .... *A. sterilis*
5	Mode of disarticulation: non-disarticulating 12
	12	Lodicule strigosa type 13
		13 Awn inserted at about middle of lemma, or absent or rudimentary ...... *A. nuda*
		13 Awn inserted at about lower third of lemma .... *A. strigosa*
	12	Lodicule sativa type 14
		14 Awn absent or rudimentary .... *A. sativa*
		14 Awn inserted at about middle of lemma 15
		15 Lemma structure resembling glumes .... *A. sativa*
		15 Lemma structure tough 16
		16 Lemma tip bisubulate; epiblast brevis type .... *A. abyssinica*
		16 Lemma tip bidenticulate; epiblast sativa type and fatua type .... *A. sativa*
	12	Lodicule fatua type 17
		17 Epiblast 0.45–0.7 mm wide .... *A. sativa* × *fatua* or *sativa* × *sterilis* (each resembling more parent)
		17 Epiblast more than 0.3 up to 0.4 mm wide 18
		18 Awn inserted between upper third to upper fourth of lemma; lemma tip bisetulate-biaristulate .... *A. hispanica*
		18 Awn inserted from upper fourth to below tip of lemma; lemma tip shortly bisubulate .... *A. brevis*
		18 Awn inserted at about middle of lemma or absent or rudimentary .... *A. nuda*
5	Mode of disarticulation: all florets disarticulating 19
19	Lodicule fatua type 20
	20	Epiblast 0.3 mm wide or less; scars narrow elliptic; periphery ring only confined to one-quarter; awn inserted at about lower third of lemma ... *A. lusitanica*
	20	Epiblast more than 0.3 up to 0.4 mm wide 21
		21 Periphery ring only confined to one-eighth to one-sixteenth .... *A. longiglumis*
		21 Periphery ring only confined to one-third to one-half .... *A. barbata*
	20	Epiblast 0.45–0.7 mm wide 22

22		Scar of third floret heart-shaped 23
	23	Periphery ring only confined to one-third to one-half; scars elliptic; epiblast septentrionalis type .... *A. occidentalis*
	23	Periphery ring all around scar .... *A. vaviloviana*
22		Scar of third floret not heart-shaped .... *A. fatua*
19		Lodicule sativa type 24
24		Epiblast 0.45–0.7 mm wide 25
	25	Scar of third floret heart-shaped; periphery ring all around scar .... *A. abyssinica* × *vaviloviana*
	25	Scar of third floret not heart-shaped .... *A. sativa fatuoid*
24		Epiblast 0.3 mm wide or less 26
	26	Periphery ring only confined to one-quarter 27
		27 Awn inserted between lower third to middle of lemma; lemma tip bisubulate; epiblast sativa type .... *A. hirtula*
		27 Awn inserted at about lower third of lemma; lemma tip bisetulate-biaristulate; epiblast septentrionalis type .... *A. damascena*
	26	Periphery ring only confined to one-third to one-half .... *A. wiestii*
24		Epiblast more than 0.3 up to 0.4 mm wide 28
	28	Awn inserted at about lower third of lemma; scars narrow elliptic; periphery ring only confined to one-quarter; epiblast brevis type .... *A. matritensis*
	28	Awn inserted at about middle of lemma 29
		29 Periphery ring only confined to one-eighth to one-sixteenth *A. longiglumis*
		29 Periphery ring only confined to one-third to one-half .... *A. hybrida*

4. WEIGHTED KEY WITH CHARACTERS 10–25 ONLY
WEIGHT (22, 25 = 0; 14, 15, 19, 23 = 1; 10, 11, 12, 13, 16, 17, 18, 20, 21, 24 = 2)

1	Lemma vestiture below awn insertion: beset densely with macrohairs 2
2	Lodicule strigosa type; mode of disarticulation: non-disarticulating; lodicule often with prickles .... *A. strigosa*
2	Lodicule fatua type 3
	3	Mode of disarticulation: lowermost floret only disarticulating 4
		4 Palea back beset with prickles .... *A. sterilis*
		4 Palea back beset with hairs 5
			5 Lemma tip bilobed; awn inserted at about middle of lemma; palea with 2 rows of cilia along edges of keels; lodicule rarely with a few prickles; epiblast 0.75–1.0 mm wide; chromosome number $2n = 28$ ..... *A. maroccana*
			5 Lemma tip bisubulate; awn inserted at about lower third of lemma; palea with 1–3 rows of cilia along edges of keels; lodicule never bearing prickles; epiblast 0.45–0.7 mm wide; chromosome number $2n = 42$ .... *A. atherantha*

- 3 Mode of disarticulation: all florets disarticulating 6
  - 6 Epiblast 0.3 mm wide or less; scars narrow elliptic; periphery ring only confined to one-quarter; awn inserted at about lower third of lemma .. **A. lusitanica**
  - 6 Epiblast 0.45–0.7 mm wide 7
    - 7 Scar of third floret heart-shaped 8
      - 8 Scars elliptic; periphery ring only confined to one-third to one-half; palea with 1–3 rows of cilia along edges of keels; epiblast septentrionalis type; chromosome number $2n = 42$ .... **A. occidentalis**
      - 8 Scars heart-shaped or slightly so; periphery ring all around scar; palea with 1–2 rows of cilia along edges of keels; epiblast brevis type; chromosome number $2n = 28$ .... **A. vaviloviana**
    - 7 Scar of third floret not heart-shaped .... **A. fatua**
  - 6 Epiblast more than 0.3 up to 0.4 mm wide 9
    - 9 Scars linear .... **A. longiglumis**
    - 9 Scars elliptic ....**A. barbata**
- 2 Lodicule sativa type 10
- 10 Mode of disarticulation: lowermost floret only disarticulating 11
  - 11 Scars heart-shaped or slightly so; periphery ring only confined to one-quarter; awn inserted at about middle of lemma; lemma tip bidenticulate; palea with 1 row of cilia along edges of keels; epiblast brevis type; chromosome number $2n = 14$ .... **A. canariensis**
  - 11 Scars elliptic .... **A. trichophylla**
- 10 Mode of disarticulation: all florets disarticulating 12
  - 12 Epiblast 0.45–0.7 mm wide .... **A. sativa fatuoid**
  - 12 Epiblast more than 0.3 up to 0.4 mm wide 13
    - 13 Scars narrow elliptic; periphery ring only confined to one-quarter; awn inserted at about lower third of lemma; epiblast brevis type .... **A. matritensis**
    - 13 Scars oval to round .... **A. hybrida**
    - 13 Scars linear .... **A. longiglumis**
  - 12 Epiblast 0.3 mm wide or less 14
    - 14 Lemma tip bisubulate; awn inserted between lower third to middle of lemma; epiblast sativa type .... **A. hirtula**
    - 14 Lemma tip biaristulate .... **A. wiestii**
    - 14 Lemma tip bisetulate-biaristulate 15
      - 15 Periphery ring only confined to one-quarter; awn inserted at about lower third of lemma .... **A. damascena**
      - 15 Periphery ring only confined to one-third to one-half .... **A. wiestii**
- 1 Lemma vestiture below awn insertion: without macrohairs or only a few often found around awn insertion 16
- 16 Lemma structure resembling glumes 17
  - 17 Lemma tip bidenticulate 18
    - 18 Lodicule fatua type .... **A. sativa** × **fatua** or **sativa** × **sterilis** (each resembling more parent)

		18	Lodicule sativa type .... *A. sativa*

- 17 Lemma tip bisubulate .... *A. nuda*
- 16 Lemma structure tough 19
- 19 Mode of disarticulation: non-disarticulating 20
  - 20 Epiblast more than 0.3 up to 0.4 mm wide 21
    - 21 Lemma tip shortly bisubulate; awn inserted from upper fourth to below tip of lemma; lodicule never bearing prickles .... *A. brevis*
    - 21 Lemma tip bisetulate-biaristulate 22
      - 22 Lodicule fatua type; awn inserted between upper third to upper fourth of lemma; palea with 1 row of cilia along edges of keels; lodicule rarely with a few prickles .... *A. hispanica*
      - 22 Lodicule strigosa type; awn inserted at about lower third of lemma; palea with 1–2 rows of cilia along edges of keels; lodicule often with prickles .... *A. strigosa*
  - 20 Epiblast 0.45–0.7 mm wide 23
    - 23 Lodicule fatua type .... *A. sativa* × *fatua* or *sativa* × *sterilis* (each resembling more parent)
    - 23 Lodicule sativa type 24
      - 24 Lemma tip bisubulate; palea with 1–2 rows of cilia along edges of keels; epiblast brevis type; chromosome number $2n = 28$ .... *A. abyssinica*
      - 24 Lemma tip bidenticulate .... *A. sativa*
- 19 Mode of disarticulation: lowermost floret only disarticulating 25
  - 25 Epiblast 0.45–0.7 mm wide 26
    - 26 Periphery ring only confined to one-half; awn inserted at about lower fourth of lemma; palea with 2 rows of cilia along edges of keels; chromosome number $2n = 28$ .... *A. murphyi*
    - 26 Periphery ring only confined to one-third; awn inserted at about lower third of lemma; palea with 1–3 rows of cilia along edges of keels; chromosome number $2n = 42$ .... *A. sterilis*
  - 25 Epiblast more than 0.3 up to 0.4 mm wide 27
    - 27 Glumes unequal 28
      - 28 Lemma tip bisubulate .... *A. eriantha*
      - 28 Lemma tip bisetulate-biaristulate .... *A. clauda* × *eriantha* (as in *eriantha* but spikelets narrower and longer)
    - 27 Glumes equal 29
      - 29 Scars linear; periphery ring only confined to one-eighth; awn inserted at about upper third of lemma; palea with 1–2 rows of cilia along edges of keels; lodicule fatua type; epiblast sativa type; chromosome number $2n = 14$ .... *A. ventricosa*
      - 29 Scars elliptic .... *A. trichophylla*
- 19 Mode of disarticulation: all florets disarticulating 30
  - 30 Scar of third floret heart-shaped 31
    - 31 Scars heart-shaped or slightly so 32
      - 32 Lodicule fatua type .... *A. vaviloviana*

                32   Lodicule sativa type .... *A. abyssinica* × *vaviloviana*
            31   Scars oval to round .... *A. hybrida*
         30   Scar of third floret not heart-shaped 33
            33   Lodicule sativa type 34
                34   Glumes unequal; scars narrow elliptic; periphery ring only confined to one-quarter; awn inserted at about upper third of lemma; palea with 1–2 rows of cilia along edges of keels; chromosome number $2n = 28$ ....
                    A. macrostachya
                34   Glumes equal .... *A. sativa fatuoid*
            33   Lodicule fatua type 35
                35   Glumes unequal; scars linear; periphery ring only confined to one-eighth; awn inserted between lower third to middle of lemma; lemma tip bisetulate-biaristulate; palea with 1–2 rows of cilia along edges of keels; lodicule rarely with a few prickles; epiblast sativa type; epiblast more than 0.3 up to 0.4 mm wide; chromosome number $2n = 14$ .... *A. clauda*
                35   Glumes equal .... *A. fatua*

5. **WEIGHTED KEY ACCORDING TO RESULTS FROM "CHARANAL"**
    (All characters)
    WEIGHT (1, 4, 7, 10, 11, 16 = 0; 3, 5, 6, 8, 13, 20, 22 = 1; 2, 18, 19, 21, 23, 25 = 2; 9, 12, 14, 15, 17, 24 = 3)

1   Lemma tip bilobed .... *A. maroccana*
1   Lemma tip shortly bisubulate; awn inserted from upper fourth to below tip of lemma .... *A. brevis*
1   Lemma tip biaristulate 2
   2   Spikelet with 2 florets .... *A. wiestii*
   2   Spikelet with 2–3 florets 3
      3   Awn inserted at about lower third of lemma; ligule acute; scars narrow elliptic; periphery ring only confined to one-quarter .... *A. matritensis*
      3   Awn inserted at about middle of lemma 4
         4   Scars elliptic; color appearance green; culm geniculate or prostrate; panicle equilateral; periphery ring only confined to one-third to one-half; palea with 1–2 rows of cilia along edges of keels; palea back glabrous; lodicule never bearing prickles; epiblast brevis type; chromosome number $2n = 28$ .. *A. barbata*
         4   Scars linear .... *A. longiglumis*
1   Lemma tip bisetulate-biaristulate 5
   5   Epiblast 0.3 mm wide or less 6
      6   Spikelet with 2 florets .... *A. wiestii*
      6   Spikelet with 2–3 florets 7
         7   Scars narrow elliptic; juvenile growth prostrate; culm geniculate or prostrate; panicle slightly flagged; spikelet indeterminate; lodicule fatua type; epiblast brevis type .... *A. lusitanica*

- 7 Scars elliptic; juvenile growth prostrate to erect; culm erect; panicle equilateral; spikelet long; lodicule sativa type; epiblast septentrionalis type .... *A. damascena*
- 5 Epiblast more than 0.3 up to 0.4 mm wide 8
  - 8 Spikelet with 2 florets; awn inserted between upper third to upper fourth of lemma .... *A. hispanica*
  - 8 Spikelet with 2–6 florets; color appearance green; awn inserted between lower third to middle of lemma .... *A. clauda*
  - 8 Spikelet with 2–3 florets 9
    - 9 Awn inserted at about lower third of lemma .... *A. strigosa*
    - 9 Awn inserted at about middle of lemma 10
      - 10 Periphery ring only confined to one-eighth; relatively low plants not exceeding 95 cm (15–95); spikelet short; glumes unequal; mode of disarticulation: lowermost floret only disarticulating; lemma vestiture below awn insertion: without macrohairs or only a few often found around awn insertion; palea with 1–2 rows of cilia along edges of keels; lodicule never bearing prickles; epiblast sativa type .. *A. clauda* × *eriantha* (as in *eriantha* but spikelets narrower and longer)
      - 10 Periphery ring only confined to one-eighth to one-sixteenth ...*A. longiglumis*
- 1 Lemma tip bidenticulate 11
- 11 Epiblast more than 0.3 up to 0.4 mm wide 12
  - 12 Spikelet with 2 florets; mode of disarticulation: lowermost floret only disarticulating; scars heart-shaped or slightly so; periphery ring only confined to one-quarter; palea with 1 row of cilia along edges of keels; lodicule rarely with a few prickles; epiblast brevis type; chromosome number $2n = 14$ .... *A. canariensis*
  - 12 Spikelet with 2–4 florets .... *A. hybrida*
- 11 Epiblast 0.45–0.7 mm wide 13
  - 13 Spikelet with 2–4 florets; juvenile growth erect; spikelet long; awn inserted at about lower fourth of lemma; palea with 2 rows of cilia along edges of keels; chromosome number $2n = 28$ .... *A. murphyi*
  - 13 Spikelet with 2–5 florets .... *A. sterilis*
  - 13 Spikelet with 2–3 florets 14
    - 14 Scars heart-shaped or slightly so .... *A. sativa fatuoid*
    - 14 Scars oval to round .... *A. fatua*
  - 13 Spikelet with 1–7 florets 15
    - 15 Awn absent or rudimentary 16
      - 16 Lodicule fatua type .... *A. sativa* × *fatua* or *sativa* × *sterilis* (each resembling more parent)
      - 16 Lodicule sativa type .... *A. sativa*
    - 15 Awn inserted at about middle of lemma 17
      - 17 Lodicule fatua type .... *A. sativa* × *fatua* or *sativa* × *sterilis* (each resembling more parent)
      - 17 Lodicule sativa type .... *A. sativa*
- 1 Lemma tip bisubulate 18
- 18 Spikelet with 2 florets .... *A. ventricosa*

18	Spikelet with 2–5 florets	*A. sterilis*
18	Spikelet with 2–4 florets	19

- 19 Periphery ring only confined to one-half; spikelet long; mode of disarticulation: lowermost floret only disarticulating; awn inserted at about lower fourth of lemma; palea with 2 rows of cilia along edges of keels; lodicule fatua type; lodicule rarely with a few prickles; epiblast fatua type; epiblast 0.45–0.7 mm wide; chromosome number $2n = 28$ .... *A. murphyi*

- 19 Periphery ring only confined to one-third to one-half .... *A. hybrida*

18	Spikelet with 3–4 florets	20

- 20 Periphery ring only confined to one-third to one-half; relatively low plants not exceeding 95 cm (15–95); panicle slightly flagged; mode of disarticulation: all florets disarticulating; awn inserted at about middle of lemma; lodicule fatua type; lodicule never bearing prickles; epiblast 0.47–0.7 mm wide ...*A. occidentalis*

- 20 Periphery ring only confined to one-third .... *A. trichophylla*

18	Spikelet with 3–6 florets	21

- 21 Awn inserted at about upper third of lemma; plant perennial; color appearance green; relatively tall plants exceeding 100 cm (30–180); spikelet indeterminate; glumes unequal; mode of disarticulation: all florets disarticulating; lemma structure tough; lodicule sativa type; chromosome number $2n = 28$ *A. macrostachya*

- 21 Awn inserted at about middle of lemma or absent or rudimentary .... *A. nuda*

18	Spikelet with 2–3 florets	22

- 22 Epiblast 0.3 mm wide or less; awn inserted between lower third to middle of lemma; palea with 1 row of cilia along edges of keels .... *A. hirtula*

- 22 Epiblast more than 0.3 up to 0.4 mm wide 23

  - 23 Scars elliptic; color appearance green; culm geniculate or prostrate; relatively tall plants exceeding 100 cm (30–180); panicle equilateral; spikelet long; glumes equal; mode of disarticulation: all florets disarticulating; periphery ring only confined to one-third to one-half; lemma vestiture below awn insertion: beset densely with macrohairs, epiblast brevis type; chromosome number $2n = 28$ .... *A. barbata*

  - 23 Scars linear; color appearance glaucous; culm erect; relatively low plants not exceeding 95 cm (15–95); panicle slightly flagged; spikelet short; glumes unequal; mode of disarticulation: lowermost floret only disarticulating; periphery ring only confined to one-eighth; lemma vestiture below awn insertion: without macrohairs or only a few often found around awn insertion; epiblast sativa type; chromosome number $2n = 14$ .... *A. eriantha*

- 22 Epiblast 0.45–0.7 mm wide 24

  - 24 Awn inserted at about lower third of lemma; ligule obtuse; spikelet long; mode of disarticulation: lowermost floret only disarticulating; palea back beset with hairs .... *A. atherantha*

  - 24 Awn inserted at about middle of lemma 25

    - 25 Palea with 1–3 rows of cilia along edges of keels 26

      - 26 Scars heart-shaped or slightly so .... *A. sativa fatuoid*

      - 26 Scars oval to round .... *A. fatua*

    - 25 Palea with 1–2 rows of cilia along edges of keels 27

      - 27 Lodicule fatua type .... *A. vaviloviana*

      - 27 Lodicule sativa type 28

28 Mode of disarticulation: non-disarticulating .... *A. abyssinica*
 28 Mode of disarticulation: all florets disarticulating ....
    A. abyssinica × *vaviloviana*

6. WEIGHTED KEY FOR CYTOLOGISTS AND BREEDERS (Characters: 1–10, 25)
WEIGHT (7, 8, 9 = 0; 4, 5 = 1; 2, 3, 6 = 2; 1, 10, 25 = 3)

1 Glumes unequal 2
  2 Plant perennial; relatively tall plants exceeding 100 cm (30–180); spikelet indeterminate; spikelet with 3–6 florets; chromosome number $2n=28$ ....
    A. macrostachya
  2 Plant annual; chromosome number $2n=14$ 3
    3 Color appearance green; spikelet long; spikelet with 2–6 florets .... *A. clauda*
    3 Color appearance glaucous .... *A. eriantha*
    A. clauda × *eriantha*
    (as in *eriantha* but spikelets narrower and longer)
1 Glumes equal 4
  4 Ligule obtuse 5
    5 Chromosome number $2n=42$ .... *A. atherantha*
    5 Chromosome number $2n=28$ 6
      6 Juvenile growth erect .... *A. wiestii*
      6 Juvenile growth prostrate .... *A. barbata*
    5 Chromosome number $2n=14$ 7
      7 Culm geniculate or prostrate .... *A. lusitanica*
      7 Culm erect 8
        8 Relatively low plants not exceeding 95 cm (15–95) 9
          9 Juvenile growth prostrate to erect .... *A. damascena*
          9 Juvenile growth prostrate .... *A. hirtula*
          9 Juvenile growth erect 10
            10 Spikelet short 11
              11 Spikelet with 2–3 florets .... *A. brevis*
              11 Spikelet with 2 florets .... *A. ventricosa*
            10 Spikelet long .... *A. nuda*
        8 Relatively tall plants exceeding 100 cm (30–180) 12
          12 Color appearance green .... *A. wiestii*
          12 Color appearance glaucous 13
            13 Panicle equilateral 14
              14 Juvenile growth prostrate to erect; spikelet with 2 florets .. *A. hispanica*

		14	Juvenile growth erect .... *A. strigosa*
		13	Panicle slightly flagged .... *A. longiglumis*

4   Ligule acute 15
15   Spikelet long 16
    16   Color appearance glaucous .... *A. maroccana*
    16   Color appearance green 17
       17   Panicle slightly flagged .... *A. occidentalis*
       17   Panicle equilateral 18
          18   Spikelet with 2–3 florets .... *A. matritensis*
          18   Spikelet with 2–4 florets .... *A. murphyi*
          18   Spikelet with 3–4 florets .... *A. trichophylla*
15   Spikelet indeterminate 19
    19   Spikelet with 2–5 florets .... *A. sterilis*
    19   Spikelet with 1–7 florets .... *A. sativa*
                                  A. sativa × *fatua* or *sativa* × *sterilis*
                                       (each resembling more parent)
15   Spikelet short 20
    20   Chromosome number $2n = 14$ .... *A. canariensis*
    20   Chromosome number $2n = 41, 40,$ or 42 .... *A. sativa fatuoid*
    20   Chromosome number $2n = 28$ .... *A. abyssinica*
                                                 A. vaviloviana
                                       A. abyssinica × *vaviloviana*
    20   Chromosome number $2n = 42$ 21
       21   Spikelet with 2–4 florets .... *A. hybrida*
       21   Spikelet with 2–3 florets .... *A. fatua*

7. WEIGHTED KEY WITH A FEW OBVIOUS CHARACTERS FOR QUICK IDENTIFICATION (Characters 1, 10, 11, 17, 25) WEIGHT (11, 17 = 0; 1, 10, 25 = 1)

1   Glumes unequal 2
   2   Plant perennial; chromosome number $2n = 28$ .... *A. macrostachya*
   2   Plant annual; chromosome number $2n = 14$ 3
      3   Mode of disarticulation: all florets disarticulating .... *A. clauda*
      3   Mode of disarticulation: lowermost floret only disarticulating 4
         4   Lemma tip bisubulate .... *A. eriantha*
         4   Lemma tip bisetulate-biaristulate .... *A. clauda* × *eriantha*
                    (as in *eriantha* but spikelets narrower and longer)
1   Glumes equal 5

5		Mode of disarticulation: lowermost floret only disarticulating 6	
	6	Chromosome number $2n = 14$ 7	
		7 Lemma tip bisubulate .... *A. ventricosa*	
		7 Lemma tip bidenticulate .... *A. canariensis*	
	6	Chromosome number $2n = 28$ 8	
		8 Lemma tip bilobed .... *A. maroccana*	
		8 Lemma tip bidenticulate, or bisubulate .... *A. murphyi*	
	6	Chromosome number $2n = 42$ .... *A. sterilis*	
			A. atherantha
			A. trichophylla
5		Mode of disarticulation: non-disarticulating 9	
	9	Chromosome number $2n = 28$ .... *A. abyssinica*	
	9	Chromosome number $2n = 42$ .... *A. sativa*	
		A. sativa × *fatua* or *sativa* × *sterilis*	
		(each resembling more parent)	
	9	Chromosome number $2n = 14$ 10	
	10	Lemma tip shortly bisubulate .... *A. brevis*	
	10	Lemma tip bisetulate-biaristulate .... *A. hispanica*	
			A. strigosa
	10	Lemma tip bisubulate .... *A. nuda*	
5		Mode of disarticulation: all florets disarticulating 11	
11		Lemma tip biaristulate .... *A. barbata*	
			A. longiglumis
			A. matritensis
			A. wiestii
11		Lemma tip bisetulate-biaristulate 12	
	12	Chromosome number $2n = 28$ .... *A. wiestii*	
	12	Chromosome number $2n = 14$ .... *A. damascena*	
			A. longiglumis
			A. lusitanica
			A. wiestii
11		Lemma tip bidenticulate 13	
	13	Chromosome number $2n = 41, 40,$ or 42 .... *A. sativa fatuoid*	
	13	Chromosome number $2n = 42$ .... *A. fatua*	
			A. hybrida
11		Lemma tip bisubulate 14	
	14	Chromosome number $2n = 14$ .... *A. hirtula*	
	14	Chromosome number $2n = 41, 40,$ or 42 .... *A. sativa fatuoid*	
	14	Chromosome number $2n = 28$ .... *A. barbata*	
			A. vaviloviana
		A. abyssinica × *vaviloviana*	
	14	Chromosome number $2n = 42$ .... *A. fatua*	
			A. hybrida
			A. occidentalis

Part 2. Systematic treatment of species

14. Introduction

The following sections have been provided for every species and various non-taxonomic entities such as F_1 hybrids and fatuoids: correct name; synonymy; description; distribution; phenology; habitat; chorotype; mode of dispersal; selected specimens; identification hints and remarks; and notes.

SYNONYMS AND NOTES

The synonyms are arranged in two groups, each in chronological order of their dates of publication. The first group includes all the homotypic synonyms together with the full citation of the correct name. The second group, which is usually larger than the first, consists of all the heterotypic synonyms.

Every basionym is referred to a note by an appropriate number in square brackets. These notes give details about various nomenclatural aspects such as detection and location of types, interpretation of protologs and various other matters related to typification. A photograph of the type of the correct name accompanies the nomenclatural notes.

DESCRIPTIONS

The descriptions are relatively short but follow a more or less uniform pattern as follows: duration of life cycle, juvenile growth, color appearance, stature of culm, height, shape of panicle, range of length of spikelet and the range of its number of florets, range of length of glumes and relative length of upper and lower glume, mode of disarticulation, shape of scars of the florets, degree of confinement of periphery ring to scar, place of awn insertion, structure of lemma, configuration of lemma tips, vestiture of lemma, range of number of rows of cilia along edge of keel; vestiture of back of palea, lodicule type and presence of prickles, epiblast type and range of width, chromosome number, and belongingness to genome group.

Figures illustrating selected parts accompany the description of each species.

DISTRIBUTION, PHENOLOGY, AND HABITAT

The different political boundaries within which a given species or unit was found are enumerated under this section. Moreover, a distribution map is provided for each.

The range of flowering time and various ecological sites recorded for every species and unit are described under the phenology and habitat sections. When available, one or several photographs of typical sites are also provided.

Introduction

CHOROTYPE

The occurrence of a particular species in terms of its presence within phytogeographically recognized floristic units is given in this section for each species. I have not rigidly followed one particular author and his definitions of the various floristic units that he recognizes. Rather, to define the chorotype for each species as briefly and accurately as possible, I based myself according to convenience on Good (1974) or Zohary (1973) or Rikli (1943–1948) or on a combination of these, sometimes taking the floristic region defined by one author and sometimes the region defined by another.

MODE OF DISPERSAL

The three basic diaspores in *Avena* are (i) the whole panicle, when the grain has a cultivated base, (ii) the entire spikelet, when only the lowermost floret is disarticulating, and (iii) the floret, when every floret is disarticulating at maturity. In the last two cases, the diaspores have awns that assist in pushing or digging themselves into cracks in the soil. The stiff hairs and setae at the base of these diaspores also enhance the process of immersion into cracks.

When I mention that dispersal is by wind, I mean that dispersal occurs only over limited distances, as has been observed in the field, that is, in the Mediterranean and Middle East areas. In these areas the diaspores are pushed by strong winds along the dry, barren, and gravelly ground; dispersal is particularly effective when the mature diaspores are very loosely held between the dry glumes on the panicle so that a strong and sudden wind can release them and blow them at varying distances. These diaspores, however, are definitely not adapted to wind dispersal. Quite frequently the diaspores get stuck on the hairs of grazing animals (Figs. 1 and 2) and will achieve telechory very effectively by this means.

SELECTED SPECIMENS

The specimens in this work are not selected from the 4996 accessions received from USDA and grown in Ottawa, nor are they selected from the 5264 accessions collected by my colleagues and myself in natural habitats. Rather, these specimens are selected from herbarium specimens only. My aim was to take specimens (preferably exsiccata) that are preserved in many institutions so that the herbaria which I have not studied will possibly possess some duplicates of these specimens.

In selecting the specimens I attempted to find at least one representative from each country or geographically defined area in which a particular species occurs. For the cases where no appropriate herbarium specimen was found, the reader is referred to the accessions of the Gene Pool Collection of *Avena* (Baum et al. 1975).

In some cases, specimens with exactly the same label content (i.e., presumably duplicate specimens) belonged to different species. This is not surprising because the oat plant, except for *A. macrostachya,* is an annual, and when duplicates are collected in the field, they may often be mistaken because the species are so similar in habit. In fact duplicate specimens may truly exist in perennials only. Thus, it is not surprising to find among the selected specimens a few with the same label data (viz., same collection number, same date, same locality, and same original identification), which come from different institutions and which I identify differently. I minimized citing such cases among the selected specimens, but frequently found this phenomenon among the herbarium specimens in general.

A word of warning about the few selected specimens cited: if you find them in herbaria that I have not consulted, check them thoroughly for identification; furthermore, in large herbaria, be alert to possible duplicates that could have identical identification and yet belong to different taxa.

IDENTIFICATION HINTS AND REMARKS

These notes complement the descriptions and can assist anyone having difficulties. I believe that they will also prove to be useful in assuring correct identification. In this section, the most useful discriminatory characters of each species are mentioned; these characters are compared to help distinguish species that closely resemble one another.

15. Description of genus

AVENA L.

Avena L., Sp. Pl. 79 (1953); Gen. Pl. ed. 5, 34 No. 85 (1754)
Lectotype species: *A. sativa* L., designated by Britton and Brown, Illustr. Fl. North. U.S. ed. 2, 218 (1913)

TAXONOMIC SYNONYMS AS TO TYPES ONLY

Section *Paniculatae* Schrader, Fl. Germ. 1:368 (1906)
    Lectotype species: *A. fatua* L.
Subgenus *Genuinae* Link, Hortus Reg. Bot. Berol. 1:109 (1827)
    Lectotype species: *A. sativa* L.
Subgenus *Verae* Link, op. cit. 110
    Lectotype species: *A. fatua* L.
Section *Avena* Gaudin, Fl. Helv. 1:329 (1828)
    Lectotype species: *A. sativa* L.
Section *Avenastrum* Duby, Bot. Gall. ed. 2, 1:512 (1828)
    Lectotype species: *A. sativa* L.

Subgenus? *Avenae genuinae* Reichenb., Fl. Germ. Excurs. 52 (1830)
    Lectotype species: *A. sativa* L.
Section *Arvenses* Gaudin, Synopsis Fl. Helv. ed. Monnard: 80 (1836)
    Lectotype species: *A. sativa* L.
Section *Avenae genuinae* Koch, Synopsis ed. 1, 794 (1837)
    Lectotype species: *A. sativa* L.
Section *Annuae* Trin., Mem. Acad. Petersb. Ser. 6, 2:23 (1838)
    Lectotype species: *A. sativa* L.
Section *Eu Avena* Griseb., Spicil. Fl. Rumel. 2:452 (1845)
    Lectotype species: *A. pilosa* M.B. (i.e., *A. eriantha* Dur.)
Section *Avenatypus* Coss. et Germ., Fl. Par. 636 (1845)
    Lectotype species: *A. sativa* L.
Section *Eu Avena* Koch, Linnaea 21:391 (1848)
    Lectotype species: *A. sativa* L.
Section *Crithe* Griseb. in Ledeb., Fl. Ross. 4:412 (1853)
    Lectotype species: *A. sativa* L.
Section *Agravena* Kirschl., Fl. Als. 2:309 (1857)
    Lectotype species: *A. sativa* L.
Section *Annuae* Husnot, Graminées, 38 (1897)
    Lectotype species: *A. sativa* L.
Subgenus *Crithe* (Griseb.) Rouy, Fl. France 14:120 (1913)
Preissia Opiz., Seznam Rostl. Ceske, 79 (1852)
    Type species: *A. strigosa* Schreb.

DESCRIPTION

Annuals or perennials *(A. macrostachya)* with diffuse, sometimes more or less contracted or one-sided panicles, and nodding spikelets. Aestivation of leaves involute. Blades with sclerenchyma strands on some of the largest vascular bundles only and with no strands in most other smaller ones (*Avena* type). Spikelets with 1–6 fertile florets and 1–2 additional male or rudimentary florets. Glumes 2, unequal, or mostly equal or nearly so, with 3–11 nerves, rounded on the back, smooth, herbaceous, persistent. Rachilla hairy or glabrous, disarticulating above the glumes and between the florets into several dispersal units, or only above the glumes into one dispersal unit, or not at all. Lemmas becoming indurated at maturity, rarely remaining membraneous; bilobed, bifid with or without long bristles (aristulae), the long bristles having up to 2 smaller shorter bristles (setulae) at the base or having none. Awns dorsal, with (or without in non-disarticulating specimens) a twisted column that is round in cross section and sometimes rudimentary. Palea bifid, 2-keeled, ciliate on keels, as long as lemmas or slightly shorter. Lodicules 2, of festucoid-type structure, entire, cylindrical at base and following into an open cone, linear-lanceolate or unequally bilobed or with a short rudimentary side lobe, this lobe, when present, is attached to the base of the lodicule as a rudder to a boat and is often partly fused to the basal part of the cone-like structure. Stamens 3. Ovaries villous. Styles rudimentary or practically none. Stigmas laterally exerted. Caryopses terete, but grooved in front; endosperm hard; embryos relatively small to very large; hila linear; epiblasts present. Chromosome numbers $2n = 14, 28, 42$.

DISTRIBUTION

Found in (Fig. 44) Afghanistan, Albania, Algeria, Austria, Azores, Belgium, Bulgaria, Canary Islands, China, Corsica, Crete, Cyprus, Czechoslovakia, Denmark, Egypt, Ethiopia, Finland, France, Germany, Greece, Hungary, India, Iran, Iraq, Israel, Italy, Japan, Jordan, Korea, Lebanon, Libya, Lithuania, Luxemburg, Madeira Islands, Morocco, Nepal, Norway, Pakistan, Poland, Portugal, Romania, Sardinia, Saudi Arabia, Sicily, Spain, Sweden, Switzerland, Syria, Taiwan, Tunisia, Turkey, United Kingdom, USSR, and Yugoslavia. Introduced in Argentina, Australia, Brazil, Canada, Chile, Cuba, USA, and Uruguay.

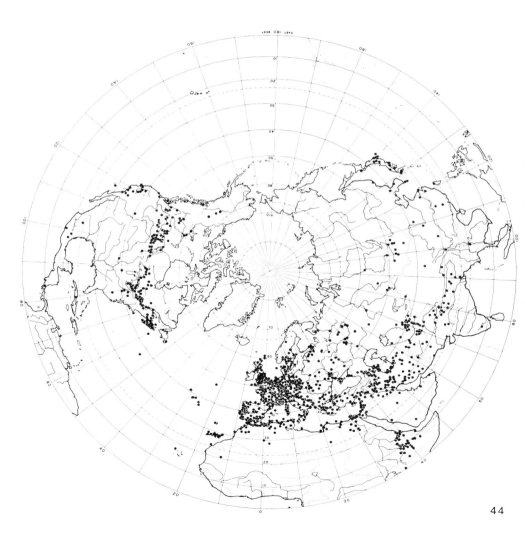

Fig. 44. World distribution of the genus *Avena,* polar projection, northern hemisphere only.

CHOROTYPE

Western and central Asiatic, the Mediterranean, and Macaronesian and northeast African highland and steppe regions; but also penetrates into other regions of the Boreal Kingdom of the Old World, and to some extent into the North African – Indian desert region. Introduced elsewhere (see above).

16. Section 1. Avenotrichon (Holub) Baum, Can. J. Bot. 52:2259 (1975)

Basionym Helictotrichon subgenus Avenotrichon Holub in B. Nèmec et al., Philipp Maxmilian Opiz und seine bedeutung für die Pflanzentaxonomie: 124 (1958).

Holotype species *A. macrostachya.*

Species included *A. macrostachya.*

Notes on taxonomy and distribution This section includes one species only. It is taxonomically isolated from the other sections (see Chapter 7 and Fig. 10), and also topographically isolated in the broad general area of northwest Mediterranean Africa. It is confined to the Djurdjura and Aurès chains of mountains in Algeria (Fig. 51) at altitudes between 1500 and 2200 m. The next closest area occupied by *Avena* at high altitudes is the east African highlands, namely Ethiopia, which is occupied by Section *Ethiopica* and by at least one species of Section *Avena*. An area at much higher altitudes is the high elevation of the Himalayan mass and its vicinity, which is occupied by members of Section *Avena* (e.g., *A. hybrida*) and at least one member of Section *Tenuicarpa*. Taxonomically it has some remarkable features, especially, the perennial habit and the special anatomical structure of the awn. It shares the unequal glumes with members of Section *Ventricosa*.

(1) A. MACROSTACHYA BAL. EX COSS. ET DUR.

Homotypic synonyms

A. macrostachya Bal. ex Coss. et Dur., Bull. Soc. Bot. France 1:318 (1855) [1]
Helictotrichon macrostachyum (Bal. ex Coss. et Dur.) Henrard, Blumea 3:430 (1940)

Figs. 45–50. Morphological and micromorphological diagnostic details of *A. macrostachya*. 45. Spikelet with the florets taken out of the glumes (left); note the 3-nerved lower glume and the 9-nerved upper glume ($\times$ 2.5). 46. Bisubulate lemma tips ($\times$ 24). 47. Scar ($\times$ 48). 48. Pair of lodicules in which the left is a fatua type and the right is a sativa type ($\times$ 50). 49. Detail of the side lobe of the sativa-type lodicule of Fig. 48 ($\times$ 120). 50. Epiblast brevis type ($\times$ 120).

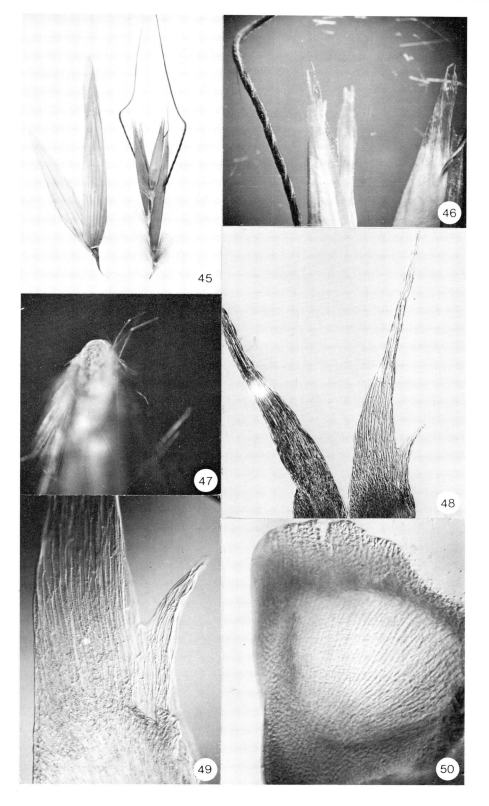

Description

Perennial plants, densely tufted. Color appearance green. Culm erect, often geniculate. Relatively tall plants, 40–100 cm high. Ligules obtuse. Panicle slightly flagged. Spikelets (Fig. 45) indeterminate in length, 20–30 mm long without the awns; each spikelet has 3–6 florets. Glumes unequal, the lower being about half the length of the upper (Fig. 45), 10–25 mm long, 3–7 nerves; all the florets are disarticulating at maturity; sometimes the upper florets remain attached and do not have scars; scars narrow elliptic (Fig. 47). Periphery ring only confined to ¼ of scar. Awns inserted at about upper ⅓ of lemma. Lemma structure tough; lemma tips (Fig. 46) bisubulate, sometimes with membranous, setula-like structure; lemmas glabrous (i.e., lacking macrohairs below awn insertion). Paleas with 1–2 rows of cilia along the edges of the keels. Lodicules (Figs. 48, 49) sativa type; sometimes the side lobe is very small, or one of the two lodicules in a pair may be fatua type; lodicules never bearing prickles. Epiblast brevis type, 0.40–0.45 mm wide (Fig. 50); some plants are seedless (i.e., sterile). Chromosome number $2n=28$. Genome unknown.

Distribution

Native to (Fig. 51) Algeria.

Phenology

Flowers from May to June.

Habitat

Mountain slopes and pastures, often rocky (Figs. 52, 53).

Chorotype

Endemic to the Mauritanian steppes province of the Irano-Turanian region.

Mode of dispersal

The diaspores are mostly florets, but these are sterile. This species is poorly understood; and one method of dispersal is vegetative through basal offshoots.

Selected specimens

ALGERIA, Djebel Tougour prope Batna, 30 June 1853, *Balansa No. 718* (BM, FI, G, LE, and P).

A. macrostachya

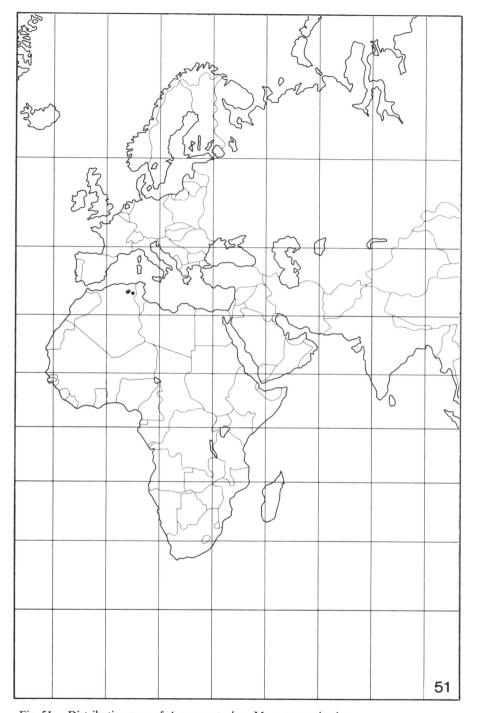

Fig. 51. Distribution map of *A. macrostachya*. Mercator projection.

A. macrostachya

Figs. 52–53. Two sites of *A. macrostachya,* on Mt. Djurdjura, Algeria.

Identification hints and remarks

This species cannot be confused with any *Avena* species because it is the only perennial species in the genus. The spikelets of *A. clauda* somewhat resemble those of *A. macrostachya* because of their unequal glumes, but the habit of the panicle is different. This species can, however, be confused with a few species of the genus *Helictotrichon,* which are all perennial species. Many authors classified *A. macrostachya* within *Helictotrichon* or among the group of perennial species of a large genus *Avena*. The controversy will obviously continue for some time (see Chapter 5). The reasons for including *A. macrostachya* in *Avena* are based on gross morphological habit, shape and structure of glumes, anatomy of the awns, histotaxis of blades, and nature of the endosperm.

Notes

[1] It was first proposed by Balansa under the name *Avena macrostachia*, in Pl. Alger. Exsicc. No. 718 (1853?) as a *nom. nud.* I have examined the holotype in P (Fig. 54); it is implied from the protolog. The paratypes from P also belong here. I was able to detect many isotypes from BM, FI, and G. Herb. LE has an isoparatype collected by La Perraudière.

17. Section 2. Ventricosa Baum, Can. J. Bot. 52:2259 (1975)

Basis for name   I have named this section after *A. ventricosa*.

Holotype species   *A. ventricosa*.

Species included   *A. clauda, A. eriantha, A. ventricosa*.

Naturally occurring hybrids   *A. clauda* × *eriantha*.

Notes on taxonomy and distribution   The species in this section are diploids with the C genome. They have unequal glumes except for *A. ventricosa*, which has mostly subinequal glumes and is similar in this respect to Section *Ethiopica*. *Avena clauda* can be a noxious plant and is of limited economic importance because it is used locally as hay. This section thrives in disjunct populations in an area between 25° and 45° latitude North, from the west Mediterranean area eastward to the Himalayan mass (Fig. 55). This distribution consists of the area that once was covered by the ancient Tethys Sea. Although we do not possess fossil evidence, I speculate that this distribution of disjunct populations is of relict nature. Phenetic similarity based on all characters suggests relationships with Section *Agraria* (see Chapter 7 and Fig. 10).

(2)   A. CLAUDA DUR.

Homotypic synonym

   A. clauda Dur., Rev. Bot. Duchartre 1:360 (1845–1846) [1]

Heterotypic synonyms

   A. clauda var. eriantha Bal. ex Coss., Bull. Soc. Bot. France 1:15 (1854) [2]
   A. clauda subv. leiantha Malz., Monogr. 232 (1930) [3]
   A. clauda subv. eriantha (Bal. ex Coss.) Malz., loc. cit.

A. macrostachya

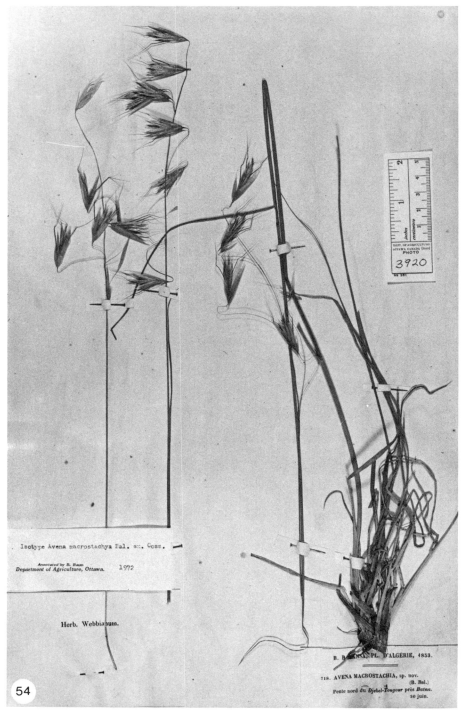

Fig. 54. Holotype of *A. macrostachya*.

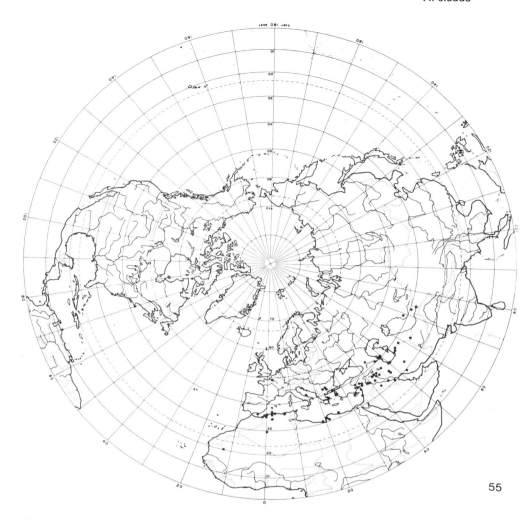

Fig. 55. Distribution map of section *Ventricosa*. Polar projection.

Description

Annual plants. Juvenile growth prostrate. Color appearance green or dark green. Culm erect. Relatively low plants, 20–70 cm high. Ligules obtuse. Panicles slightly flagged. Spikelets (Figs. 56, 57) long, 20–28 mm long without the back awn; each spikelet has 2–6 florets. Glumes (Fig. 56) unequal in length, the lowest being ⅓–½ the size of the upper; the lowest have 3–5 nerves and the upper have 7 nerves; all the florets are disarticulating at maturity; scars linear (Fig. 59). Periphery ring only confined to ⅛ of scar. Awn inserted between lower ⅓–½ of lemma. Lemma structure tough; lemma tips (Fig. 58) bisetulate-biaristulate, the setulae sometimes very small and in

A. clauda

close proximity to the aristulae; lemma without macrohairs below awn insertion or only a few hairs found around the place of the awn insertion. Paleas with 1–2 rows of cilia along the edges of the keels. Lodicules fatua type (Fig. 60), rarely with a few prickles (Fig. 61). Epiblast sativa type (Fig. 62), 0.3 mm wide. Chromosome number $2n = 14$. Genome CC.

Distribution

Native to (Fig. 63) Algeria, Bulgaria, Crete, Greece, Iran, Iraq, Israel, Italy, Lebanon, Morocco, Syria, Turkey, and USSR (Azerbaidzhan, Georgia, Tadzhik, Turkmen, and Uzbek).

Phenology

Flowers from March to July but mostly from April to May.

Habitat

The populations of this species (Figs. 64–66) normally occupy undisturbed habitats but they are also found on ditch sides and edges of cultivated fields, and in rare instances as serious weeds. The plant grows on brownish and yellowish brown rocky soils, thin soils in garrigues or maquis at various stages of degeneration, or is confined to ungrazed, deforested areas of steppe forest (e.g., *Amygdalus-Pistacia* or *Quercus-Pistacia*). It successfully colonizes new planted forests, orchards, disturbed land around building sites, and road verges, provided that these places are not grazed. This species is becoming more common now than a few years ago and this trend will continue as overgrazing continues to decrease in its area of distribution. From this one can conclude that *A. clauda* once successfully occupied the maquis and other not too dense forests of these areas.

Chorotype

This is a typical Irano-Turanian species, that is, the Armenian-Persian highlands of western and central Asiatic region and penetrating into the southeast parts and the Moroccan-Tunis areas of the Mediterranean region.

Mode of dispersal

The diaspore is the floret; some florets may be glabrous, some hairy. The diaspores drop readily at maturity; they may be transported slightly by wind, or else by ants, or very restrictively by man.

Figs. 56–62. Morphological and micromorphological details of *A. clauda*. 56. Spikelet including glumes; note the unequal glumes ($\times$ 2.5). 57. Spikelet without the glumes ($\times$ 2.5). 58. Detail of a bisetulate-biaristulate lemma tip ($\times$ 40). 59. Linear scar; note the periphery ring confined only to ⅛ of scar ($\times$ 100). 60. Lodicule fatua type ($\times$ 120). 61. Lodicule apex with a few prickles ($\times$ 120). 62. Sativa-type epiblast ($\times$ 120).

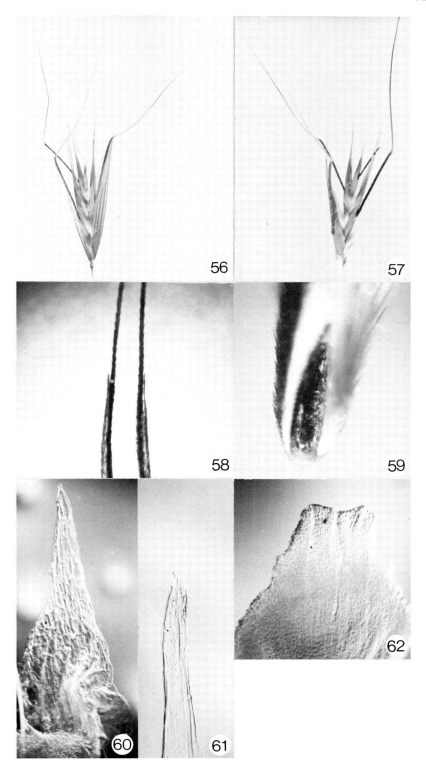

A. clauda

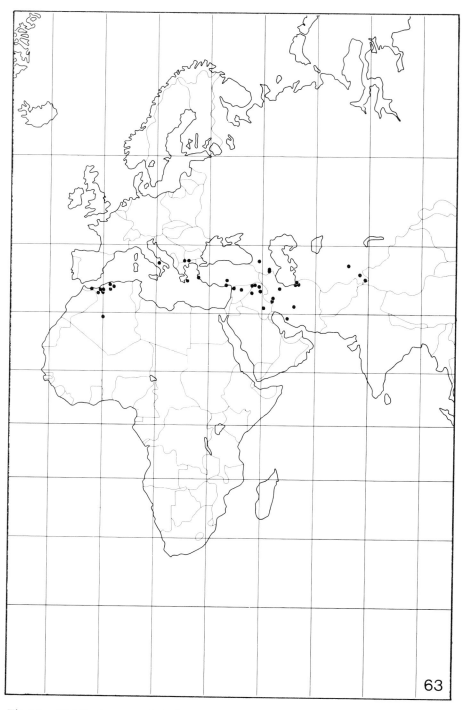

Fig. 63. Distribution map of *A. clauda*. Mercator projection.

Figs. 64–66. Three sites of *A. clauda*. 64. In natural habitat together with Hordeum, Secale, etc . . . Zagros area near Sanandaj. 65. Weed in a cherry orchard, collected there as hay. See also Fig. 66, overleaf.

Three sites of *A. clauda,* continued from page 143. Fig. 66. Hill slope near Songhor, edge of field.

Selected specimens

ALGERIA, Oran dans les champs incultes, 8 April 1852, *Balansa No. 553* (K and W); BULGARIA, Pcelina Pripecene, 6 May 1959, *Velcev-Bondev No. 509* (BR, C, COI, G, LE, S, UPS, and W); GREECE, Hymettos, June 1898, *Heldreich* (CGE and FI); IRAN, Gorgan Gonbad-Kabous Malik-Tappe, 23 April 1956, *Sharif No. 8098E* (K); IRAQ, Mosul Tel Kotchek Central Steppe, *Gillett No. 10824* (BAG and K); ITALY, Apulien Mte Gargano Manfredonia, 20 April 1964, *Seitz* (M); MOROCCO, Beni-Bu-Yahi Djebel Kerker, 22 April 1934, *Sennen and Mauricio* (MPU); SYRIA, Aleppum Dschebel Nahas, March 1841, *Kotschy No. 59* (FI, G, K, LE, P, and W); TURKEY, Smyrne Izmir coteaux calcaires, 30 April 1854, *Balansa No. 9* (C, G, K, P, and W); USSR *Azerbaidzhan,* Near Terter Khanage station, 2 June 1911, *Omelchenko No. 274* (WIR); USSR *Georgia,* Iberia, 1822, *Fischer* (G); USSR *Tadzhik,* E-Tair-Su River Valley Khak, 26 May 1932, *Gontscharov No. GR77* (LE); USSR *Turkmen,* Near Bambard-L-Bharm, 27 April 1958, *Merton No. 329i* (W); USSR *Uzbek,* Tashkent Trasnovo, 20 May 1924, *Vavilov No. 696* (WIR).

Identification hints and remarks

This is the only species in *Avena* with the following combination of morphological characters that can be discerned by the naked eye: inequal glumes and all florets disarticulating. The perennial *A. macrostachya* has these same two features but *A. clauda* is annual. Two entities that resemble *A. clauda* are *A. eriantha* and *A. clauda* × *eriantha* F_1 hybrids, but these two can readily be distinguished from the former by the mode of disarticulation.

Notes

[1] According to the protolog there are many syntypes. I have recovered a great part of the type collection in various herbaria. Baudrimont, in Jeanjean, Catalogues des Plantes Vasculaires de la Gironde, Actes Soc. Linn. Bordeaux 99:5 (1961), says that Durieu's herbarium is in Bordeaux. I was unable to examine these specimens. It seems to me that a great number of Durieu's specimens are in Paris. I have designated as lectotype a specimen in P (Fig. 67) annotated by Durieu "Avena clauda DR/Mascara 16 Mai 44." The paralectotype of FI bears two interesting labels in Durieu's handwriting: (i) "Avena/sp. n. Durieu" and (ii) "Avena clauda DR/In collibus prope Oran/majo 1842," and a citation of the other syntypes cited in the protolog such as Aucher 2929 and Kotschy 59. The K herbarium has another paralectotype annotated by Durieu; it has a letter written to Gay by Durieu pinned to it. The FI herbarium has the two paralectotypes (Aucher 2929 and Kotschy 59) pinned on the same sheet. These paralectotypes are found also in K, P, HAL, and G. In addition Paris has two sheets collected by Durieu but not labeled in his writing, one from Oran and the other from Mascara; the dates on which they were collected correspond almost exactly with those of the lectotype and the paratype from FI respectively. They may be considered as paralectotypes also.

[2] I have examined the holotype in P and an isotype in K, and it is Balansa No. 554. The specimen is also annotated by Cosson.

[3] Subv. *leiantha* seems to be the typical subvariety, in the sense of the Rules, according to the protolog. It is, therefore, equivalent to var. *clauda* subv. *clauda*. Because No. 514 has been described as a new variety, I chose it as lectotype (WIR).

(3) A. ERIANTHA DUR.

Homotypic synonym

A. eriantha Dur., Rev. Bot. Duchartre 1:360 (1845–1846) [1]

Heterotypic synonyms

Trisetum pilosum Roem. et Schult., Syst. Veg. 2:662 (1817) [2]
A. pilosa Marsch.-Bieb. ex Roem. et Schult., loc. cit. pro syn. [2]
A. pilosa (Roem. et Schult.) Marsch.-Bieb., Fl. Taur. Cauc. 3:84 (1819) nom. illegit. [2]

A. clauda

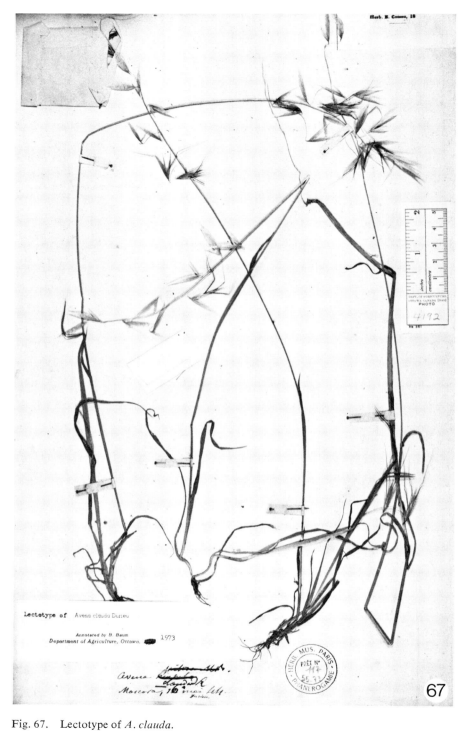

Fig. 67. Lectotype of *A. clauda*.

A. pilosa var. glabriflora Grun., Bull. Soc. Nat. Moscou 40:459 (1867) [3]
A. pilosa var. pubiflora Grun., loc. cit. [4]
A. clauda var. solida Hausskn., Mitt. Thür. Bot. Ver. N.F. 6:43 (1894) [5]
A. pilosa subv. glabriflora (Grun.) Malz., Monogr. 236 (1930)
A. pilosa subv. pubiflora (Grun.) Malz., op. cit. 235

Description

Annual plants. Juvenile growth prostrate. Color appearance glaucous. Culm erect. Relatively low plants, 20–80 cm high. Ligules obtuse. Panicle slightly flagged. Spikelets (Fig. 68) short, 18–21 mm long without the awns; each spikelet has 2–3 florets, the third normally lacking an awn. Glumes unequal in length, the lower ½–⅔ shorter than the upper; the lower has 3–5 nerves, the upper has 7 nerves and is 20–25 mm long; only the lowermost floret disarticulating at maturity; scars linear (Fig. 70). Periphery ring only confined to ⅛ of scar. Awn inserted at about middle of lemma. Lemma structure tough; lemma tips bisubulate (Fig. 69) and densely beset with macrohairs or sometimes glabrous; lemmas without macrohairs below the awn insertion. Paleas with 1–2 rows of cilia along the edges of the keels; palea back beset with hairs. Lodicules fatua type (Fig. 71), never bearing prickles. Epiblast sativa type (Fig. 72), 0.3 mm wide. Chromosome number $2n=14$. Genome $C_p C_p$.

Distribution

Native to (Fig. 73) Algeria, Bulgaria, Greece, Iran, Iraq, Israel, Lebanon, Morocco, Syria, Turkey, USSR (Azerbaidzhan and Turkmen).

Phenology

Flowers from March to July, although mostly from April to May.

Habitat

It seems to prevail (Fig. 74) in the *Quercus brantii* climax areas, but is found in general in places protected from grazing, on hills, calcareous slopes, and sandy areas or in steppes such as *Rhamnus-Artemisia*. It is also found in semi-steppe habitats dominated by *Crotalaria-Paliurus*, together with *A. sterilis* and *A. barbata*. It is found in maquis-type habitats and nowadays is in the process of recolonizing these habitats as overgrazing becomes decreasingly common.

Chorotype

Irano-Anatolian East-Mediterranean and in the Atlas area of the West Mediterranean subregion.

A. eriantha

Figs. 68–72. Morphological and micromorphological diagnostic details of *A. eriantha*. 68. Spikelets: left, without the glumes; right, including glumes (× 2.5). 69. Bisubulate lemma tips (× 40). 70. Linear scar (× 80). 71. Fatua-type lodicule (× 120). 72. Sativa-type epiblast (× 120).

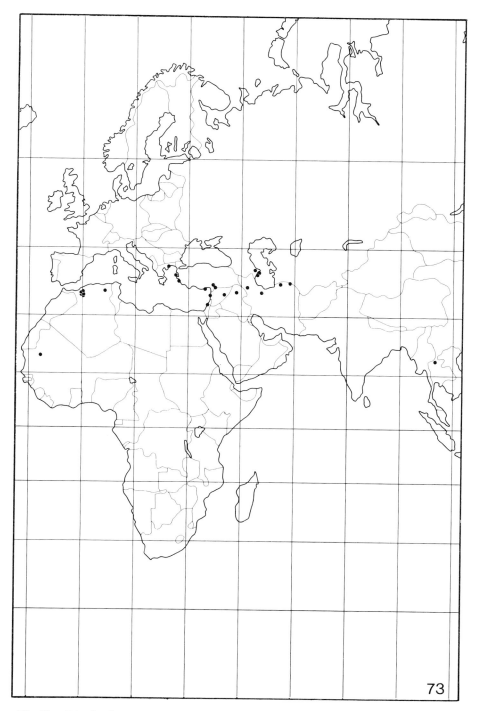

Fig. 73. Distribution map of *A. eriantha*. Mercator projection.

A. eriantha

Fig. 74. A site of *A. eriantha*, near Qasr-Shirin, Iran, edge of *Quercetum brantii*.

Mode of dispersal

The diaspore is the spikelet without the glumes; it is strongly pointed at the base and therefore suitable for dispersal by grazing animals. It is probably also, to a limited extent, transferred by wind and by ants. It seems to be locally restricted because of grazing (Figs. 1, 2).

Selected specimens

ALGERIA, Mostagamen coteaux incultes, 15 April 1851, *Balansa No. 88* (C, G, and K); BULGARIA, Monte Um-Dagh, 30 April 1883, *Sintenis No. 67* (BM, BR, and LE); IRAN, Mt. Elburs Keredj prope Khur et Pasha, 3 June 1937, *Rechinger No. 715A* (W); IRAQ, Mesopotamia Shayet, 7 March 1919, *Calder No. 1952* (BM); SYRIA? Djebel Tabrent, *Joly* (MPU); TURKEY, Maros Pazorcik Narli-Karabiyikl, 11 May 1957, *Davis No. 27803* (BM); USSR *Azerbaidzhan*, Baku hort. bot., 15 May 1955, *Karjagin No. 4354* (C, CAN, LE, M, S, and W); USSR *Turkmen,* Ashkabad W. of Newtonowa, 19 April 1897, *Litvinov No. 2206* (LE and W).

Identification hints and remarks

This is the only species in the genus *Avena* that has both unequal glumes and only the lowermost floret disarticulating at maturity. It cannot be conclusively identified this way because *A. clauda* $\times$ *eriantha* F_1 hybrids have the same combination. The F_1 hybrids or F_1-like phenotypes are distinguished by their bisetulate-biaristulate lemma tips, whereas in this species individuals have bisubulate lemma tips (Fig. 69).

Notes

[1] The protolog mentions no specific locality except that it refers to the same as that of *A. clauda* (i.e., "Oran, Mascara, etc . . ."). I have found three pertinent specimens in herb. P. Of these I designate as lectotype (Fig. 75) the specimen labeled in Durieu's handwriting "Avena eriantha DR./Cult. de graines/d. Mascara juillet 1845." One paralectotype "Oran, plateau du Djebel Santo/21 avril 1842," also collected by Durieu, is an *A. clauda* × *eriantha* F_1 hybrid.

[2] *Trisetum pilosum* and *A. pilosa* are based on the same type. The LE herbarium has a specimen labeled "Avena pilosa Iberia/D. Steven misit"; it is a very immature specimen, but it is obviously identical with *A. eriantha*. I regard this specimen as holotype. There is another specimen of great importance; it is labeled by three different people: "Avena strigosa/Iberia Stev./1812" by Fischer probably; "nova species/A. pilosa M." by Bieberstein; and "Trisetum" by Schultes. Two different species are affixed on this isotype sheet; one is *A. eriantha* and the other is *A. barbata*. In the Trinius herbarium, also at LE, there is a fragment of the holotype. The P herbarium has a specimen labeled "Avena pilosa MB ex Iberia Fischer"; it is *A. barbata* and should also be regarded as isotype. Obviously the type collection is made of discordant elements, and fortunately there is no argument about the holotype. Since the protolog states "calycinis glumis inequalibus" there should not be any confusion (see Note 3 below). The name *A. pilosa* is illegitimate because of the earlier *A. pilosa* Scopoli (1772).

[3] This is equivalent to *A. pilosa* var. *pilosa* as implied from Gruner (loc. cit.) who based his var. *glabriflora* on the only specimen that he found and examined in M.B.'s herbarium.

[4] The holotype of this variety is conserved in BM. The label reads "Pyralagai 24 Apr. 64" (i.e., 1864) and coincides with the protolog. Its identity is also *A. eriantha*.

[5] Haussknecht regarded *A. pilosa* M.B. as a variety of *A. clauda* and gave this new name at the varietal rank.

A. eriantha

Fig. 75. Lectotype of *A. eriantha*.

(4) A. VENTRICOSA BAL. EX COSS.

Homotypic synonyms

A. ventricosa Bal., Plantes d'Algérie 1852 Exsicc. No. 557 (1852) nom. nud. [1]
A. ventricosa Bal. ex Coss. Bull. Soc. Bot. France 1:14 (1854) [2]
A. ventricosa ssp. ventricosa (Bal. ex Coss.) Malz. Monogr. 241 (1930)

Heterotypic synonyms

A. bruhnsiana Gruner, Bull. Soc. Nat. Moscou 41:458 (1867) [3]
A. ventricosa ssp. bruhnsiana (Gruner) Malz., op. cit. 242
A. beguinotiana Pamp., Arch. Bot. Forli 12:18 (1936) [4]

Description

Annual plants. Juvenile growth semi-prostrate to erect. Color appearance glaucous. Culms erect. Relatively low plants, 25–70 cm high. Ligules obtuse with a point. Panicle flagged or nearly so. Spikelets (Fig. 76) short, 1.7–2.4 cm long without the awns; each spikelet has 2 florets. Glumes slightly unequal, the lowest ⅛–¼ shorter than the upper, each has 7–9 nerves and is 25–40 mm long; only the lowermost floret is disarticulating at maturity; scars linear (Fig. 78). Periphery ring only confined to ⅛ of scar. Awns inserted between upper ⅓–½ of lemma, and often the column is not diverging from the back of the lemma but continues on the same plane. Lemma structure tough; lemma tips (Fig. 77) bisubulate with two nerves ending in each lobe; lemmas without macrohairs below the insertion of the awn, but the tips often densely covered with hairs. Paleas with 1–2 rows of cilia along the edges of the keels. Lodicules fatua type (Fig. 79), very rarely bearing a few prickles. Epiblast sativa type (Fig. 80), 0.35 mm wide. Chromosome number $2n=14$. Genome C_vC_v.

Distribution

Native to (Fig. 81) Algeria, Cyprus, Iraq, Libya, Saudi Arabia, and USSR (Azerbaidzhan).

Phenology

Flowers from March to May.

Habitat

It is found (Fig. 82) on rocky plateaus, in interior desert sands, in disturbed grasslands, at the edge of forests, and in shallow calcareous soils. It often forms tiny colonies mixed with *A. clauda, A. eriantha, A. barbata,* and *A. sterilis*.

A. ventricosa

Figs. 76–80. Morphological and micromorphological diagnostic details of *A. ventricosa*. 76. Spikelet: left, including glumes; and right, without the glumes; note the slightly unequal glumes (× 2). 77. Bisubulate lemma tips (× 30). 78. Linear scar (× 40). 79. A pair of fatua-type lodicules (× 50). 80. Sativa-type epiblast (× 120).

A. ventricosa

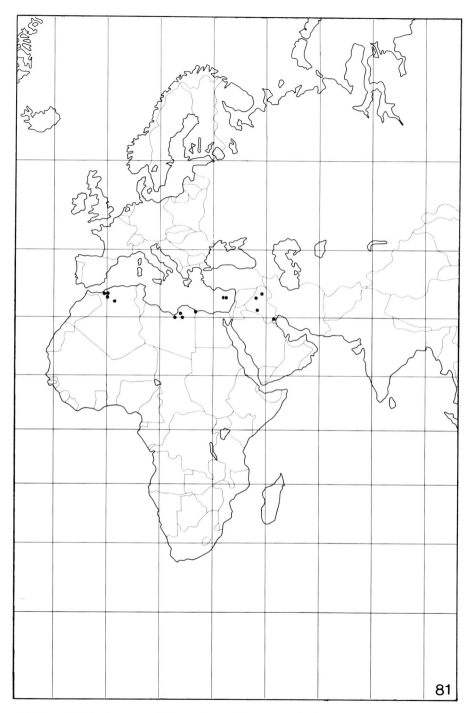

Fig. 81. Distribution map of *A. ventricosa*. Mercator projection.

Fig. 82. A site of *A. ventricosa*, 3 km east of Oran, Algeria.

Chorotype

Today it is primarily a species of the South Mediterranean, but it appears to have a relict distribution and is thus found in areas of South Mediterranean, Saudi Arabian, and the Irano-Turanian regions.

Mode of dispersal

The diaspore is the spikelet without the glumes. It has a pointed base and is usually hairy at the tip only. The spikelet falls down at maturity; it may be slightly dispersed by wind or by ants or, to longer distances, by grazing animals. The species, however, is partly hindered by overgrazing and by its presumably limited ecological tolerance.

Selected specimens

ALGERIA, Oran Batterie Espagnole champs, 5 April 1852, *Balansa No. 557* (FI, G, K, P, and W); CYPRUS, near Dhiorios 900 ft edge of pine forest, 25 March 1962, *Meikle No. 2340* (K); IRAQ, Mesopotamia Sharyat, 7 March 1919, *Calder No. 1952* (BM); LIBYA, Cirenaica Agheila steppa, 14 March 1833, *Pampanini No. 360* (FI); SAUDI ARABIA, NE Jabal Al-Amudah, 22 March 1968, *Mandaville No. 1737* (BM).

Identification hints and remarks

The appearance of the spikelet of this species (Fig. 76) is unique in the genus. Furthermore, the species cannot be confused with any other species in the genus *Avena* because of the following combination of characters, which can easily be depicted: linear scar (Fig. 78), only the lowermost floret disarticulating, and glumes slightly unequal. When the glumes are more unequal than usual, specimens of this species may be confused with *A. eriantha*. A close look at the glumes of *A. eriantha,* which has also linear scars, will show that the lower glume is ½–⅔ shorter than the upper. For practical purposes the glumes of *A. ventricosa* are regarded as being equal in the various keys to species.

Notes

[1] This is the first effective publication of *A. ventricosa,* but it is not valid according to the Rules because of lack of description. Cosson validated *A. ventricosa* two years later (see Note 2 below).

[2] Cosson quoted Balansa No. 557 in the protolog, and this specimen also could be regarded as the holotype. I have examined this specimen in P. But the label on this specimen quotes Balansa No. 89, and in addition there are four specimens No. 557 in P. Furthermore, I found in B a specimen annotated by Durieu: "Avena ventricosa spec. nov. ined./ Oran. Coteaux arides 19 avril 1842," and Cosson (op. cit. p. 11) stated "Les études auxquelles M. Durieu et moi nous nous sommes livrés . . ." meaning that Durieu participated with Cosson. I chose a specimen No. 557 in P as lectotype and regard the 3 others as isolectotypes (Fig. 83). Consequently Balansa No. 89 is a paralectotype and so is also the Durieu specimen in B. I also found isolectotypes at FI, G, K, and W, and an isoparalectotype at FI.

[3] I found the holotype in BM and it belongs here. The label reads: "Pyralagai 24 Apr. 1864" and coincides with the information in the protolog. According to Lanjouw and Stafleu (1954: 102) Bruhns' specimens are at BM.

[4] I have seen a number of syntypes in FI. I selected "No. 361, Libia-Cirenaica, Amseat a sud di Bardia, 24.3.1933, R.Pampanini" as lectotype. The paralectotypes were collected nearby on different dates by the same collector; they are Nos. 360, 362, 363, and 364, and they belong here.

A. ventricosa

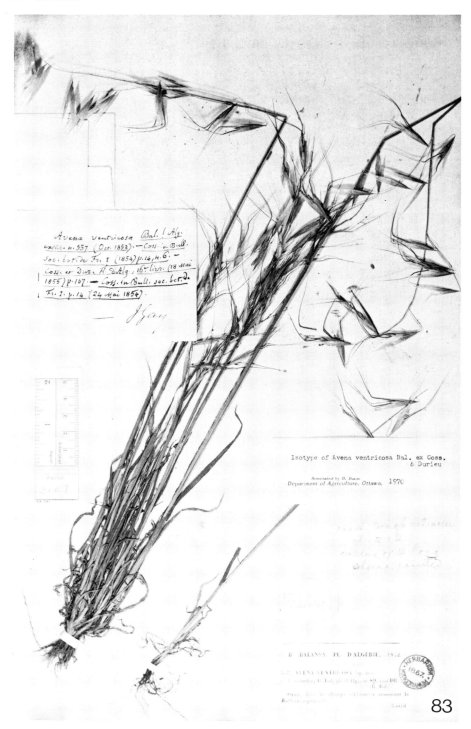

Fig. 83. Isolectotype of *A. ventricosa*.

(5) A. CLAUDA × ERIANTHA F₁ HYBRID

Synonyms

A. eriantha var. acuminata Coss et Dur., Bull. Soc. Bot. France 1:14 (1854) [1]

This species is similar to *A. eriantha* but the spikelets are narrower and more elongate (Fig. 84) because the tips of lemmas are biaristulate-bisetulate as in *A. clauda* but the setae and aristulae are not as long as in the latter species.

Distribution

Native to (Fig. 85) Algeria, Syria, Turkey, and USSR, but to be expected anywhere *A. clauda* and *A. eriantha* overlap in their distribution.

Phenology

Flowers from April to May, and rarely in August.

Fig. 84. Morphological diagnostic details o *A. clauda* × *eriantha* F₁ hybrid. Spikele without the glumes; note the similarity wit the spikelet of *A. eriantha* but the elongate shape (× 3.5).

A. clauda × eriantha F₁ hybrid

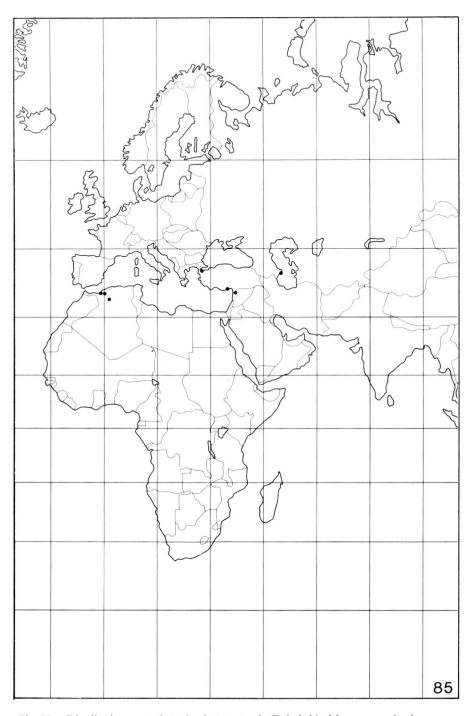

Fig. 85. Distribution map of *A. clauda* × *eriantha* F$_1$ hybrids. Mercator projection.

A. clauda × eriantha F₁ hybrid

Habitat

It was found in calcareous slopes, and even in a botanical garden where probably the two parents were grown together.

Mode of dispersal

Diaspores similar to those of *A. eriantha* (i.e., the spikelet without the glumes falls as a dissemule).

Selected specimens

ALGERIA, Oran Le Grand Lac de la Senia, 10 April 1852, *Balansa* (G); SYRIA, Aleppum Dschebel Nahas, No. 14337 (HAL); TURKEY, Tekir, May 1893, *Velenovsky* (PRC); USSR *Azerbaidzhan,* Apsheron Bot. Gardens, 6 May 1968, *Musayev* (LE).

Notes

[1] This was also published in Coss. et Dur., Expl. Sci. Alger. 2:109 (1854-1855). There, the syntypes are listed. Furthermore, Balansa published the name only in his exsiccata No. 556. In P, I found much material of the type collection. I have designated Balansa No. 88 as the lectotype. P has also an isolectotype, a paralectotype: Balansa No. 556 and an isoparalectotype, a paralectotype: Balansa No. 8, and a paralectotype collected by Durieu in Mascara, 12.5.1844. All the P material is also annotated by Cosson. In C, W, and FI I found isoparalectotypes Balansa No. 8; in W and FI I found the isoparalectotypes Balansa No. 556; and in G and K I found isolectotypes. All these specimens belong here.

18. Section 3. Agraria Baum, Can. J. Bot. 52:2259 (1975)

Basis for name I have named this section after *A. agraria,* which is now a synonym of *A. hispanica.*

Holotype species *A. hispanica*

Species included *A. brevis, A. hispanica, A. nuda, A. strigosa.*

Naturally occurring hybrids I have not found spontaneous hybrids among the material examined. Artificial hybrids were reported in the literature, and I have obtained some produced by the Welsh Plant Breeding Station (i.e., *A. brevis × strigosa*). If hybrids occur spontaneously, they must be rare; among these the most likely to expect are *A. hispanica* × *strigosa* in the southeast of France and the vicinity in the Iberian peninsula where the two species overlap in their distribution.

Notes on the taxonomy and distribution

All the species included in this section have a cultivated base; thus they were used as cereal crops to a greater degree than today. These species share the A genome, which is shared also by members of Section *Tenuicarpa*. The distribution area of the section (Fig. 86) is essentially Atlantic Europe and the Canary Islands, between 25° and 65° latitude North. Phenetic similarity based on all characters (see Chapter 7 and Fig. 10) suggests relationships with Section *Ventricosa* and with Section *Tenuicarpa*.

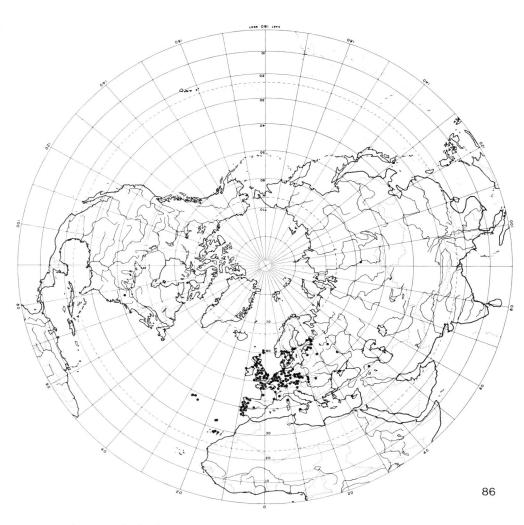

Fig. 86. Distribution map of section *Agraria*. Polar projection.

(6) A. BREVIS ROTH

Homotypic synonyms

A. brevis Roth, Bot. Abh. Beob. 42 (1787) [1]
A. sativa var. brevis (Roth) Kcke. in Kcke. et Werner, Handb. Getreidb. 1:213 (1885)
A. sativa var. brevis (Roth) Fiori et Paoletti, Icon. Fl. Ital. 1:29 (1895)
A. strigosa var. brevis (Roth) Hausskn., Mitt. Thür. Bot. Ver. N.F. 6:45 (1894)
A. strigosa ssp. brevis (Roth) Husnot, Gram. 2:38 (1897)
A. sativa ssp. brevis (Roth) Aschers. et Graeb., Synopsis 2:237 (1899)
A. strigosa ssp. strigosa prol. brevis (Roth) Thell., Vierteljahrs. Nat. Ges. Zürich 56:332 (1911)
A sativa ssp. sativa var. brevis (Roth) Fiori, Nuga. Fl. Anal. Ital. 1:105 (1923)
A. strigosa ssp. brevis (Roth) Hayek, Repert. Sp. Nov. Fedde Beih. 30:321 (1933)

Heterotypic synonyms

A. uniflora Parl., Pl. Nov. 84 (1842) [2]
A. brevis var. uniflora (Parl.) Drouet, Mém. Soc. Acad. Dept. Aube (Troyes) 30:206 (1866)
A. mandoniana Coss. et Bal., Bull. Soc. Bot. France 15:185 (1868) nom. nud. [3]
A. strigosa var. abbreviata Hausskn., Mitt. Thür. Bot. Ver. N.F. 6:44 (1894) [4]
A. strigosa ssp. strigosa prol. brevis var. glabrata subv. turgida Vav. in Malz., Monogr. 265 (1930) [5]
A. strigosa ssp. strigosa prol. brevis var. semiglabra Malz., op. cit. 265 [6]
A. strigosa ssp. strigosa prol. brevis var. trichophora Malz., op. cit. 264 [7]
A. strigosa ssp. strigosa var. glabrescens subv. uniflora (Parl.) Malz., op. cit. 262
A. strigosa ssp. strigosa prol. brevis var. candida Mordv. in Wulff, Kult. Fl. SSSR 2:429 (1936) [8]
A. strigosa ssp. strigosa prol. brevis var. candida subv. euuniflora Mordv. in Wulff, op. cit. 430 [9]
A. strigosa ssp. strigosa prol. brevis var. candida subv. turgida (Vav). Mordv. in Wulff, op. cit. 429
A. strigosa ssp. brevis var. glabrata subv. turgida (Vav). Tab. Mor., Bol. Soc. Brot. 2, 12:244 (1937)
A. strigosa ssp. brevis var. semiglabra (Malz.) Tab. Mor., loc. cit.
A. strigosa ssp. brevis var. trichophora (Malz.) Tab. Mor., loc. cit.
A. strigosa ssp. mandoniana Tab. Mor., op. cit. 245 [10]
A. strigosa ssp. mandoniana subv. açoreana Tab. Mor., loc. cit. [11]

Description

Annual plants. Juvenile growth erect. Color appearance glaucous. Culm erect. Relatively low plants, 40–70 cm high. Ligules obtuse. Panicles slightly flagged. Spikelets short, 1–1.5 cm long without the awn; each spikelet (Fig. 87) has 2–3 florets or sometimes only one floret. Glumes are nearly equal in length, with 7–8 nerves, and are 10–16 mm long; the florets are non-disarticulating at maturity. The awn is inserted just below the tip (Fig. 88) to ¼ below the tip of the lemma (Fig. 89); the awn column is not diverging but continues on the same plane of the lemma. Lemma structure tough; lemma tips (Figs. 90–92) shortly biaristulate-bisetulate to quadrimucronate, or shortly biaristulate-bilobed (Fig. 90); the lemma has no macrohairs below the awn insertion and only a few hairs grow sometimes around the awn insertion (Fig. 89). Palea with 1 row of cilia along edges of keels, and only near the tip of the palea, sometimes there are 2 rows of cilia; the palea back is beset with prickles. Lodicules essentially fatua type (Fig. 93), but they are small (about 0.5 mm long), short, more or less triangular, and mostly bent (Fig. 93, especially the one on the right) or occasionally straight (Fig. 94), and never bearing prickles. Epiblast brevis type (Figs. 95, 96), 0.3–0.4 mm wide. Chromosome number $2n=14$. Genome AA.

Distribution

Native to (Fig. 97) Austria, Azores, Belgium, Canary Islands, Czechoslovakia, France, Germany, Hungary, Italy, Luxemburg, Madeira, Netherlands, Poland, Portugal, Spain, Sweden, United Kingdom, USA, USSR, and Yugoslavia.

Phenology

Flowers from May to August.

Habitat

It was once cultivated in western Europe; now it is only rarely cultivated as a grain crop, but it seems to do well in poor upland soils. It is a weed of cultivated fields in the Azores, the Canaries, Madeira, and Portugal.

Figs. 87–96. Morphological and micromorphological diagnostic details of *A. brevis*. 87. Spikelet (× 5). 88. Spikelet without the glumes showing the awn insertion just below the tip of the lemma (× 8). 89. Spikelet without the glumes showing the awn insertion at ¼ below tip of lemma, and showing also a few hairs around the awn insertion (× 6). 90. Half of lemma tip showing an aristula in close proximity to the awn, and a setula that is broader and longer than the aristula and is lobe shaped (× 70). 91. Lemma tip quadrimucronate (× 50). 92. Shortly biaristulate-bisetulate tip of lemma (× 50). 93. Pair of lodicules essentially fatua type, but with a peculiar configuration and size found only in this species (× 50). See Figs. 94–96 overleaf.

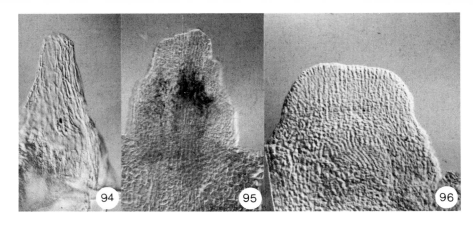

Figs. 94–96. 94. Lodicule as in Fig. 93 but with a straight configuration (× 120). 95–96. Brevis-type epiblasts (× 120).

Fig. 97. Distribution map of *A. brevis*. Mercator projection.

Chorotype

Macaronesian, West North Mediterranean and has been introduced to Eurosiberian subregion of the Euro-Siberian region.

Mode of dispersal

Chiefly by man through propagation of seeds. The whole panicle is a man-made diaspore.

Selected specimens

AUSTRIA, Oesterreich Kultiviert, *Hohenacker* (COI); AZORES ISLANDS, Ilha de S. Miguel Lagoa, July 1903, *Carreiro No. 1003* (COI); BELGIUM, Gembloux, *Scheidweiler* (BR); CANARY ISLANDS, Teneriffa in campis Tegueste, 7 July 1845, *Bourgeau No. 1035* (K); CZECHOSLOVAKIA, Tabor Bohemia Sloupnice, 3 August 1907, *Fleischer* (PRC); FRANCE, Cote d'Or Flavigny sur Ozerain, 11 August 1925, *Desplantes No. 5595* (MT); GERMANY, Hannover Bassum Agris culta, July 1875, *Beckmann No. 73-67-31* (K); HUNGARY, in Frumdorf, 1822, *Schmidely* (G); ITALY, Palermo, August 1864, *LeJolis* (COI); LUXEMBURG, *De Cloet* (BR); MADEIRA ISLAND, Prope S. Roque inter segetes 3 July 1865, *Mandon No. 271* (B, BM, BR, C, G, K, LE, P, PRC, S, and W); NETHERLANDS, Hort. Musae Utrecht (BR); POLAND, Kamienna Gora, *Romer No. 6810* (M); PORTUGAL, Viesen in crops *T. turgidum*, 6 July 1927, *Vavilov No. 1099* (WIR); SPAIN, sponte occurit ex horto, *Schreber* (W); SWEDEN, Skane Malmo, 22 July 1920, *Holmberg* (COI); *United Kingdom,* Aberystwyth Welsh Plant Breeding Station, August 1929, No. *CN 937-4* (K); USA, Pennsylvania Jardin Gaudy, *Moricand* (G); USSR grown at Voronezh Kamennaya Experiment Station, August 1927, *Malzew* (WIR); YUGOSLAVIA, Laibach, *No. 169* (WIR).

Identification hints and remarks

This species has often been confused with *A. strigosa*. Although Malzew (1930) did not regard it as a species, he recognized it as a taxon of *A. strigosa* (i.e., "proles brevis"). Malzew distinguished "proles brevis" from *A. strigosa* by its shorter spikelets and its approximately 12-mm long glumes. Many authors followed Malzew but regarded "proles brevis" as species and in addition to Malzew's characters they attributed to *A. brevis* shorter lemma tips as a characteristic. The descriptions and the keys, in this work, show that at least three taxa, which I called species, can be recognized among Malzew's *A. strigosa*: *A. brevis*, *A. hispanica* and *A. strigosa*. The configurations of the lemma tips (Figs. 90–92) and of the lodicules (Fig. 93) are unique to *A. brevis* and are not found in any other species in the genus. Furthermore, the cilia along the edges of the keels are found only near the tip of the palea; this is another feature unique to this species. These characters may provide markers for conclusive identification in cases of doubt.

Notes

[1] According to De Candolle (Phytographie p. 444 (1880)), Roth's herbarium was in Oldenburg. W. T. Stearn (British Museum) told me that the Oldenburg herbarium was transferred to B just before World War II and suffered, therefore, great damage. I was unable to find the holotype in B. I found, however, an isotype at LE in herb. Trinius, and it has the following label: "Avena brevis/mihi" written by Roth and "mis. auct. Roth" written by Trinius (Fig. 98). A specimen in W is closely associated in time with the type and may even be part of the type collection; this is an *A. brevis* sent to Wulfen by Schreber who knew Roth very well.

[2] The holotype of *A. uniflora* is in FI; it was collected by Webb.

[3] The "type" here is important because it is the basis for *A. strigosa* spp. *mandoniana* Tab. Mor. (see Note 10 below). Paris (P) has the original and 4 duplicate sheets; other duplicates are in LE (1 sheet), S (2 specimens), C (3 spec.), B (1 spec.), W (2 spec.), BM, K, G, PRC, and BR (1 sheet each).

[4] It is another name given by Haussknecht for *A. brevis* but at the variety level.

[5] The holotype and isotype are conserved in WIR and belong here.

[6] I have seen several syntypes and I designate No. 146 as lectotype; it is conserved in WIR.

[7] I was unable to find the type specimen; therefore, I designate Plate 27 (Fig. 1) as lectotype. The description and the identity of the figure belong here.

[8] Two subvarieties are described under this variety. I designate the type of subv. *turgida* (Vav.) Mordv. as the type of that variety and, therefore, equal to *A. brevis*. See also Note 9 below.

[9] I was unable to find the type of subvar. *eu. uniflora* but from the description, which may be designated as lectotype according to Article 9 Note 1, it also belongs here.

[10] I designate as lectotype the original specimen in P (see Note 3 above) because Taborda de Morais did not mention a specific sheet as his type; instead he cited the exsiccata only, and I was unable to find one in his collection in COI.

[11] The holotype is in COI and labeled by its author also "Especimen tipo."

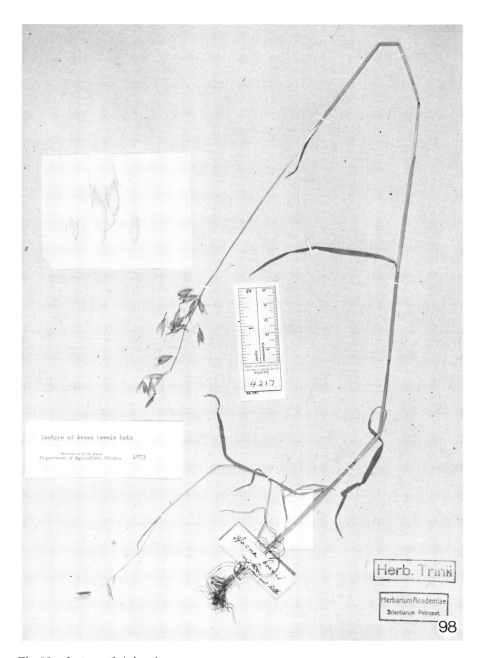

Fig. 98. Isotype of *A. brevis*.

(7) A. HISPANICA ARD.

Homotypic synonyms

A. hispanica Ard. ex Saggi, Accad. Padov. 2:112 (1789) [1]
A. fusca Ard. ex Saggi, op. cit. Tab. 4 [2]

Heterotypic synonyms

A. freita Orteg. ex Spreng., Bot. Gart. Halle 14 (1800) [3]
A. agraria Brot., Fl. Lusit. 1:105 (1804) [4]
A. agraria var. sesquialtera Brot., op. cit. 106 [5]
Danthonia strigosa var. elatior Roem. et Schult., Syst. 2:691 (1817) [6]
A. alta Cavan. ex Roem. et Schult. loc. cit. [7]
A. cavanillesii Hort. ex Roem. et Schult. loc. cit. [7]
A. hispanica Hort. ex Roem. et Schult. loc. cit. [7]
A. strigosa var. elatior Kunth, Rev. Gram. 1:103 (1829) [8]
A. sativa var. melanosperma Reichenb., Fl. Germ. 1:52 (1830) nom. illegit [9]
A. sativa var. brevistrigosa Kcke., Syst. Uebers. Cereal. 17 (1873) [10]
A. strigosa var. uniflora Hack., Oesterr. Bot. Zeitschr. 27:125 (1877) nom. nud. [11]
A. strigosa var. sesquialtera (Brot.) Hack., Cat. Gram. Portug. 19 (1880)
A. strigosa var. glabrata subv. secunda Mordv., Tr. Sb. Gen. Sel. Semen. 3:367 (1929) [12]
A. strigosa ssp. strigosa prol. brevis var. glabrata subv. secunda Mordv., loc. cit.
A. strigosa ssp. strigosa var. glabrescens subv. elatior (Roem. et Schult.) Malz., Monogr. 260 (1930)
A. strigosa ssp. strigosa var. glabrescens subv. sesquialtera (Brot.) Malz., loc. cit.
A. strigosa ssp. strigosa var. glabrescens subv. subbrevis Malz., op. cit. 261 [13]
A. strigosa var. agraria (Brot.) Sampaio, Bol. Soc. Brot. 2, 7:116 (1931)
A. strigosa ssp. brevis var. albobrevis Vasc., Ens. Sem. Melhor. Pl. Bol. No. 20, Ser. A, 39 (1935) [14]
A. strigosa ssp. brevis var. nigrescens Vasc., loc. cit. [15]
A. strigosa ssp. strigosa prol. brevis var. secunda (Mordv.) Mordv. in Wulff, Kult. Flora SSSR, 2:431 (1936)
A. strigosa ssp. strigosa var. kewensis Vav. ex Mordv. in Wulff, op. cit. 426 [16]
A. strigosa ssp. strigosa var. melanocarpa Mordv. in Wulff, op. cit. 428 [17]
A. strigosa ssp. strigosa var. typica subv. elatior f. divaricata Mordv. in Wulff, op. cit. 427 [19]
A. strigosa ssp. strigosa var. typica subv. sesquialtera (Brot.) Mordv. in Wulff, loc. cit.
A. strigosa ssp. strigosa prol. brevis var. nigricans Mordv. in Wulff, op. cit. 430 [18]

A. strigosa ssp. agraria (Brot.) Tab. Mor., Bol. Soc. Brot. 2, 12:240 (1937)
A. strigosa ssp. agraria subv. subbrevis (Malz.) Tab. Mor., op. cit. 241
A. strigosa ssp. agraria subv. subbrevis f. albula Tab. Mor., loc. cit. [20]
A. strigosa ssp. agraria subv. subbrevis f. obscura Tab. Mor., loc. cit. [21]
A. strigosa ssp. agraria var. totiglabra subv. sesquialtera (Brot.) Tab. Mor., loc. cit.
A. strigosa ssp. agraria subv. sesquialtera f. albobrevis (Vasc.) Tab. Mor., op. cit. 242
A. strigosa ssp. agraria subv. sesquialtera f. nigrescens (Vasc.) Tab. Mor., loc. cit.
A. strigosa ssp. strigosa var. glabrescens subv. unispermica Tab. Mor., op. cit. 239 [22]
A. strigosa ssp. strigosa var. glabrescens subv. unispermica f. lucida Tab. Mor., op. cit. 240 [23]
A. strigosa ssp. strigosa var. glabrescens subv. unispermica f. nigra Tab. Mor., loc. cit. [24]
A. strigosa ssp. agraria var. totiglabra Tab. Mor., Bol. Soc. Brot. 2, 13:640 (1939) [25]
A. strigosa ssp. agraria var. totiglabra subv. subbrevis (Malz.) Tab. Mor., loc. cit.
A. strigosa ssp. agraria var. totiglabra subv. subbrevis f. albula (Tab. Mor.) Tab. Mor., op. cit. 641
A. strigosa ssp. agraria var. totiglabra subv. subbrevis f. obscura (Tab. Mor.) Tab. Mor., loc. cit.
A. strigosa ssp. agraria var. totiglabra subv. sesquialtera (Brot.) Tab. Mor., loc. cit.
A. strigosa ssp. agraria var. totiglabra subv. sesquialtera f. albobrevis (Vasc.) Tab. Mor., op. cit. 642
A. strigosa ssp. agraria var. totiglabra subv. sesquialtera f. nigrescens (Vasc.) Tab. Mor., loc. cit.
A. strigosa ssp. agraria var. agrarisubpilosa Tab. Mor., op. cit. 640 [26]
A. strigosa spp. strigosa var. glabrescens subv. unispermica f. nigella Tab. Mor., op. cit. 637 [27]

Description

Annual plants. Juvenile growth prostrate to erect. Color appearance glaucous. Culms erect. Relatively tall plants, 70–110 cm high. Ligules obtuse. Panicle equilateral. Spikelets short, 1.3–2.4 cm long without the back awns, variable in shape (Figs. 99, 100) and each with two florets. Glumes nearly equal, 12–20 mm long; all the florets are non-disarticulating at maturity. Awns inserted between upper $1/3$–$1/4$ of lemma, and rarely close to middle of lemma. Lemma structure tough; the lemma tips are bisetulate-biaristulate (Figs. 103, 104), but often the setulae are in close proximity to the aristulae, and sometimes the setulae have a very faint nerve or are membranous without a nerve; a few macrohairs may be present around the insertion of the awn

Figs. 99–107. Morphological and micromorphological diagnostic details of *A. hispanica*. 99–100. Two types of spikelets found in this species; the kind in Fig. 99 (× 3) resembles that of *A. strigosa* to some extent, and the other and more common type in Fig. 100 (× 4) is shorter with plump seeds. 101–102. Same as Figs. 99 and 100 but without the glumes and slightly magnified; note the special structure of the plump type in Fig. 102 (× 6) which is unique in the genus, and note in Fig. 101 (× 8) the few macrohairs scattered around the awn insertion. 103. Close-up of the lemma tip of Fig. 102; note the setula with the nerve protruding into it (× 16). 104. Close-up of the lemma tip of Fig. 101; note here also the setula (× 12). 105. Fatua-type lodicules; note the obtuse apex and the number of hydathodes (× 50). 106. Detail of the lower part of the right-hand lodicule of Fig. 105 (× 120). 107. Brevis-type epiblast (× 120).

A. hispanica

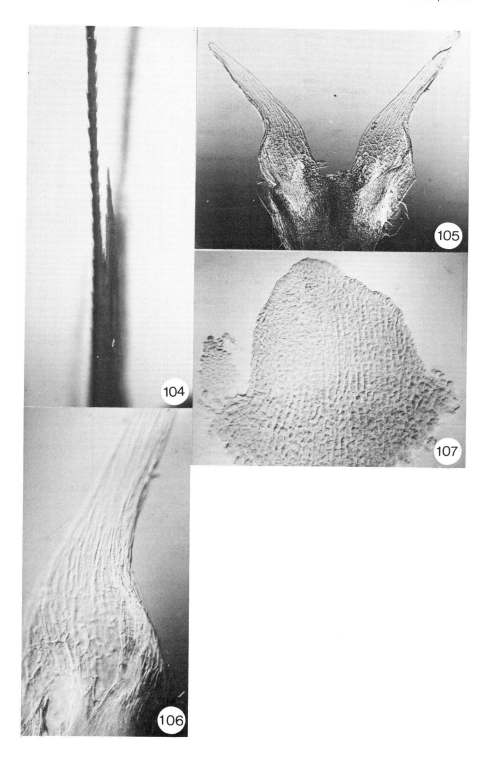

(Fig. 101), but they are always absent below the awn insertion (Figs. 101, 102). Palea with one row of cilia along edges of keels, but only on the upper ½ of the palea or sometimes on the upper ⅙ only whereas the lower parts of the palea do not have cilia along the edges of the keels; palea back beset with prickles or short verrucae. Lodicules fatua type (Figs. 105, 106) and often the body is similar to that of the strigosa type (i.e., with many hydathodes and with an obtuse apex); the lodicule very rarely bears one prickle. Epiblast brevis type (Fig. 107), approximately 0.4 mm wide. Chromosome number $2n = 14$. Genome presumably AA.

Distribution

Native to (Fig. 108) Portugal, Atlantic Spain, and Atlantic France. Introduced also in Belgium, Brazil, Czechoslovakia, Estonia, Germany, Ireland, and the United Kingdom.

Phenology

Flowers from March to September, but mostly from May to July.

Habitat

Cultivated until recently in northwest Spain and in Portugal, and often found as a weed in cereals. Cultivated to a very limited extent in other distribution areas. Thrives best on light or sandy soils.

Chorotype

Lusitanian part of the Mediterranean region and the southwest fringes of the western European part of the Euro-Siberian region; introduced into the European subregion of the Euro-Siberian region.

Mode of dispersal

Chiefly dispersed by man for cultivation or inadvertently through impurities with other cereal grains; very restricted though. The whole panicle in this species is a man-made diaspore.

Selected specimens

BELGIUM, Hainaut Obourg, July 1865, *Martinis No. 80* (CGE); BRAZIL, Avityba Parana ruderatis, 3 January 1916, *Dusen No. 17482* (G); CZECHOSLOVAKIA, Olomouc, July 1931, *Lans* (COI); ESTONIA, Near Dorpat Nemiko Experimental Farm, 1911, *Ratlef No. 366* (WIR); FRANCE, Champagney Champs d'Avoine, 25 July 1879, *Vendrely No. 369* (MT); GERMANY, Erlangen inter segetes, 1817, *Zuccarini* (M); PORTUGAL, Barcoico cultivada, June 1931, *Sousa No. 3447* (COI); SPAIN, Algarve Faro Hovel, *Witman No. 1845* (FI); UNITED KINGDOM, Scotland Inverness Nairn, 29 July 1898, *Marshall* (CGE).

A. hispanica

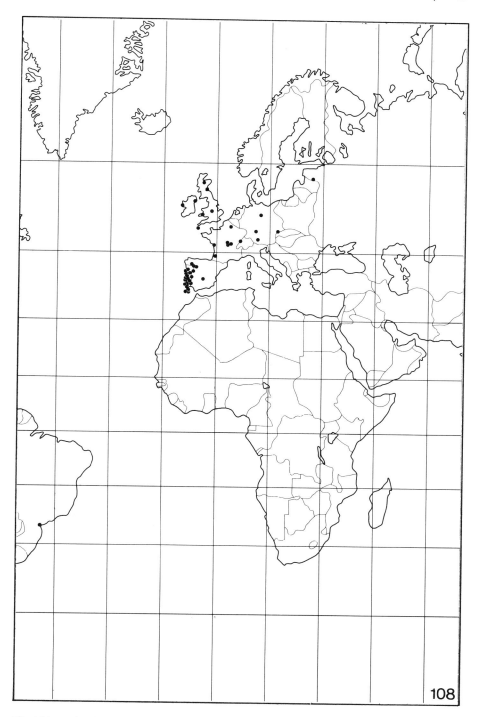

Fig. 108. Distribution map of *A. hispanica*. Mercator projection.

A. hispanica

Identification hints and remarks

This species can be confused with *A. strigosa* and *A. brevis*. The latter can be readily confirmed or eliminated by checking for micromorphologic characters that are unique to it (see Identification hints of *A. brevis*). Although there is great similarity between *A. strigosa* and *A. hispanica,* they can be conclusively identified on the basis of the lodicule, which is essentially fatua type in *A. hispanica* (Figs. 105, 106) and strigosa type in *A. strigosa*. Although the lodicule of this species is fatua type, it is different from the regular type in that the tip has many hydathodes and is obtuse as in the strigosa type. In addition to these differences, *A. hispanica* has awns inserted in the upper $1/3-1/4$ of the lemma, whereas in *A. strigosa* the same are inserted at about the lower $1/3$; although the point of insertion varies, it can be used for definite identification. Quite often, the special configuration of the tip of the lemma (Figs. 99, 100) is found among specimens.

Notes

[1] There is no doubt in my mind about the identity of the drawing and the protolog of *A. hispanica*. The description is remarkably detailed and clear. I was unsuccessful, however, in obtaining the type of this species; Dr. G. Barioli informed me that the types of Arduino were destroyed in World War II. I obtained from FI a dubious isotype with the annotation "Avena strigosa nigra/Hispania Witman." The other annotations I could not interpret, but the identity of the species is clear and it belongs here. According to Article 9 Note 1, it is permissible to recognize a description or figure as type. In this case both the description and the figure, which complement each other, should be regarded as type until the specimen type is recovered.

[2] The plate of *A. hispanica* has been named *A. fusca,* which is probably a slip. So *A. fusca* should be regarded as a superfluous name.

[3] The original collection of Sprengel is lost (Stafleu 1967:455) but I was able to recover pertinent specimens in HAL, LE, and C. The identity of the three specimens is clear and they belong here. The two specimens from HAL are annotated "Avena freita Ortegi" and "Avena Freita Ortegae" and both belonged to Herb. C. Schkuhr. Schkuhr was a contemporary of Sprengel so that these two specimens may be regarded as isotypes. The LE specimen was obtained from Schkuhr but the dates are somewhat contradictory; one date is 1790 and the other is "Januario 1801." This is also a probable isotype. The C specimen comes from Herb. Viborg and is annotated "Profess. Märtens 1799" on the back of the sheet, and on the front "A. freita Ortega." It is doubtful if the C specimen has anything to do with the type collection; it is important because it is contemporaneous with the other, also has the same identity, and perhaps may have been obtained from seeds sent by Ortega himself to various botanical gardens. Another pertinent specimen is in M and is annotated "A. freitas du Royaume de Galice" and was also seen by Schreber around 1800.

[4] According to Taborda de Morais (1937) the type was lost at LISU, but his careful interpretation of the protolog leaves no doubt about the identity of this taxon. This identification can also be reinforced by the distribution of the species. I select COI No. 3459 as neotype.

[5] I have the same remark as in Note 4 above. I select COI No. 3444 as neotype.

[6] This variety is based on five synonyms given in the protolog, and all belong here.

[7] See Note 6 above. I found one specimen in BR annotated "Avena alta Cavan" but cannot ascertain its identity with the type.

[8] Since it is based on *A. freita* Orteg., *A. agraria* Brot., and *A. alta* Cav., its identity is obvious. But, because this taxon is based on those synonyms which are also quoted by Roemer and Schultes, it may not be a coincidence, so it is quite probable that this is a new combination *A. strigosa* var. *elatior* (Roem. et Schult.) Kunth.

[9] It is illegitimate because it postdates *melanosperma* Sweet. It belongs here because *A. fusca* Ard. is cited. I was, however, unable to detect any related specimen.

[10] I was unable to find the type because it was lost during World War II. Because many authors misidentified *A. hispanica* as either *A. brevis* or *A. strigosa,* it is quite conceivable that this taxon belongs here because Koernicke stated: "uebergang von brevis zu strigosa."

[11] I examined the "type" in W. The label reads: "In arvis prope Bussaco, 14/5/1876" among others.

[12] I have selected No. 2394 from WIR as lectotype. This specimen bears also the No. 5226/1.

[13] I have examined all the syntypes at WIR and designated No. 1106 as lectotype; this specimen was collected in Lissabon (Portugal) by Vavilov on 2 July 1927. All the paralectotypes belong here.

[14] Of the three syntypes examined, I designate No. 7941 as lectotype. All belong here and are conserved in ELVAS.

[15] I have examined five syntypes from ELVAS; all belong here, and I chose No. 3864 as lectotype.

[16] I have designated No. 2380 as lectotype; the specimen is in WIR. According to Vavilov this name should replace the earlier *A. strigosa* ssp. *orcadensis* var. *flava* Marqu., which was split into so many varieties.

[17] I was unable to find any syntype material, but I think that it belongs here because of the distribution.

[18] I have selected No. 2395 from WIR as lectotype; it also bears No. 5257, and is labeled by Mordvinkina.

[19] I have selected No. 2452 from WIR as lectotype; the paralectotype No. 7432 also belongs here. Both types mention, in Mordvinkina's handwriting, "Pied de Mouche," which is also stated in the protolog including the figure.

[20] The protolog does not mention any specimen, but the 1939 publication lists some. I have chosen No. 3459 from COI as lectotype.

[21] Same as Note 20 above, but I chose No. 3438 of COI as lectotype.

[22] This is a new name for "A. strigosa b. sesquialtera Hackel, Catal, Gram. Portugal 19 (1880). non Brotero." No specimens are mentioned but Hackel's specimen is obviously implied, and it belongs here (see Note 11 above).

[23] No specimens are mentioned in this 1937 publication, but later in the 1939 publication (p. 638), 13 specimens are mentioned. I have selected No. 3447 as lectotype. All the paralectotypes also belong here.

[24] No specimens mentioned in this publication; later, in the 1939 publication, 9 are mentioned but under f. *nigella;* see Note 27 below for the lectotype.

[25] It is the typical var. of ssp. *agraria* because according to Taborda de Morais's method of working, var. α is the typical variety. Furthermore, it includes subv. *subbrevis,* which is designated as "Avena agraria Brot. sensu stricto" by the author. Thus this variety is equal to var. *agraria* according to the present Rules.

[26] I was unable to examine the holotype designated by its author, or the paratype. However, the picture (20 f. 3) and the geographic distribution confirm that it belongs here.

[27] Of the specimens cited, I was able to examine only two syntypes from COI. I designated No. 3451 as lectotype; this and the paralectotype No. 3446 belong here.

(8) A. NUDA L.

Homotypic synonyms

> A. nuda L., Demonstr. Pl. 3 (1753) [1]
> A. sativa ssp. nuda (L.) Gillet et Magne, Nouv. Fl. France ed. 3:532 (1873)
> A. sativa var. nuda (L.) Kcke. in Kcke. et Werner, Handb. Getreidb. 1:218 (1885)

A. strigosa var. nuda (L.) Hausskn., Mitt. Thür. Bot. Ver. N.F. 6:45 (1894)
A. sativa var. nuda (L.) Schmalh., Fl. Central and S. Russia 2:618 (1897) comb. illegit.
A. sativa ssp. nuda (L.) Aschers. et Graebn., Synopsis 2:237 (1899)
A. fatua ssp. nuda (L.) Thell., Vierteljahrs. Naturf. Ges. Zürich 56:328 (1912)
A. strigosa ssp. strigosa prol. nuda (L.) Malz., Monogr. 266 (1930)
A. strigosa ssp. strigosa var. nuda (L.) Tab. Mor., Bol. Soc. Brot. 2, 13:639 (1939)

Heterotypic synonyms

*A. nuda var. biflora Haller, Nov. Comm. Soc. Sci. Göttingen 6: Tab. 6 fig. 35 (1775) [2]
*A. nuda var. triflora Haller, op. cit. fig. 34 [2]
*A. albicans Lestib., Bot. Belg. 2:35 (1827) [3]
A sativa var. biaristata Alefeld, Landw. Fl. 322 (1866) [4]
A. sativa ssp. nuda var. biaristata (Alefeld) Aschers. et Graebn. op. cit. 238
A. nudibrevis Vavilov, Bull. Appl. Bot. Pl. Breed. 16:48, 176 (1926) [5]

Description

Annual plants. Juvenile growth erect. Color appearance glaucous. Culm erect. Relatively low plants, 60–80 cm high. Ligules obtuse. Panicle slightly flagged. Spikelets (Fig. 109) long, 20–28 mm without the awns; each spikelet has 3–6 florets. Glumes slightly unequal, the lower being ¼–⅕ shorter than the upper, and each has 7–9 nerves and is about 20–22 mm long; the florets are non-disarticulating at maturity. Lemmas awnless or awns inserted at about middle of lemmas. Lemma structure same as glumes; lemma tips biaristulate with a short and weak aristula (Fig. 110) to bisubulate (Fig. 111); lemmas without macrohairs below the insertion of the awn. Palea back beset with prickles. Lodicules strigosa type (Figs. 112, 114) or fatua type (Figs. 113, 115). Epiblast (Fig. 116) brevis type, 0.35 mm wide. Chromosome number $2n=14$. Genome AA.

Distribution

Native to (Fig. 117) Austria, Belgium, Czechoslovakia, Denmark, Germany, Greece, and the United Kingdom.

Phenology

Flowers from June to September, and rarely from April.

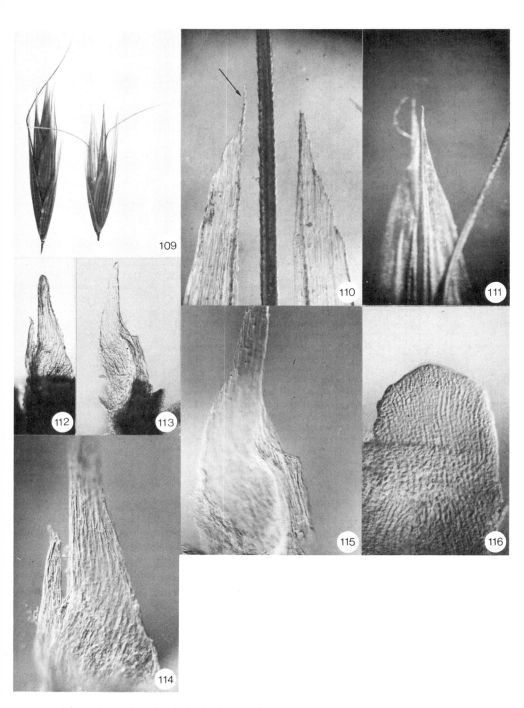

Figs. 109–116. Morphological and micromorphological diagnostic details of *A. nuda*. 109. Spikelets of two sizes; note the slightly unequal glumes and the varing number of florets, and that although the left spikelet is longer the glumes are essentially the same size as in the shorter one (× 2.5). 110. Biaristulate lemma tip; the aristula (arrow) is short (× 40). 111. Bisubulate lemma tip (× 40). 112. Strigosa-type lodicule (× 50). 113. Fatua-type lodicule (× 50). 114. Detail of the lodicule in Fig. 112 (× 120). 115. Detail of the side lobe of the lodicule in Fig. 113 (× 120). 116. Brevis-type epiblast (× 120).

Fig. 117. Distribution map of *A. nuda*. Mercator projection.

A. nuda

Habitat

Restrictively cultivated, and used to a limited extent in breeding programs. It was probably grown to a larger extent in the past.

Chorotype

European subregion of the Euro-Siberian region.

Mode of dispersal

Only by cultivation. The caryopses fall at maturity from the papery lemmas, and are in fact the diaspores. The plants are harvested before the seeds fall. The whole panicle in this species is a man-made diaspore.

Selected specimens

AUSTRIA, Neosstadt, (PRC); BELGIUM, Cappellen moissons, 2 September, 1881, Hennen (BR); CZECHOSLOVAKIA, Tabor Bohemia Sloupnice, 25 August 1909, *Fleischer* (PRC); DENMARK, Goteborg somentes vindas, 29 June 1946, *Rainha No. 223121* (COI); GERMANY, Bonn Poppelsdorf, 1889 *Scheppig No. 73-67-13* (K); UNITED KINGDOM, Cornwall Dicksons, *Babington No. 4* (CGE).

Identification hints and remarks

This species can occasionally be confused with the naked type of *A. sativa*. It certainly was confused with *A. sativa* until Vavilov's work (1926) and Malzew's monograph (1930). Morphologically, it can conclusively be distinguished from the naked type of *A. sativa* by inspection of the lodicules and epiblasts. In *A. nuda* the lodicules are strigosa type (Figs. 112, 114) or fatua type (Figs. 113, 115) and the epiblast is brevis type, whereas in *A. sativa* the lodicules are sativa type and the epiblasts are sativa or fatua type.

Except for the hexaploid naked oats, *A. nuda* is unique in the structure of the lemmas, which are similar to glumes. The existence of two kinds of lodicules is an unsolved problem; it may perhaps point to two different species but I have not sufficient evidence to confirm this. The same situation also exists in *A. longiglumis*.

Notes

[1] The typification of this name is not simple, but fortunately taxonomic identity is not a problem because all the material pertinent to the protolog belongs to one and the same species, often called by oat breeders and oat workers "small naked oat."

A. nuda

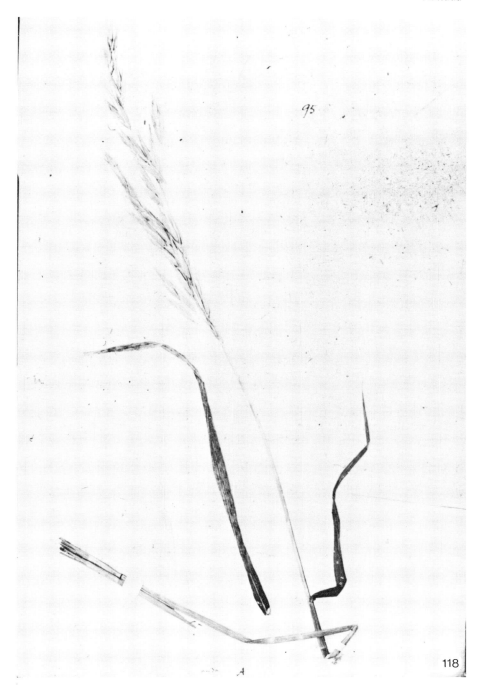

Fig. 118. Authentic specimen of *A. nuda*. This is not the type, but it is a Linnaean specimen related to the Systema Naturae 10th ed. (see text for details).

The protolog reads: "AVENA nuda similis A. sativae, a qua differt receptaculo flosculorum elongato et calycem excedente, floculis glabris ultra calycem prominentibus."

The Linnaean herbarium in London (LINN) has one specimen No. 95.8 (Savage 1945, p. 19) annotated by Linnaeus "nuda A" (Fig. 118). In Stockholm (S) I found several pertinent specimens. One is annotated similarly "nuda A" by Linnaeus, but what is most important is Dahl's handwriting on the back of the sheet "à Linné P." (see Lindman 1909, p. 6, for the significance of the statement). Although these two specimens were annotated by Linnaeus, they appear to be more closely linked in time and place with the publication of Systema Naturae ed. 10 than with the protolog. In Systema Naturae Linnaeus added all the newly discovered species since 1753 by inserting them between species arranged according to the sequence in the Species Plantarum. Instead of assigning a number to these new species, he assigned a letter: A, B, C ... (Baum, 1968). Thus *A. nuda* appears after "sativa 5" as "nuda A" and before "fatua 6" (Linnaeus 1759, p. 878). Obviously, these two specimens coincide with *Systema Naturae* and not with *Demonstrationes*. There is, however, a possibility that the sign "A" was added later by Linnaeus himself to the specimens he had examined earlier.

In addition to these specimens I found in Stockholm (S) a specimen annotated "A. nuda specimen ex horto Uppsal: communicavit Hortul. Nietzel." According to Lindman (1909, p. 12), Nietzel "was employed at Uppsala botanical garden from 1739 to 1756." Because this specimen is so closely linked with the protolog of *A. nuda* that was published in an account of the Uppsala Botanic Garden, I designate it as lectotype even though it does not bear Linnaeus' own handwriting. Its identity is not questionable; it is the diploid naked-type oat.

I was able to examine additional sheets from that time and probably from the same source: in Stockholm (S) by Casström and Swartz; in Stockholm (SBT) by Bergius; and in Uppsala (UPS) by Thunberg annotated "cult. in horto Upsaliense. Thunberg." These specimens have the same identity, and may be regarded as paralectotypes.

[2] I have placed these two varieties among the synonyms here on the basis of my judgment of the fit of the figures only, and on the text (p. 19 and 20). I was unable to find related authentic specimens in herb. P (although for other Haller types I was more successful).

[3] According to the description, it is very likely to belong here because Lestiboudois' description of *A. nuda* L. clearly applies to the hexaploid naked oat, whereas the protolog of *A. albicans* is very similar to the description of the real *A. nuda*. The protologue also compares this species and the former, so clearly Lestiboudois knew two kinds of naked oats. The types of Lestiboudois are unknown (Stafleu 1967:265).

[4] The types of Alefeld were lost during World War II. I have not been able to find specimens annotated by him in various herbaria. I designate the protolog as lectotype; the description is clear enough and it belongs

here according to my taxonomic concepts. Moreover, Alefeld quotes in the protolog "Nackter Hafer Metzger's," which is the real *A. nuda* L., and indeed in Metzger (1824:54) *A. nuda* L. is mentioned as the correct name for that taxon.

[5] Vavilov (loc. cit.) thought that *A. nuda* L. was the naked hexaploid oat and, therefore, he called the small naked oat (a diploid species) *A. nudibrevis* Vav. Later Mansfeld (1953, p. 480) seemed to have suspected that *A. nuda* L. was the diploid species, and placed *A. nudibrevis* in the synonymy of *A. nuda* L. Vavilov did not work according to conventional taxonomic practice and so he did not designate a type, nor did he conserve a type collection, although he used an extensive herbarium. Probably *A. nudibrevis* may be based on seed collections and plants grown from them for morphological, cytological and immunological research by Vavilov and his disciples, and type collections were never kept. According to Mordvinkina (personal communication), who was one of his disciples, the descriptions of new taxonomic entities were done from panicles or seeds, or both, and notes. I found that the specimens in the herbarium (WIR) were multiplications, and that the original material is not extant.

Additional note

[6] Sometimes *A. nuda* Thuill., Fl. Par. ed. 2, 1:59 (1799) is cited. This is an erroneous citation because Thuillier quoted *A. nuda* L. I found Thuillier's specimens in G, K, and BR and all are *A. brevis* and do not belong here.

(9) A. STRIGOSA SCHREB.

Homotypic synonyms

A. strigosa Schreb., Spic. Fl. Lips. 52 (1771) [1]
Danthonia strigosa (Schreb.) P. Beauv., Agrostogr. 92 (1812)
Preissia strigosa (Schreb.) Opiz, Seznam Rostl. České 79 (1852)
A. preissia Opiz, op. cit. 20
A. sativa var. strigosa (Schreb.) Kcke. in Kcke. et Werner, Handb. Getreidb. 1:213 (1885)
A. sativa var. strigosa (Schreb.) Fiori in Fiori and Paoletti, Flora Anal. Ital. 1:72 (1896) comb. illegit.
A. sativa ssp. strigosa (Schreb.) Aschers. et Graebn., Synopsis 236 (1899)
A. strigosa ssp. strigosa (Schreb.) Thell., Vierteljahrs. Nat. Ges. Zürich 56:331 (1912) comb. illegit.
A. sativa ssp. sativa var. strigosa (Schreb.) Fiori, Nuov. Fl. Anal. Ital. 1:109 (1923)

Heterotypic synonyms

A. nervosa Lam., Tabl. Encycl. 1:201 (1791) [2]
A. alta Cavan. ex Roem. et Schult., Syst. Veg. 2:691 (1817) pro syn. Danthonia strigosa
A. hispanica Hort. ex Roem. et Schult., loc. cit. pro syn. Danthonia strigosa
A. arduensis Lej. ex Steud., Nom. Bot. ed. 2, 1:171 (1840) pro syn. A. strigosa [3]
A. ambigua Schönh. Fl. Thür. 517 (1850) [4]
A. fatua var. ambigua (Schönh.) Hausskn., Mitt. Geogr. Ges. Thür. Jena 3:237 (1885) [5]
*A. strigosa ssp. glabrescens Marquand, Rep. Bot. Soc. Exch. Club Brit. Isles 6:324 (1922) [6] [7]
*A. strigosa ssp. glabrescens var. albida Marquand, loc. cit. [6] [8]
*A. strigosa ssp. glabrescens var. cambrica Marquand, loc. cit. [6]
*A. strigosa ssp. orcadensis Marquand, loc. cit. [6]
*A. strigosa ssp. orcadensis var. flava Marquand, op. cit. 325 [6] [9]
*A. strigosa ssp. orcadensis var. intermedia Marquand, loc. cit. [6] [10]
*A. strigosa ssp. orcadensis var. nigra Marquand, loc. cit. [6]
*A. strigosa ssp. pilosa Marquand, op. cit. 323 [6]
*A. strigosa ssp. pilosa var. alba Marquand, op. cit. 324 [6]
*A. strigosa ssp. pilosa var. fusca Marquand, loc. cit. [6]
A. strigosa var. tricholepis Holmb., Bot. Not. 1926:182 (1926) [11]
A. strigosa ssp. strigosa var. glabrescens (Marq.) Thell., Rec. Trav. Bot. Neerland. 25:435 (1929)
A. strigosa ssp. strigosa var. orcadensis (Marqu.) Thell., loc. cit.
A. strigosa ssp. strigosa prol. brevis var. glabrata Malz., Monogr. 265 (1930) [12]
*A. strigosa ssp. strigosa var. glabrescens subv. unilateralis Malz., op. cit. 260 [13]
A. strigosa ssp. strigosa var. solida subv. tricholepis (Holmb.) Malz., op. cit. 256
A. strigosa ssp. strigosa var. subpilosa Malz., op. cit. 257 [14]
A. strigosa ssp. strigosa var. subpilosa subv. orcadensis (Marq.) Malz., loc. cit.
A. strigosa ssp. strigosa prol. brevis var. tephrea Mordv. in Wulff, Fl. Cult. Pl. 2:430 (1936) [15]
*A. strigosa ssp. strigosa prol. brevis var. tephrea subv. epruinosa Mordv., loc. cit. [16]
*A. strigosa ssp. strigosa prol. brevis var. tephrea subv. longistrigs Mordv., loc. cit. [16]
*A. strigosa ssp. strigosa prol. brevis var. tephrea subv. rachipubescens Mordv., loc. cit. [16]
A. strigosa ssp. strigosa var. alba (Marq.) Mordv., op. cit. 425
A. strigosa ssp. strigosa var. albida (Marq.) Mordv., loc. cit.
A. strigosa ssp. strigosa var. fusca (Marq.) Mordv., loc. cit.
A. strigosa ssp. strigosa var. gilva Mordv., op. cit. 428 [17]
A. strigosa ssp. strigosa var. intermedia (Marq.) Mordv., op. cit. 425
A. strigosa ssp. strigosa var. nigra (Marq.) Mordv., loc. cit.
A. strigosa ssp. strigosa var. typica Vav. ex Mordv., op. cit. 427 [18]
A. strigosa ssp. strigosa var. unilateralis (Malz.) Mordv., op. cit. 428

A. strigosa ssp. brevis var. glabrata (Malz.) Tab. Mor., Bol. Soc. Brot. 2, 12:244 (1937)
A. strigosa ssp. strigosa var. glabrescens f. albida (Marq.) Tab. Mor., op. cit. 239
A. strigosa ssp. strigosa var. glabrescens f. cambrica (Marq.) Tab. Mor., loc. cit.
A. strigosa ssp. strigosa var. subpilosa subv. tricholepis (Holmb.) Tab. Mor., Bol. Soc. Brot. 2, 13: Tab. 1 (1939)
A. glabrescens (Marq.) Herter, Revist. Sudamer. Bot. 6:141 (1940)

Description

Annual plants. Juvenile growth erect. Color appearance glaucous. Culms erect. Relatively tall plants, 80–120 cm high. Ligules obtuse, rarely with a point. Panicle equilateral. Spikelet (Fig. 119) short, 2.0–2.5 cm long without the awns; each spikelet has 2–3 florets. Glumes equal in length or nearly so, 16–24 mm long, with 5–9 nerves; all the florets are non-disarticulating at maturity. Awns inserted at about lower 1/3 of lemma (Fig. 119). Lemma structure tough; lemma tips bisetulate–biaristulate (Figs. 120–122) or sometimes biaristulate only; lemmas with macrohairs present or absent below the insertion of the awn or only a few hairs around the insertion. Paleas with 1–2 rows of cilia along the edges of the keels; palea back beset with prickles or rarely without. Lodicules (Figs. 123, 124) strigosa type and often bearing prickles (Figs. 125, 126); the side lobe may sometimes be fused to the upper part (Fig. 127). Epiblast brevis type (Figs. 128, 129), 0.3–0.35 mm wide. Chromosome number $2n = 14$. Genome $A_s A_s$.

Distribution

Found in (Fig. 130) Austria, Belgium, Corsica, Czechoslovakia, Denmark, Finland, France, Germany, Hungary, Lithuania, Luxemburg, Norway, Portugal, Spain, Sweden, Switzerland, the United Kingdom, and USSR.

Phenology

Flowers from June to September, and rarely to October also.

Habitat

Cultivated to a limited extent, chiefly in Wales where it does well on poor upland soils. This species was cultivated to a slightly greater extent in Germany, the United Kingdom, and in various other parts of Eastern Europe. It is a weed among fields, especially cereal crops, but is also found on waste grounds. It thrives well and is more common on sandy soil.

Chorotype

European part of the Euro-Siberian region, and probably native to the Atlantic provinces of that region.

Mode of dispersal

Chiefly dispersed by man in cultivation or inadvertently as a weed in cultivation, especially of cereals. The whole panicle in this species is a man-made diaspore.

Selected specimens

AUSTRIA, Kreutzen inter segetes, *Hackel No. 291* (C, COI, K, LE, M, P, PRC, S, W, and WU); BELGIUM, Hainaut Obourg, July 1866, *Martinis No. 247* (BR, CGE, and K); CORSICA, Corte, 1841, *Leo of Metz* (CGE); CZECHOSLOVAKIA, Lomnice, 7 July 1884, *Weidmann* (COI); DENMARK, Silkeborg, July 1897, *Lorenzen* (HAL); FINLAND, Isthmus Karelicus par Sakkola, 5 August 1897, *Lindberg No. 53* (C, CAN, COI, G, K, LE, P, W, and WU); FRANCE, Wissembourg, 6 July 1859, *Duval-Jouve No. 2768* (CGE, G, LE, P, and S); GERMANY, Bassum Hannover agris arenosis, July 1887, *Beckmann* (B, BR, GB, LE, PRC, and W); HUNGARY, Eger sandfelder Voitersrenble, 19 August 1913, *Schleicher No. 251436* (CAN); LITHUANIA, Kovno Penevezh Birzhay Smaili, 1912, *Kezhelis No. 447* (WIR); LUXEMBOURG, Vielsalm, 27 July 1878, *Everu* (BR); NORWAY, Tjomo, 18 August 1878, *Bryhn* (LD); PORTUGAL, Camarido Caminha, June 1885, *Cunha No. 412* (COI); SPAIN, Galicia Lugo, 3 August 1932, *Mordvinkina No. 2513* (WIR); SWEDEN, Skallsjo in Hulan, 11 August 1892, *Bagge* (CGE); SWITZERLAND, Samen Kontrol Station Zürich, 15 July 1890, *Stebler and Schroter No. 163* (C, COI, G, S, and W); UNITED KINGDOM, Pitts-Hill near Petworth Sussex, 2 September 1918 (CGE and K); USSR, Krim in Palen, 20 July 1869, *Raentz* (PRC); USSR *Leningrad* Serteisk Bykhtiya, 14 August 1913, *Shaganov No. 527* (WIR); USSR, prope urbem Pskow inter segetes, 8 August 1902, *Andrejew No. 1997* (C, G, LE, PRC, S, and WU).

Identification hints and remarks

Like *A. sterilis*, this species will rarely be confused with other species, but species that are closely similar to it (*A. hispanica* and *A. brevis*) may be misidentified as *A. strigosa*. Conclusive identification can be effected by inspection of the lodicules because *A. strigosa* has strigosa-type lodicules (Figs. 123–127), which are unique to this species (see also "Identification hints" under *A. hispanica*). There may be some confusion with some species of section *Tenuicarpa*, but careful inspection will show whether the spikelets have cultivated bases or if they disarticulate at maturity. If the specimen to be identified has a cultivated base it obviously cannot belong to Section *Tenuicarpa*.

Figs. 119–129. Morphological and micromorphological diagnostic details of *A. strigosa*. 119. Spikelet: left, including glumes; right, without the glumes ($\times$ 2). 120. Bisetulate-biaristulate lemma tips ($\times$ 25). 121. Detail of one aristula with one setula (arrow) of Fig. 120 ($\times$ 60). 122. Detail of a setula fused with the aristula, note the nerve (arrow) in the setula ($\times$ 60). 123. A pair of strigosa-type lodicules ($\times$ 50). 124. Strigosa-type lodicule; note the side lobe (arrow) fused to upper part of the body, and note the many hydathodes at the apex of the lodicule ($\times$ 120). 125. Strigosa-type lodicule with prickles ($\times$ 50). 126. Detail of the prickly area of Fig. 125 ($\times$ 120). 127. Strigosa-type lodicule with side lobe (arrow) almost completely fused ($\times$ 50). 128–129. Brevis-type epiblasts ($\times$ 120).

A. strigosa

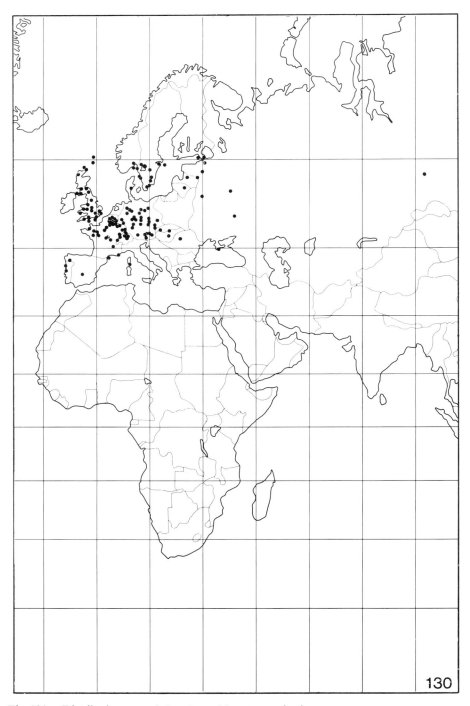

Fig. 130. Distribution map of *A. strigosa*. Mercator projection.

Notes

[1] I found the holotype in M, where it is to be expected (Stafleu 1967: 434). The holotype (Fig. 131) is labeled "Avena strigosa Spicil. Lips./Circa Lipsiam in/agris lecta." It is fragmented into three parts, and in addition, on the same sheet are affixed two other specimens that do not belong to the type collection and also do not belong here because of their identity. I have numbered these on that sheet and so Nos. 2, 3, 5 are the holotype, No. 1 is an *A. barbata,* and No. 4 is an *A. wiestii* from Egypt. At the end of the protolog, it is stated "Similis a Tournefortio in Creta lecta est, sed flosculis . . ." I have found in M a sheet pertinent to this part of the protolog; it is labeled "Creta. T/ legit Tournefort." This specimen from Crete is an *A. barbata.* It is quite possible that specimen No. 1 (of the holotype sheet, mentioned above) is part of the Tournefort collection.

[2] I have examined the holotype in herb. Lamarck in P. The label reads "Avena pennsylvanica L. ex Pourret, A. strigosa Schreb., Avena nervosa Lam. ill. gen., Avena dubia Leers. No. 89. t.9. f.3." The identity of the holotype is *A. strigosa,* but Leers' entry refers to *Ventenata dubia.* In the general herbarium in P, I found a fragment of *A. nervosa* in an envelope that says "Herb. Poiret in Herb. Moquin-Tandon." In herb. C, I found a specimen that must be an isotype because on the back of the sheet it reads "Avena nervosa dedit LaMark"; this specimen belonged to Vahl.

[3] In BR I found two specimens that may be part of the "type" collection. One states "Avena strigosa arduennensis Lejeune in litteris"; the other "Avena strigosa var. β. arduennensis." Both belong to Lejeune's herbarium, and belong here. In LE, I found a specimen annotated "A. strigosa β. arduennensis houppe de poils à la base de l'arête, D. Lejeune 1820." K has a similar specimen sent to Gay by the author. See also Note 1 of *A. sativa* × *fatua* F_1 hybrid.

[4] In herb. Hausknecht in JE, I found a specimen annotated "Avena ambigua mihi" by Schoenheit himself. I chose this one as lectotype, and it belongs here.

[5] This is often quoted (see Chose and Niles 1962) as *A. fatua* var. *ambigua* Hausskn., without Schoenheit. There are two reasons why it should be regarded as a combination. In the protolog Hausknecht states: "dieselbe Form . . . stellt *A. ambigua* Schönh. dar." Also, in herb. W, I found a specimen annotated by Hausknecht "A. fatua var. ambigua Schönh. pr. sp." According to the description, however, Hausknecht meant *A. fatua,* and other specimens annotated by him in JE as var. *ambigua* are *A. fatua* and *A. sativa* × *fatua* F_1 hybrid.

[6] I found many specimens by Marquand at K but the majority postdate the 1922 publication. These may be authentic specimens, but not types. It is difficult to be conclusive about the identity of these taxa; most if not all are likely to belong here, but there is a possibility that some may be *A. hispanica* or even *A. brevis.* I have placed them here in the synonymy with some doubt. According to the descriptions they should be here.

A. strigosa

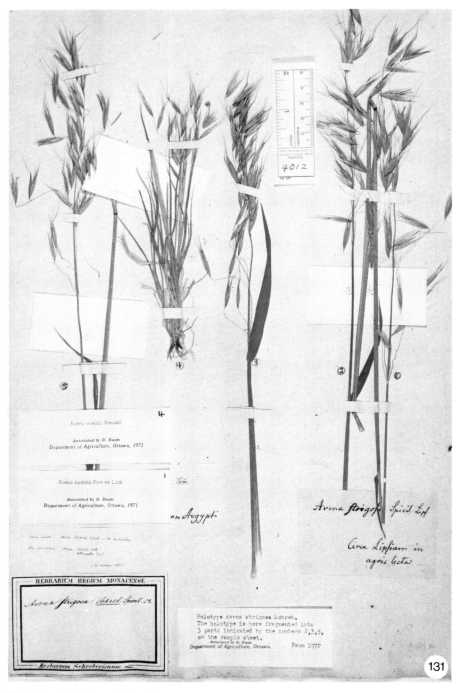

Fig. 131. Holotype of *A. strigosa*.

The Welsh Plant Breeding Station does not keep a herbarium with Marquand's specimens; the closest material is at K, as mentioned above.

[7] In BM there are 2 specimens annotated by Marquand, but the annotation slip bears no date; they belong here.

[8] In BM I found one specimen annotated by Marquand, again with no date on the annotation slip.

[9] In BM I have seen one specimen annotated by Marquand, which might be taken as type; it belongs here.

[10] Same remark as in Note 9 above.

[11] I have obtained two syntypes from LD. I chose the specimen collected by Thedenius in August 1880 as lectotype, and it is also annotated by Holmberg. The paralectotype was collected by Bryhn, and bears on the label the words "cf. tricholepis Holmb." in Holmberg's own handwriting. Both types belong here.

[12] I was unable to see the type collection in WIR, but according to the protolog and the distribution it could belong here.

[13] I did not find the type in WIR; furthermore no specimens are listed in the protolog. It may well belong here because of its distribution.

[14] I have not seen any specimen in WIR that could be from the type collection. According to the distribution, it should belong here. Moreover, one of the synonyms given by Malzew is *A. arduenensis* Lejeune ex Trinius, and it is probably the same as that of Steudel. I saw Lejeune's material in BR (see Note 3) and it belongs here.

[15] I have found two syntypes in WIR and have designated as lectotype No. 2406 (which bears also No. 4481). This specimen and the paralectotype, No. 2406/4482 belong here.

[16] I have not seen types of these subvarieties, but I presume that they belong here, judging from the descriptions alone. The other possibility is *A. hispanica*.

[17] Among the material in WIR, I have selected specimen No. 2513 as lectotype; it is from Lugo (Spain) as stated in the protolog and belongs here.

[18] It belongs here because it appears to be the typical variety (i.e., it is equal to var. *strigosa* according to the Rules) and because ssp. *glabrescens* var. *cambrica* Marquand and var. *glabrescens* (Marquand) Thell. are quoted among the synonyms.

19. Section 4. Tenuicarpa Baum, Can. J. Bot. 52:2259 (1975)

Basis for name   I have given this name to allude to the thin and narrow spikelets of the species in this section.

Holotype species   *A. wiestii*

Species included   *A. barbata, A. canariensis, A. damascena, A. hirtula, A. longiglumis, A. lusitanica, A. matritensis, A. wiestii.*

Naturally occurring hybrids   I have described the artificially produced *A. lusitanica* × *longiglumis* F_1 hybrid (p. 245) because I think that there is some chance that they occur spontaneously. They are to be expected, although I have not encountered them in the material examined; furthermore I have not encountered any hybrids from this section in that material.

Notes on taxonomy and distribution   The members of this section have relatively narrow spikelets; they are in this respect similar to those of Section *Agraria* but they do not have florets with a cultivated base. The florets have small and more or less elongate scars as opposed to the more or less rounded scars and thick spikelets of Section *Pachycarpa* and Section *Avena*. The species are diploids except for *A. barbata,* which is tetraploid, and also except for *A. wiestii,* which is diploid and tetraploid. All the species possess variants of the A genome. Section *Tenuicarpa* is distributed (Fig. 132) at about 25–40° latitude North in a belt from the Atlantic islands and Portugal in the west Mediterranean eastward to the Himalayan mass. This is the area occupied by the ancient Tethys Sea. Some species are extremely scarce but may be locally abundant (e.g., *A. canariensis*) or some may be abundant throughout the area (e.g., *A. barbata*). The territory occupied by this section is the same as that of Section *Ventricosa*. Phenetic similarity, however, suggests that this section occupies a position between Section *Agraria* and Section *Avena* (see Chapter 7 and Fig. 10). Some species of this section were introduced in the Americas.

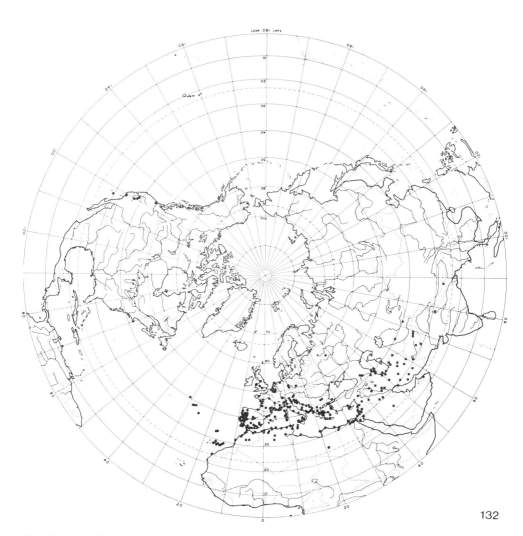

Fig. 132. Distribution map of section *Tenuicarpa*. Polar projection.

(10) A. BARBATA POTT EX LINK

Homotypic synonyms

- A. barbata Pott ex Link, Jour. Bot. Schrader 2:314 (1799) [1] [2]
- A. fatua var. barbata (Pott ex Link) Fiori et Paoletti, Icon. Fl. Ital. 1:29 (1855)
- A. sterilis ssp. barbata (Pott ex Link) Gillet et Magne, Nouv. Fl. France ed. 3:532 (1873)
- A. strigosa ssp. barbata (Pott ex Link) Thell., Vierteljahrs. Nat. Ges. Zürich 56:330 (1912)
- A. sativa ssp. fatua var. barbata (Pott ex Link) Fiori, Nouv. Fl. Anal. Ital. 1:109 (1923)

A. barbata ssp. barbata (Pott ex Link) Tab. Mor., Bol. Soc. Brot. 2, 13:617 (1939)
A. alba var. barbata (Pott ex Link) Maire et Weiller, Fl. Afr. Nord 2:275 (1953)

Heterotypic synonyms

A. barbata Brot., Fl. Lusit. 1:108 (1804) [3]
A. hirsuta Hornem., Hort. Hafn. 1:102 (1813) nom. illegit. [4]
A. sesquitertia Hort. ex Steud., Nom. Bot. ed. 1, p. 95 (1821) pro syn.
A. hirsuta var. humilis Nees, Nov. Act. Acad. Caes. Leop. Carol. 19 suppl. 1:158 (1843) [5]
A. hoppeana Scheele, Flora 27:57 (1844) [6]
A. hirsuta De Moor, Traité Gram. 165 (1854) nom. illegit. [7]
A. deusta Ball, Jour. Linn. Soc. London 16:719 (1878) pro syn.
A. hirsuta var. racemosa Lojac., Natural. Sicil. 4:138 (1885) [8]
A. barbata var. hoppeana (Scheele) Richt., Pl. Eur. 1:62 (1890)
A. barbata var. caspica Hausskn., Mitt. Thür. Bot. Ver. N.F. 6:41 (1894) [9]
A. barbata var. solida Hausskn., loc. cit. [10]
A. almeriensis Gdgr., Bull. Soc. Bot. France 52:443 (1905) [11]
A. sterilis var. micrantha Trabut, Bull. Agron. Alg. Tunisie 16:354 fig. d (1910) nom. nud. [12]
A. strigosa ssp. barbata var. solida (Hausskn.) Thell., Vierteljahrs. Nat. Ges. Zürich 56:331 (1912)
A. barbata var. triflora Trab., 4^e Conf. Genet. Paris:433, f. 9 (1913)
A. fatua ssp. fatua var. pilibarbis Thell., Repert. Sp. Nov. Fedde 13:54 (1913) [13]
A. sallentiana Pau, Bol. Soc. Aragon Cienc. Nat. 1918:133 (1918) [4]
A. fatua ssp. fatua f. pilibarbis (Thell.) Thell., Rec. Trav. Bot. Neerl. 25:426 (1929)
A. wiestii var. glabra Nabelek, Publ. Fac. Sci. Univ. Masaryk (Brno) No. 111:10 (1929) nom. illegit. [15]
A. fatua ssp. fatua var. pilosissima subv. pilibarbis (Thell.) Malz., Monogr. 317 (1930)
A. strigosa ssp. barbata var. typica Malz., op. cit. 270
A. strigosa ssp. barbata var. typica subv. triflora Malz., op. cit. 271 [16]
A. strigosa ssp. strigosa var. solida (Hausskn.) Malz., op. cit. 256
A. strigosa ssp. wiestii subv. caspica (Hausskn.) Malz., op. cit. 278
A. barbata var. eubarbata Maire, in Jahand. et Maire, Cat. Pl. Maroc 1:50 (1931)
A. barbata var. sallentiana (Pau) Jahand. et Maire, loc. cit.
A. hirsuta var. sallentiana (Pau) Senn. et Mauric., Cat. Fl. Rif. Or. 129 (1933)
A. alba var. barbata subv. fallax Maire et Weiller in Maire, Cat. Pl. Afr. Nord No. 2857 (1939) [17]
A. barbata ssp. barbata var. typica (Malz.) Tab. Mor., Bol. Soc. Brot. 2, 13:617 (1939)
A. barbata ssp. barbata var. typica subv. glabritriflora Tab. Mor., op. cit. 621 [18]

A. barbata ssp. hirtula var. calva Tab. Mor., op. cit. 626 [19]
A. barbata ssp. hirtula var. caspica (Hausskn.) Tab. Mor., op. cit. Tab. 1
A. barbata ssp. hirtula var. malzewii subv. trifloriaristulata Tab. Mor., op. cit. 623 [20]
A. alba var. barbata f. fallax (Maire et Weiller) Maire et Weiller, Fl. Afr. Nord 2:276 (1953)
A. alba var. barbata f. genuina (Aschers. et Graeb.) Maire et Weiller, loc. cit.
A. alba var. barbata f. triflora Maire et Weiller, Bull. Soc. Sci. Nat. Phys. Maroc 37:145 (1957) [21]

Description

Annual plants. Juvenile growth prostrate. Color appearance dark green to green. Culms usually geniculate to erect, sometimes prostrate. Low to tall plants, usually 60–80 cm high, sometimes up to 150 cm. Ligules obtuse. Panicle equilateral. Spikelets (Fig. 133) long, 21–29 mm without the awns; each spikelet has 2–3 florets. Glumes nearly equal with 9–10 nerves, and 15–30 mm long; all the florets are disarticulating at maturity; scars elliptic (Fig. 134); the scar of the third floret is not heart-shaped. The periphery ring is only confined to ⅓–½ of the scar. Awns inserted at about middle of lemma. Lemma structure tough; lemma tip bisubulate or biaristulate (Fig. 135), sometimes appearing bisetulate (Figs. 136, 137) but the setulae are lacking nerves and are only membranous; the lemmas are densely beset with macrohairs below the insertion of the awn. Palea with 1–2 rows of cilia along the edges of the keels; palea back glabrous or beset with prickles. Lodicules fatua type (Fig. 138) and usually narrow, elongate and often more involute than the other species of the genus, never bearing prickles. Epiblast brevis type (Fig. 139), 0.3–0.4 mm wide. Chromosome number $2n=28$. Genome AABB.

Distribution

Native to (Fig. 140) Afghanistan, Albania, Algeria, Austria, Azores, Belgium, Bulgaria, Canary Islands, Corsica, Crete, Cyprus, Egypt, France, Germany, Greece, India, Iran, Iraq, Israel, Italy, Lebanon, Libya, Madeira, Morocco, Nepal, Pakistan, Poland, Portugal, Sardinia, Saudi Arabia, Sicily, Spain, Syria, Tunisia, Turkey, the United Kingdom, USSR, and Yugoslavia. Introduced in Argentina, Australia, Brazil, Chile, Uruguay, and USA.

Phenology

Flowers from February to May but also until September in some areas (e.g., Italian Riviera area). In the Canary Islands, it also flowers from October to March. In the Southern Hemisphere, where it was introduced, the period of flowering is from September to February, but it flowers mainly from September to November.

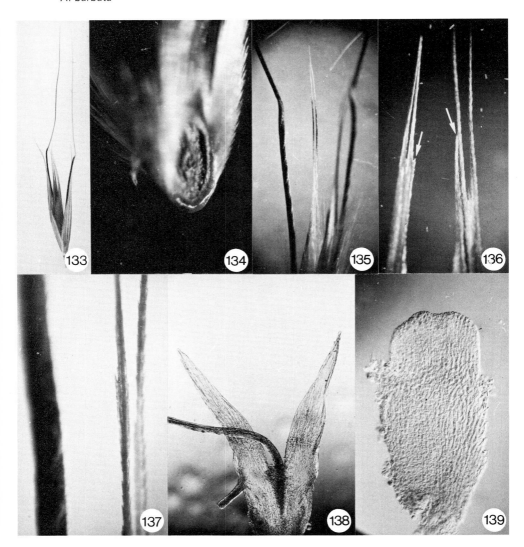

Figs. 133–139. Morphological and micromorphological diagnostic detail of *A. barbata*. 133. Spikelet (× 2.5). 134. Scar; note the shiny periphery ring here confined to ½ of the elliptical scar (× 40). 135. Biaristulate lemma tips (× 35). 136. Two kinds of lemma tips found in this species: left arrow, a nerveless setula-like structure; and right arrow, a true and nerved setula (× 40). 137. Detail of a nerveless setula-like structure (× 75). 138. A pair of fatua-type lodicules; note their narrow and elongate shape (× 50). 139. Epiblast (× 120).

Habitat

It is probably the most frequently occurring species after *A. sterilis*. It is very successful in disturbed and undisturbed sites. In its natural habitat, it populates shallow, stony hillsides and open parklands and pasture and seems to be confined to deforested or shrubby sites. It grows at the edges of salty

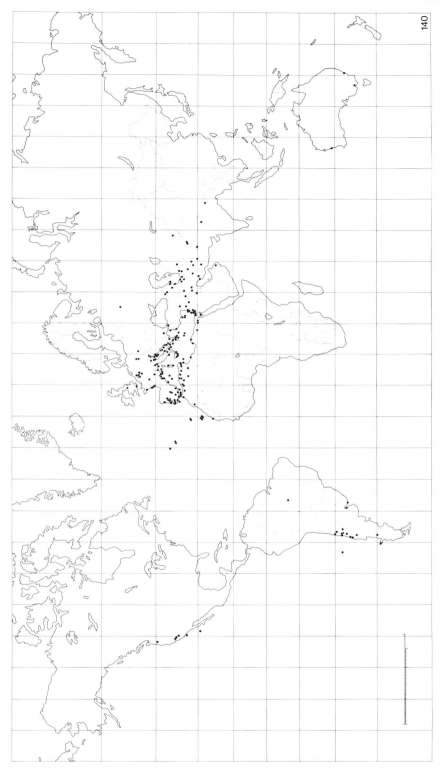

Fig. 140. Distribution map of *A. barbata*. Mercator projection.

marshes, on limestone formations, and even in flooded paddy fields. It is chiefly found on rocky or gravelly slopes of mountains and valleys. It occurs together with *A. wiestii* on volcanic soils, or with *A. longiglumis* on sandy soils, or with *A. lusitanica* in steppe habitats. *Avena barbata* is essentially a weed of poor shallow soils and often occurs there with *A. sterilis,* which has difficulty in competing with the former in those habitats. It is common as a weed in cultivated fields, roadsides, waste places in villages or towns, and along walls and excavation sites.

Chorotype

All Mediterranean, western Asiatic (i.e., the Armenian-Persian Highlands) and South Russia, an area sometimes called the Mediterranean–Irano-Anatolian and North African – Indian desert regions. It is adventitious or naturalized in the Euro-Siberian region (mainly the European subregion), in the flanks of the Himalayas (i.e., the Indian region), the Pacific North American region (i.e., mainly the Californian coast), the South Brazilian region, and the Andean and Pampas regions. It is not common in the last three regions.

Mode of dispersal

By wind and various animals (e.g., ants). The diaspore is the floret, and is very effective.

Selected specimens

AFGHANISTAN, Kabul Tangi-Gharu 1600 m, 21 October 1935, *Scheibe No. 59A* (W); ALBANIA, Valiona Capo Luiquetta, *Baldacci No. 98* (BR); ALGERIA, Alger Kouba ditione urbis, 1879, *Gandoger No. 504* (PRC); ARGENTINA, Buenos Aires the harbour, 5 September 1946, *Sparre No. 362* (S); AUSTRALIA, Broadmeadows near Melbourne, November 1943, *Smith No. 85* (BRI); AZORES, Murio Tectior-Insul., May 1838, *Hothot No. 150* (FI); BELGIUM, Liège Nessonvaux, *Crepin* (BR); BRAZIL meridionalis, *Sellow No. 14233* (HAL); BULGARIA, Trojanum Dardanchi Versuras, 25 May 1883, *Sintenis No. 870* (BR, LE, M, PRC, UPS, W, and WU); CANARY ISLANDS, Gran Canaria Teror 500 m, 22 April 1971, *Kunkel No. 14131* (C); CHILE, La-Leona-Ran montis, October 1828, *Bertero No. 68* (FI); CORSICA, Boniface: Ste Manga, 6 April 1938, *Meylan No. 2399* (G); CRETE, Kissamos Ins. Grabusa agria, June 1942, *Rechinger No. 12123* (G, K, and S); CYPRUS, Lakkovounera Fo 600 ft, 6 April 1953, *Merton No. 964* (K); EGYPT, Kairo prope Turra deserti, 20 April 1908, *Bornmüller No. 11105* (B, LE, S, W, and WU); FRANCE, Toulon endroits incultes, 11 April 1839, *Durieu de Maisonneuve No. 81* (BR, FI, and LE); GERMANY, Mosel Wininger, July 1866, *Huter* (CGE); GREECE, Valle Persei Pindus Tymphaeus 15 June 1896, *Sintenis No. 601* (G, LE, S, and W); IRAN, Persepolis ruinas rupestribus, *Kotschy No. 269/839* (FI, G, LE, M, P, S, UPS, and W); INDIA, Kumaon Almora, *Strachey and Winterbottom No. 4* (CAL); IRAQ, Kurdistan Mt. Auroman, 15 June 1957, *Rechinger No. 12370* (W); ISRAEL, Jerusalem Mt. Scopus, 12 April 1931, *Zohary and Jaffe No. 107* (BR, C, COI, HUJ, LE, M, S, UPS, and W); ITALY, Ligurien Ospedaletti hugeln, 27 May 1905, *Kneucker No. 1193* (B, C, COI, G, GB, W, WIR, and WU); LEBANON, Coelesyria Chtora, 26 May 1889, *Peyron No. 1301* (G); LIBYA, Cyrenaica Faidia humosis calcareis, 26 April 1938, *Maire and Weiller No. 1525* (MPU); MADEIRA IS., Funchal Ribeira Joan Gomez, April 1900, *Bornmüller No. 1355* (BR, G, and PRC); MOROCCO, Segangan Atlaten Beni-Sidel, *Sennen and Mauricio No.*

960 (MA); NEPAL, Shiar Khala cultivated ground, *Gardener No. 859* (BM); PAKISTAN, Baluchistan Quetta, 6 May 1956, *Norris No. 115* (K); POLAND, Kamienna Gora, *Romer No. 6835* (M); PORTUGAL, Olyssiponensis, 30 May 1851, *Welwitsch No. 429* (FI, K, and S); SARDINIA, Cagliari in collibus, April 1827, *Müller* (CGE, HAL, S, and UPS); SAUDI ARABIA, Ash-Shaygit Al-Atshan, 23 February 1968, *Mandaville No. 1336* (BM); SICILY, Palermo, *Parlatore* (FI); SPAIN, Malacae collibus, April 1837, *Boissier* (CGE, FI, G, HAL, LE, P, and W); SYRIA, 38 km a Damascus, 24 May 1925, *Rechinger No. 13227* (W); TUNISIA, Sousse Arad et Nefzaoua cultis, April 1909, *Pitard No. 1074* (G); TURKEY, Smyrnae (Izmir) collibus, 1827, *Fleischer* (BR); UNITED KINGDOM, SW Yorks Bradford city sidings, *McCallum No. 2777* (K); URUGUAY S. Montevideo, *Horneman* (C); USA, *California*, Santa Barbara, 3 July 1913, *Hitchcock No. 637* (BR); USA *Oregon*, Salem dry ground railroad, *Nelson No. 845* (BR); USSR *Azerbaidzhan*, Lenkoran, 1868, *Haussknecht* (JE); USSR *Georgia* Tiflis, 13 June 1911, *Koslowsky No. 261* (WIR); USSR *Turkmen*, Ashkabad Newtonowa, 19 April 1898, *Litwinow No. 2206* (W); YUGOSLAVIA, Fiume Rijeka, *Noë No. 1309* (BR, CGE, LE, P, S, UPS, and W).

Identification hints and remarks

Avena barbata has been confused, on morphological grounds, with *A. wiestii* and *A. lusitanica* (formerly called *A. hirtula* by some authors). In this work I recognize, in addition to these three species, *A. matritensis,* which can also be confused with them. Malzew (1930) and recently Ladizinsky and Zohary (1971) regard these four species, together with *A. abyssinica* and *A. vaviloviana,* as one "biological" species, *A. strigosa.* On morphological grounds, this is unacceptable, and I have demonstrated in this work that a number of taxa can be recognized at the rank of species.

Of the first four species mentioned, *A. wiestii* and *A. matritensis* have a sativa-type lodicule, which can serve as a marker to eliminate them conclusively when identifying *A. barbata*. The third species *A. lusitanica* has lemma tips that are bisetulate-biaristulate, whereas those of *A. barbata* are biaristulate or more often bisubulate (Figs. 135–137). In addition *A. lusitanica* has a much narrower epiblast than *A. barbata.* See also the note under *A. wiestii* which I regard as a diploid and tetraploid species.

Notes

[1] I have designated as lectotype a specimen (Fig. 141) in Trinius' herbarium (LE) with the label "Avena barbata Pott/Lissabon/Lk, herb!" Furthermore, in the general herbarium in Leningrad (LE) there is another specimen that may be considered as paralectotype. On the label it says "Avena barbata Pott/Olyssipone lecta a cf. Link/A. hirsuta Roth"; its identity is *A. lusitanica*. I was unable to find better candidates elsewhere; Link's herbarium in B would be the logical place. I tried also in the Willdenow herbarium (B – Willd.) because sometimes specimens were added to this herbarium later. The specimens that presumably were in B have been lost or destroyed during World War II.

[2] Various authors thought that *A. alba* Vahl, Symb. Bot. 2:24 (1791) should be the correct name instead of *A. barbata*. Mansfeld, in Die Kulturpflanze Beiheft 2, p. 478 footnote (1959), rightly pointed out that

A. barbata

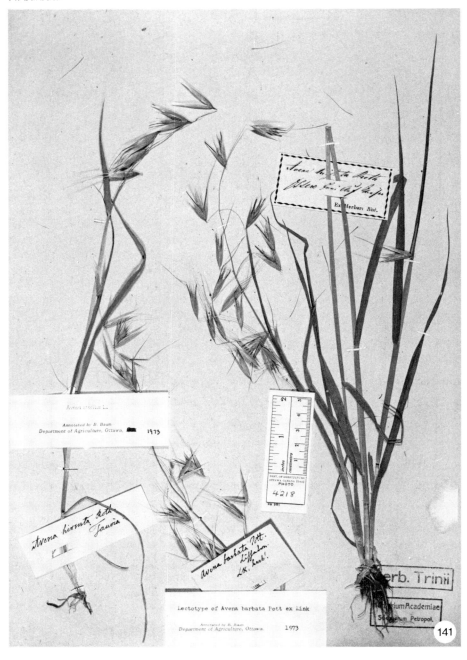

Fig 141.   Lectotype of *A. barbata*.

A. alba Vahl is probably *Arrhenatherum erianthum* Boiss. I have examined two pertinent specimens in C and I agree with Mansfeld. Of these two, I have designated as lectotype the specimen with Vahl's handwritten remark "A. elatior L. var. sec. Trin./Legi Tuneti." The other specimen was collected in "Borbona" and may be considered as paralectotype.

[3] The name *A. barbata* Brot. is often used in the literature even though it is not legitimate. Many authors thought that *A. barbata* Pott was different from *A. barbata* Brotero, but there is evidence to suggest that they were the same. According to W. T. Stearn (personal communication) Link, Pott, and Brotero were very good friends; furthermore, Link and Brotero botanized together, and, therefore, one should not regard the choice of that name by Brotero as coincidence. It probably was meant to be Link's name. A confirmation to this opinion can be found in Link H.F. Hort. Bot. Berol 1:111 (1827): "nomen A. barbata olim Pott imposuit, a nobis traditum est." Moreover, Brotero mentioned the help of Link in his introduction.

[4] Hornemann probably meant *A. hirsuta* Roth; in any case the two specimens that I have observed, both annotated "Roth" by him (one from Montevideo, the other from the Canaries; and both from C) are *A. barbata* Pott ex Link.

[5] I have examined a specimen from S collected in Chile in 1831 and sent by Jussieu in 1834 to Nees and identified by the latter as var. *humilis;* it belongs here. Furthermore, I saw in W two specimens collected in Santiago de Chile by Phillippi that may be pertinent here, and a similar one in HAL. All are *A. barbata*. The holotype, of course, was lost in B during World War II. I, therefore, designate the S specimen as neotype.

[6] I did not find the holotype, but according to Hausknecht (Mitt. Thür. Bot. Ges. 6:41 (1894)) it is an *A. barbata*.

[7] I have not seen the type, but judging from the description it should belong here.

[8] I have examined a specimen from the BM annotated "Avena hirsuta Roth var racemosa mihi/Limosa" in Jacono's handwriting. I regard this specimen as a probable isotype, and it belongs here.

[9] I have designated the specimen from Lenkoran collected by Hausskecht in 1868 as lectotype and that collected in 1861 in Baku (probably not by Hausknecht) as paralectotype. Both specimens are in JE.

[10] I have designated the specimen from Eleusis collected by Hausknecht in 1885 as lectotype. Furthermore, I examined one paralectotype collected in "Genua," but was unable to see the other paralectotype. The first two are in JE and their identity is *A. barbata*.

[11] I have seen the holotype and an isotype in LY and another isotype in B.

[12] According to the illustration it might be *A. barbata* or *A. sterilis*. In addition *A. barbata* var. *triflora* Trabut is mentioned in the 1913 publication; it is not a new taxon according to the Rules (but see also Note 16), and from the description and the probable specimens (see note 16) it appears to be the most common form of *A. barbata*.

[13] On the basis of the description and distribution I decided that this is likely to belong here, not to *A. fatua*.

[14] I have not examined the type, but judging from the description it belongs here.

[15] I have obtained the holotype from Brno (BRA); it is *A. barbata* according to my circumscription.

[16] I have designated as lectotype the specimen "Trabut No. 510, Sersou," which is conserved in WIR. Because *A. barbata* var. *triflora* was never legitimately published, this taxon was authored by Malzew in spite of the mention of Trabut as basionym. Also it is quite possible that Trabut based his *triflora* (see Note 12) on the lectotype and a paralectotype, that I examined, "Trabut 928 Mostaganem" (WIR).

[17] The holotype is Maire and Weiller No. 1525, which is conserved in MPU.

[18] I have selected No. 3932 as lectotype. The paralectotype No. 3667 is an *A. matritensis,* whereas the paralectotype No. 3956 belongs here. The three types are conserved in COI.

[19] The holotype No. 3930 and the two paratypes, Nos. 3929 and 3422, are all *A. barbata* and are conserved in COI.

[20] I have designated No. 3961 as lectotype; it is conserved in COI.

[21] Maire and Weiller thought that their combination was based on Trabut's taxon, but (for the same reason that I have mentioned in Notes 12 and 16) the name for this taxon ought to be attributed to Maire and Weiller only. Furthermore, their name for the taxon does not conflict with that of Malzew because according to the Rules the two are at different ranks.

(11) A. CANARIENSIS BAUM, RAJHATHY ET SAMPSON

Homotypic synonym

A. canariensis Baum, Rajhathy et Sampson, Can. J. Bot. 51:759 (1973) [1]

Description

Annual plants. Juvenile growth prostrate to erect. Color appearance green. Culm erect. Relatively low plants, 30–80 cm high. Ligules acute. Panicles equilateral. Spikelets short, 1.2–1.6 cm long without the awns; each spikelet has 2 florets (Fig. 142). Glumes almost equal in length; 1.3–1.9 cm long; with 7–9 nerves; only the lowermost floret is disarticulating at maturity; scars heart-shaped or slightly so (Fig. 144). Periphery ring only confined to ¼ of scar. Awn inserted at about middle of lemma. Lemma structure tough; lemma tips bidenticulate (Fig. 143); lemma densely covered with macrohairs below awn insertion. Palea with 1 row of cilia along the edges of the keels; palea back covered with verrucae or prickles. Lodicules sativa type (Fig. 145) but sometimes reminiscent of strigosa type (Fig. 146) and rarely with some prickles. Epiblast brevis type, 0.3–0.4 mm wide. Chromosome number $2n=14$. Genome AA.

Distribution

Native to (Fig. 147) Canary Islands and Morocco.

Phenology

Flowers from February to May.

Habitat

Among xeric grasslands, rocky slopes (Fig. 148), and essentially basaltic soil or volcanic debris. It usually grows above 200 m elevation and is not found in the lowlands. It is found on terrace banks of irrigated fields, roadsides, and open steppes or high exposed mountain sides; it occupies natural sites as well as areas disturbed by man. It is not a weed of field crops.

Chorotype

Endemic in the Macaronesian region; and adventitious in the Moroccan Coast adjacent to that region.

Mode of dispersal

The diaspore is the entire spikelet without the glumes; it tends to remain for a while between the glumes when mature because of the dense and relatively long hairs on the back of the lemmas. Dispersal is effected by wind, by grazing animals, and by various other means of zoochory.

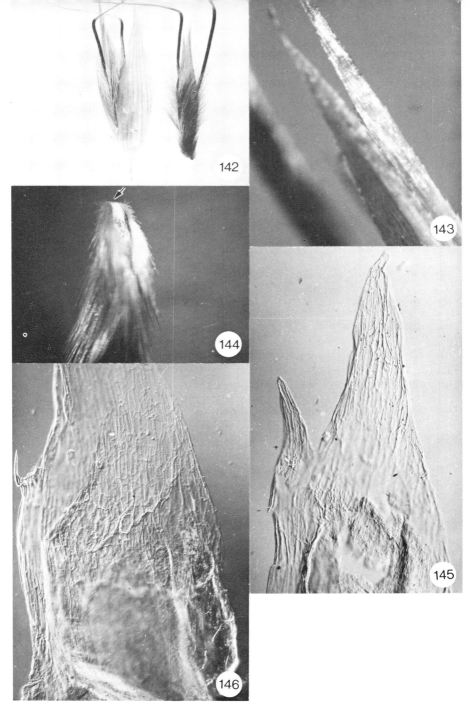

Figs. 142–146. Morphological and micromorphological diagnostic details of *A. canariensis*. 142. Spikelets: left, with glumes; right, without glumes (× 2.5). 143. Bidenticulate tip of lemma; only one tooth is focused to show more details (× 60). 144. A slightly heart-shaped scar; the arrow points to the emarginate part (× 16). 145. Lodicule essentially sativa type, but somewhat reminiscent of the strigosa type because of the relatively high insertion of the side lobe (× 120). 146. Lodicule more reminiscent of the strigosa type than that in Fig. 145, and with a prickle on the apex of the side lobe (× 120).

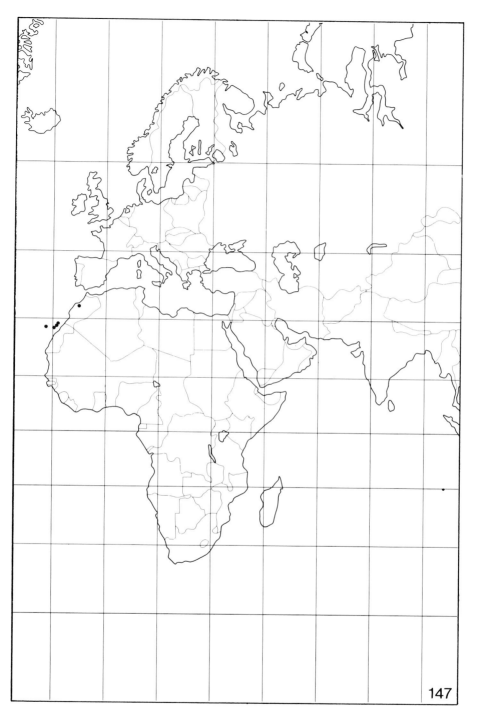

Fig. 147. Distribution map of *A. canariensis*. Mercator projection.

Selected specimens

CANARY ISLANDS, Fuerteventura Oliva, February 1846, *Bourgeau No. 1038* (FI, K, and P); MOROCCO, Casablanca, 20 March 1883, *Melleria* (P).

Identification hints and remarks

The spikelets of this species resemble, in gross morphology, a miniature *A. magna* or a miniature *A. sterilis*. At maturity, the hairs on the lemmas spread in a peculiar manner holding the diaspore for a while between the dry glumes. The appearance of the spikelets is so typical that *A. canariensis* is readily recognizable among other species. A conclusive identification can be effected by checking the chromosome number because there is no other *Avena* species that has sterilis-type spikelets and is also a diploid.

Note

[1] The name *A. canariensis* was used only once by Steudel, Nomenclator Bot. 1:171 (1840), and there it is a *nomen nudum*. Thus the present name is legitimate according to the Code. Steudel's plant is *Pentameris macrantha* (Schrad.) Nees. The holotype (Fig. 149) is "Fuerteventura, ad spiracula de Oliva, Bourgeau Pl. Canariensis No. 1038" (FI). The paratype, and many specimens collected that can also be regarded as paratypes, are mentioned in Baum, Rajhathy, and Sampson, op. cit. 759 and in the *Wild Oat Gene Pool* listing by Baum et al. (1975). I also found isotypes at P and K herbaria.

Fig. 148. A site of *A. canariensis, A. occidentalis,* and *A. barbata,* on talus slope above Rio de Palmas, Fuerteventura, Canary Islands.

Fig. 149. Holotype of *A. canariensis*.

(12) A. DAMASCENA RAJHATHY ET BAUM

Homotypic synonym

A. damascena Rajhathy et Baum, Can. J. Genet. Cytol. 14:646 (1972) [1]

Description

Annual plants. Juvenile growth prostrate to erect. Color appearance green. Culm erect. Relatively low plants, 20–90 cm high. Ligules obtuse. Panicles equilateral. Spikelets long (Fig. 150), 20–26 mm; each spikelet has 2–3 florets. Glumes equal or nearly so, each has 7–9 nerves and is 21–26 mm long; all the florets are disarticulating at maturity. Scars (Fig. 152) round to mostly elliptic; periphery ring only confined to ¼ of scar. Awn inserted at about lower ⅓ of lemma. Lemma structure tough; lemma tips (Fig. 151) bisetulate–biaristulate, in fact each lemma is deeply cleft into two attenuate lobes ending in aristulae and with two nerves ending in each of these lobes, one nerve protruding into the aristula and the other protuding into the short setula but close to the apex of the lobe, that is, on the upper part of the aristula (Fig. 151); lemmas densely beset with macrohairs below awn insertion. Paleas with 1 row of cilia along the edges of the keels; palea back beset with a few prickles. Lodicules sativa type (Fig. 153) with small side lobes; lodicules never bearing prickles. Epiblast septentrionalis type (Fig. 154), 0.25 mm wide. Chromosome number $2n=14$. Genome A_dA_d.

Distribution

Native to (Fig. 155) Syria.

Phenology

Flowers from April to May.

Habitat

Known only from one locality on a dry wadi covered with pebbles in the Syrian desert.

Chorotype

Endemic to the southwest fringe of the Armenian-Persian Highlands within the western and central Asiatic region, in other words the western Irano-Armenian area.

Figs. 150–154. Morphological and micromorphological diagnostic details of *A. damascena*. 150. Spikelet including glumes ($\times$ 2.2). 151. Deeply clefted lemma tip; the arrow points to the place where the setula emerges; note that each lobe resembles an aristula but has two nerves until the setula diverges from it ($\times$ 25). 152. Elliptic scar with the shiny periphery ring confined to ¼ of the scar ($\times$ 30). 153. Sativa-type lodicule; arrow points to the side lobe ($\times$ 120). 154. Septentrionalis-type epiblast ($\times$ 120).

A. damascena

A. damascena

Fig. 155. Distribution map of *A. damascena*. Mercator projection.

Mode of dispersal

The diaspores are the florets, which seem to be successful, but this species is restricted to one site only.

Selected specimens

See Note [1].

Identification hints and remarks

This species is very similar to *A. wiestii*. An external morphologic character that might serve to distinguish this species from *A. wiestii* is the place of insertion of the awn. In *A. damascena* the awn is inserted in the lower ⅓ of the lemma, whereas in *A. wiesti* it is inserted about the middle. Another character that is useful is the periphery ring, which is confined to ⅓–½ in *A. wiestii* and only to ¼ in *A. damascena*. The typical feature of this species is the structure of the lemma tips; they are deeply cleft into two attenuate lobes with two nerves in each. Within each lobe, one nerve protrudes into an aristula and the other into a setula. This structure differs from that of *A. wiestii* because the setula is on the lobe, not on the body of the lemma; compare Fig. 151 with Figs. 193 and 194 of *A. wiestii*.

Note

[1] Holotype (Fig. 156) conserved in DAO with two isotypes: Martens and Rajhathy, Syria, 60 km N. of Damascus.

A. damascena

Fig. 156. Holotype of *A. damascena*.

(13) A. HIRTULA LAG.

Homotypic synonyms

A. hirtula Lag., Gen. et Sp. Nov. 4 (1816) [1]
A. barbata var. hirtula (Lag.) Perez-Lara, Anal. Soc. Españ. Hist. Nat. 15:398 (1886)
A. strigosa ssp. hirtula (Lag.) Malz., Monogr. 247 (1930)
A. barbata ssp. hirtula (Lag.) Tab. Mor., Bol. Soc. Brot. 2, 13:622 (1939)
A. barbata ssp. hirtula var. malzevii Tab. Mor., loc. cit. [2]
A. alba var. hirtula (Lag.) Emb. et Maire, Bull. Soc. Sci. Nat. Phys. Maroc 37:145 (1957)

Heterotypic synonyms

A. barbata var. minor Lange, Naturhist. For. Kjøbenhavn Vid. Medd. 2, 1:39 (1860) [3]
A. barbata var. triflora Willkomm. in Willkomm. et Lange, Prodr. Fl. Hisp 1:68 (1862) nom. illegit [4]
A. barbata var. fuscescens Batt. et Trab., Fl. Alger. Monoc. 62 (1884) [5]
A. lagascae Sennen, Plantes d'Espagne 1926 No. 5980 (1926) [6]
A. strigosa ssp. hirtula subv. minor (Lange) Malz., op. cit. 249
A. serrulatiglumis Sennen et Mauricio in Sennen, Diagn. Nouv. Pl. Espagne et Maroc 1928–35:248 (1936) [7]
A. barbata ssp. barbata var. typica subv. triflora (Willk.) Tab. Mor., op. cit. 620
A. barbata ssp. hirtula var. malzevii subv. minor (Lange) Tab. Mor., op. cit. 625
A. alba var. hirtula subv. minor (Lange) Emb. et Maire, Bull. Soc. Sci. Nat. Phys. Maroc 37:145 (1957)
A. prostrata Ladizinsky, Israel J. Bot. 20:297 (1971) [8]

Description

Annual plants. Juvenile growth prostrate. Culm prostrate or erect. Relatively low plants, 10–50 cm high. Ligules obtuse. Panicle equilateral. Spikelets short, 15–17 mm long without the awns; each spikelet (Fig. 157) has 2–3 florets. Glumes equal in length or nearly so, each about 15 mm long and with 7–9 nerves. All the florets are disarticulating at maturity; scars elliptic (Fig. 159). Periphery ring only confined to ¼ of scar. Awns inserted between lower ⅓–½ of lemma. Lemma structure tough; lemma tip bisubulate (Fig. 158) and sometimes biaristulate; lemmas densely beset with macrohairs below the insertion of the awn. Paleas with 1 row of cilia along the edges of the keels; palea back beset with prickles, which are sometimes scanty. Lodicules (Figs. 160, 161) sativa type, small, about 0.7 mm long, and never bearing prickles. Epiblast (Figs. 162, 163) essentially sativa type but varies in type (e.g., brevis type, Fig. 163), always with a notched or emarginate apex. Chromosome number $2n = 14$. Genome $A_p A_p$.

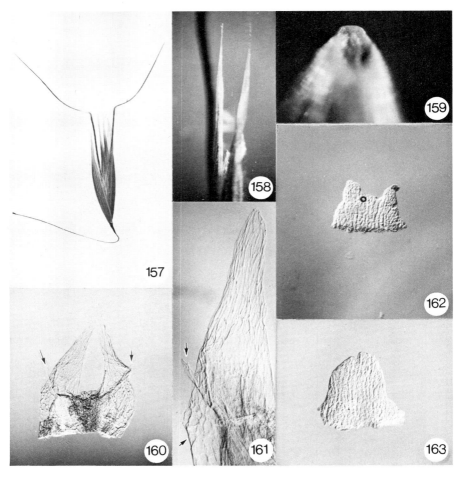

Figs. 157–163. Morphological and micromorphological diagnostic details of *A. hirtula*. 157. Spikelet (× 3). 158. Bisubulate lemma tip (× 33). 159. Elliptic scar; note the periphery ring confined to ¼ of scar (× 76). 160. Pair of sativa-type lodicules; arrows point to the side lobes (× 50). 161. Detail of the base of the left-hand lodicule of Fig. 166; upper arrow points to side lobe, lower arrow points to membranous wing (× 120). 162. Sativa-type epiblast; this is the most commonly found configuration (× 120). 163. Brevis-type epiblast (× 120).

Distribution

Native to (Fig. 164) Algeria, Morocco, Spain.

Phenology

Flowers from April to May.

A. hirtula

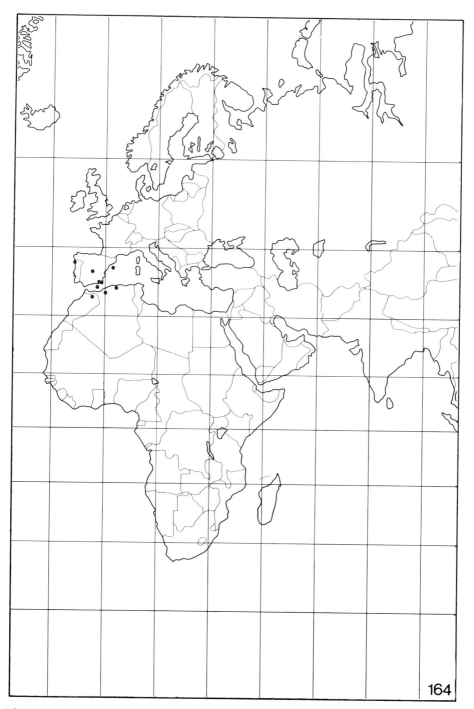

Fig. 164. Distribution map of *A. hirtula*. Mercator projection.

Habitat

Hills, sandy places, rocky places, normally in arid places, and sometimes in edges of plantations and along roadsides; locally quite common and often found with *A. barbata* and *A. lusitanica.*

Chorotype

West Mediterranean.

Mode of dispersal

The diaspores are the individual florets. These drop readily at maturity and may also be carried for a limited distance by the wind or by various means of zoochory. The ecological requirements probably limit the dispersal of this species.

Selected specimens

ALGERIA, Mostaganem sur coteaux incultes, 19 April 1851, *Balansa No. 86* (LE, MT, P, and W); MOROCCO, Gueznaia hauteurs Talmagait, 28 May 1934, *Sennen and Mauricio No. 9690* (G); SPAIN, Murcie Sierras Pantano sites arides, 24 May 1926, *Jeronimo No. 5980* (COI, LE, MA, P, and W).

Identification hints and remarks

This species could easily be confused with *A. damascena,* except that the two are largely disjunct according to the present state of our knowledge. *A. hirtula* differs from both *A. damascena* and *A. wiestii* by the bisubulate lemma tips (Fig. 158), which are rarely biaristulate and by the sativa-type epiblast configuration (Figs. 162, 163). It can also be confused with *A. barbata* because the spikelets are similar, but *A. barbata* has fatua-type lodicules and is a tetraploid.

Notes

[1] The typification of *A. hirtula* is somewhat complicated because the original herbarium of Lagasca was destroyed (D.C. Phytogr. 426, Stafleu 1967:253). According to Willkomm in Willk. et Lge., Prodr. Fl. Hisp. 1:68 (1862), the type of *A. hirtula* might still be in MA because under *A. barbata* var. *genuina* Wk. he stated *"A. hirtula* Lag. ex autopsia specim. orig. in herb. Madritensi!" I have obtained from MA two specimens. One specimen is labeled "Isotypus" and it is a duplicate of the "Typus" which I did not obtain. I cannot consider these as types at all because they date 22 July 1820. Furthermore, there are two different species on this "Isotypus" sheet: *A. barbata* and *A. lusitanica.* The sec-

ond specimen, which I have obtained on loan, I designate as lectotype because it is closest to the protolog, although it is not dated, and the label reads "Avena hirtula Lag./Circa Orcelim" in Lagasca's handwriting (Fig. 165). I found an authentic specimen in M also labeled by Lagasca, but it had been sent to that herbarium in 1820. Another similar specimen is in BM and is annotated by J. Gay "Ex hort. meo, seminib., ab auct. missis." Willkomm (loc. cit.) commented on the confusion between *A. hirtula* Auct. and *A. hirtula* Lag., and it is quite possible that this confusion started with Lagasca himself because, according to my taxonomic concepts, at least three different species were found in his annotated material (i.e., *A. hirtula, A. barbata,* and *A. lusitanica*). Sennen was aware of this confusion and in a note under his Exsiccata No. 5980 (see A. lagascae among the synonyms) he stated that *A. lagascae* is identical with "*A. hirtula* Lag., non auct." Sennen's view supports my choice of the lectotype and was probably based on similar observation.

[2] According to Taborda de Morais's method of working, var. *malzewii* is equal to var. *hirtula* because it is the typical variety of that subspecies; see p. 623 where he stated "tantum subspeciei characteribus."

[3] The type collection consists of discordant elements because the species is difficult to distinguish from its close allies; from C, I obtained a type sheet that contained two syntypes. One syntype collected in "Pontevedra Galicia Aug. 23, 1852" equals *A. hirtula,* whereas the second collected in "La Carolina 11 Mai, 1852" equals *A. lusitanica*. The former specimen I designate as lectotype.

[4] This is a new name for *A. barbata* var. *minor* Lange, and is superfluous.

[5] I have examined the holotype from P. Trabut, among others, wrote on the label "Avena barbata minor Lge.?/Sourjouab mai 83."

[6] The holotype label from MA says "A. lagascae Sennen, Plantes d'Espagne 1926 No. 5980/=A. barbata var. minor Lge., =A. hirtula Lag. non auct./Murcie: Sierras, Pantano, sites arides. 24.5. Leg. Hno. Jeronimo." The specimen is an exsiccata and was distributed to many herbaria. Thus, in COI, MT, LE, W, and P I found many isotypes, which all belong here. *Avena lagascae* can be retained as legitimate because its author would have rejected *A. hirtula* on the basis of Article 69 of the Code.

[7] I have seen the holotype from herb. MA and an isotype from G, and both belong here.

[8] I have examined an isotype at K and have also obtained from its author seeds from the original collection. The specimen and seeds belong here, and *A. prostrata* was collected in the same area where *A. lagascae* was found many years before.

A. hirtula

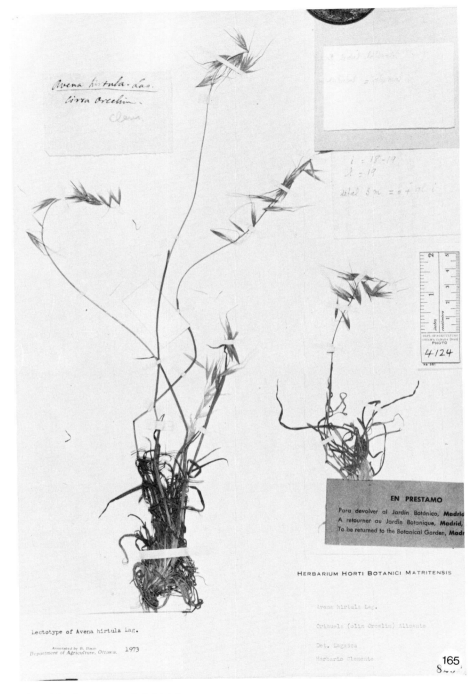

Fig. 165. Lectotype of *A. hirtula*.

(14) A. LONGIGLUMIS DUR.

Homotypic synonyms

A. longiglumis Dur., Rev. Bot. Duchartre 1:359 (1846) [1]
A. barbata ssp. longiglumis (Dur.) Lindb., Act. Soc. Sci. Fenn. N. Ser. B, 1:12 (1932)

Heterotypic synonyms

A. longiglumis subv. glabrifolia Malz., Monogr. 240 (1930) [2]
A. longiglumis subv. pubifolia Malz., op. cit. 239 [3]
A. longiglumis var. genuina Maire et Weiller, Fl. Afr. Nord 2:272 (1953) [4]
A. longiglumis var. genuina subv. australis Maire, loc. cit. [5]
A. longiglumis var. tripolitana Maire et Weiller in Maire, loc. cit. [6]

Description

Annual plants. Juvenile growth prostrate. Color apearance glaucous. Culm erect. Relatively tall plants, 30–120 cm high. Ligules obtuse, often with a point. Panicle equilateral sometimes slightly flagged. Spikelets (Fig. 166) long, 20–35 mm long without the awns; each spikelet has 2–3 florets. Glumes equal or nearly so, 25–40 mm long with 9–11 nerves; all the florets are disarticulating at maturity; scars (Fig. 168) linear. Periphery ring confined to $1/8$–$1/16$ of scar, often pointed towards the base. Awns inserted at about the middle of the lemma. Lemma structure tough; lemma tips biaristulate or bisetulate-biaristulate (Fig. 167) with long and well-developed aristulae; lemmas densely beset with macrohairs below the awn insertion. Paleas with 1 row of cilia along the edges of the keels; palea back often with prickles in proximity of the keels and at the tip only, sometimes completely lacking any prickles. Lodicules fatua type (Fig. 169) or sativa type (Fig. 170), with prickles (Figs. 170, 171) or without prickles (Fig. 169). Epiblast (Fig. 172) septentrionalis type, 0.3–0.35 mm wide. Chromosome number $2n=14$. Genome A_1A_1.

Distribution

Native to (Fig. 173) Algeria, Egypt, Israel, Jordan, Libya, Morocco, Portugal, Sardinia, and Spain.

Phenology

Flowers from March to July.

Figs. 166–172. Morphological and micromorphological diagnostic details of *A. longiglumis*. 166. Spikelet including glumes, and one floret to the left; note that the glumes are much longer than the florets and that the florets are covered with macrohairs below the insertion of the awn (× 2.5). 167. One side of a bisetulate-biaristulate lemma tip; arrow points to setula (× 48). 168. Linear scar; note the pointed periphery ring, which is confined to $\frac{1}{16}$ of scar (× 70). 169. A pair of fatua-type lodicules without prickles (× 50). 170. Sativa-type lodicule also bearing prickles (× 50). 171. Detail of the prickly part of Fig. 170 (× 120). 172. Septentrionalis-type epiblast (× 120).

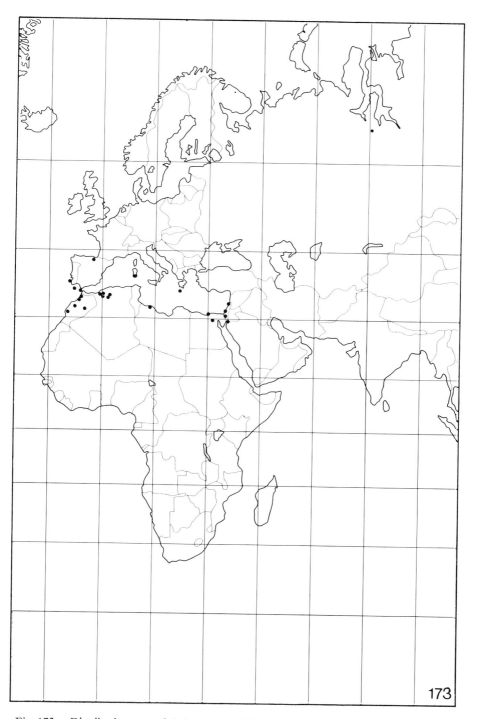

Fig. 173. Distribution map of *A. longiglumis*. Mercator projection.

A. longiglumis

Habitat

Hills, mountain slopes, fields (Fig. 174), cliffs, calcareous substrate, seaside sands, oases, sandy loam, and interior desert sand. It grows also in disturbed habitats, in irrigated soil, and on ditch sides.

Chorotype

West-South-East Mediterranean and Saharo-Arabian.

Mode of dispersal

The diaspores are the florets; they are long with a pointed base and are normally hairy. These dissemules drop readily after maturity and may be carried by zoochory (e.g., ants) to nearby places. This inefficient mode of dispersal may account for the scarceness of this species.

Selected specimens

ALGERIA, Oran Batterie Espagnole champs, 5 April 1852, *Balansa No. 552* (FI, K, P, and W); EGYPT, Abusir-Amria seaside road, 26 March 1961, *Täckholm* (G and K); ISRAEL, Palestine between Fenan and Tlach, 21 March 1908, *Aaronson No. 1178* (AAR and WIR); JORDAN, Wadi Ram, 14 April 1945, *Davis No. 10451;* LIBYA, Tripolitania

Fig. 174. A site of *A. longiglumis,* along the road between Rabat and Meknes 30 km E. of Rabat, Morocco.

Karabouli arenosis, 15 April 1938, *Maire and Weiller No. 1522* (MPU); MOROCCO, Mamora Dar-Salem sables, 1 May 1924, *Jahandiez No. 253* (BM); PORTUGAL, Algarve Faro champs incultes, 18 April 1853, *Bourgeau No. 2065* (FI, K, and P); SARDINIA, 27 June 1926, *Vavilov No. 54201* (WIR); SPAIN, Santa Cathalina propre Puerto S. Maria, 14 April 1849, *Bourgeau No. 500* (CGE, FI, K, P, and UPS).

Identification hints and remarks

At first glance the spikelets, if large, may appear similar to those of *A. sterilis* or, if small, they may appear similar to those of *A. barbata* or *A. lusitanica*. For this reason, this species has often been overlooked in early floristic accounts. It can readily and conclusively be identified by a combination of three characters that can be seen by the naked eye: glumes equal, all florets disarticulating at maturity, and scars linear (Fig. 168). This combination is unique in the genus. The existence of two lodicule types (i.e., sativa type and fatua type) is certainly not consistent with the situation in other species in the genus *Avena* that have only one type, except for *A. nuda*. Some future research, at the populational level, may throw more light on this.

Notes

[1] I have seen a number of syntypes in various herbaria and also recovered the main collection in P. The protolog, among others, states "Oran, Arzew, Mostaganem." I designate as lectotype (Fig. 175) the specimen labeled in Durieu's handwriting "Avena longiglumis DR./Mostaganem 22 avril 44." Two interesting paralectotypes exist in P; Durieu labeled one "Oran (Algerie)/Juin 1842/legit DR.!" and the other "Avena longivalvis. n. sp./Oran. Plateau du côteaux/maritime et côteaux sablon-/neux, à el Oudja/6 mai 42." The latter demonstrates the intent of Durieu in naming it *A. longivalvis*. In B, I have seen two isolectotypes. FI possesses one sheet with two different specimens on it: one is annotated "Avena longiglumis/specimen cultivé, 1845" and can be considered as a paralectotype; the other, judging from the label, could be considered an isolectotype but it is an *A. ventricosa,* although it is annotated *longiglumis* by Durieu. The K specimen is a plant grown from seeds of the specimen labeled *A. longivalvis* above, and can be considered as a paralectotype also. It is worth noting that I did not find any Arzew specimens by Durieu; these may be in herb. BORD (see Note 1 under *A. clauda*).

[2] According to the Code the correct name should be *A. longiglumis* var. *longiglumis* subv. *glabrifolia* (but see next Note 3). I was able to see two out of three syntypes. I have designated No. 1178 as lectotype. This specimen and the paralectotype belong here; they are conserved in WIR.

[3] Malzew (Monogr. 491) says: "There exists finally a whole series of distinctions characterizing the smaller hereditary forms within varieties themselves; such forms we call subvarieties." Furthermore, he stated that this subvariety is "forma typica." Consequently it is equivalent to *A. longiglumis* var. *longiglumis* subvar. *longiglumis* and the type of the species is also the type here. I have seen, in WIR, the specimens cited under this subvariety; all belong here.

A. longiglumis

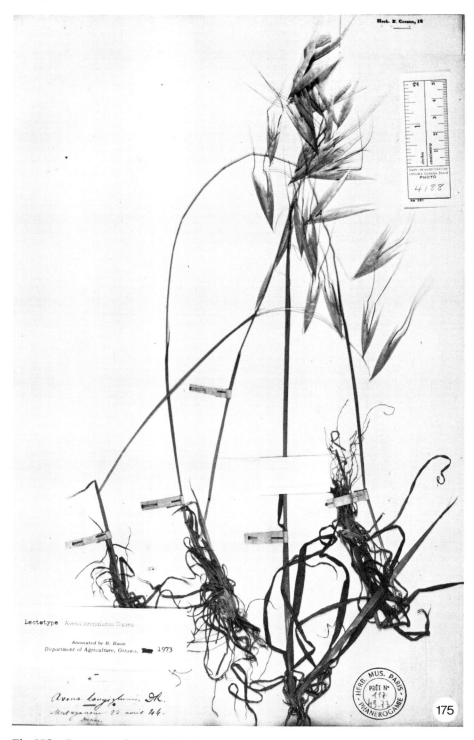

Fig. 175. Lectotype of *A. longiglumis*.

[4] This is the typical variety and is equivalent to var. *longiglumis*.

[5] I have obtained from MPU two specimens (Balansa 2323 from Mogador), which are syntypes and belong here. The Maire specimen, which is quoted in the protolog and which would be the obvious choice for a lectotype, I could not see.

[6] I have examined the holotype from MPU; it is Maire and Weiller 1522.

(15) A. LUSITANICA (TAB. MOR.) BAUM COMB. NOV. ET STAT. NOV.

Homotypic synonyms

> A. barbata ssp. hirtula var. malzevii subv. lusitanica Tab. Mor., Bol. Soc. Brot. 2, 13:624 (1939) [1]

Heterotypic synonyms

> A. barbata var. genuina Willk. in Willk. et Lange, Prodr. Fl. Hisp. 1:68 (1862) [2]
> A. wiestii var. solida Hausskn., Mitt. Thür. Bot. Ver. N.F. 6:42 (1894) [3]
> A. strigosa ssp. wiestii var. solidiflora Thell., Vierteljahrs. Nat. Ges. Zürich 56:335 (1912) [4]
> A. strigosa ssp. abyssinica var. solidiflora (Thell.) Malz., Monogr. 284 (1930)
> A. strigosa ssp. barbata var. typica subv. genuina Malz., op. cit. 272 [5]
> A. barbata ssp. barbata var. typica subv. genuina (Willk.) Tab. Mor., op. cit. 618
> A. barbata ssp. hirtula var. malzevii subv. subaristulata Tab. Mor., op. cit. 625 [6]
> A. barbata ssp. hirtula var. subcalva Tab. Mor., loc. cit. [7]
> A. barbata ssp. barbata var. subtypica subv. triflori-subtypica Tab. Mor., op. cit. 622 [8]

Description

Annual plants. Juvenile growth prostrate. Culm geniculate or prostrate. Ligules obtuse or bluntly acute (i.e., with a few longer teeth at about the middle). Panicle equilateral or more often slightly flagged. Spikelets with indeterminate length, 18–33 mm long (Fig. 176) without the awns; each spikelet has 2–3 florets, the third normally without an awn. Glumes nearly equal in length, 17–30 mm long with 7–9 nerves; all the florets disarticulating at maturity; scars (Fig. 178) narrow elliptic. The periphery ring confined to

¼ of scar or less. Awns inserted at about lower ⅓ of lemma. Lemma structure tough; lemma tips bisetulate-biaristulate with normally well developed setulae and aristulae (Fig. 177); lemmas densely beset with macrohairs below awn insertion. Paleas with 1 row of cilia along the edges of the keels; palea back beset with prickles. Lodicules (Figs. 179, 180) fatua type, usually narrow with an elongate appearance; lodicules never bearing prickles. Epiblast brevis type (Fig. 181), 0.3 mm wide. Chromosome number $2n=14$. Genome $A_s A_s$ presumably.

Distribution

Native to (Fig. 182) Algeria, Canary Islands, Corsica, Crete, France, Greece, Israel, Italy, Libya, Madeira, Morocco, Netherlands, Portugal, Sardinia, Sicily, Spain, Switzerland, Turkey, and Yugoslavia.

Phenology

Flowers from March to July, and rarely at the end of February also.

Habitat

Shallow rocky soils, arid hill slopes and plains, sand dunes, pastures, parklands, roadsides, excavation sites, cultivated fields, and waste places. Often forms mixed stands with *A. barbata* and is rare on good fertile soil. It seldom forms large populations and is well adapted to undisturbed and disturbed habitats.

Chorotype

Essentially a West Mediterranean species penetrating into the Northeast Mediterranean and the adjacent southern fringes of the Euro-Siberian region.

Mode of dispersal

The diaspore is the floret. It is usually hairy. It may be carried some distance by various animals and occasionally by man, when this species is gathered for hay together with *A. barbata*.

Selected specimens

ALGERIA, Ghardaia Beni-Isghen collibus, March 1902, *Chevallier No. 528* (BR); CRETE, *Sieber* (W); FRANCE, Lot et Garonne bord des champs, 4 June 1851, *Debeaux No. 882* (CGE, LE, P, and S); GREECE, Insel Mikra Kaimeni Thera on lava, 13 April 1911, *Hayek* (GB); ITALY, Ripas Arrone Balneocaball, 1 June 1814, *Bubany* (PRC); LIBYA, Cyrenaica Benghasi, 2 March 1883, *Ruhmer No. 368* (JE); MADEIRA ISLAND, July 1862, *Clarke No. 194* (K); MOROCCO, Chaonia Titmellil aridis, 9 May 1912, *Pitard No. 1281* (G); NETHERLANDS, *Schleicher* (FI); PORTUGAL, Seixo Chaves, 17 May 1945, *Garcia No. 858* (COI); SPAIN, Murcia Sierrae de Caruscoy pascuis, 22 May 1891, *Porta et Rigo* (PRC); SWITZERLAND, Zürich, Guterbahnhof, 3 June 1919, *Thellung* (BAS); YUGOSLAVIA, Montenegro Boka Kosorska, May 1900, *Rohlena* (PRC).

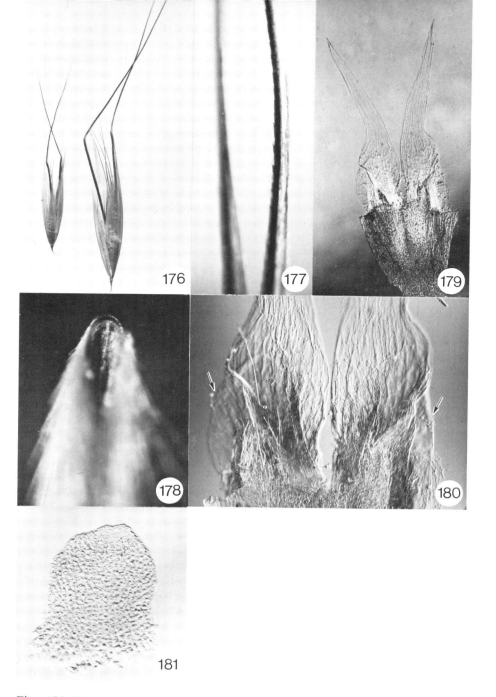

Figs. 176–181. Some morphological and micromorphological diagnostic details of *A. lusitanica*. 176. Spikelets of various sizes (× 2.5). 177. Bisetulate-biaristulate tip of lemma; focus is on one aristula and its adjacent setula; note the nerve protruding into the setula (× 63). 178. Narrow elliptic scar (× 76). 179. Fatua-type lodicule pair; note the narrow and elongate shape peculiar to this species and to *A. barbata*. 180. Detail of the bases of the lodicules shown in Fig. 179; note the absence of any side lobe, and note the thin membranous wings (arrows), which are present in all lodicules of *Avena* (× 120) 181. Brevis-type epiblast (× 120).

A. lusitanica

Fig. 182. Distribution map of *A. lusitanica*. Mercator projection.

A. lusitanica

Identification hints and remarks

This species has been confused with *A. wiestii* and *A. barbata* mostly because they often grow together and also because they are similar in spikelet configuration and structure. It can be readily distinguished from *A. barbata* by careful inspection of the lemma tips: in *A. lusitanica* they are bisetulate-biaristulate (Fig. 177), whereas in *A. barbata* they are biaristulate or bisubulate. Furthermore *A. lusitanica* has a much smaller epiblast than *A. barbata*. From *A. wiestii* it can be distinguished by inspection of the lodicule: it is sativa type in *A. wiestii*, whereas in *A. lusitanica* it is fatua type. Moreover it can also be confused with *A. matritensis*, but *A. matritensis* has a sativa type lodicule.

Notes

[1] I have obtained all the type collection from COI. All have the same identity except No. 3966, which is an *A. barbata*. I have designated No. 3481 as lectotype (Fig. 183).

[2] Although it might be implied as the typical variety, I did not place this name in the synonymy of *A. barbata* because it seems that most weight should be given to the statement by Willkown in the protolog that "A. hirtula Lag. ex autopsia specim. orig. in herb. Madritensi!" This would be the obvious choice of lectotype for this taxon, and it belongs here (see Note 1 under *A. hirtula* Lag.). This specimen is not the type of *A. hirtula* Lag.

[3] The protolog reads: "aus Nordafrika liegt sie mir vor von Benghasi, Cyrenaica (leg. Ruhmer No. 368) in *var. solida* mit festsitzenden Ährchen." I have obtained the holotype from JE, and it belongs here.

[4] This is a new name substituted for *A. wiestii* var. *solida* Hausskn.

[5] It is *genuina* Malz. not Willk., because Malzew excludes the Willk. type. Many cited specimens, especially the Eastern Mediterranean ones, belong here though. Because Malzew meant the typical subvariety, it could as well be placed in the synonymy of *A. barbata*.

[6] I have examined the holotype from COI; it is No. 3980 and it belongs here.

[7] I have examined the type collection from COI. The holotype No. 3981 and paratype No. 3982 belong here, but the paratype No. 3986 is an *A. barbata*.

[8] I have obtained the holotype No. 3923 from COI. It belongs here but the plant is sterile, that is, the florets are empty and no seeds are present.

A. lusitanica

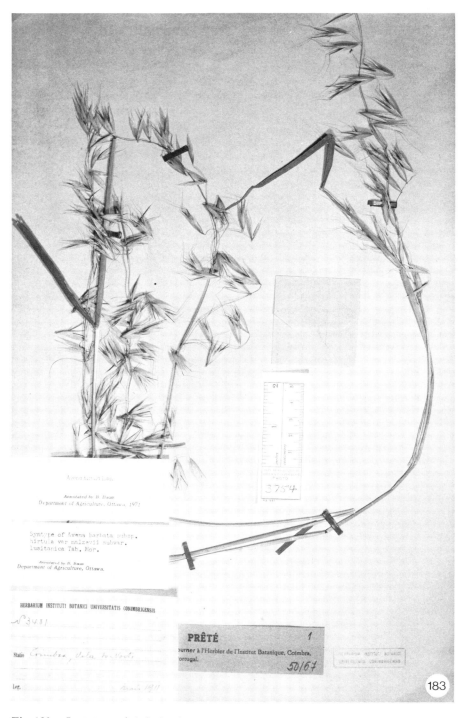

Fig. 183. Lectotype of *A. lusitanica*.

(16) A. MATRITENSIS BAUM SP. NOV.*

Heterotypic synonyms

*A. hirsuta Moench, Meth. Pl. Suppl. 64 (1802) nom. ambig., nom. dub. [1]
*A. hirsuta Roth, Catal. Bot. 3:19 (1806) nom. illegit. [2]
A. villosa Lag. ex Roem. et Schult., Syst. Veg. 2:670 (1817) pro syn. A. hirsuta Moench [3]
A. barbata var. saxatilis Lojac., Fl. Sicul. 3:302 (1909) [4]
A. bolivaris Sennen, Butll. Inst. Catal. Hist. Nat. 18:176 (1918) nom. nud. [5]
A. strigosa ssp. hirtula subv. pseudostrigosa Malz., Monogr. 251 (1930) [6]
A. barbata ssp. hirtula var. malzevii subv. pseudostrigosa (Malz.) Tab. Mor., Bol. Soc. Brot. 2, 13:623 (1939)
A. barbata ssp. barbata var. typica subv. hirsuta (Moench) Tab. Mor., op. cit. 619

Description

Annual plants. Color appearance green. Culm erect. Ligules acute. Panicle equilateral. Spikelets (Fig. 184) long, 25–28 mm long without the awns; each spikelet has 2–3 florets. Glumes equal in length or nearly so, 25–30 mm long, with 7–9 nerves; all the florets are disarticulating at maturity; scars (Fig. 186) narrow elliptic. The periphery ring is confined to ¼ of the scar or less. Awns inserted at about lower ⅓ of lemma. Lemma structure tough; lemma tips (Fig. 185) biaristulate, and they may sometimes appear as bisetulate also but only because of membranous structures, not real setulae; lemmas densely beset with macrohairs below the awn insertion. Paleas with 1 row of cilia along the edges of the keels; palea back beset with prickles. Lodicules (Figs. 187, 188) sativa type but usually with a small side lobe and approaching a strigosa-type configuration (Fig. 188); lodicules rarely bearing a few prickles. Epiblast (Fig. 189) brevis type, 0.3–0.35 mm wide. Chromosome number probably $2n = 14$. Genome unknown.

Distribution

Native to (Fig. 190) Algeria, Belgium, Corsica, France, Italy, Morocco, Portugal, Sardinia, Sicily, Spain, the United Kingdom, and Yugoslavia.

Phenology

Flowers from April to August, rarely to September and even October.

---
*See Note 7 for typification.

A. matritensis

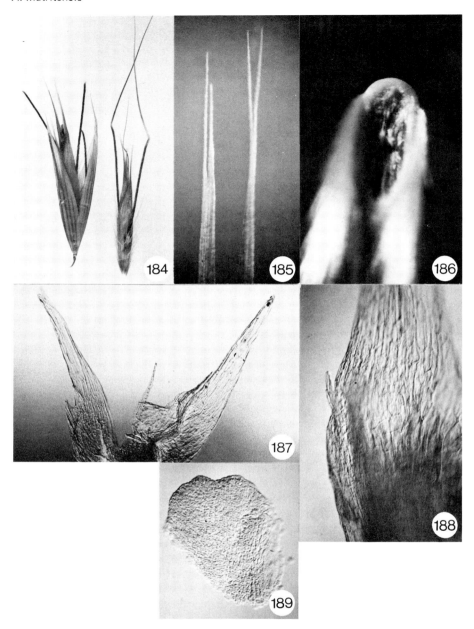

Figs. 184–189. Morphological and micromorphological diagnostic details of *A. matritensis*. 184. Spikelets: left, including glumes; right, without the glumes, and note the lemmas covered with macrohairs below the awn insertion (× 2.5). 185. Two biaristulate lemma tips (× 33). 186. Narrow elliptical scar; note the periphery ring confined to ¼ of the scar (× 76). 187. Two sativa-type lodicules showing the variability of the configuration of the side lobe; here they somewhat resemble a strigosa-type lodicule (× 50). 188. Detail of part of a small side lobe (× 120). 189. Brevis-type epiblast (× 120).

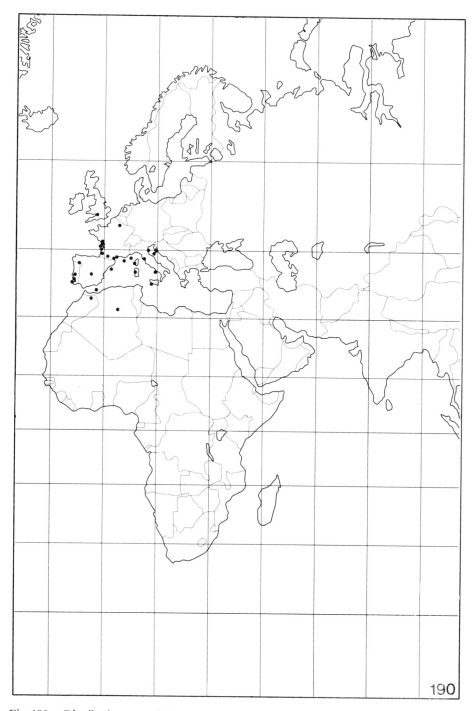

Fig. 190. Distribution map of *A. matritensis*. Mercator projection.

Habitat

Roadsides, waste places, grassy hills, clearings in maquis, and sandy hills. This is locally a rare species and its ecological requirements are practically unknown.

Chorotype

West Mediterranean, penetrating and adventitious in the Atlantic province of the Euro-Siberian region.

Mode of dispersal

The diaspores are the florets. These are relatively long and hairy. The florets drop at maturity and may be carried short distances by small animals or rarely by man.

Selected specimens

ALGERIA, Mzab Metlili, 11 June 1898, *Cosson* (P); BELGIUM, Haute-Crotte décombres, 19 October 1902 (BR); CORSICA, Calvi les champs, *No. 4622* (FI); FRANCE, Charente Inferieure Glacis de Rochelle, *Letourneux No. 229* (B, BR, C, G, LE, P, and W); ITALY, Neapolitani Granatelli, May 1841, *Heldreich* (FI); MOROCCO, Gueznaia hauteurs Talamagait, 28 May 1934, *Sennen and Mauricio No. 9690* (MA); PORTUGAL, Serra d'Arrabida, May 1951, *Walters* (CGE); SARDINIA, Tempio dans les maquis, *Reverchon No. 179* (BR); SICILY, Palermo, May, *Todaro No. 416* (CGE); SPAIN, Madrid, *Dufour* (BM); UNITED KINGDOM, Vice Gloucestershire Tewkesbur, September 1951, *Bannister No. 73-67-7* (K); YUGOSLAVIA, Porto Re Grabrovo, 9 June 1902, *Degen No. 127* (CGE, K, LE, W, and WU).

Identification hints and remarks

This species can be confused with *A. barbata, A. lusitanica,* and *A. wiestii*. The species can be conclusively identified by the following unique combination of characters: the scars are typically narrow elliptic, the peculiar lodicules (Figs. 187, 188) are sativa type, and the lemma tips are biaristulate. Besides *A. matritensis,* only *A. damascena* and *A. lusitanica* have awns inserted at about the lower ⅓ of the lemma, but both have bisetulate-biaristulate lemma tips.

Notes

[1] In "Early Marburg Botany: An Introduction to Moench's Methodus horti Botanici et agri Marburgensis," W. T. Stearn (1966) stated that "Moench's herbarium no longer exists." In the Trinius herbarium in LE I found two interesting specimens: one is labeled "A. hirsuta Roth e Gallia australi," and another with the inscriptions: "Avena hirsuta Roth

s.n. Avenae barbata Pott culta in hto. Goetting." in Trinius's handwriting. Neither specimen has any relation to the type but both are close in time. The former does not belong here but the latter does. For this reason I have put *A. hirsuta* here, dubiously (see also Note 2 below).

[2] According to the descriptions *A. hirsuta* Moench and *A. hirsuta* Roth are completely different. Thus it is possible, but very unlikely, that Roth did not know about Moench's *A. hirsuta*. Roth's specimens in LE (see Note 1 above) may be pertinent here. For the same reason as above I have put this name here in the synonymy, dubiously.

[3] Because this name has been published as a synonym, it is illegitimate. I have examined its "type"; one of its labels reads: "Ab amiciss. Dufour nobiscum comunicata sub nomine: Avena villosa Lagasc." It belongs here, and is from BM.

[4] Among others, the protolog mentions: "A. wiestii Steud. fide Degen." I have seen a number of specimens of "Degen No. 127" and I designate the specimen at K as lectotype; it belongs here.

[5] I have examined the "type" in MA and duplicates from COI and W. All belong here.

[6] I was unable to find the syntypes. According to Tab. 13 and 14 in Malzew and according to the distribution, it seems to belong here; particularly characteristic are the long spikelets.

[7] The formal latin description of this taxon is implied in the synonymy, and reference is particularly made to *A. strigosa* ssp. *hirtula* subv. *pseudostrigosa* Malz., Monogr. 251 (1930). I have designated as holotype the original specimen of *A. bolivaris* (see note 5), that is, Sennen No. 3597 from MA (Fig. 191).

A. matritensis

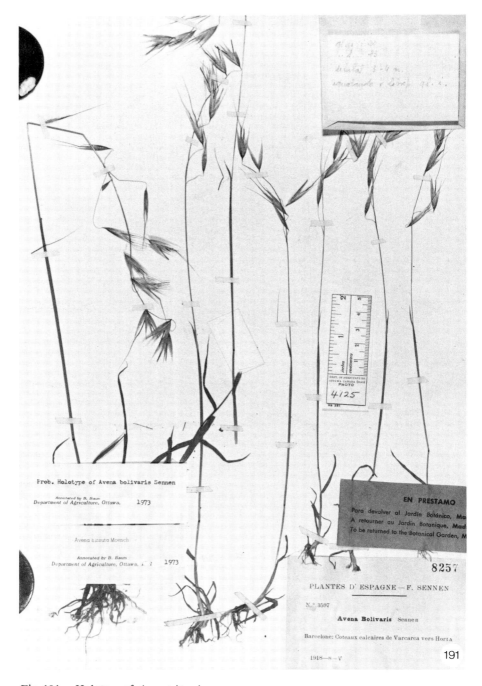

Fig. 191. Holotype of *A. matritensis*.

(17) A. WIESTII STEUDEL

Homotypic synonyms

A. wiestii Steud. Syn. Pl. Glum. 1:231 (1854) [1]
A. barbata var. wiestii (Steud.) Hausskn., Mitt. Thür. Bot. Ver. N.F. 6:45 (1894)
A. strigosa ssp. wiestii (Steud.) Thell., Vierteljahrs. Nat. Ges. Zürich 56:333 (1912)
A. alba var. wiestii (Steud.) Maire et Weiller, Fl. Afr. Nord 2:277 (1953)
A. barbata ssp. wiestii (Steud.) Mansf., Kulturpfl. Beih. 2:479 (1959)
A. nuda ssp. wiestii (Steud.) Löve and Löve, Bot. Notiser 114:50 (1961)

Heterotypic synonyms

A. strigosa ssp. barbata var. subtypica Malz., Monogr. 275 (1930) [2]
A. strigosa ssp. hirtula subv. aristulata Malz., op. cit. 252 [3]
A. strigosa ssp. hirtula subv. glabrifolia Malz., op. cit. 251 [4]
A. strigosa ssp. wiestii subv. deserticola Malz., op. cit. 277 [5]
A. barbata ssp. barbata var. subtypica (Malz.) Tab. Mor., Bol. Soc. Brot. 2, 13:621 (1939)
A. barbata ssp. hirtula var. deserticola (Malz.) Tab. Mor., op. cit. Tab. 1
A. alba var. hirtula subv. aristulata (Malz.) Maire et Weiller, Fl. Afr. Nord 2:277 (1953)
A. alba var. hirtula subv. glabrifolia (Malz.) Maire et Weiller loc. cit.

Description

Annual plants. Juvenile growth erect. Color appearance green or dark green. Culms erect. Low or tall plants, 40–120 cm high. Ligules obtuse with a point. Panicle equilateral. Spikelets (Fig. 192) indeterminate, 1.6–3.0 cm long without the awns; each spikelet has 2 florets. Glumes equal in length or slightly unequal, 15–30 mm long, with 5–7 nerves; all the florets are disarticulating at maturity; scars oval elliptic (Fig. 195) to narrow elliptic. Periphery ring only confined to ⅓–½ of scar. Awns inserted at about middle of lemma. Lemma structure tough; lemma tips (Figs. 193, 194) bisetulate-biaristulate or biaristulate only; lemmas densely beset with macrohairs below the insertion of the awn. Paleas with 1 row of cilia along the edges of the keels; palea back with prickles or sometimes glabrous. Lodicules (Figs. 196–198) sativa type, usually with very small side lobes, and these need special care to discern; lodicules never bearing prickles. Epiblast septentrionalis type (Figs. 199, 200), 0.2–0.3 mm wide. Chromosome numbers $2n=14$ and $2n=28$. Genome A_sA_s and presumably AABB, but not yet fully interpreted.

Distribution

Native to (Fig. 201) Afghanistan, Algeria, Bulgaria, Egypt, Greece, Iran, Iraq, Israel, Italy, Libya, Saudi Arabia, Sicily, Spain, Turkey, and USSR (Azerbaidzhan, Ukraine, and Turkmen).

A. wiestii

Phenology

Flowers from March to July.

Habitat

Occurs sporadically in fairly dense colonies but populations are of variable size although mostly small (Figs. 202, 203). It grows in sandy loess, sandy hills, sandstone, volcanic debris, among rocks, on gravelly slopes, and in *Artemisietea* steppes. The tetraploid forms have similar habitat preferences to *A. barbata*.

Chorotype

East Mediterranean, East Saharo-Arabian, and Irano-Turanian, penetrating to the adjacent fringes of the West Mediterranean region (Algeria interior) and to the southern fringes of the Pontic province.

Mode of dispersal

The diaspore is the floret. The diaspores fall readily at maturity. They are definitely dispersed relatively short distances by the wind, but grazing animals and ants also contribute substantially to their dispersal. The weedy types (i.e., the tetraploids) are extremely adaptable to various habitats.

Selected specimens

AFGHANISTAN, Kandehar gebiet Germaub, 1 May 1935, *Kerstan No. 409* (W); ALGERIA, Alger collines arides, May 1837, *Bové* (CGE); EGYPT, Cairo-Alexandria road 60 km from Alexandria, 20 April 1962, *Bochantsev* (LE); GREECE, Insel Thera Phira-Pyrgos, 14 April 1911, *Hayek* (GB); IRAN, Halate-Mehran Luristan Pushti-Kuh, 27 March 1948, *Behboudi No. 224* (W); IRAQ, Biredjik Seitun, 16 April 1883, *Sintenis No. 364* (BR, LE, M, P, and WU); ISRAEL, Jericho Ain-i-Sultan Wadi Kilt, 30 March 1897, *Bornmüller No. 1650* (B, BR, and PRC); ITALY, Stromboli S. Vincenzo, 8 April 1913, *Hayek* (GB); LIBYA, Cirenaica Amseat a sud Bardia, 24 March 1933, *Pampanini No. 288* (K); SAUDI ARABIA, Musland, 1909, *Musil* (PRC); SICILY, Messina unkultivierte felder, *Zodda No. 498* (GB); TURKEY, Cilicie montagne d'Anamour, May 1872, *Peronin No. 115* (PRC); USSR *Azerbaidzhan*, Apsheron, 22 April 1968, *Musayev* (LE); USSR *Turkmen*, Bukhara Amu Darya R. valley, 27 April 1915 (LE); USSR *Ukraine*, Salyan Bankovskii Steppe, 20 April 1925, *Brzhezitskii No. 77* (WIR).

Figs. 192–200. Morphological and micromorphological diagnostic details of *A. wiestii*. 192. Spikelet including the glumes ($\times$ 3). 193. Bisetulate-biaristulate tips of lemmas ($\times$ 33). 194. Detail of one setula (arrow) adjacent to the aristula; the setula seems almost fused to the aristula; note also the presence of the nerve in the setula ($\times$ 60). 195. An oval elliptic scar; note the shiny periphery ring confined to ⅓ of scar ($\times$ 76). 196. A pair of sativa-type lodicules with very small side lobes, which are typical of this species ($\times$ 50). 197. Detail of a portion of the right lodicule of Fig. 196 showing the side lobe (arrow) magnified ($\times$ 120). 198. Detail of a portion of the left lodicule of Fig. 196 showing the side lobe (arrow) magnified ($\times$ 120). 199–200. Septentrionalis-type epiblasts ($\times$ 120).

A. wiestii

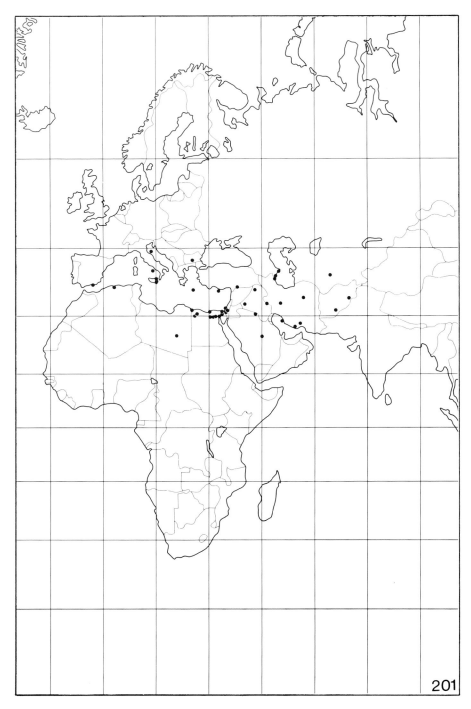

Fig. 201. Distribution map of *A. wiestii*. Mercator projection.

A. wiestii

Figs. 202–203. Two sites of *A. wiestii*. 202. Slopes facing Lar River, Elburs, Iran. 203. In a fenced area 2 km E of Mahmud Abad, Iran.

Identification hints and remarks

This species is often confused with *A. barbata* and *A. lusitanica*, and can also be confused with *A. damascena* and with other species of section *Tenuicarpa*. Furthermore, I recognize *A. wiestii* as a species with diploid and tetraploid forms on phenetic grounds. A character that can serve for conclusive identification is the lodicule. In *A. wiestii* the lodicule is sativa type, whereas in *A. barbata* and *A. lusitanica* the lodicules are fatua type. For discrimination from *A. damascena,* the place of insertion of the awn is a useful marker (see "Identification hints" under *A. damascena).* The tetraploid forms of *A. wiestii* seem to be confined to the Middle East, whereas *A. barbata* covers practically the entire area of Section *Tenuicarpa* (Fig. 132). I was unable to detect a taxonomic discrimination between the diploid and tetraploid forms that would suggest the recognition of different species. Instead, I found good discrimination between *A. wiestii* as defined here, and *A. barbata* an exclusively tetraploid species.

Notes

[1] I have examined the holotype (Fig. 204) in P. It is annotated by Steudel and was collected by Wiest in 1835. According to Stafleu (1967:458) Steudel's original collection is in P. I have found a specimen in LE which may be an isotype, and its label reads: "549 Avena arundinacea Del., in Aegypto inferiori Dr. A. Wiest, 1835."

[2] I was unable to examine the holotype. According to the description, distribution, and picture, it seems to be the tetraploid form of *A. wiestii* (that is, it belongs here).

[3] I have not found in WIR any type material, but in GB I found an isosyntype (Zodda No. 498), which I designate as lectotype, and it belongs here.

[4] I found only one syntype in WIR; it is No. 994, which was collected in El Arisch on 2 May 1925, and I designate it as lectotype. Malzew quoted 944 but it is 994, and it belongs here.

[5] Because this is the typical subvariety, it belongs here. Malzew states in the protolog. "Forma typica," and according to the Rules this is equal to subv. *wiestii.*

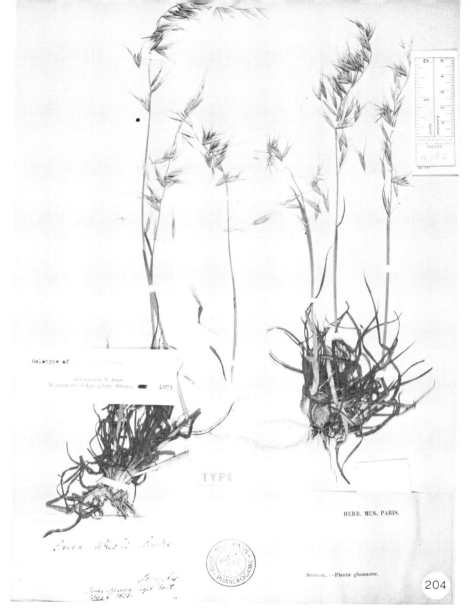

Fig. 204. Holotype of *A. wiestii*.

(18) A. LUSITANICA × LONGIGLUMIS F_1 HYBRID

Description

Resembling *A. longiglumis* except: sterile (i.e., does not set seeds); scars slightly broader; spikelets (Fig. 205) 20–22 mm long (i.e., shorter than most of those found in *A. longiglumis;* and lemmas often with a few hairs around the insertion of the awn.

Distribution

It has never been observed but might be expected in the area of overlap of *A. lusitanica* and *A. longiglumis*.

Mode of dispersal

Irrelevant because the diaspore is always sterile.

Fig. 205. Morphological details of *A. lusitanica* × *longiglumis* F_1 hybrid. Spikelets: left, without the glumes; right, with the glumes, and note the similarity to *A. longiglumis* (× 3).

20. Section 5. Ethiopica Baum, Can. J. Bot. 52:2259 (1975)

Basis for name I have named the section after the country where the species occurs almost exclusively.

Holotype species *A. abyssinica.*

Species included *A. abyssinica, A. vaviloviana.*

Naturally occurring hybrids The occurrence of *A. abyssinica* × *vaviloviana* hybrids is relatively frequent in Ethiopia.

Notes on taxonomy and distribution Each species has glumes that are quite often unequal but not to the extent of *A. clauda* or *A. eriantha;* they are similar in length to those of *A. ventricosa.* The broad wing-like auricle of the lodicules is unique to this section. The species are distributed between 0° and 20° latitude north; they are confined to the highlands of Ethiopia and also occur restrictively in the highlands of adjacent Yemen and Aden (Fig. 206). All the species are tetraploids and carry the AB genome. Phenetically speaking, Section *Ethiopica* appears to be close to Section *Avena,* when based on all characters (Fig. 10 of Classification chapter) although with respect to the genome it is close to *A. barbata* of Section *Tenuicarpa* as is shown when the phenetic relationships are based on a restricted number of characters (Fig. 9).

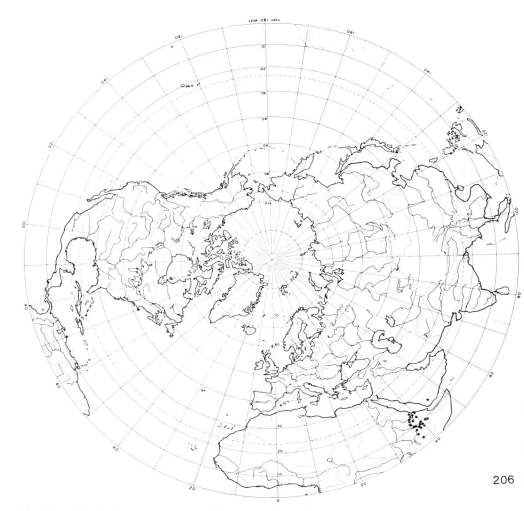

Fig. 206. Distribution map of section *Ethiopica*. Polar projection.

(19) A. ABYSSINICA HOCHST.

Homotypic synonyms

A. abyssinica Hochst. in Schimper, Iter Abyss. Sect. 3, No. 1877 (1846) [1] [2]
A. sativa var. abyssinica (Hochst.) Kcke. in Kcke. et Werner, Handb. Getreidb. 1:216 (1885)
A. sativa var. abyssinica (Hochst.) Engl., Abhandl. Preuss. Akad. Wiss. 2:129 (1891) comb. illegit.
A. strigosa var. abyssinica (Hochst.) Hausskn., Mitt. Thür. Bot. Ver. N.F. 6:45 (1894)
A. strigosa ssp. abyssinica (Hochst.) Thell., Vierteljahrs. Nat. Ges. Zürich 56:335 (1912)
A. alba ssp. abyssinica (Hochst.) Löve and Löve, Bot. Not. 114:50 (1961)

Heterotypic synonyms

A. sativa var. braunii Kcke. in Kcke. et Werner, loc. cit. [3]
A. sativa var. hildebrantii Kcke. in Kcke. et Werner, loc. cit. [3]
A. sativa var. schimperi Kcke. in Kcke. et Werner, loc. cit. [3]
A. barbata var. abbreviata Hausskn., Mitt. Thür. Bot. Ver. N.F. 13/14:51 (1899) [4]
A. abyssinica f. glaberrima Chiov., Ann. Ist. Bot. Roma 8:343 (1907) [5]
A. wiestii var. pseudoabyssinica Thell., Vierteljahrs. Nat. Ges. Zürich 56:307 (1912) [6]
A. strigosa ssp. wiestii var. pseudoabyssinica Thell., op. cit. 334 [6]
A. strigosa ssp. abyssinica var. glaberrima (Chiov.) Thell., op. cit. 336
A. strigosa ssp. abyssinica var. glaberrima (Chiov.) Malz., Monogr. 285 (1930)
A. strigosa ssp. abyssinica var. subglaberrima Malz., loc. cit. [7]
A. strigosa ssp. vaviloviana var. pseudoabyssinica (Thell.) Malz., op. cit. 280
A. abyssinica var. chiovendae Mordv. in Wulff, Kult. Fl. SSSR 2:421 (1936) [8]
A. strigosa ssp. abyssinica var. glaberrima (Chiov.) Tab. Mor., Bol. Soc. Brot. 2, 12:243 (1937)
A. vaviloviana var. pseudoabyssinica (Thell.) C. E. Hubbard in Hill, Fl. Trop. Afr. 10:119 (1937)
A. barbata ssp. vaviloviana var. pseudoabyssinica (Thell.) Tab. Mor., Bol. Soc. Brot. 2, 13: Tab. 1 (1939)
A. abyssinica var. baldratiana Cif., Atti Ist. Bot. Univ. Pavia Ser. 5, 2:152 (1944) [9]
A. abyssinica var. neoschimperi Cif., loc. cit. [9]

Description

Annual plants. Juvenile growth erect. Color appearance glaucous, sometimes light green. Culm erect. Relatively low plants, 40–95 cm high. Ligules acute. Panicle equilateral. Spikelets short, 2.0–2.5 cm long without the back awn; each spikelet (Fig. 207) has 2 florets, and rarely 3. Glumes almost equal in length, 20–25 mm long, with 7–8 nerves. Florets non-disarticulating at maturity. Awn inserted at about the middle of the lemma, sometimes reduced in size or rarely rudimentary. Lemma structure tough; lemma tip long bisubulate and it may sometimes appear to be bisetulate because of membranous nerveless lobes (Fig. 208); lemma without macrohairs below awn insertion, or with only a few hairs around the awn insertion. Palea with 1–2

Figs. 207–211. Morphological and micromorphological diagnostic details of *A. abyssinica*. 207. Spikelet with 2 subequal glumes, 2 florets, and 2 fully developed awns (× 6). 208. Tip of lemma is bisubulate but appears bisetulate; note that the setula (arrow) is a false one because there is no nerve (× 60). 209. Typical sativa-type lodicule of this species; note the broad wing (arrow) and the small side lobe (× 50). 210. Detail of a small side lobe of a sativa-type lodicule (× 120). 211. Brevis-type epiblast (× 120).

rows of cilia along edges of keels; back of palea beset with prickles. Lodicules (Figs. 209, 210) sativa type and with a broad wing at the base of the body, and with a side lobe, which is often small (Fig. 210); lodicules never bearing prickles. Epiblast (Fig. 211) brevis type, 0.5 mm wide. Chromosome number $2n=28$. Genome AABB.

Distribution

Native to (Fig. 212) Ethiopia and Saudi Arabia. Introduced to a very limited extent in Algeria for research purposes.

Phenology

Flowers from August to May, but mainly from October to March.

Habitat

Grows on elevated basaltic plateaus above 2400 m, in cereal fields (mainly wheat and barley as a "tolerated weed"), and quite often gathered as a cultivated plant or even rarely cultivated (Figs. 213, 214).

Chorotype

Endemic to the northeast African highland region.

Mode of dispersal

Chiefly by man, either actively by growing this species (and this is done very restrictively) or passively by harvesting with other cereals (e.g., rye, barley, and wheat). The whole panicle in this species is a man-made diaspore.

Selected specimens

ETHIOPIA, promiscue in agris, 1844, *Schimper No. 1877* (B, BM, CGE, K, LE, P, PRC, and W); SAUDI ARABIA, Beihan among wheat, March 1957, *Congdon No. H5-2-57* (K).

Identification hints and remarks

This species can very easily be confused with *A. sativa*, especially by the appearance of the spikelets. Specimens of *A. abyssinica* often have a size and shape comparable with those of *A. sativa*. Conclusive identification can be obtained by examining the epiblast. The epiblast of *A. sativa* is fatua or sativa type, whereas that of *A. abyssinica* is brevis type (Fig. 211). If no mature

A. abyssinica

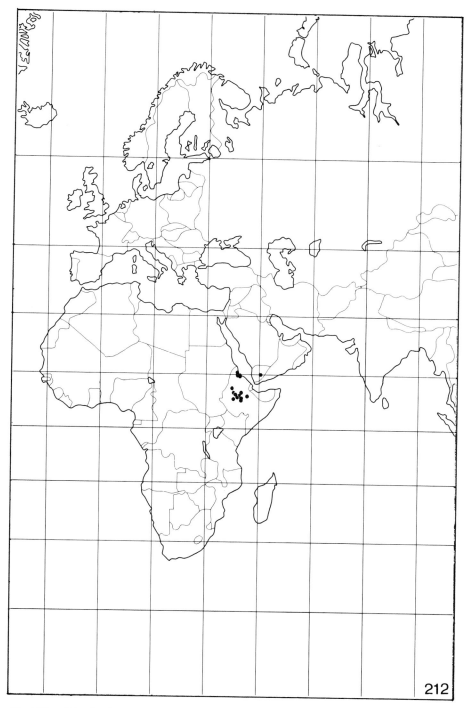

Fig. 212. Distribution map of *A. abyssinica*. Mercator projection.

Fig. 213. A site of *A. abyssinica*, 10 km NE Aksum, Ethiopia.

Fig. 214. A site of *A. sterilis*, *A. abyssinica*, and *A. vaviloviana* growing together, 5 km N Kembolcha, Ethiopia.

seeds are present, one can distinguish between the two species by means of the lodicules. In *A. abyssinica* the broad wing at the base of the body of the lodicule (Fig. 209) is typical, and no comparable structure exists in *A. sativa*.

Notes

[1] The original label of *A. abyssinica* Hochst. is: "Schimperi iter Abyssinicum. Sectio tertia. 1877. Avena abyssinica Hochst. A. hirtula Lag.? Inter A. fatuam et strigosam intermedia flosculis basi pilosis triaristatis (subquinque aristatis). Promiscue in agris. U.t. 1844." According to Hochstetter, C. F., and Steudel, E. G., "Unio Itineraria. Abyssinian Plants." London Jour. Bot. Hooker 5:7-9 (1846) the exsiccata were written at Esslingen in August 15, 1845, but according to a note from Gay in an isotype (Fig. 215) in the herbarium at Kew (K) the effective date of publication seems to be 1846. Hubbard, C. E., in A. W. Hill, Flora of Tropical Africa 10(1):120 (1937) is apparently the first authority to cite the correct name of this taxon.

[2] I did not chose a lectotype of *A. abyssinica* Hochst. because I could not find out which collection can be regarded as the original. According to Stafleu, Taxonomic literature (1967:458) the original collection could be in Paris because "Steudel founded the Esslingen *Unio itineraria* together with Hochstetter," and Steudel's collections are now in Paris. On the other hand, Clayton (personal communication) told me that Hochstetter's original collection is at TUB, and I was not able to see this collection. I examined many isotypes of *A. abyssinica* though: two from B; one from LE, UPS, S, W, and M; many from P; and one from K, G, BM, HAL, PRC, FI, and CGE.

[3] The basis for assessing the synonymy of *A. sativa* var. *braunii,* var. *hildebrantii* and var. *schimperi* is the description only. Koenicke based these varieties on the lemma color only.

[4] *A. barbata* var. *abbreviata* is a new name given by Hausknecht for *A. abyssinica* (see pp. 50–51).

[5] I was able to examine all the syntypes of *A. abyssinica* f. *glaberrima* and I have designated No. 6541 "Medri od Tesfa: Adi Ghebsus m. 1600 c.s.m. 15.10.1905" (FI) as lectotype. All the paralectotypes have the same identity except for No. 751, which is an *A. vaviloviana*.

[6] Not only is Thellung inconsistent in placing his var. *pseudo abyssinica* under the species *A. wiestii* and then under *A. strigosa* in the same work, but the type of *A. abyssinica* is part of the syntype of his var. *pseudo abyssinica*. Thellung, however, recognizes the taxon *A. abyssinica* Hochst. and according to the rules the type cannot be separated from the name of this taxon, even if the taxon is placed at the subspecies level. To further confuse the issue, I have examined two syntypes from the British Museum (BM) (i.e., Schimper No. 603), that were

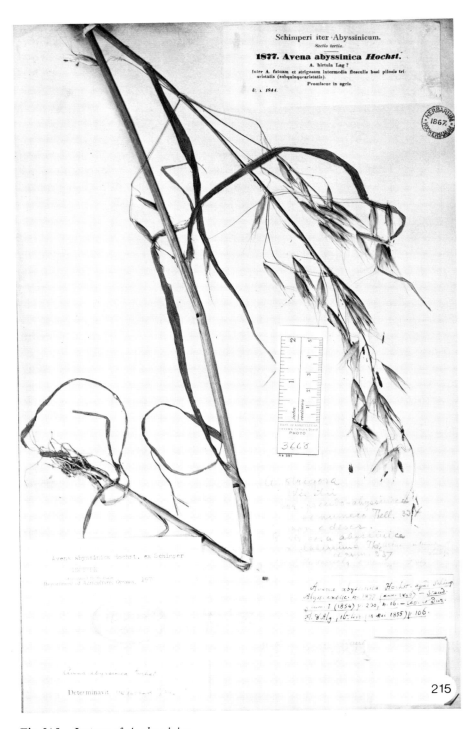

Fig. 215. Isotype of *A. abyssinica*.

identified as *A. vaviloviana.* In any case, because the type of *A. abyssinica* is part of the syntype collection, and because I did not examine the identity of the Schimper No. 603 specimen from the Zürich University herbarium (quoted by Thellung), I regard this taxon to be synonymous to *A. abyssinica;* and obviously, likewise, all the combinations that use this name in various combinations.

[7] I have designated No. 1038 as lectotype: it is conserved in WIR with the 2 other paralectotypes (Nos. 1020 and 1174).

[8] I have designated as lectotype a specimen grown and collected by Mordvinkina from seeds collected by Vavilov. The label on this specimen has two numbers, 1678 and 4969; it is from Ethiopia, Mullat 2100 m alt., dated 5 August 1927 and it has an additional label attached to it with the same date and another number (846). It is conserved in WIR.

[9] Professor R. Tomaselli (PAV) wrote me that Ciferri's specimens were "destroyed by the SS's incursion in the Institute" during World War II.

(20) A. VAVILOVIANA (MALZ.) MORDV.

Homotypic synonyms

> A. strigosa ssp. vaviloviana Malz., Bull. Appl. Bot. Pl.-Breed. 20:138 (1929) [1]
> A. vaviloviana (Malz.) Mordv. in Wulff, Fl. Cult. Pl. 2:422 (1936)
> A. barbata ssp. vaviloviana (Malz.) Tab. Mor., Bol. Soc. Brot. 2, 13:626 (1939)

Heterotypic synonyms

> A. strigosa ssp. abyssinica var. pilosiuscula Thell., Vierteljahrs. Nat. Ges. Zürich 56:336 (1912) [2]
> A. strigosa ssp. vaviloviana var. pilosiuscula (Thell.) Malz., Monogr. 281 (1930)
> A. vaviloviana var. pilosiuscula (Thell.) Hubb. in Hill, Fl. Trop. Afr. 10:119 (1937)
> A. barbata ssp. vaviloviana var. pilosiuscula (Thell.) Tab. Mor., op. cit. 627

Description

Annual plants. Juvenile growth erect. Color appearance green. Culms erect. Relatively low plants, 40–95 cm high. Ligules acute. Panicle equilateral. Spikelets (Fig. 216) short, 1.6–2.1 cm long without the awns; each spikelet has 2–3 florets (Fig. 217). Glumes almost equal in length, 20–25 mm long, with 7–9 nerves; all the florets are disarticulating at maturity; scars (Figs. 219, 220) heart-shaped or slightly so, sometimes elliptic to narrow elliptic. Periphery ring all around the scar (Figs. 219, 220). Awns inserted at about the middle

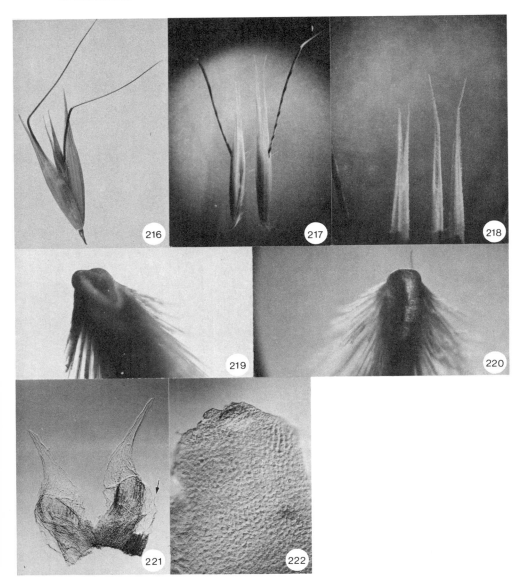

Figs. 216–222. Morphological and micromorphological details of *A. vaviloviana*. 216. Spikelet including glumes (× 3.5). 217. Two florets taken out of the glumes; note the few hairs around the insertion of the awn (× 7). 218. Bisubulate lemma tips of the two florets of Fig. 217 (× 40). 219–220. Heart-shaped scars. Note the periphery ring all around each scar; this configuration is typical of the species (× 25). 221. Fatua-type lodicules; note the broad wing (arrow) at the base (× 50). 222. Brevis-type epiblast (× 120).

of the lemma. Lemma structure tough; lemma tips narrowly bisubulate (Fig. 218), and a subule like an aristula may sometimes appear but in this case it is membranous only and without a nerve; lemmas with macrohairs below the insertion of the awn or glabrous or only a few hairs around the insertion (Fig. 217). Paleas with 1–2 rows of cilia along the edges of the keels; palea back beset with prickles. Lodicules (Fig. 221) fatua type and often at the base of the body there is a broad wing shaped like an auricle. Epiblast (Fig. 222) brevis type, 0.4–0.45 mm wide. Chromosome number $2n=28$. Genome AABB.

Distribution

Native to Ethiopia (Fig. 223).

Phenology

Flowers from September to April.

Habitat

Noxious weed in cereal fields, especially wheat and barley on the elevated basaltic plateaus of Ethiopia; sometimes harvested, although the seeds shatter quite early (Figs. 224, 225). Scattered here and there in small populations in natural habitats (e.g., exposed slopes).

Chorotype

Endemic to the Northeast African highland region, primarily in Ethiopia.

Mode of dispersal

The diaspores are the florets. They are glabrous or hairy. They readily drop down after maturity and from there may be dispersed short distances by various means of zoochory or even by wind. The chief dispersal vector is man (see under habitat section).

Selected specimens

ETHIOPIA, Adoae inter segetes, 1842, *Schimper No. 950* (BM, CGE, G, HAL, K, LE, M, P, and PRC).

Identification hints and remarks

When the origin of the specimen is unknown, *A. vaviloviana* can be confused with *A. barbata* and other similar species of section *Tenuicarpa*.

A. vaviloviana

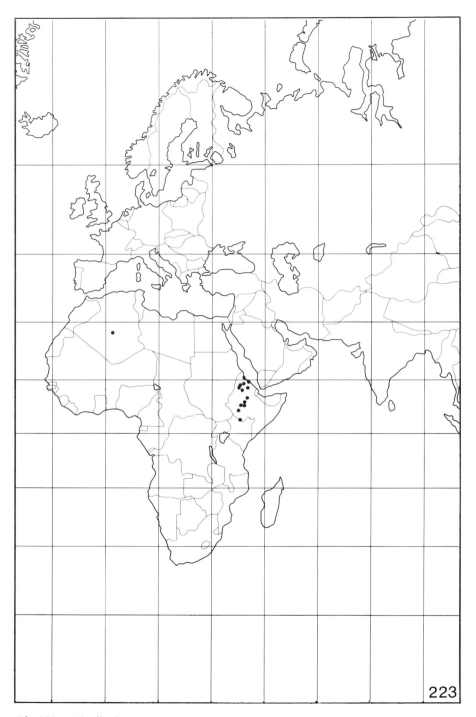

Fig. 223. Distribution map of *A. vaviloviana*. Mercator projection.

A. vaviloviana

Figs. 224–225. Two sites of *A. vaviloviana,* 20 km NE Gonder, and 20 km N of Adis Ababa, Ethiopia.

Furthermore it certainly can be confused with *A. abyssinica* × *vaviloviana* F_1 hybrids. For conclusive identification it is enough to examine the scars of all the florets because *A. vaviloviana* and the hybrid are the only species in the genus with heart-shaped scars and with a periphery ring all around the scar (Figs. 219–220). *A. vaviloviana* can be definitely distinguished from an *A. abyssinica* × *vaviloviana* F_1 hybrid because the former has a fatua-type lodicule, whereas the latter has a sativa-type lodicule.

Notes

[1] No specimens are listed in the protolog. In the Monograph (Malzew 1930:280) the subspecies is detailed with groupings at the varietal level and each contains a list of specimens that can be regarded as syntypes of this subspecies. The first variety var. *pseudo-abyssinica,* by implication, is the typical variety because Malzew quotes "Forma typica." To my mind, this narrows down the syntypes for subspecies *vaviloviana* to the syntypes of that variety. I have seen all these syntypes in WIR and have selected specimen No. 1027 as lectotype (Fig. 226); Vavilov collected it on 9 April 1927 and gave it the number 52286. I have also examined an isolectotype that has two identical labels (one by Malzew and the other by Vavilov) except that the latter bears the number 52201.

Fig. 226. Lectotype of *A. vaviloviana*.

The paralectotypes Nos. 917 and 1033 are also in WIR and belong here. Another paralectotype is Schimper No. 950, which I did not see in WIR, but which I did find in BM, CGE, G (2 specimens), HAL (2 specimens), K, PRC, LE, P, and M (2 sheets). All the isoparalectotypes belong here except one sheet of G, which is *A. abyssinica* × *vaviloviana* F_1 hybrid and one sheet of HAL, which is *A. sterilis*. "A. vaviloviana (Malzew) Malzew, Bull. Applied Bot. Leningrad Suppl. 38, 135 (1930)" is quoted by Index Kewensis. Malzew obviously never intended to make this combination in the same work where he published this taxon as subspecies; the combination is even not implied. The first combination was done by Mordvinkina (1936).

[2] Thellung meant that this variety is the typical variety, not only because it is the first variety but because he based it on Steudel's epithet, which is the same as *A. abyssinica* Hochst. Furthermore, on p. 337 he admits that he did not see the type of *A. abyssinica* (Schimper No. 1877, see Notes 1 and 2 of *A. abyssinica*), and that he placed it under var. *glaberrima*. The only specimen listed under var. *pilosiuscula* is Schimper No. 950, and it is a paralectotype of *A. strigosa* ssp. *vaviloviana* (see Note 1 above). I have put var. *pilosiuscula* among the synonyms here because of the mix-up inadvertently created by Thellung, and of course, on the basis of the protolog. Strictly speaking, I could have used the first argument above and placed var. *pilosiuscula* together with *A. abyssinica* by ignoring the type. Malzew and Hubbard were aware of this mix-up by Thellung and, to rectify the situation, they transferred var. *pilosiuscula* to *vaviloviana* as can be seen among the synonyms.

(21) A. ABYSSINICA × VAVILOVIANA F_1 HYBRID

Synonyms

- A. wiestii var. glabra Hausskn., Mitt. Thür. Bot. Ver. N.F. 13/14:49 (1899) [1]
- A. strigosa ssp. wiestii var. glabra (Hausskn.,) Thell., Vierteljahrs. Nat. Ges. Zürich 56:334 (1912)
- A. strigosa ssp. wiestii var. intercedens Thell., loc. cit. [2]
- A. strigosa ssp. vaviloviana var. glabra (Hausskn.) Malz., Monogr. 280 (1930)
- A. strigosa ssp. vaviloviana var. intercedens (Thell.) Malz., op. cit. 282
- A. vaviloviana var. glabra (Hausskn.) Hubbard in Hill, Fl. Trop. Afr. 10:119 (1937)
- A. vaviloviana var. intercedens (Thell.) Hubbard, op. cit. 120
- A. barbata ssp. vaviloviana var. glabra (Hausskn.) Tab. Mor., Bol. Soc. Brot. 2, 13: Tab. 1 (1939)
- A. barbata ssp. vaviloviana var. intercedens (Thell.) Tab. Mor., loc. cit.

Description

Similar to *A. vaviloviana* (Fig. 227) except: lodicule sativa type, similar to that of *A. abyssinica,* sometimes only partially expressed, that is, with side

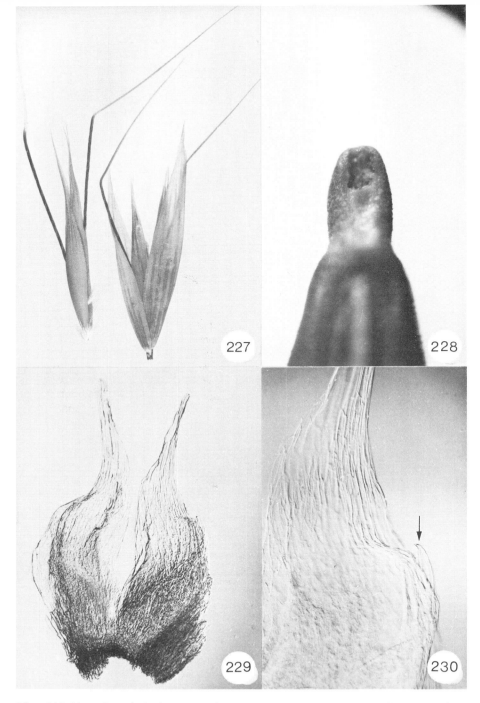

Figs. 227–230. Morphological and micromorphological diagnostic details of *A. abyssinica* × *vaviloviana* F_1 hybrids. 227. Spikelets: left, without glumes; right, with glumes; note the similarity with *A. vaviloviana* (× 3.2). 228. 'Semidisarticulating' scar occurring under cultivation conditions in *A. vaviloviana* or in *A. abyssinica* × *vaviloviana* F_1 hybrids; note that the periphery ring is obvious and all around the scar while the cavity of the scar is half covered (× 45). 229. A pair of sativa-type lodicules (× 50). 230. Detail of a portion of the right lodicule of Fig. 229 showing the side lobe (arrow) (× 120).

lobe occurring on one lodicule only (Figs. 229, 230); lemmas often with a few hairs around the insertion of the awn; sometimes the plants are sterile (i.e., no seeds are produced); and in cultivation, the scars are somewhat deteriorated, that is, they become slightly 'semi-disarticulating' (Fig. 228).

Distribution

Native to Ethiopia (Fig. 231).

Phenology

Flowers from August to May.

Habitat

Commonly found where the two parents coexist (Fig. 214).

Mode of dispersal

Dissemules are similar to those of *A. vaviloviana* and are also dispersed similarly.

Selected specimens

ETHIOPIA, Adoae inter segetes, 1842, *Schimper No. 950* (G); ETHIOPIA Eritrea Asmara 55 km NE, 11 May 1927, *Vavilov No. 1035* (WIR).

Notes

[1] The holotype is a small fragment of a panicle collected by "Schweinfurth, valle Mogod, Eritrea 8/4/92," and duly annotated by Haussknecht. It is conserved in JE, and belongs here. Appended to this specimen is a letter to Haussknecht by Schweinfurth, dated 15 July 1899.

[2] I was unable to see any type material labeled as such by Trabut and in accordance with the data specified in the protolog. But, according to the protolog, this species must belong here because the description fits Figures 1 and 2 and my own description of these hybrids, especially with respect to the scars. In W, I found a sheet with two specimens mounted on it; these were collected by Trabut from Ethiopian plants grown as *A. abyssinica* in Algeria (probably near Oran). One is an *A. abyssinica* and the other is an *A. vaviloviana*. Although Thellung probably did not see this sheet, it indicates that Trabut grew both *A. abyssinica* and *A. vaviloviana* in the same plot, and consequently what Thellung described as var. *intercedens* was likely a hybrid of these two species. This circumstantial evidence supports my assumption.

A. abyssinica × vaviloviana F₁ hybrid

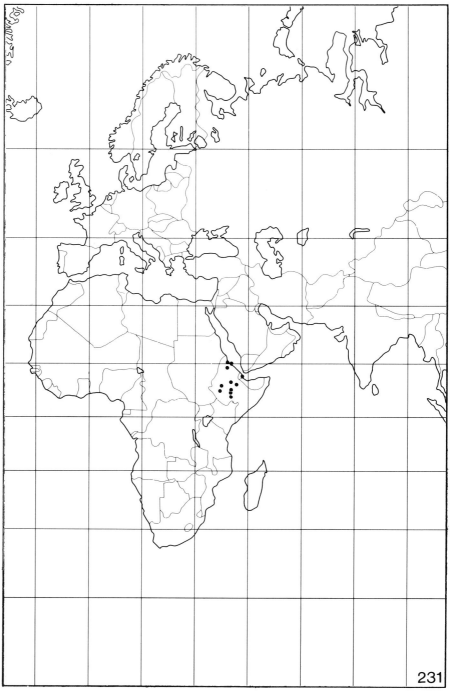

Fig. 231. Distribution map of *A. abyssinica* × *vaviloviana* F₁ hybrids. Mercator projection.

21. Section 6. Pachycarpa Baum, Can. J. Bot. 52:2260 (1975)

Basis for name The name of this section implies the large spikelets of the species. In section *Avena*, *A. sterilis* and *A. atherantha* have forms with spikelets similar in size.

Holotype species *A. maroccana*

Species included *A. maroccana, A. murphyi*

Note on the taxonomy and distribution This section is restricted to a small area in the southern tip of Spain and in Morocco (Fig. 232). The species have

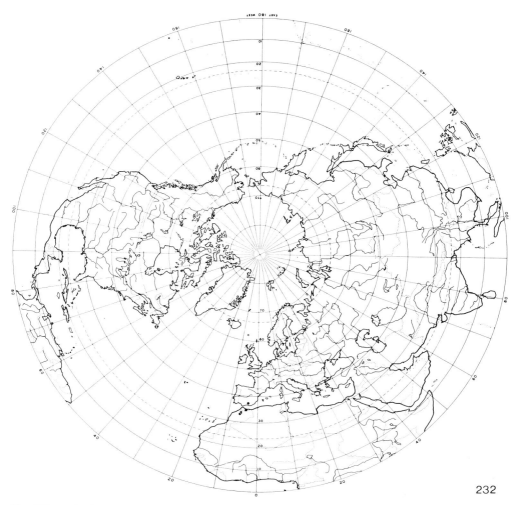

Fig. 232. Distribution map of section *Pachycarpa*. Polar projection.

large, thick, and heavy spikelets, with broad round scars at the base of the first floret only; also they share basically the AACC genome. Phenetically (Figs. 9, 10) they are close to section *Avena*. Did these species used to be distributed more widely? One answer is yes, if it is assumed that they once occupied a larger area of the Tethyan orogeny. Another answer would be no, and then it could be construed that they are incipient species. See Chapter 8.

(22) A. MAROCCANA GDGR.

Homotypic synonym

A. maroccana Gdgr., Bull. Soc. Bot. France 55:658 (1908) [1]

Heterotypic synonym

A. magna Murphy et Terrell, Science 159:103 (1968) [2]

Description

Annual plants. Juvenile growth prostrate. Color appearance glaucous. Culm erect. Relatively tall plants, 50–100 cm high. Ligules acute. Panicle equilateral. Spikelets (Fig. 233) long, 2–3 cm without the awns; each spikelet has 2–3 florets. Glumes nearly equal, 3–4 cm long, with 7–9 nerves; only the lowermost floret is disarticulating at maturity; scars oval to round (Fig. 236). Periphery ring only confined to ¼ of scar. Awns inserted at about middle of lemma. Lemma structure tough; lemma tips bilobed (Fig. 234) or the lobes attenuate (Fig. 235), and always with two nerves ending in each lobe; lemmas very densely beset with macrohairs below the insertion of the awn and often above them also (Figs. 233, 234). Paleas with 2 rows of cilia along the edges of the keels; palea back (Fig. 237) beset with hairs, especially towards the apex. Lodicules fatua type similar to those of *A. sterilis,* and sometimes with one or a few prickles. Epiblast (Fig. 238) fatua type, 0.8–1.0 mm wide. Chromosome number $2n = 28$. Genome AACC.

Distribution

Native to Morocco (Fig. 239).

Phenology

Flowers from April to May.

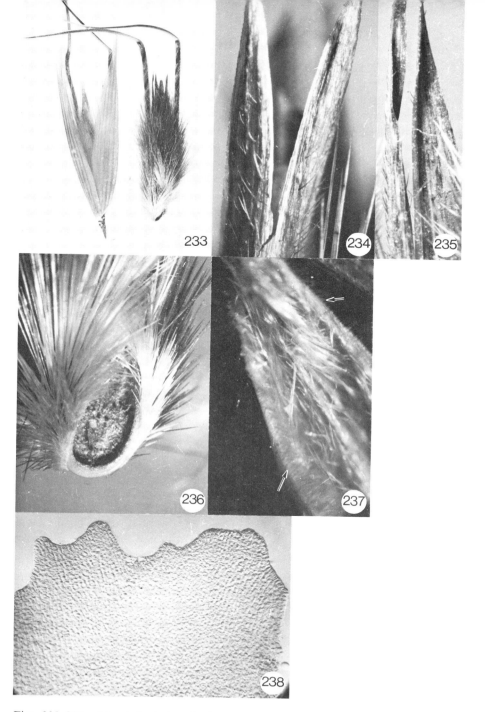

Figs. 233–238. Morphological and micromorphological diagnostic details of *A. maroccana*. 233. Spikelets: left, with glumes; right, without the glumes; note the dense vestiture of the lemmas (× 2.5). 234. Bilobed lemma tip (× 60). 235. Lemma tip with attenuate lobes (× 60). 236. Oval scar; note the periphery ring confined to ¼ of scar (× 40). 237. Hairy back of palea; arrow shows the cilia along edges of keel; the hairs are much longer than the cilia (× 70). 238. Fatua-type epiblast (× 120).

A. maroccana

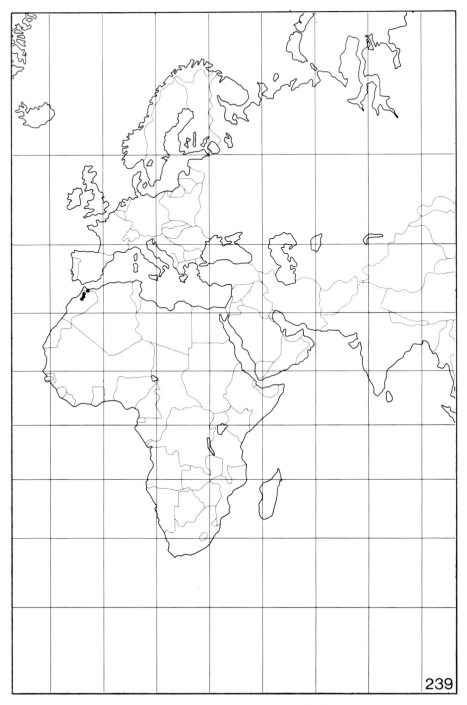

Fig. 239. Distribution map of *A. maroccana*. Mercator projection.

Habitat

It grows (Fig. 240) on fertile alluvial soil, on reddish brown clay soils, on ditch sides, in and around cultivated fields, and on roadsides. It frequently forms mixed stands with *A. sterilis*.

Chorotype

Endemic to Morocco, that is, the southern part of the west Mediterranean subregion.

Mode of dispersal

The diaspore is large, thick and densely hairy, and consists of the entire spikelet without the glumes. It is similar to that of *A. sterilis* in shape and size but *A. maroccana* is considerably less successful because it has, for unknown reasons, only a very restricted distribution. The diaspore falls readily after maturity and may be carried some distance by wind or by animals, including man.

Selected specimens

MOROCCO, 30 km SE of Tiflet, roadside, 27 May 1964, *Zillinsky* (DAO, K, NA, and US).

Fig. 240. A site of *A. maroccana*, 8 km W of Maazia, Morocco.

Identification hints and remarks

This recently rediscovered species was confused with *A. sterilis*. In the present work it may be confused with other species having spikelets similar to those of *A. sterilis* (i.e., *A. atherantha* and *A. trichophylla*). Conclusive identification can be effected by inspection of the vestiture of the back of the palea. It is hairy in *A. maroccana* and only covered with prickles in *A. sterilis*. *Avena atherantha* also has hairs on the back of the palea, but it can be reliably distinguished from this species by an examination of the lemma tips or, better, by cytological examination: *A. maroccana* is tetraploid whereas *A. atherantha* is hexaploid. The hairy spikelets are unique to *A. maroccana*; the hairs are (Fig. 233) grayish white and very dense in most cases. I would state that in the habit of the spikelets alone, *A. maroccana* can readily be recognized.

Notes

[1] I found the type (Fig. 241) in LY; it is labeled "Maroc in sterilibus herbosis maritimis circa Ceuta 25.4.1903." It is the right name for *A.*

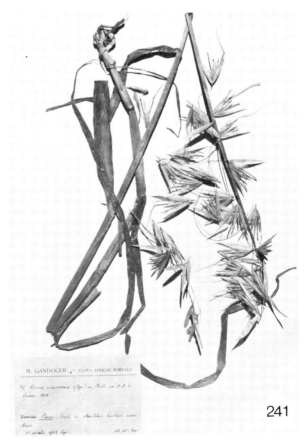

Fig. 241. Holotype of *A. maroccana*.

magna, the name about which there was so much controversy (see Science 163:594-595 (1969)).

[2] I have examined the holotype at K and also isotypes from NA and US; they all belong here.

(23) A. MURPHYI LADIZ.

Homotypic synonym

A. murphyi Ladiz., Israel J. Bot. 20:24 (1971) [1]

Description

Annual plants. Juvenile growth erect. Color appearance green. Culm erect. Relatively tall plants, 50–100 cm high. Ligules acute. Panicle equilateral. Spikelets (Fig. 242) long, 2–3 cm without the awns; each spikelet has 2–4 florets. Glumes equal in length or nearly so, 3–4 cm long, with 9–11 nerves; only the lowermost floret is disarticulating at maturity; scars oval to round, elliptic or broad elliptic (Fig. 244). Periphery ring confined to ½ of scar or less. Awns inserted at about lower ¼ of lemmas (Fig. 242). Lemma structure tough; lemma tips bidenticulate (Fig. 243) to bisubulate, with two nerves ending in each tooth; lemmas without macrohairs below the awn insertion or only a few present around the place of insertion. Paleas with 2 rows of cilia along the edges of the keels; palea back beset with prickles. Lodicules (Fig. 245) fatua type; lodicules with prickles (Fig. 246) or without prickles. Epiblast fatua type (Fig. 247) 0.6–0.7 mm wide. Chromosome number $2n = 28$. Genome AACC.

Distribution

Native to (Fig. 248) a restricted area in southern Spain between Tarifa and Vejer de la Frontera.

Phenology

Flowers from April to June.

Habitat

Grows in heavy alluvial soil, on field edges, roadsides, and also undisturbed land patches.

Chorotype

West Mediterranean, endemic to South Spain.

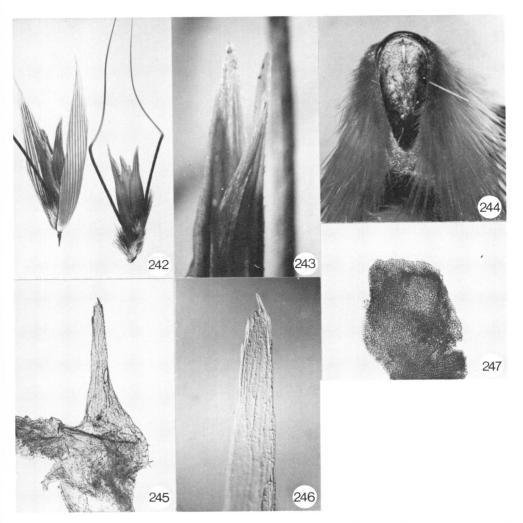

Figs. 242–247. Morphological and micromorphological diagnostic details of *A. murphyi*. 242. Two spikelets: left, including glumes; right, without the glumes; note the low insertion of the awns and the peculiar shape of the florets ($\times$ 2). 243. Bidenticulate lemma tip ($\times$ 30). 244. An oval elliptic scar; note the shiny periphery ring confined to almost ½ of the scar ($\times$ 30). 245. Fatua-type lodicule with a few prickles ($\times$ 50). 246. Detail of Fig. 245; apex of lodicule with the prickles ($\times$ 120). 247. Fatua-type epiblast ($\times$ 50).

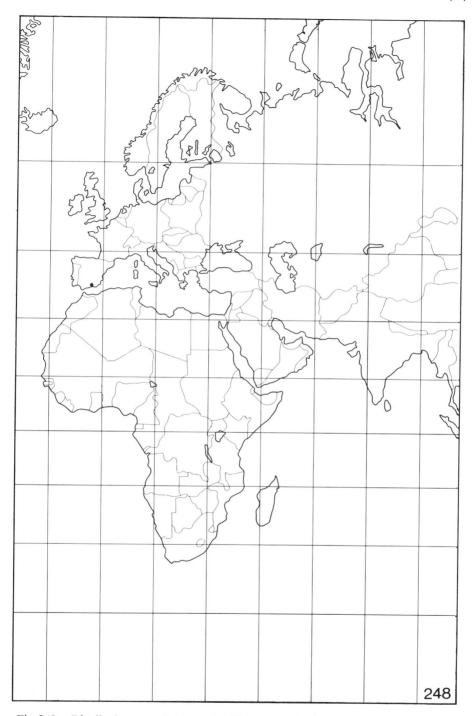

Fig. 248. Distribution map of *A. murphyi*. Mercator projection.

Mode of dispersal

The diaspore is the spikelet without the glumes. It is similar in size to *A. sterilis* but is glabrous. This species is found together with *A. sterilis* in the same habitat, but its distribution is very restricted, and therefore there must be other factors that determine its lack of success in colonizing habitats. The same situation exists between *A. maroccana* and *A. sterilis*. The diaspores drop at maturity and may be transported by various animals, including man.

Selected specimens

See Note [1].

Identification hints and remarks

This species can be confused primarily with *A. sterilis* because of the habit of the spikelets, and because *A. murphyi*, in its restricted habitat and range of distribution, grows together with *A. sterilis* (Ladizinsky 1971). A character useful for discrimination that can be seen with the naked eye is the insertion of the awn. The awn is inserted at about the lower ¼ of the lemma in *A. murphyi* and at about the lower ⅓ in *A. sterilis*. The shape and the structure of the spikelets is typical of this species (Fig. 242), that is, the glabrous lemmas, which are much shorter than the glumes, and the very prominent nerves at the diverging lemma tips. The periphery ring can provide another useful marker because it is confined to ½ in *A. murphyi*, and to ⅓ in *A. sterilis*. The conclusive check obviously is the chromosome number.

Note

[1] I have examined the isotype at K. I did not see the holotype but I obtained seed material from the author.

22. Section 7. Avena (see Baum, Can. J. Bot. 52:2259 (1975))

Basis for name The name of this section follows the requirements of the *Code of Nomenclature* (Stafleu et al. 1972), namely, that the section containing the type species of the genus have the same name as the genus.

Holotype species *A. sativa*.

Species included *A. atherantha, A. fatua, A. hybrida, A. occidentalis, A. sativa, A. sterilis, A. trichophylla*.

Naturally occurring hybrids *A. sativa* × *fatua, A. sativa* × *sterilis*. These hybrids are quite common. I have not yet found other hybrids, but they are to be expected in the areas of hybrid habitats. I was unable to study artificially produced material (such as *A. sativa* × *hybrida* and *A. fatua* × *hybrida*) because of lack of viable material of *A. hybrida;* perhaps spontaneous hybrids have passed undetected by me because of lack of markers.

Notes on taxonomy and distribution The species in this section have relatively large spikelets and all are hexaploids with basically the AACCDD genome. Many species have epiblasts of the fatua type and of larger size than species in other sections, except for those of section *Pachycarpa*. Two species, *A. fatua* and *A. sterilis,* have a wide distribution, are very noxious weeds, and are sometimes used for hay. The cultivated oats, *A. sativa,* is one of the most important cereal crops; it often competes economically with barley in some areas. The section as a whole occupies most of the areal of the genus, that is, from the Equator to the Arctic Circle (Fig. 249). The species of this section are probably of more recent origin than those of the other sections.

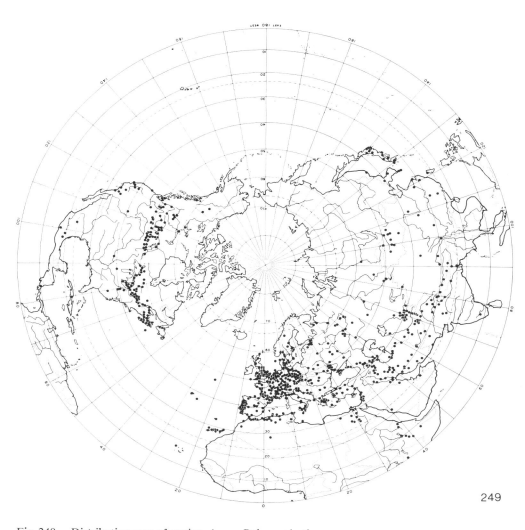

Fig. 249.   Distribution map of section *Avena*. Polar projection.

(24) A. ATHERANTHA PRESL

Homotypic synonyms

A. atherantha Presl, Cyp. et Gram. Sicul. 30 (1820) [1]
A. barbata var. atherantha (Presl) Grossh., Trudy Bot. Inst. Azerbaidzh. Fil. Akad. Nauk SSSR 8:207 (1939)
A. strigosa ssp. barbata var. typica subv. atherantha (Presl) Malz., Monogr. 274 (1930) [1]

Heterotypic synonyms

A. maxima Presl, Fl. Sicul. 44 (1826) [2]
A. sterilis var. maxima Perez-Lara, Anal. Soc. Espan. Hist. Nat. 15:398 (1886) [3]
A. sterilis ssp. macrocarpa var. maxima (Perez-Lara) Thell., Vierteljahrs. Nat. Ges. Zürich 56:315 (1912)
A. sterilis ssp. macrocarpa var. setosissima subv. maxima (Perez-Lara) Malz., op. cit. 389
A. sterilis ssp. macrocarpa var. setosissima subv. hirsutimaxima Tab. Mor., Bol. Soc. Brot. 2, 13:576 (1939) [4]
A. sterilis ssp. macrocarpa var. setosissima subv. trichomaxima Tab. Mor., op. cit. 575 [5]

Description

Annual plants. Juvenile growth prostrate to erect. Ligules obtuse and often with a point. Panicle equilateral. Spikelets (Fig. 250) long, 3–4 cm without the awns; each spikelet has 2–3 florets. Glumes equal in length, 3–5 cm long, with 7–9 nerves; only the lowermost floret disarticulating at maturity; scars (Fig. 252) oval to round, mostly broad elliptic. Periphery ring only confined to ¼ of scar. Awns inserted at about lower ⅓ of lemma. Lemma structure tough; lemma tips (Fig. 251) bisubulate, with a very long subule to almost biaristulate; lemmas densely beset with macrohairs below the awn insertion and reminiscent of *A. maroccana,* and sometimes there are a few hairs only. Paleas with 1–3 rows of cilia along the edges of the keels; palea back (Fig. 253) beset with hairs at least toward the apex. Lodicules fatua type (i.e., similar to those of *A. sterilis*); lodicules never bearing prickles. Epiblast (Fig. 254) fatua type, 0.5–0.6 mm wide. Chromosome number $2n = 42$. Genome AACCDD.

Distribution

Native to (Fig. 255) Algeria, Canary Islands, Corsica, Cyprus, Egypt, France, Greece, Italy, Kenya, Portugal, Sicily, Spain, Switzerland, Tunisia, the United Kingdom, USSR (Crimea), and Yugoslavia.

Phenology

Flowers from March to May, sometimes in June and July, and rarely in August and September. In Kenya, it flowers in November only.

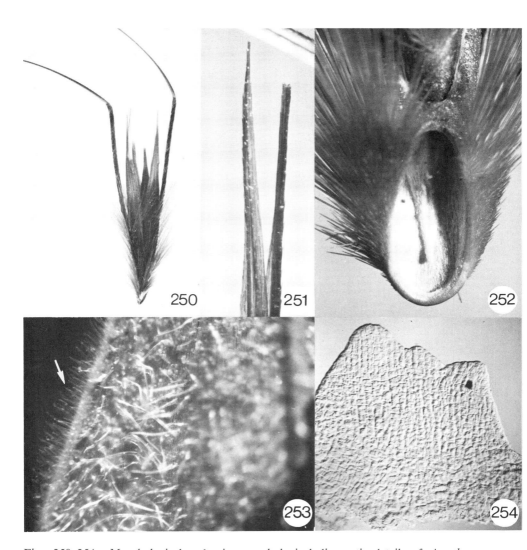

Figs. 250–254. Morphological and micromorphological diagnostic details of *A. atherantha*. 250. Spikelet excluding the glumes (× 2). 251. Bisubulate lemma tip; the apex of the right subule is broken (× 30). 252. Oval scar; note the periphery ring confined to ¼ of scar (× 30). 253. Palea back beset with hairs that are longer than the cilia (arrow) along the edges of the keel (× 76). 254. Fatua-type epiblast (× 120).

A. atherantha

Fig. 255. Distribution map of *A. atherantha*. Mercator projection.

Habitat

It is locally rare, although its range of distribution is large. It grows in edges of cultivated fields and various post-segetal habitats. Although its diaspore resembles that of *A. sterilis,* it is not an aggressive weed and definitely does not constitute a threat to agriculture. Its natural or primary habitat seems to be on stony soil in park forests and in garigues, or low shrub Mediterranean vegetation.

Chorotype

All Mediterranean, adventitious in the European subregion of the Euro-Siberian region. Its status in Kenya is not clear, that is, it may be adventitious there or of relict distribution.

Mode of dispersal

By wind, and chiefly by grazing animals such as sheeps. Its diaspore is the entire spikelet without the glumes (Fig. 250). The diaspore is rather heavy and large.

Selected specimens

ALGERIA, Miserghin prairies bord du lac, 12 April 1852, *Balansa No. 551* (G, P, and W); CANARY ISLANDS, Teneriffe, Santa Ursula Muro, 18 March 1933, *Asplund No. 391* (G); CORSICA, Balagne bei Algajola, 27 June 1916, *Kukenthal No. 1656* (G); CYPRUS, Larnaka, 4 May 1893, *Deschamps No. 499* (PRC); EGYPT, *Presl* (PRC); FRANCE, Montpellier, June 1864, *Duval-Jouve* (BR); GREECE, *Delphi,* 9 September 1938, *BWNC* (CGE); ITALY, Liguria San Remo Olive garden, May 1901, *Kuntze* (GB); KENYA, 15 m N.W. of Timboroa edge of wheat field, 11 November 1961, *Bogdan* (DAO); PORTUGAL, Coimbra Calhabe, June 1937, *Matos No. 3669* (COI); SICILY, Palermo, 10 April 1855, *Huet du Pavillon* (CGE and LE); SPAIN, Andalousie Puerto St. Maria, 14 April 1852, *Bourgeau* (CGE); SWITZERLAND, Zürich Guterbahnhof, 1 August 1917, *Thellung* (BAS); TUNISIA, Sebkha Es Seldjoum, April 1881, *Roux* (P); UNITED KINGDOM, Cambridge, 25 June 1828, *Henslow* (CGE); USSR, Tauria Jaltiensis Gurzuf Aju-Dag, 17 May 1958, *Kotov* (KW); YUGOSLAVIA, Fiume Rijeka, *Noë No. 529* (BR, CGE, LE, PRC, and W).

Identification hints and remarks

This species can be easily confused with *A. sterilis, A. maroccana,* and *A. trichophylla.* The latter can be ruled out conclusively by examination of the lodicule. If the lodicule is fatua type it is not *A. trichophylla.* Furthermore, *A. sterilis* can be conclusively ruled out by examination of the back of the palea, because both *A. maroccana* and *A. atherantha* have hairs on the palea (Fig. 253), whereas in *A. sterilis* the palea is beset with prickles. The distinction between *A. maroccana* and *A. atherantha* can be made in many cases, but not in all cases, by a naked-eye inspection of the lemma tips and the vestiture of the spikelet. In *A. maroccana* the lemma tips are bilobed whereas in *A. atherantha* they are bisubulate or even almost biaristulate; and the hairiness of the spikelet is very dense and whitish in *A. maroccana*. See also additional remarks under *A. sterilis* and *A. maroccana*.

Notes

[1] I was unable to see the holotype of *A. atherantha*, but I found an isotype or fragment of the holotype in Trinius's herbarium (LE). The inscription on the label is "Avena atherantha Presl, mis. in Sicil. a se lectam Presl." Its identity is unquestionable. Furthermore, Malzew (Monogr. p. 274, footnote 383) stated that according to Willkomm (Oesterr. Bot. Zeitschr. 40:147 (1890)) *A. atherantha* Presl in herb. Prag (presumably PR) should be attributed to *"A. sterilis maxima."* The observation by Willkomm might support my assumption of the identity of the specimen in the Trinius's herbarium. Malzew (loc. cit.), however, decided to exclude the type and by the description alone included the name of that taxon as part of *A. barbata*.

[2] The holotype is conserved in PRC and is labeled "Avena maxima Pr./ A. pennsylvanica. Cyp. Gram. sic. excl. syn./ In arvis Panormi, inter segetes./Maj. 1817," presumably in Presl's handwriting. Another specimen with the same identity and a similar appearance may be a duplicate, that is, an isotype; although the label is not written by Presl, it bears a similar date and locality and seems to have been written by someone later because of the word "original."

[3] I found in W a specimen that agrees with the information given in the protolog. Its number is 113 and it is presumably annotated by Perez-Lara, collected on "8 Mai 1876," and it belongs here. However, Perez-Lara also mentions "in lusitania Meridionali (Bourg.)"; this is *A. longiglumis* of which I saw three specimens in herb. FI, K, and P (i.e., Bourgeau No. 2065). I designate the Perez-Lara specimen as lectotype and regard the three Bourgeau specimens as paralectotypes. See also Note 10 of *A. sterilis* L.

[4] I was able to obtain the holotype from COI; it is specified in Taborda de Morais's publication. I was unable to examine the paratype.

[5] I have seen the holotype (No. 3463) and one paratype (No. 3669); both belong here and are conserved in COI.

(25) A. FATUA L.

Homotypic synonyms

> A. fatua L., Sp. Pl. 80 (1753) [1]
> A. fatua ssp. fatua (L.) Thell., Vierteljahrs. Nat. Ges. Zürich 56:319 (1912)
> A. sativa ssp. fatua (L.) Fiori, Nuov. Fl. Anal. Ital. 1:109 (1923)

A. sativa ssp. fatua var. fatua (L.) Fiori, loc. cit.
A. pilosa Scopoli, Fl. Carn. ed. 2, 1:86 (1772), nom. ambig. [2]
A. lanuginosa Gilib., Exerc. Phyt. 2:539 (1792), nom. illegit. [3]
A. patens St. Lager in Cariot, Étude des Fleurs, ed. 8, 2:921 (1889) nom. illegit. [3]

Heterotypic synonyms

A. fatua var. pilosissima S.F. Gray, Nat. Arr. Brit. Pl. 2:131 (1821) [4]
A. nigra Wallroth, Linnaea 14:544 (1840) [5]
A. fatua var. glabrata Peterm., Fl. Bienitz 13 (1841) [6]
*A. fatua var. grandiflora Scheele, Flora 27:57 (1844)
A. fatua var. glabrescens Coss., Bull. Soc. Bot. France 1:15 (1854) [7]
A. fatua var. hirsuta Neilr., Fl. Nieder-Oesterr. 59 (1859) [8]
A. fatua var. flavescens Schur, Oesterr. Bot. Zeitschr. 10:71 (1860) [9]
A. fatua var. glabrescens Schur, loc. cit., nom. illegit. [9]
A. fatua var. nigrescens Schur, loc. cit. [10]
A. fatua var. genuina Ducom., Taschenb. Schweiz. Bot. 863 (1869)
A. fatua var. abbreviata Hausskn., Mitt. Geogr. Ges. Thür. Jena 3:239 (1885) [11]
A. fatua var. glabrescens Hausskn., loc. cit., nom. illegit. [12]
A. fatua var. nigrescens Hausskn., loc. cit., nom. illegit. [13]
A. fatua var. nigrescens f. albescens Hausskn., loc. cit. [14]
A. fatua var. nigrescens f. cinerascens Hausskn., loc. cit. [15]
A. sativa var. sericea Hook. f., Fl. Brit. Ind. 7:275 (1896) [16]
A. fatua var. glabrata Stapf in Rendle, Jour. Linn. Soc. Bot. 36:401 (1904) nom. illegit. [17]
A. fatua var. vulgaris Tourlet, Cat. Pl. Vasc. Indre Loire 568 (1908)
A. fatua ssp. fatua f. glabrata (Peterm.) Thell., Rec. Trav. Bot. Neerl. 25:426 (1929)
A. fatua ssp. fatua f. pilosissima (S.F. Gray) Thell., loc. cit.
A. fatua ssp. fatua f. pseudotransiens Thell., op. cit. 427 [18]
A. fatua ssp. cultiformis var. trichocarpa Malz., Monogr. 346 (1930) [19]
A. fatua ssp. fatua var. glabrata (Peterm.) Malz., op. cit. 320
A. fatua ssp. fatua var. glabrata subv. flocculosa Malz., op. cit. 322 [20]
A. fatua ssp. fatua var. pilosissima (S.F. Gray) Malz., op. cit. 316
A. fatua ssp. fatua var. pilosissima subv. scabrida Malz., op. cit. 318 [21]
A. fatua var. pilosa Syme ex Wolley-Dod, Fl. Sussex 507 (1937) [22]
A. fatua ssp. fatua var. intermedia subv. minima Tab. Mor., Bol. Soc. Brot. 2, 13:593 (1939)
A. fatua ssp. fatua var. pilosissima subv. biflora Tab. Mor., op. cit. 591 [23]
A. fatua ssp. fatua var. pilosissima subv. biflora f. cinerasecens (Hausskn.) Tab. Mor., op. cit. Tab. 1
A. fatua ssp. fatua var. pilosissima subv. scabrida f. albescens (Hausskn.) Tab. Mor., op. cit. 590
A. fatua ssp. fatua var. pilosissima subv. scabrida f. nigrescens (Hausskn.) Tab. Mor., op. cit. 591
A. fatua ssp. eufatua Hyland., Nord. Karlvax. 1:286 (1953) [24]
A. fatua var. pilosissima subv. scabrida (Malz.) Cabrera, Rev. Invest. Agricolas 11:378 (1957)

Description

Annual plants. Juvenile growth prostrate to erect. Color appearance green or light green. Culm erect. Relatively tall plants, 80–160 cm high. Ligules acute. Panicle equilateral. Spikelets (Fig. 256) short, 15–22 mm long without the awns; each spikelet has 2–3 florets. Glumes almost equal in length, 18–25 mm long, with 9–11 nerves; all the florets are disarticulating at maturity; scars (Fig. 257) oval to round, rarely elliptic; scar on third floret (Fig. 258) similar in configuration to scar on first floret or occasionally slightly flattened at the base but definitely not heart-shaped, and often there is no third floret at all. The periphery ring (Fig. 257) is confined to ⅓–¼ of scar. The awn is inserted at about the middle of the lemma. Lemma structure tough; the lemma tips are bidenticulate, or long or short bisubulate, or almost bilobed; macrohairs below awn insertion densely to sparsely present or totally absent. Paleas with 1–3 rows of cilia along edges of the keels, often 2 rows only; palea back beset with prickles. Lodicules (Fig. 259) fatua type, and never bearing prickles. Epiblast (Figs. 260, 261) fatua type, 0.45–0.6 mm wide. Chromosome number $2n = 42$. Genome AACCDD.

Distribution

Native to (Fig. 262) Afghanistan, Algeria, Austria, Belgium, Bulgaria, China, Czechoslovakia, Denmark, Egypt, Ethiopia, France, Germany, Hungary, India, Iran, Iraq, Italy, Japan, Kashmir, Korea, Lebanon, Luxemburg, Nepal, Netherlands, Norway, Pakistan, Poland, Portugal, Romania, Saudi Arabia, Spain, Sweden, Switzerland, Tibet, the United Kingdom, USSR, and Yugoslavia. Introduced in Argentina, Australia, Canada, Union of South Africa, and USA.

Phenology

Flowers from February to October depending on the area, but mostly from May to August in the northern hemisphere; and from October to January in the southern hemisphere.

Habitat

It is one of the three most important weeds in this genus; and is probably one of the most noxious weeds of cultivation in temperate and north-temperate areas. *Avena fatua* grows among field crops, in waste places, along

---

Figs. 256–261. Morphological and micromorphological diagnostic details of *A. fatua*. 256. Spikelet; note the lemmas are sparsely beset with macrohairs below the awn insertion (× 3). 257. Scar of the first floret; it is almost round; note the shiny periphery ring, which is confined to ⅓ of scar (× 40). 258. Scar of third floret; note that it (arrow) is slightly flattened at base (× 40). 259. A fatua-type lodicule; note the membranous wing (arrow), which is present in all the lodicules in the genus (although sativa-type lodicules have a side lobe in addition to the wing) (× 120). 260. Fatua-type epiblast (× 50). 261. Fatua-type epiblast with upper margin differently shaped (× 120).

A. fatua

283

A. fatua

Fig. 262. Distribution map of *A. fatua*. Mercator projection.

disturbed river banks, in orchards, along shoulders of highways, beside railroad tracks, on sand, in oases, beside docks, on limestone rocky soil, in disturbed wood clearings, and in valleys. It is normally locally abundant. It thrives particularly well in cultivated oat fields and among cereals in general.

Chorotype

Confined to the temperate and north-temperate regions of the world in the northern hemisphere and to similar climatic conditions in the southern hemisphere. It is therefore primarily Euro-Siberian, west and central Asiatic, and has been introduced into the Sino-Japanese, the flanks of the Himalayas of the Indian region, the west Mediterranean, S.W. Australian, Atlantic North American, South African and the Pampas Regions.

Mode of dispersal

The diaspores are the florets. They may be hairy or glabrous and are dispersed by man as he harvests them along with cereal grains, or else they drop at maturity; they may also be dispersed by various means of zoochory. The diaspores are known to have a differential dormancy and winterhardiness. All these factors contribute to the success of this species.

Selected specimens

AFGHANISTAN, Kataaghan 10 km Khumri, *Rechinger No. 33706* (W); ARGENTINA, San Juan Conception, *Cuzzo No. 1112* (BR); AUSTRALIA, Moreton Petrie weed of cult., 3 December 1931, *Blake No. 188* (BR); AUSTRIA, Aistersheim, *Keck No. 3494* (B, C, GB, LE, P, PRC, S, W, and WU); BELGIUM, Vise Bord de la Meuse, 1901, *Hardy No. 1892* (BR); BULGARIA, Radomir, August 1887, *Velenovsky* (PRC); CANADA *Alberta*, Medicine Hat Roadside, 22 July 1950 (DAO); CANADA *British Columbia*, Creston reclamation area, 5 August 1943, *Fastham* (UBC and V); CANADA *Manitoba*, Portage La Prairie, July 1894, *McMorine* (DAO); CANADA *New Brunswick*, Westmorland Point-du-chêne, 6 August 1964, *Roberts No. 64-2763A* (CAN); CANADA *Nova Scotia*, Kings Port Williams Railway, 14 August 1955, *Erskine No. 55906* (ACA and CAN); CANADA *Ontario*, Thunder Bay Longlac, 29 July 1952, *Baldwin and Breitung No. 3599* (CAN and WIN); CANADA *Prince Edward Island*, Charlottetown Railway yards, 9 August 1953, *Erskine No. 2333* (ACA and DAO); CANADA *Quebec*, St. Jean St. Valentin 4 mi du Lac, 29 July 1951, *Cinq-Mars* (DAO and QFA); CANADA *Saskatchewan*, Assiniboia Tableland Champs, 7 August 1951, *Boivin and Gillett No. 8566* (DAO and SASK); CHINA Kwangtung Yan-Fou, 2 February 1928, *Wang No. 566* (W); CZECHOSLOVAKIA, Bohemia Litomysl, July 1911, *Obdrzalek* (PRC); DENMARK, Valley Kjobenhavn, 14 September 1867, *Lange* (COI); EGYPT, Nazlet Ileiyan opp. El Billeida, 25 February 1927, *Täckholm* (S); ETHIOPIA, Goba 9300 ft elevation, *Kuckuk No. AB4* (DAO); FRANCE, Puy de Dome Clermont-Ferrand, 3 July 1882, *Fretleribaud No. 3923* (MT); GERMANY, Thuringia Weimar Feldern, July 1885, *Haussknecht* (BM and PRC); HUNGARY, Budapest vallis Lipotmezo, 13 June 1903, *Degen No. 161* (CGE, GB, W, and WU); INDIA, Punjab Mt. Buchara to Parmir Kekhg, 16 September 1916, *Vavilov No. 682* (WIR); IRAN, Kerman in ruderatis, 24 April 1948, *Rechinger No. 2943* (W); IRAQ, Basra Insula Shatt-Al-Arab, 16 March 1957, *Rechinger No. 8446* (W); ITALY, Bozene segetes, *Haussman No. 512* (BR); JAPAN, Honshu Yokohama, 15 June 1913, *Hisauchi No. 1125* (TI); KASHMIR, NW Himalaya Pangel, *Stolorzke No. 537894* (CAL); KOREA, Quelpaert Honguo in agris, 20 May 1908, *Taquet No. 1852* (G); LEBANON, Shemlan, 4 February 1955, *No. 1613* (G); LUXEMBOURG Torgny moissons, August

1919, *Dolisy* (BR); NEPAL, North of Tukucha Chhairogaon 9000 ft, 30 May 1954, *Stainton and Syke No. 808* (K); NETHERLANDS, Near Utrecht roadside, 7 July 1954, *Leeuwenberg* (COI); NORWAY, Tonsberg, August 1892, *Lewin No. 217934* (CAN); PAKISTAN, Baluchistan, 1891, *Elliott,* (K); POLAND, Lublin Novo Alexandria segetes, 11 July 1904, *Maximov No. 2819* (C, G, LE, S, and W); PORTUGAL, Souselas, May 1932, *Taborda de Morais No. 3763* (COI); ROMANIA, Transilvania Hermanstadt, July 1860, *Schur No. 7507* (P); SAUDI ARABIA, 40 km NE of Abragal Kabrit, 22 March 1968, *Mandaville No. 1725;* SPAIN, Leon Toreno del Sil prope Tonferrad, 14 July 1927, *Vavilov No. 1139* (WIR); SWEDEN, Goteborg Vid Kvillegatan, 5 August 1927, *Ohlsen No. 17449* (MT); SWITZERLAND, Zürich Guterbahnhof, 26 September 1917, *Thellung* (BAS); TIBET, Kam Yan-Tsy-Tsyan Basin, 25 July 1900, *Ladygin No. 367* (LE); UNION OF SOUTH AFRICA, Good Hope Loevenschwanz, September 1827, *Ecklon No. 925-14229* (HAL); UNITED KINGDOM, Norfolk between Hillington and Roydon, 4 July 1936, *Hubbard No. 20* (CGE, G, and S); USA *Alaska,* Juneau old barnyard, 5 September 1926, *Anderson No. 442* (CAN); USA *California,* Santa Cruz mountains, 7 July 1913, *Hitchcock No. 638* (BR); USA *Missouri,* Sheffield, 22 June 1905, *Bush No. 3025* (CGE); USA *Utah,* near Provo, 8 June 1959, *Anderson No. 248* (COI); USA *Washington,* S. Pullman Palouse, 17 July 1912, *Gaines No. 456* (WIR); USSR *Armenia,* Erevan, 1926, *Stoletova No. 903* (WIR); USSR *Azerbaidzhan,* Baku Mardakhany Vine, 26 May 1926, *Malzew No. 997* (WIR); USSR *Georgia,* Gori Chala lowlands, 1923, *Zhukovsky No. 703* (WIR); USSR *Kazakh,* Atbasar valley station, 2 September 1911, *Obraztsov No. 294* (WIR); USSR, *RSFSR Leningrad* Luga weed in orchard, 3 August 1918, *Ganeshin No. 1207* (WIR); USSR *Mongolia,* along Boro-Gol river, 1922, *Pissarev No. 1238* (WIR); USSR *Tadzhik,* Ura-Tyube, 10 August 1918, *Balabaev No. 832* (WIR); USSR *Turkmen,* Tach-Kepri Murghab, 26 August 1925, *Malzew No. 628* (WIR); USSR *Ukraine* Kherson Gurevka in crops, 13 June 1899, *Paczoski No. 1217* (WIR); USSR *Uzbek* Kokhan Khanate Kekh, 27 June 1871, *Fedchenko* (LE); YUGOSLAVIA, *Tojnica,* 13 July 1892, *Beck* (PRC).

Identification hints and remarks

The circumscription of this species in this work excludes many plants that, according to Malzew's system, would have been included into one broad taxon. My circumscription of *A. fatua* is essentially phenetic and is at variance with the biological species concept (e.g., Ladizinsky and Zohary 1971), which regards *A. fatua, A. sativa,* and practically all the hexaploid species of this work (Section Avena) as one biological species. In this work, there can be some confusion between *A. fatua* and the closely similar species, *A. hybrida* and *A. occidentalis.* Conclusive identification can be effected by means of the scars of the third floret, the lodicules, and the epiblasts. The sativa-type lodicule exists in *A. hybrida,* whereas *A. fatua* and *A. occidentalis* have a fatua-type lodicule (Fig. 259). The scar of the third floret is heart-shaped in *A. hybrida* and *A. occidentalis,* whereas in *A. fatua* it is rounded (Fig. 258). The epiblast in *A. hybrida* and *A. occidentalis* is septentrionalis type, and in *A. fatua* it is fatua type (Figs. 260–261).

Notes

[1] Strictly speaking I am violating the Code by using the name *A. fatua* L. for this taxon. According to typification procedure and following the taxonomic concept adopted in this work, *A. fatua* should be the correct name for *A. hybrida;* however, I feel *A. fatua* should be conserved because during the last two hundred years, and especially the last twenty years, this name has been used for Common Wild Oats in many scientific and semipopular articles. We do not have provision in the Code

for conservation of specific names, but there is a committee on stabilization of names of plants of economic importance. The preliminary report was published in Regnum Vegetabile 36 (1964) and additions were published in Regnum Vegetabile 60:180-114 (1969). In none of these lists does *A. fatua* appear. Obviously the correct application of the name depends on the taxonomic concept. In this monograph a number of taxa are recognized at the specific rank, namely *A. occidentalis, A. hybrida,* and *A. fatua;* these species used to be recognized as only one single species called *A. fatua.* The typification of *A. fatua* follows:

Avena fatua L., Sp. Pl. ed. 1:80 (1753).

"6. AVENA panicula patente, calycibus trifloris, flosculis basi pilosis. Fl. Suec. 97.
Avena seminibus hirsutis. Fl. lapp. 30. Roy. lugdb. 65.
Festuca utriculis lanugine flavescentibus. Bauh. pin. 10
Gramen avenaceum, locustis lanugine flavescentibus. Scheuch. gram. 239.
Habitat in Europae agris inter segetes."

Linnaeus repeated unchanged the phrase name from Flora Suecica; therefore, the specimens associated with it must first be considered for typification. I found only one specimen in LINN, No. 95.9 (Savage 1945, p. 19) that is closely linked with the phrase name and that appears to be closely associated with the publication of the Species Plantarum; this specimen is labeled "fatua 6 Fl. Suec." in Linnaeus' handwriting and I designate it as the lectotype (Fig. 263).

This specimen is identifiable as *A. hybrida* Peterm. because of the heart-shaped scar of the third floret, nearly subulate tips of lemmas, and a sativa-type lodicule, among other traits. *A. fatua* L., therefore, is the correct name for this species.

The first synonym listed after the phrase name is that of Flora Lapponica. This work is based upon specimens that are conserved, as was customary in the 18th century and earlier, in the form of a book, and this book is filed in the library at the Institut de France, Paris (Stearn 1957:115). I have examined the specimen of *Avena seminibus hirsutis,* which is glued on p. 13 of this book and which bears the following words in Linnaeus's handwriting "Avena seminibus basi hirsutis. Fl. Lapp. p. 21," "Gramen avenaceum, utriculis lanugine flavescentibus Tournef. 525," and "30." The identity of this paralectotype is the same as that of the lectotype.

I was unable to see the Van Royen specimen in Leiden (L), but I suspect that its identity is different from the above two because *A. hybrida* or the real *A. fatua* is very rare in Leiden. I also searched in the Burser herbarium in Uppsala, Sweden (UPS) for the specimen linked with Bauhin's synonym, but could not find such a specimen.

Finally, in the Vienna Museum herbarium (W) I discovered a specimen by Scheuchzer, which is the last-mentioned paralectotype in the protolog; its identity is *A. sterilis* L.

A. fatua

Fig. 263. Lectotype of *A. fatua*.

From the foregoing typification it is apparent that the type collection consists of two, or maybe three, discordant elements. The lectotype, which I have chosen and documented in Taxon 23:581 (1974), attaches the name to the taxon that I call *A. hybrida* in this work. I have decided for the reasons mentioned above to retain the name *A. fatua* for the taxon here circumscribed, with, of course, exclusion of the type, and in violation of the Code. At a later date, I shall also propose conservation of the name to the Committee for Stabilization of Plants of Economic Importance.

[2] From the description alone, it is difficult to know what species this is. Furthermore, many species of *Avena*, annuals and perennials (now *Helictotrichon*), are mentioned in the same description. In his Flora Carniolica, 2nd ed., Scopoli stated at the beginning of his "Plantarum Carnioliae": "Species Plantarum in Syst. Nat. Ill. Linnaei non numeratae, Asterisco (*) notantur..." All the species in this list bear the same number as those of the "Flora." Since *A. pilosa* does not have an asterisk it was probably not considered as a new species by Scopoli (?). The only specimen that Scopoli explicitly refers to is *A. pensylvanica* of Mygind's herbarium. I have obtained this plant from BP, and identified it as *A. brevis*. I was unable to locate any other original material, so that this name still remains obscure.

[3] These are new names for *A. fatua* L., which is cited as synonym.

[4] In spite of the fact that the types of S.F. Gray are unknown I tried to locate some authentic specimens in BM, but in vain. According to the description it belongs here.

[5] I did not find the type, but according to the description it is either *A. fatua* or *A. hybrida* and most likely the former.

[6] I tried to detect the type among the specimens in the W herbarium, but could not find it. (I found the type of *A. hybrida* Peterm. there!) According to the description it belongs here.

[7] The type collection is given in detail in a subsequent publication: Coss. et Dur., Expl. Sci. Alger 2:113 (1855). I designate the specimen from P labeled: "Cult. Hort. D.R. e seminibus prope Biskra a cl. Dr Gujon lectis" as lectotype. The specimen collected by Balansa "gravier de l'oued Biskra, à Biskra/ avril/1853" is a paralectotype and is also in P.

[8] This specimen, because it is described as the typical variety, is equivalent to *A. fatua* var. *fatua*.

[9] This is a composed diagnosis for the two var.; the first var. (see note 10) is the typical var. The two belong here according to the description; var. *glabrescens* is illegitimate because of var. *glabrescens* Coss.

[10] It seems that the first var., in other species also, was meant to be the typical var. (see also Schur, op. cit. p. 70: *A. sativa* var. *mutica*) and is,

therefore, equal to *A. fatua* var. *fatua*. In this case I was able to find an authentic specimen deposited in P, but because it was collected in 1876, it has no relation to the type. It is, however, labeled by Schur "Avena fatua L. var. nigrescens microspicula" and "No. 13125."

[11] I have examined the holotype and isotype in JE and found an isotype in LE and in W; they were collected in Ettersberg in August 1883.

[12] It is illegitimate because of the earlier var. *glabrescens* Cosson (1854). I have examined the type collection in JE and designated as lectotype the specimen annotated by Haussknecht, among others, "f. glabrescens nigra"; it belongs here. The paralectotype has also many annotations and one reads "f. nigra subglabra," but its identity is *A. fatua* × *sativa* F_1 hybrid.

[13] It is illegitimate because of the earlier *nigrescens* Schur (1860). I obtained from JE four pertinent specimens and have designated as lectotype the specimen labeled "nigresc. subpilosa" by Haussknecht. One paralectotype is annotated "nigresc. subglabresc." and there is a duplicate, that is, an isoparalectotype. Another paralectotype is annotated "nigra valde pilosa." All belong here.

[14] I have examined the holotype from JE; it is annotated "albescens pilosa," among others.

[15] I have designated the specimen labeled "mit hellbraunen, nach vorn hin Kahl werdenden Samen/Saatfelder b. Weimar" as lectotype. The paralectotype, among other annotations, is labeled "brunnea pilosa." Both types are in JE and belong here.

[16] Of the three syntypes at K, I chose as lectotype the specimen collected by Hooker in Sikkim on Sept. 7, 1849. The two paralectotypes and the lectotype belong here.

[17] It is illegitimate because of the earlier *A. fatua* var. *glabrata* Peterm. (1841). I have not seen the type collection; from the description it is difficult to decide if it is *A. fatua* or *A. hybrida*.

[18] The only specimens I have found in connection with the protolog were in K. I designate: "Bouly de Lesdain, Nord, Dunkerque décembre près du port, 17.6.1925" as lectotype. A possible paralectotype collected in the same place by the same collector, but on Oct. 11, 1925, does not belong here but in *A. sativa*.

[19] I have examined most of the syntypes from WIR. I designated No. 1139 as lectotype; most paralectotypes belong here except No. 1141, which is an *A. sativa fatuoid*.

[20] Of the two syntypes in WIR I designate No. 371 as lectotype.

[21] Many syntypes are listed on pp. 318–319 and are conserved in WIR; of these I chose as lectotype No. 1207, which is an *A. fatua*. Some

paralectotypes are *A. sativa fatuoid* (such as No. 905) and quite a few are *A. hybrida* (such as No. 902); some are even *A. trichophylla* (such as No. 1221), or even *A. sterilis* (such as No. 915).

[22] The type probably never existed, but it belongs here according to the description, which can serve as type.

[23] According to the protolog the type is the same as that of *A. fatua* ssp. *fatua* var. *pilosissima* subv. *biflora* f. *cinerascens* (Hausskn.) Tab. Mor. because Taborda de Morais established the subvariety to accommodate the forma. This approach is implied in his work where he also systematically established varieties to accommodate his subvarieties. (Incidently, specimen No. 3854 is an *A. sterilis* and does not belong here.)

[24] This is equal to the typical variety according to the Rules, and it was common practice in the 1950's to designate them with the prefix *eu*.

(26) A. HYBRIDA PETERM.

Homotypic synonyms

A. hybrida Peterm., Fl. Bienitz 13 (1841) [1]
A. fatua var. hybrida (Peterm.) Aschers., Fl. Brandenb. 1:828 (1864)
A. fatua ssp. fatua var. hybrida (Peterm.) Thell., Vierteljahrs. Nat. Ges. Zürich 56:322 (1912)
A. fatua ssp. fatua var. hybrida subv. petermanni Thell., op. cit. 324 [2]

Heterotypic synonyms

*A. intermedia Lestib., Bot. Belg. 3:36 (1827) [3]
A. fatua var. intermedia (Lestib.) Lej. et Court., Comp. Fl. Belg. 1:71 (1828)
A. fatua ssp. fatua var. intermedia (Lestib.) Lej. et Court., loc. cit.
*A. vilis Wallr., Linnaea 14:543 (1840) [4]
A. intermedia Lindgr., Bot. Not. 1841:151 (1831) nom. illegit. [5]
A. fatua var. intermedia (Lindgr.) Hartm., Handb. Skand. Fl. ed. 4:30 (1843)
A. japonica Steud., Syn. Pl. Gram. 1:231 (1854) [6]
A. fatua var. glabra Ducomm., Taschenb. Schweiz. Bot. 863 (1869) [7]
A. fatua var. intermedia (Lindgr.) Ducomm., loc. cit.
A. fatua var. contracta Hausskn., Mitt. Geogr. Ges. Thür. 3:239 (1885) [8]
A. fatua var. vilis (Wallr.) Hausskn., Mitt. Thür. Bot. Ver. N.F. 6:39 and 45 (1894)
A. fatua var. intermedia (Lindg.) Husnot, Gram. Fl. Belg. 39 (1896)
A. fatua var. glabrescens Tourlet, Cat. Pl. Vasc. Indre et Loire 568 (1908) nom. illegit. [9]
A. septentrionalis Malz., Bull. Angewandte Bot. 6:915 (1913) [10]

A. fatua ssp. fatua f. hybrida (Peterm.) Thell., Rec. Trav. Bot. Neerl. 25:426 (1929)
A. fatua ssp. fatua f. intermedia (Lestib.) Thell., loc. cit.
A. fatua ssp. meridionalis Malz., Bull. Appl. Bot. Pl.-Breed. 20:140 (1929) [11]
A. fatua ssp. septentrionalis (Malz.) Malz., loc. cit.
A. fatua ssp. fatua var. vilis (Wallr.) Malz., Monogr. 326 (1930)
A. fatua ssp. meridionalis var. elongata Malz., op. cit. 309 [12]
A. fatua ssp. meridionalis var. grandis Malz., op. cit. 306 [13]
A. fatua ssp. meridionalis var. grandis subv. scabriuscula Malz., loc. cit. [14]
A. fatua ssp. meridionalis var. longiflora Malz., op. cit. 308 [15]
A. fatua ssp. meridionalis var. longispiculata Malz., loc. cit. [16]
A. fatua ssp. septentrionalis var. glabella Malz., op. cit. 298 [17]
A. fatua ssp. septentrionalis var. glabripaleata Malz., op. cit. 295 [18]
A. fatua ssp. septentrionalis var. sparsepilosa Malz., op. cit. 297 [19]
A. fatua ssp. septentrionalis var. valdepilosa Malz., op. cit. 294 [20]
A. fatua var. intermedia (Lindgr.) Vasc., Rev. Agron. 19:19 (1931) comb. illegit.
A. meridionalis (Malz.) Roshev. in Fedtsch., Fl. Turkom. 1:105 (1932)
A. meridionalis var. grandis (Malz.) Roshev., op. cit. 106
A. fatua ssp. fatua var. intermedia subv. longispiculata (Malz.) Tab. Mor., Bol. Soc. Brot. 2, 13: Tab. 1 (1939)
A. fatua ssp. fatua var. intermedia subv. sparsepilosa (Malz.) Tab. Mor., loc. cit.
A. fatua ssp. fatua var. pilosissima subv. scabriuscula (Malz.) Tab. Mor., loc. cit.
A. fatua ssp. fatua var. pilosissima subv. valdepilosa (Malz.) Tab. Mor., op. cit. 591
A. fatua ssp. fatua var. vilis subv. elongata (Malz.) Tab. Mor., op. cit. Tab. 1
A. fatua ssp. fatua var. vilis subv. glabella (Malz.) Tab. Mor., loc. cit.

Description

Annual plants. Culms erect, nodes often hairy (Fig. 268). Ligules acute or obtuse with a point. Panicle equilateral, sometimes slightly secund. Spikelets (Fig. 264) short, 1.5–2.4 cm long without the awn; each spikelet has 2–4 florets, and most commonly 3. Glumes equal in length or nearly so, 16–22 mm long, with 7–9 (–11) nerves; all the florets disarticulating at maturity; scars oval elliptic to round, and scars on the 3rd floret (Fig. 266) and sometimes the 2nd (Fig. 267) are heart-shaped. The periphery ring is only confined to ⅓–½ of the scar. Awns inserted at about middle of lemma. Lemma structure tough; lemma tips (Fig. 265) irregularly bidenticulate to bisubulate; macrohairs absent or present below the insertion of the awn or only a few found around the insertion. Paleas usually with 2 rows of cilia along the edges of the keels, but sometimes with 1 or rarely with 3; palea back beset with prickles. Lodicules (Figs. 269, 270) sativa type, and usually the side lobe is very small; the lodicules never bear prickles. Epiblast (Figs. 271–273) septentrionalis type, 0.35 mm wide. Chromosome number $2n=42$. Genome AACCDD.

Distribution

Native to (Fig. 274) Afghanistan, Austria, Belgium, China (west), Denmark, France, Germany, India, Iraq, Japan, Luxemburg, Nepal, Pakistan, Sweden, Tibet, Turkey, the United Kingdom, and the USSR. Introduced in Canada.

Phenology

Flowers chiefly from July to August, but a number of plants in flowering stages have been observed from March to October.

Habitat

Mountain slopes, hills, valleys, and in *Quercus* forest clearings. Its primary habitat is in high elevations of central Asia, such as Nepal, Tibet, Tadzhikistan, and northeast to Mongolia. It grows also on road shoulders, as a weed in cultivation, and in various waste places. It is well adapted to cold climate, and is not a serious weed.

Chorotype

West and central Asiatic, and penetrating as a weed into the Euro-Siberian region and very slightly into the northern Atlantic – North American region.

Mode of dispersal

The diaspore is the floret. A great many of the diaspores are glabrous and relatively small. The diaspores either drop at maturity, are transferred a short distance by the wind, or are probably dispersed by various means of zoochory. To a very small extent the diaspores pass as impurities in cereal grains. The diaspores are presumably winter hardy.

A. hybrida

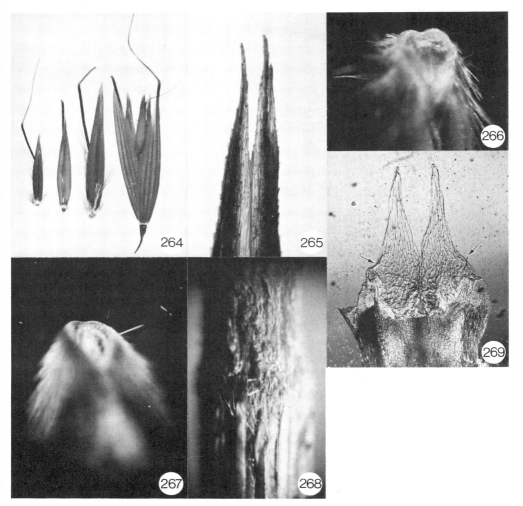

Figs. 264–273. Morphological and micromorphological diagnostic details of *A. hybrida*. 264. Spikelet and 3 florets removed from another spikelet (× 3.5). 265. Shortly bisubulate lemma tip; note the membranous, setula-like structure on the right subule (× 50). 266. 3rd floret with a heart-shaped scar (× 76). 267. 2nd floret with a slightly heart-shaped scar (× 40). 268. Hairy nodes; note that the hairs are actually at the edge of the node (× 24). 269. Sativa-type lodicules; arrows indicate very small side lobes (× 50).

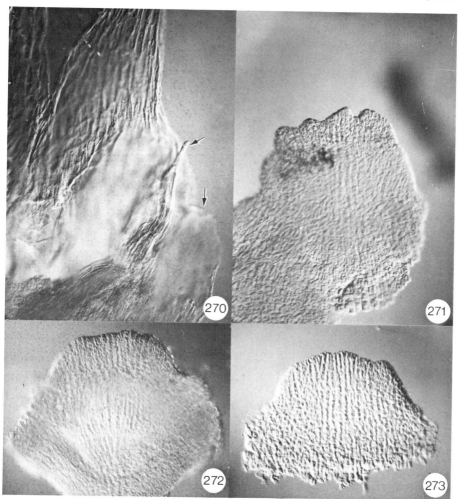

Figs. 270–273. 270. Lower part of the right lodicule magnified with upper arrow showing the small side lobe; lower arrow points to the membranous wing (× 120). 271–273. Septentrionalis-type epiblasts; note the convex upper margin, which is a characteristic feature of this type (× 120).

A. hybrida

Fig. 274. Distribution map of *A. hybrida*. Mercator projection.

Selected specimens

AFGHANISTAN, Ghazni Salzab Okak-Behzud, 6 July 1962, *Rechinger No. 17949* (W); AUSTRIA, Aistersheim, *Keck No. 3494* (G); BELGIUM, Eeckeren champs cultivés, 1 August 1887, *Van-de-Put* (BR); CANADA, Manitoba Aweme, 1913, *Criddle No. 537* (WIR); CHINA, Turkestan Arpa Khan Yarkaud, *Scully* (G); DENMARK, Valley agris, September 1861, *Lange* (BR); FRANCE, Compiegne champs potagers, 20 August 1844, *Crepin* (BR); GERMANY, Hamburg, 1841, *Sender* (CGE); INDIA, Kumaon Almora, *Strachey and Winterbottom* (BR and P); IRAQ, 60 km S.E. of Bagdad Hafriya, 10 April 1957, *Rechinger No. 9084* (W); JAPAN, Chiba Higashikai Sushika-Ku, 28 June 1936, *Makino No. 45101* (CAN); LUXEMBOURG, Bettembourg, *Reichling No. 21442* (BR); NEPAL, Balangra pass 12000 ft. in *Quercus* forest, 19 July 1952, *Polunin and Syke No. 2505* (K); PAKISTAN, Dacca, *Wett No. 537887* (CAL); SWEDEN, Scania Skurup, 15 July 1907, *Schultz* (MT); TIBET, Zaidam 9200 ft., 30 August 1843, *Przewalski* (LE); TURKEY, Konya, 1925, *Zhukovsky No. 802* (WIR); UNITED KINGDOM, Cloygate Beanfield, 1871, *Watson No. 1307* (CGE); USSR, *RSFSR*, Arkhangelsk Onega Vongudy, 18 August 1910, *Egorov No. 38* (WIR); USSR *Armenia*, near Gezeldar, 1926, *Stoletova No. 890* (WIR); USSR *Georgia*, Ksan river valley, 1922, *Zhukovsky No. 715* (WIR); USSR *Mongolia*, Aymak Bain-Khongor, 25 July 1948, *Grubov No. 6904* (LE); USSR, *RSFSR*, Semipolatinsk Bukhtarma Altai, 1924, *Sinskaya No. 639* (WIR); USSR *Tadzhik*, Ura Tjube Yangi-Aryk, 5 August 1918, *Balahaev No. 838* (WIR); USSR *Kirghiz*, Fergana Kokand Sokh, 1926, *Kobelev No. 1003* (WIR); USSR *Urol* Saranul Petropavlovsk, 27 August 1912, *Papkov No. 374* (WIR).

Identification hints and remarks

This species can be confused with *A. fatua*, and if the geographical origin of the specimen is unknown it can also be confused with *A. occidentalis*. Conclusive identification can be effected by examination of the lodicules and epiblasts. *Avena hybrida* has sativa-type lodicules (Figs. 269, 270) and septentrionalis-type epiblasts (Figs. 271–273), whereas *A fatua* and *A. occidentalis* have fatua-type lodicules. The epiblast of *A. fatua* is of the fatua type, but *A. occidentalis* also has septentrionalis-type epiblasts, which are nevertheless wider than those in *A. hybrida*. Furthermore, scars of the third florets are heart-shaped in *A. hybrida* and *A. occidentalis*, but not in *A. fatua*.

Notes

[1] I have detected a specimen in W, annotated by Petermann "Avena hybrida Peterm.! Lips. fl." signed "Peterm."; I designate this specimen as lectotype (Fig. 275).

[2] This is a new name for *A. hybrida* Peterm. and is based on the same type.

[3] According to the description, it could well belong here or could be an *A. sativa fatuoid*. Stafleu (1967:265) states that Lestiboudois' specimens are unknown.

[4] Judging from the description it might belong here; the other probable alternative is *A. fatua*. I tried unsuccessfully to find Wallroth's specimens in JE, G, and BR.

A. hybrida

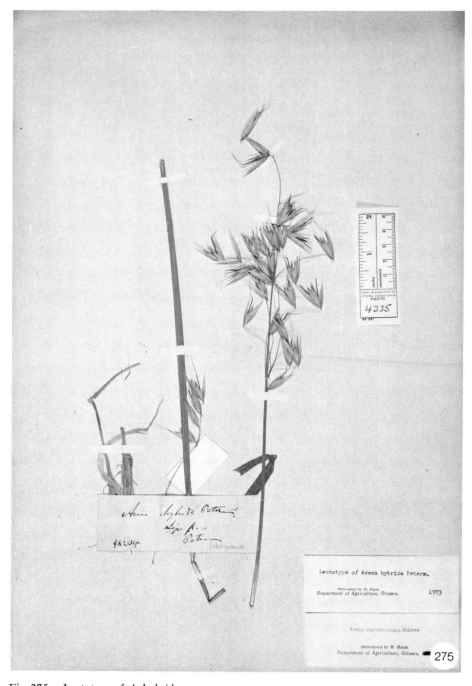

Fig. 275. Lectotype of *A. hybrida*.

[5] The name is superfluous because of the earlier *A. intermedia* Lestib. The type collection is scattered in various herbaria. The holotype is in C and is annotated by Lindgreen "Avena intermedia Lindgr.!/Degeberg Vestrogothiae Lindgreen." The isotype, also in C, is similar. A specimen in B annotated in Lindgreen's hand "Avena intermedia Lindgr. nov. spec." may be an isotype. LE, P, and W have specimens that may be isotypes; these are annotated "Vestrogothia circa Lidköping."

[6] I have examined the holotype from L; its accession number is 100190. The isotype is in P with Steudel's annotations similar to those of the holotype. M has a specimen (No. 6978); it may also be an isotype but it does not belong here.

[7] This is a new name for *A. hybrida* Peterm. at the variety level.

[8] The holotype is labeled so by Haussknecht and is conserved in JE.

[9] It is superfluous to the former var. *glabrescens* Cosson (1854). I was unable to examine the types but because this is a new name for *A. hybrida* Peterm. the type collection also includes it. For this reason it appears here in the synonymy.

[10] The protolog does not mention particular collections or specimens; instead it states: from Archangelsk through Perm to West-Siberia. All these are listed in Malzew's monograph on pp. 294 and 295 in his variety α, and from these the lectotype could be chosen provided that the date of these potential syntypes will not postdate the publication of *A. septentrionalis*. Consequently, I have designated No. 17 as lectotype and the date on this one is 1.8.1910. Some paralectotypes are Nos. 37, 434, 217, and 215.

[11] The protolog offers few guidelines for the typification of this subspecies. In Malzew's monograph, published one year later, we get a clue. Among the syntypes that I have seen (and these are listed under the different subvarieties of var. *grandis*), some belong here, some to *A. fatua,* and some to *A. occidentalis.* I have chosen No. 628 as lectotype and it belongs here. All the type material is in WIR. It is worthwhile to mention that Malzew's concept of this subspecies completely loses its meaning under my present taxonomy.

[12] I have designated No. 552 from WIR as lectotype. Among the paralectotypes No. 594 belongs to *A. fatua,* whereas Nos. 599 and 615 belong here.

[13] In the protolog Malzew stated that this variety is "forma typica" of *A. fatua* ssp. *meridionalis.* Thus the lectotype is the same (see Note 11 above).

[14] Because the lectotype of ssp. *meridionalis* is one of the syntypes of this subvariety, it automatically becomes the lectotype here too. This is No. 628 and, therefore, by lectotypification this subvariety is equivalent to *A. fatua* ssp. *meridionalis* var. *meridionalis* subv. *meridionalis.*

[15] I have found all the syntypes in WIR, and designated No. 1002 as lectotype. The specimen No. 1001 lacks seeds, so no positive identification is possible, but all the other paralectotypes belong here.

[16] I have searched in WIR for the syntypes, but was unable to find any. I decided to place this variety under synonymy here because the description and the picture point to great similarity with other varieties that belong here.

[17] I have found all the syntypes in WIR. Most belong here and I designate No. 1245 as lectotype. The paralectotypes Nos. 380, 832, 833, and 1238 are *A. fatua*.

[18] Of the few syntypes that I was able to examine from WIR, I have designated No. 38 as lectotype. Malzew cites also *A. sativa* var. *sericea* Hooker pro parte but it is *A. fatua* and does not belong here (see Note 16 under *A. fatua*).

[19] I have seen many syntypes in WIR and I chose No. 1237 as lectotype.

[20] This is equivalent to var. *septentrionalis* according to the Rules because Malzew stated here "forma typica."

Supplementary note: *A. fatua* ssp. *meridionalis* (Malz.) A. W. Hill, Ind. Kew. Suppl. 9:31 (1939) is erroneously cited by Chase (1962) because the combination was never made by Hill.

(27) A. OCCIDENTALIS DUR.

Homotypic synonym

A. occidentalis Dur., Cat. Graines Bordeaux 25 (1864) [1]

Heterotypic synonyms

A. fatua f. deserticola Hausskn., Mitt. Thür. Bot. Ges. 13/14:46 (1899) [2]
A. fatua ssp. fatua var. pilosissima subv. deserticola (Hausskn.) Malz., Monogr. 320 (1930)
A. fatua ssp. meridionalis var. grandis subv. puberula Malz., op. cit. 306 [3]
A. fatua ssp. meridionalis var. grandis subv. villosa Malz., op. cit. 307 [4]
A. fatua ssp. fatua var. pilosissima subv. parva Tab. Mor., Bol. Soc. Brot. 2, 13:592 (1939) [5]
A fatua ssp. fatua var. pilosissima subv. puberula (Malz.) Tab. Mor., op. cit. Tab. 1
A. fatua ssp. fatua var. pilosissima subv. villosa (Malz.) Tab. Mor., op. cit. Tab. 1

Description

Annual plants. Juvenile growth prostrate to erect. Color appearance green. Culm erect. Relatively low plants, 50–80 cm high. Ligules acute. Panicle slightly flagged or unilateral. Spikelets (Fig. 276) long, 2–3 cm without the awns; each spikelet has 3–4 florets. Glumes equal or nearly so, 30–35 mm long, with 7–9 nerves; all the florets are disarticulating at maturity; scars round-elliptic, the third and fourth are heart-shaped (Fig. 278) and sometimes the second may also be heart-shaped. Periphery ring confined to ⅓ of scar. Awns inserted at about middle of lemma. Lemma structure tough; lemma tips (Fig. 277) bisubulate, or sometimes ending into very short aristulae; lemmas densely beset with macrohairs below the insertion of the awn, rarely glabrous or only a few hairs present. Paleas with 1–3 rows of cilia along edges of keels, most with only 2 rows; palea back beset with prickles. Lodicules (Figs. 279, 280) fatua type never bearing prickles. Epiblast (Fig. 281) septentrionalis-type, 0.6 mm wide. Chromosome number $2n = 42$. Genome AACCDD.

Distribution

Native to (Fig. 282) Azores, Canary Islands, Egypt, Ethiopia, Madeira, Portugal, and Saudi Arabia. Introduced in Mexico and the USA (California and Oregon).

Phenology

Flowers chiefly from March to June, but the range may be from January to October.

Habitat

Found (Fig. 148) on alluvial soils, in oases, along banks of rivers, in valleys, on rocky slopes, and among crops, and also on basaltic soil. Moisture seems to be more of a factor than type of soil.

Chorotype

Macaronesian and south Mediterranean regions and the northern parts of the east Saharo-Arabian subregion. Introduced and naturalized in the southern part of the Pacific North American region and in the northern parts of the Caribbean region.

Mode of dispersal

The diaspores are the florets. They are glabrous or more often hairy. The florets shed at maturity and may be dispersed by the wind or by animals

A. occidentalis

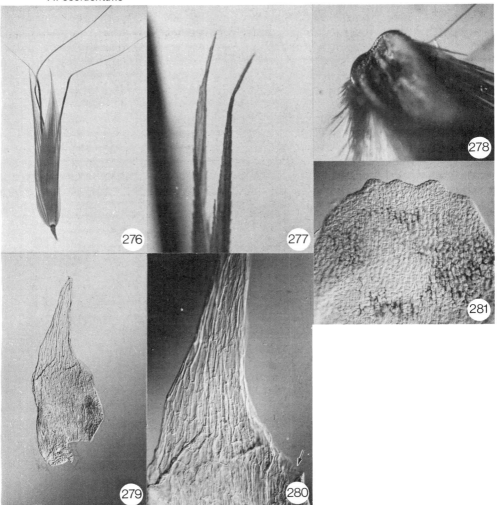

Figs. 276–281. Morphological and micromorphological diagnostic details of *A. occidentalis*. 276. Entire spikelet; note the resemblance between this spikelet and that of *A. longiglumis* (× 2). 277. Bisubulate tip of lemma; the awn is in the background out of focus (× 50). 278. Heart-shaped scar of a 3rd floret; note the shiny periphery ring confined to ⅓ of scar (× 50). 279. Fatua-type lodicule (× 50). 280. Detail of the lower part of the body of the lodicule in Fig. 279; note the folded membranous wing and the absence of a side lobe (× 120). 281. Septentrionalis-type epiblast (× 120).

(e.g., ants and sheep). It could be locally dispersed by man at least within the Canary Islands because it is found there among crops, but the crops are not transported to other places. In addition *A. occidentalis* is restricted to a warmer climate than *A. fatua*. It appears, therefore, that man contributes very little to dispersal. Man introduced *A. occidentalis* to Mexico and the USA during the Spanish Period by carrying sheep, other animals, and feed from the Canaries to the New World.

A. occidentalis

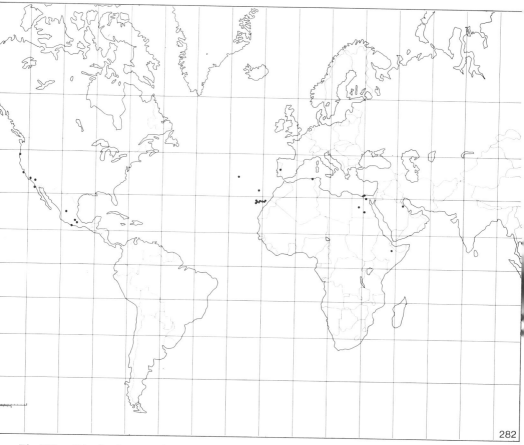

Fig. 282. Distribution map of *A. occidentalis*. Mercator projection.

Selected specimens

AZORES ISLANDS, St. Michael's Islands, 1845, *Hunt No. 302* (CGE); CANARY ISLANDS, Gomera San Sebastian, 17 April 1964, *Lid* (O); EGYPT, Farafrah oasis Libyan desert, 2 July 1874, *Ascherson No. 2515* (JE); ETHIOPIA, Fafan River Benhuai, 28 January 1886, *Schweinfurth* (JE); MADEIRA ISLAND, Eastern part, May 1971, *Kjellsson* (C); MEXICO, Posadas propre Puebla Rancho, 4 March 1909, *Nicolas* (BR); PORTUGAL, Covilha arridores da Cidada, 10 June 1939, *Taborda de Morais No. 3856* (COI); SAUDI ARABIA, Qaryat Al Ulya 25 km SE, 7 May 1964, *Mandaville No. 216* (K); USA *California,* San Diego, 8 May 1884, *Orcutt* (COI); USA *Oregon,* Sauvies Island, July 1883, *Howell* (BR).

Identification hints and remarks

Plants that belong to this species are often misidentified as *A. fatua* or even *A. sterilis*. Simple inspection of the mode of disarticulation will readily eliminate the possibility of *A. sterilis*. The scars of the third and fourth florets are heart-shaped in *A. occidentalis* (Fig. 278), but round in *A. fatua* (or else the spikelet of *A. fatua* has only two florets). The spikelets of *A. occidentalis* resemble those of *A. longiglumis* in external habit only (Fig. 276), that is, in the number of florets and in the shape of the glumes. Conclusive identification can be effected by inspection of the epiblast; it is septentrionalis type in this species, but fatua type in *A. fatua*.

Notes

[1] This species has been described in 1865 in the 1864 Catalogue. The latter date is often mistakenly quoted. Simultaneously the protolog was reproduced in Bull. Soc. Bot. France 12: Bibl. 78 (1865). According to the protolog H. de La Perraudière is the collector and the specimen was in J. Gay herbarium; it is now at K. Furthermore, one seed has been taken by Durieu, which then "devint l'origine de l'existence de la plante dans le Jardin de Bordeaux"; I found this specimen in P. The two specimens, K and P, obviously constitute the type collection. I designate the P specimen as lectotype (Fig. 283) because it bears Durieu's annotation "Avena occidentalis DR/HDR. 20 juin 1857/De l'Ile de Fer (Laperraudière)"; I regard the K specimen as a paralectotype, and it is annotated "41 Avena Hierro (=île de Fer) 6 mai 1855," presumably by La Perraudière himself. The paralectotype has also J. Gay's notes appended to it, mentioning the correspondence with Durieu.

[2] According to the protolog there are many syntypes. I have obtained all of them from JE. I have selected the specimen labeled "A fatua fusca longiglumis vegeta/Benhusi (Faja)/Schweinfurth 28/1.86" as lectotype. Some of the paralectotypes belong here and others are *A. sterilis* or even *A. fatua*.

[3] I was unable to find the holotype at WIR. It is likely to belong here according to the description and distribution, but its true identity has not been confirmed.

[4] I was unable to find in WIR any of the syntypes mentioned in the protolog. Judging from the description and the distribution it is likely to belong here, but this remains to be confirmed.

[5] I have examined the holotype from COI; it is No. 524. Because the identification was not conclusive, I cannot be certain that it belongs here.

A. occidentalis

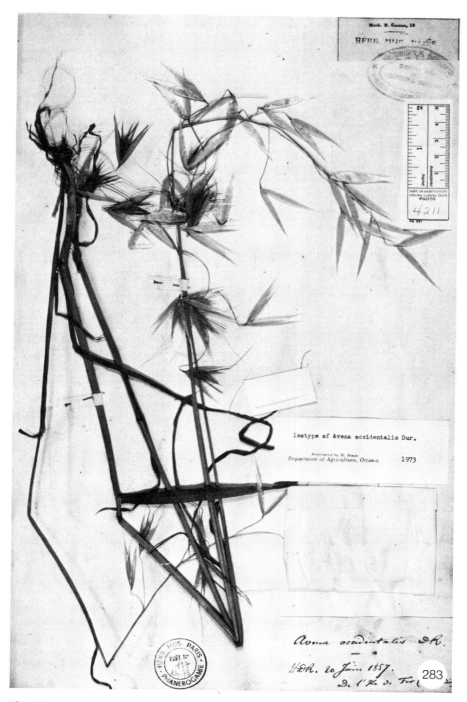

Fig. 283. Lectotype of *A. occidentalis*.

(28) A. SATIVA L.

Homotypic synonyms

A. sativa L., Sp. Pl. 79 (1753) [1]
A. dispermis Mill., Gard. Dict. ed. 8, n. 1 (1768) nom. illegit. [2]
A. fatua var. sativa (L.) Hausskn., Mitt. Geogr. Ges. Thür. 3:238 (1885)
A. fatua ssp. sativa (L.) Thell., Vierteljahrs. Nat. Zürich 56:325 (1911)
A. sativa ssp. sativa (L.) Fiori, Nuov. Fl. Anal. Ital. 1:108 (1923)
A. sativa ssp. sativa (L.) Tab. Mor., Bol. Soc. Brot. 2, 13:598 (1939)

Heterotypic synonyms

A. orientalis Schreb., Spicil. Fl. Lips. 52 (1771) [3]
A. nuda var. multiflora Haller, Nov. Comm. Soc. Sci. Göttingen 6: Tab. 6, fig. 36 (1775) [4]
A. heteromalla Haller, op. cit. figs. 32, 33 [5]
A. sativa var. alba Haller, op. cit. fig. 29 [6]
A. sativa var. nigra Haller, op. cit. fig. 31 [7]
A. sativa var. variegata Haller, op. cit. fig. 30 [8]
A. sativa var. nigra Schrank, Baier. Fl. 1:375 (1789) [9]
A. tatarica Ard. ex Saggi, Accad. Padov. 2:101, Tab. 1 (1789) [10]
A. pendula Gilib., Exerc. Phyt. 2:539 (1792) nom. illegit. [11]
A. heteromalla Moench, Meth. Pl. 195 (1794) [5]
A. racemosa Thuill., Fl. Env. Paris, ed. 2, 59 (1799) [12]
*A. sativa var. alba Koel., Descr. Gram. 290 (1802) nom. illegit. [13]
*A. sativa var. nigra Koel., loc. cit. nom. illegit. [13]
*A. sativa var. nigra subv. glabra Koel., loc. cit. [13]
*A. sativa var. nigra subv. pilosa Koel., loc. cit. [13]
A. agraria var. mutica Brot., Fl. Lusit. 1:106 (1804) [14]
A. georgica Zuccag., Obs. Bot. Cent. 1:14 (No 31) (1806), reprinted in Roemer, Coll. Bot. 126 (1809) [15]
A. rubra Zuccag., Obs. Bot. Cent. 1:14 (No 30) (1806), reprinted in Roemer, Coll. Bot. 126 (1809) [16]
A. podolica Pascal ex Zuccagni in Roem., Coll. Bot. 126 (1809) pro syn. A. rubra Zuccagni [17]
A. sativa var. mutica Stokes, Bot. Mat. Med. 1:158 (1812) [18]
A. anglica Hort. ex Roem. et Schult., Syst. Veg. 2:669 (1817) pro syn.
A. chinensis Fisch. ex Roem. et Schult., loc. cit., pro syn. A. nuda L. [19]
A. nuda var. chinensis Fisch. ex Roem. et Schult., loc. cit.
A. cinerea Hort. ex Roem. et Schult., loc. cit., pro syn.
A. flava Hort. ex Roem. et Schult., loc. cit., pro syn.
A. georgiana Roem. et Schult., loc. cit., pro syn. [20]
A. trisperma Auct. ex Roem. et Schult., loc. cit., pro syn.
A. unilateralis Brouss. ex Roem. et Schult., loc. cit. pro syn. A. orientalis Schreb.
A. orientalis var. mutica S. F. Gray, Nat. Arr. Brit. Pl. 2:130 (1821) [21]
A. sativa var. mutica S. F. Gray, loc. cit. [21]
A. sativa var. nigra S. F. Gray, loc. cit., nom. illegit. [21]
*A. hungarica Lucé, Topogr. Nachr. Insel Oesel 20 (1823) nom. nud.

*A. sativa var. aristata Schlecht., Fl. Berolin. 1:51 (1823)
A. sativa var. mutica Schlecht., op. cit. 52 nom. illegit. [22]
A. chinensis Metzg., Eur. Cereal. 53-54 (1824) [23]
A. sativa var. praecox Metzg., op. cit. 71 nom. nud.
A. trisperma Schuebl. ex Schüz, Diss. Inaug. Bot. 8 (1825) [24]
A. sativa var. leucosperma Sweet, Hort. Brit. 452 (1826) nom. nud.
A. sativa var. melanosperma Sweet, loc. cit., nom. nud.
A. sativa var. mutica Lestib., Botan. Belg. 2:35 (1827) nom. illegit.
A. nuda var. chinensis Fisch. ex Link, Hort. Berol. 1:109 (1827) pro syn. A. chinensis Metzg.
A. nuda var. chinensis Kunth, Rev. Gram. 1:103 (1829) [25]
A. chinensis Link, Handb. Gewächs. 1:43 (1829) nom. illegit. [26]
A. orientalis var. nuda Link, Hort. Berol. 2:254 (1833) [27]
A. orientalis var. mutica Spenner ex Schuebler et Martens, Fl. Würtemb. 71 (1834), nom. illegit. [28]
A. sativa var. aristata Schuebler et Martens, loc. cit., nom. illegit. [29]
A. sativa var. aristata f. alba Schuebler et Martens, loc. cit. [30]
A. sativa var. aristata f. nigra Schuebler et Martens, loc. cit. [30]
A. sativa var. mutica Schuebler et Martens, loc. cit., nom. illegit. [28]
*A. sativa var. aristata Krause, Abbild. Beschr. Getr. Heft 7:13 (1837) nom. illegit. [29]
*A. sativa var. nigra Krause, op. cit. 15, nom. illegit. [31] see [32]
A. sativa var. praegravis Krause, op. cit. 7, nom. illegit. [32]
*A. sativa var. rubida Krause, op. cit. 13, see [32]
*A. sativa var. turgida Krause, op. cit. 7-8, see [32]
A. orientalis var. mutica Peterm., Fl. Lips. 105 (1838) nom. illegit. [33]
A. sativa var. melanosperma Peterm. Fl. Bienitz 13 (1841) nom. illegit. [34]
A. sativa var. mutica Peterm. loc. cit., nom. illegit. [35]
A. sativa var. trisperma (Schuebler) Koch in Peterm., loc. cit.
A. sativa var. chinensis Doell, Rhein. Fl. 99 (1843) [36]
A. sativa var. genuina Godr., Fl. Lorr. 3:150 (1844) [37]
A. sativa var. involvens Godr., loc. cit. [38]
A. sativa var. nigra Wood, Class-book ed. 2, 610 (1847) nom. illegit. [39]
A. sativa var. secunda Wood, loc. cit. [40]
A. byzantina Koch, Linnaea 21:392 (1849) [41]
A. nuda var. quadriflora Opiz, Seznam Rostl. Ceske 20 (1852) nom. nud. [42]
*A. orientalis var. aristata Maly in Opiz, loc. cit., nom. nud. [44]
A. orientalis var. mutica Maly in Opiz, loc. cit., nom. nud. [43]
A. sativa var. normalis Opiz, loc. cit., nom. nud. [44]
A. sativa var. normalis Opiz, loc. cit., nom. nud. [44]
*A. sativa var. submutica Opiz, loc. cit., nom. nud. [44]
A. sativa var. contracta Neilr., Fl. Nieder-Oesterr. 58 (1859) [45]
A. sativa var. diffusa Neilr., loc. cit. [46]
A. sativa var. aristata Schur, Oesterr. Bot. Zeitschr. 10:70 (1860) nom. illegit. [47]
A. sativa var. fusciflora Schur, op. cit. 71 [48]
A. sativa var. mutica Schur, op. cit. 70, nom. nud. [49]
A. sativa var. nigra Schur, op. cit. 70 [50]
A. sativa var. rufa Schur, op. cit. 70
A. sativa var. nigra Provancher, Fl. Canad. 689 (1862) nom. illegit. [51]

A. sativa var. secunda Provancher, loc. cit., nom. illegit. [52]
A. sativa var. chinensis (Metzger) Aschers., Fl. Brand. 1:827 (1864)
A. sativa var. trisperma (Schuebler) Aschers., loc. cit., nom. illegit.
A. fusca Schur, Enum. Pl. Transsilv. 756 (1866) nom. illegit. [53]
A. fuscoflora Schur, loc. cit., pro. syn. A. fusca Schur [54]
A. sativa var. aristata Alefeld, Landwirth. Flora 321 (1866) nom. illegit. [55] [59]
A. sativa var. arundinacea Schur, loc. cit. [56]
A. sativa var. culinaris Alefeld, op. cit. 322 [57]
A. sativa var. fusca Alefeld, op. cit. 321, nom. illegit. [58]
A. sativa var. hyemalis Alefeld, loc. cit. [59]
A. sativa var. leucocarpa Schur, loc. cit. [60]
A. sativa var. montana Alefeld, op. cit. 320 [59]
A. sativa var. mutica Schur, loc. cit., nom. illegit. [61] [60]
A. sativa var. mutica Alefeld, loc. cit., nom. illegit. [61] [59]
A. sativa var. niger Alefeld, op. cit. 321 [59]
A. sativa var. obtusata Alefeld, loc. cit. [59] [62]
A. sativa var. pugnax Alefeld, loc. cit. [59]
A. sativa var. tristis Alefeld, op. cit. 322 [59] [62]
A. sativa var. vulgaris Alefeld, op. cit. 320 [59]
A. sativa var. aristata Ducomm., Taschenb. Schweiz. Bot. 863 (1869) nom. illegit. [63]
A. sativa var. mutica Ducomm., loc. cit., nom. illegit. [64]
A. orientalis var. mutica Ducomm., loc. cit., nom. illegit. [64]
A. sativa var. trisperma (Schuebler ex Schüz) Ducomm., loc. cit., comb. illegit.
A. distans Schur, Oesterr. Bot. Zeitschr. 20:22 (1870) [65]
A. nigra Heuzé, Pl. Alimentaires 1:505 (1873) [66]
A. verna Heuzé, op. cit. 504 [66]
A. sterilis f. parallela Hausskn., Mitt. Thür. Bot. Ver. 3:240 (1884) [67]
A. fatua var. sativa f. secunda Hausskn., Mitt. Geogr. Ges. Thür. 3:239 (1885) [68]
A. sativa var. affinis Kcke. in Kcke. et Werner, Handb. Gretreidb. 1:208 (1885) [69]
A. sativa var. aurea Kcke. in Kcke. et Werner, op. cit. 207 [69]
A. sativa var. brunnea Kcke. in Kcke. et Werner, loc. cit. [69]
A. sativa var. chinensis (Fisch.) Kcke. in Kcke. et Werner, op. cit. 208 [69]
A. sativa var. cinerea Kcke. in Kcke. et Werner, op. cit. 207 [69]
A. sativa var. flava Kcke. in Kcke. et Werner, loc. cit. [69]
A. sativa var. grisea Kcke. in Kcke. et Werner, op. cit. 210 [69]
A. sativa var. gymnocarpa Kcke. in Kcke. et Werner, op. cit. 208 [69]
A. sativa var. inermis Kcke. in Kcke. et Werner, op. cit. 217 [69]
A. sativa var. krausei Kcke. in Kcke. et Werner, op. cit. 210 [69]
A. sativa var. macrantha Hack., Bot. Jahrb. Engler 6:244 (1885) [70]
A. nuda var. elegantissima Hort., Wien Ill. Gart.-Zeit. 11:379 (1886) [71]
A. orientalis var. turgida Eriks., Bot. Centralbl. 38:787 (1889) [72]
A. sterilis var. degenerans Hausskn., Mitt. Thür. Bot. Ver. N.F. 6:40 (1894) [73]
A. sativa var. orientalis (Schreb.) Hook. f., Fl. Brit. India 7:275 (1896)

A. sativa var. chinensis (Metzger) Vilm., Blumengartn. ed. 3 by Sieb. et Voss. 1:1205 (1896) comb. illegit.
A. sativa var. orientalis (Schreb.) Vilm., loc. cit.
A. sativa var. orientalis (Schreb.) Schmalh., Fl. Central and S. Russia 2:617 (1897)
A. sativa ssp. diffusa (Neilr.) Aschers. et Graebn., Synopsis Mitt. Fl. 2:234 (1899)
A. sativa ssp. diffusa var. aristata (Krause) Aschers. et Graebn., loc. cit.
A. sativa ssp. diffusa var. aristata subv. trisperma (Schuebler) Archers. et Graebn., loc. cit.
A. sativa ssp. diffusa var. aurea (Kcke.) Aschers. et Graebn., op. cit. 235
A. sativa ssp. diffusa var. aurea subv. krausei (Kcke.) Aschers. et Graebn., loc. cit.
A. sativa ssp. diffusa var. brunnea (Kcke.) Aschers. et Graebn., loc. cit.
A. sativa ssp. diffusa var. brunnea subv. montana (Alefeld) Aschers. et Graebn., loc. cit.
A. sativa ssp. diffusa var. grisea (Kcke.) Aschers. et Graebn., loc. cit.
A. sativa ssp. diffusa var. grisea subv. cinerea (Kcke.) Aschers. et Graebn., loc. cit.
A. sativa ssp. diffusa var. mutica (Alefeld) Aschers. et Graebn., op. cit. 234
A. sativa ssp. diffusa var. mutica subv. praegravis Aschers. et Graebn., loc. cit. [74]
A. sativa ssp. diffusa var. nigra Aschers. et Graebn., op. cit. 235 [74]
A. sativa ssp. diffusa var. rubida Aschers. et Graebn., loc. cit. [74]
A. sativa ssp. nuda var. affinis (Kcke.) Aschers. et Graebn., op. cit. 238
A. sativa ssp. nuda var. chinensis Aschers. et Graebn., loc. cit.
A. sativa ssp. nuda var. gymnocarpa (Kcke.) Aschers. et Graebn., loc. cit.
A. sativa ssp. nuda var. inermis (Kcke.) Aschers. et Graebn., loc. cit.
A. sativa ssp. orientalis (Schreb.) Aschers. et Graebn., op. cit. 235
A. sativa ssp. orientalis var. flava (Kcke.) Aschers. et Graebn., op. cit. 236
A. sativa ssp. orientalis var. tartarica (Ard.) Aschers. et Graebn., op. cit. 235
A. sativa ssp. orientalis var. tristis (Alefeld) Aschers. et Graebn., op. cit. 236
A. algeriensis Trab., Bull. Soc. Hist. Nat. Afr. Nord 2:151 (1910) nom. illegit. [75]
A. fatua var. subuniflora Trabut, Bull. Agr. Alg. Tunisie Maroc 16:360 (1910) [76]
A. sativa var. subuniflora (Trab.) Thell., Vierteljahrs. Nat. Ges. Zürich 56:227 (1911)
A. fatua ssp. fatua var. intermixta Thell., op. cit. 325 (1912) [77]
A. fatua ssp. sativa var. contracta (Neilr.) Thell., op. cit. 326
A. fatua ssp. sativa var. diffusa (Neilr.) Thell., loc. cit.
A. fatua ssp. sativa var. subuniflora (Trab.) Thell., op. cit. 327
A. ponderosa L. ex Jackson, Ind. Linn. Herb. in Suppl. Proceed. Linn. Soc. London 124th Session 42 (1912) nom. nud., [1] last paragraph.
A. sterilis ssp. byzantina (Koch) Thell., op. cit. 316
A. sterilis ssp. byzantina var. biaristata (Hack.) Thell., op. cit. 316
A. sterilis ssp. byzantina var. culta Thell., op. cit. 317 [78]
A. trabutiana Thell., Repert. Sp. Nov. Fedde 13:53 (1913) [79]

A. fatua ssp. sativa f. brachytricha Thell., op. cit. 55 [80]
A. fatua ssp. sativa f. glaberrima Thell., op. cit. 54 [80]
A. fatua ssp. sativa f. macrathera Thell., loc. cit. [80]
A. fatua ssp. sativa f. pseudosubuniflora Thell., op. cit. 55 [80]
A. fatua ssp. sativa f. setulosa Thell., loc. cit. [80]
A. fatua ssp. sativa f. subuniflora (Trab.) Thell., loc. cit.
A. sterilis ssp. byzantina f. pseudosativa Thell., op. cit. 53 [81]
A. sterilis var. algeriensis (Trab.) Trab., Jour. Hered. 5:77 (1914)
A. praecoqua Litw., Bull. Appl. Bot. 8:564 (1915) [82]
A. praecocioides Litw., loc. cit. [82]
A. shatiloviana Litw., loc. cit. [82]
A. sterilis f. solidissima Thell., Naturw. Wochenschr. N.F. 17:455 (1918) [83]
A. sativa f. chlorathera Thell., Ber. Schweiz. Bot. Ges. 26/29:172 (1920) [84]
A. sativa f. subpilosa Thell., loc. cit. [85]
A. sativa ssp. autumnalis Marquand, Welsh Pl. Br. St. Bul. Ser. C. 2:8 (1922) [86]
A. sativa ssp. sativa var. diffusa (Neilr.) Fiori, Nuov. Fl. Anal. Ital. 1:109 (1923)
A. sativa ssp. sativa var. orientalis (Schreb.) Fiori, loc. cit.
A. 6-flora Larrañaga, Escritos D.A. Larrañaga 2:49 (1923) (Inst. Hist. Geog. Urug.) [87]
A. diffusa var. asiatica Vav., Bull. Appl. Bot. Pl. Breed. 16:94 (1926) [88]
A. diffusa var. bachkirorum Vav., loc. cit. [88] [89]
A. diffusa var. iranica Vav., loc. cit. [88] [90]
A. diffusa var. kasanensis Vav., loc. cit. [88] [91]
A. diffusa var. persica Vav., loc. cit. [88] [92]
A. diffusa var. segetalis Vav., loc. cit. [88] [93]
A. diffusa var. volgensis Vav., loc. cit. [88] [94]
A. nuda var. mongolica Vav., op. cit. 47, 175 [95]
A. sativa f. pilosiuscula Vav., op. cit. 91, 210 [96]
A. fatua ssp. macrantha (Hack.) Malz., Bull. Appl. Bot. Pl. Breed. 20:141 (1929)
A. fatua ssp. nodipilosa Malz., op. cit. 140 [97]
A. fatua ssp. praegravis Malz., op. cit. 142 [98]
A. fatua ssp. sativa f. chlorathera (Thell.) Thell., Rec. Trav. Bot. Neerland. 25:428 (1929)
A. fatua ssp. sativa f. pseudosubpilosa Thell., loc. cit., nom. illegit. [99]
A. fatua ssp. sativa f. subdecidua Thell., op. cit. 427 [100]
A. fatua ssp. sativa f. subpilosa (Thell.) Thell., op. cit. 428
A. sterilis ssp. byzantina var. hypatricha Thell., op. cit. 432 [101]
A. sterilis ssp. byzantina var. hypomelanathera Thell., op. cit. 431 [102]
A. sterilis ssp. byzantina var. pseudosativa (Thell.) Thell., op. cit. 433
A. sterilis ssp. byzantina f. solidissima (Thell.) Thell., op. cit. 432
A. byzantina var. induta Thell. in Parodi, Rev. Fac. Agron. Vet. Univ. Buenos Aires 7:165 (1930) [103]
A. fatua ssp. macrantha proles nudata Malz., Monogr. 313 (1930) [104]
A. fatua ssp. macrantha var. asiatica (Vav.) Malz., op. cit. 312
A. fatua ssp. macrantha var. asiatica subv. iranica (Vav.) Malz., op. cit. 313

A. fatua ssp. macrantha var. calva Malz., loc. cit. [105]
A. fatua ssp. macrantha var. longipila Malz., op. cit. 312 [106]
A. fatua ssp. macrantha var. pilifera Malz., op. cit. 311 [107]
A. fatua ssp. macrantha var. pilifera subv. homomalla Malz., op. cit. 312 [108]
A. fatua ssp. macrantha var. pilifera subv. subpilifera Malz., op. cit. 311 [109]
A. fatua ssp. nodipilosa prol. decortica Malz., op. cit. 303 [110]
A. fatua ssp. nodipilosa prol. decortica subv. mongolica (Pissar. ex Vav.) Malz., op. cit. 304
A. fatua ssp. nodipilosa var. glabra Malz., op. cit. 303 [111]
A. fatua ssp. nodipilosa var. glabra subv. pilosiuscula (Vav.) Malz., loc. cit.
A. fatua ssp. nodipilosa var. glabra subv. pseudoligulata Malz., loc. cit. [112]
A. fatua ssp. nodipilosa var. glabra subv. secunda Malz., loc. cit. [112]
A. fatua ssp. nodipilosa var. glabricuscula Malz., op. cit. 301 [113]
A. fatua ssp. nodipilosa var. glabricuscula subv. kasanensis (Vav.) Malz., op. cit. 302
A. fatua ssp. nodipilosa var. piligera Malz., op. cit. 301 [114]
A. fatua ssp. nodipilosa var. subglabra Malz., op. cit. 302 [115]
A. fatua ssp. nodipilosa var. subglabra subv. speltiformis (Vav.) Malz., loc. cit.
A. fatua ssp. praegravis prol. grandiuscula Malz., op. cit. 357 [116]
A. fatua ssp. praegravis prol. grandiuscula subv. affinis (Kcke.) Malz., loc. cit.
A. fatua ssp. praegravis var. leiantha Malz., op. cit. 355 [117]
A. fatua ssp. praegravis var. leiantha subv. subeligulata Malz., op. cit. 356 [118]
A. fatua ssp. praegravis var. leiantha subv. turgida (Erikss.) Malz., op. cit. 355
A. fatua ssp. praegravis var. macrotricha Malz., op. cit. 354 [119]
A. fatua ssp. praegravis var. macrotricha subv. arundinacea (Schur) Malz., op. cit. 355
A. fatua ssp. praegravis var. macrotricha subv. norvegica Malz., op. cit. 354 [120]
A. fatua ssp. sativa var. brachytricha (Thell.) Malz., op. cit. 338
A. fatua ssp. sativa var. brachytricha subv. pseudotransiens Malz., op. cit. 339 [121]
A. fatua ssp. sativa var. brachytricha subv. spelticola Malz., loc. cit. [122]
A. fatua ssp. sativa var. glaberrima (Thell.) Malz., op. cit. 340
A. fatua ssp. sativa var. glaberima subv. contracta (Neilr.) Malz., op. cit. 341
A. fatua ssp. sativa var. glaberrima subv. culinaris (Alefeld) Malz., op. cit. 344
A. fatua ssp. sativa var. glaberrima subv. eligulata (Vav.) Malz., op. cit. 342
A. fatua ssp. sativa var. pilosa (Koel.) Malz., op. cit. 336
A. fatua ssp. sativa var. pilosa subv. subpilosa (Thell.) Malz., op. cit. 337
A. sterilis ssp. byzantina var. hypatricha (Thell.) Malz., op. cit. 402
A. sterilis ssp. byzantina var. macrotricha Malz., op. cit. 398 [123]

A. sterilis ssp. byzantina var. macrotricha subv. culta (Thell.) Malz., op. cit. 400
A. sterilis ssp. byzantina var. macrotricha subv. hypomelanathera (Thell.) Malz., op. cit. 399
A. sterilis ssp. byzantina var. macrotricha subv. solidissima (Thell.) Malz., op. cit. 401
A. sterilis ssp. byzantina var. solida subv. secunda Malz., op. cit. 398 [124]
A. sterilis ssp. nodipubescens Malz., op. cit. 383 [125]
A. sterilis ssp. nodipubescens var. longiseta subv. subculta Malz., op. cit. 386 [126]
A. sterilis ssp. nodipubescens var. solidiflora Malz., op. cit. 384 [127]
A. sterilis ssp. nodipubescens var. solidiflora subv. pilosiuscula Malz., op. cit. 385 [128]
A. sterilis ssp. pseudosativa (Thell.) Malz., op. cit. 376
A. sterilis ssp. pseudosativa var. subsolida Malz., op. cit. 378 [129]
A. sterilis ssp. pseudosativa var. subsolida subv. transietissima Malz., loc. cit. [130]
A. ludoviciana var. transietissima Thell. ex Malz., loc. cit., pro syn. subv. transietissima Malz. [130]
A. sterilis ssp. pseudosativa var. thellungiana Malz., loc. cit. [131]
A. sativa var. lusitanica Sampaio, Bol. Soc. Brot. 2, 7:117 (1931) [132]
A. sativa ssp. nodipilosa (Malz.) Vasc., Rev. Agron. 19:19 (1931)
A. sativa ssp. orientalis var. armata Petrop., Bull. Appl. Bot. Pl. Breed. Suppl. 45:25 (1931) [133]
A. sativa ssp. orientalis var. borealis Petrop., loc. cit. [134]
A. sativa ssp. diffusa (Neilr.) Hiitonen, Suonen Kasvis 198 (1933)
A. praegravis Roshev. in Komarov, Fl. URSS 2:268 (1934) [135]
A. praegravis var. arundinacea (Schur) Roshev., loc. cit.
A. praegravis var. macrotricha (Malz.) Roshev., loc. cit.
A. grandis Nevski, Acta Univ. Asiae Med. 8b. Bot. 17:6 (1934) [136]
A. macrantha (Hack.) Malz. in Keller et al. editors, Sorn. Rast. URSS 1:206 (1934) (Feb.)
A. macrantha (Hack.) Nevski, loc. cit. (April) comb. illegit.
A. persarum Nevski, loc. cit. [137]
A. sativa var. bashkirorum (Vav.) Nevski, op. cit. 5, pro syn. A. volgensis (Vav.) Nevski
A. sativa var. segetalis (Vav.) Nevski, loc. cit.
A. sativa var. volgensis (Vav.) Nevski, loc. cit., pro syn. A. volgensis (Vav.) Nevski
A. thellungii Nevski, op. cit. 6 [138]
A. volgensis (Vav.) Nevski, op. cit. 5
A. nodipilosa (Malz.) Malz. in Keller et al. editors, Sorn. Rast. URSS 1:205 (1934)
A. byzantina var. rubida (Krause) Vasc., Ens. Sem. Melhor. Pl. Bol. 20, Ser. A, 40 (1935)
A. sativa ssp. genuina Vasc., loc. cit. [139]
A. sativa ssp. genuina var. brunnea (Kcke.) Vasc., loc. cit.
A. sativa ssp. genuina var. mutica (Alefeld) Vasc., loc. cit.
A. sativa ssp. grandiglumis var. hirsuta Vasc., op. cit. 39 (140)
A. sativa ssp. grandiglumis var. cinerea (Kcke.) Vasc., loc. cit.
A. sativa ssp. praegravis (Langethal) Vasc., op. cit. 40 (141)

A. byzantina var. anopla Mordv. in Wulff, Fl. Cult. Pl. 2:415 (1936) [142]
A. byzantina var. anopla subv. anatolica Mordv., loc. cit. (143)
A. byzantina var. cinnamomea Mordv., op. cit. 416 [144]
A. byzantina var. cremea Mordv., loc. cit. [145]
A. byzantina var. culta (Thell.) Mordv., op. cit. 413
A. byzantina var. culta f. Zhukovskii Mordv., op. cit. 415 [146]
A. byzantina var. graeca Mordv., op. cit. 417 [147]
A. byzantina var. incana Mordv., op. cit. 416 [148]
A. byzantina var. maroccana Mordv., loc. cit. [149]
A. byzantina var. nigra Mordv., loc. cit. [150]
A. byzantina var. solida f. praecox Mordv., op. cit. 413 [151]
A. byzantina var. ursina Mordv., op. cit. 417 [152]
A. sativa var. cinerea f. scabra Mordv., op. cit. 380 [153]
A. sativa var. eligulata Vav. ex Mordv., op. cit. 384 [154]
A. sativa var. homomalla (Malz.) Mordv., op. cit. 385
A. sativa var. inermis f. elegantissima Mordv. et Rodina in Mordv., op. cit. 386 [155]
A. sativa var. inermis f. tardiflora Mordv. et Rodina, loc. cit. [156]
A. sativa var. krausei f. baydarica Mordv., op. cit. 379 [157]
A. sativa var. krausei f. citirna Mordv., loc .cit. [158]
A. sativa var. ligulata Vav. ex Mordv., op. cit. 384 [159]
A. sativa var. maculata Mordv., op. cit. 387 [160]
A. sativa var. mutica f. gigantea Mordv., op. cit. 376 [161]
A. byzantina ssp. byzantina (Koch) Tab. Mor., Bol. Soc. Brot. 2, 13: Tab. 1 (1939)
A. byzantina ssp. byzantina var. hypatricha (Thell.) Tab. Mor., loc. cit.
A. byzantina ssp. byzantina var. induta subv. pilosiuscula (Malz.) Tab. Mor., loc. cit.
A. byzantina ssp. byzantina var. induta subv. secunda (Malz.) Tab. Mor., loc. cit.
A. byzantina ssp. byzantina var. macrotricha (Malz.) Tab. Mor., loc. cit.
A. byzantina ssp. byzantina var. macrotricha subv. biaristata (Hack. ex Trab.) Tab. Mor., loc. cit.
A. byzantina ssp. byzantina var. macrotricha subv. culta (Thell.) Tab. Mor., loc. cit.
A. byzantina ssp. byzantina var. macrotricha subv. hypomelanathera (Thell.) Tab. Mor., loc. cit.
A. byzantina ssp. byzantina var. macrotricha subv. solidissima (Thell.) Tab. Mor., loc. cit.
A. byzantina ssp. byzantina var. macrotricha subv. subculta (Malz.) Tab. Mor., loc. cit.
A. byzantina ssp. byzantina var. solida subv. solidiflora (Malz.) Tab. Mor., loc. cit.
A. byzantina ssp. pseudosativa (Thell.) Tab. Mor., op. cit. 610
A. byzantina ssp. pseudosativa var. thellungiana (Malz.) Tab. Mor., op. cit. 611
A. byzantina ssp. pseudosativa var. transietissima (Thell.) Tab. Mor., loc. cit.
A. sativa ssp. praegravis var. grandiuscula (Malz.) Tab. Mor., op. cit. Tab. 1.

A. sativa ssp. praegravis var. grandiuscula subv. affinis (Kcke.) Tab. Mor., loc. cit.
A. sativa ssp. praegravis var. leiantha (Malz.) Tab. Mor., op. cit. 606
A. sativa ssp. praegravis var. leiantha subv. subeligulata (Malz.) Tab. Mor., op. cit. Tab. 1.
A. sativa ssp. praegravis var. leiantha subv. turgida (Erikss.) Tab. Mor., loc. cit.
A. sativa ssp. praegravis var. macrotricha (Malz.) Tab. Mor., op. cit. 604
A. sativa ssp. praegravis var. macrotricha subv. arundinacea (Schur) Tab. Mor., op. cit. Tab. 1.
A. sativa ssp. praegravis var. norvegica Tab. Mor., loc. cit.
A. sativa ssp. sativa var. brachytricha (Thell.) Tab. Mor., loc. cit.
A. sativa ssp. sativa var. brachytricha subv. asiatica (Vav.) Tab. Mor., loc. cit.
A. sativa ssp. sativa var. brachytricha subv. iranica (Vav.) Tab. Mor., loc. cit.
A. sativa ssp. sativa var. brachytricha subv. spelticola (Malz.) Tab. Mor., loc. cit.
A. sativa ssp. sativa var. brachytricha subv. speltiformis (Vav.) Tab. Mor., loc. cit.
A. sativa ssp. sativa var. brachytricha subv. subglabra (Malz.) Tab. Mor., loc. cit.
A. sativa ssp. sativa var. glaberrima (Thell.) Tab. Mor., loc. cit.
A. sativa ssp. sativa var. glaberrima subv. calva (Malz.) Tab. Mor., loc. cit.
A. sativa ssp. sativa var. glaberrima subv. contracta (Neilr.) Tab. Mor., loc. cit.
A. sativa ssp. sativa var. glaberrima subv. eligulata (Vav. ex Mordv.) Tab. Mor., loc. cit.
A. sativa ssp. sativa var. glaberrima subv. glabra (Malz.) Tab. Mor., loc. cit.
A. sativa ssp. sativa var. glaberrima subv. pilosiuscula (Vav.) Tab. Mor., loc. cit.
A. sativa ssp. sativa var. glaberrima subv. pseudoligulata (Malz.) Tab. Mor., loc. cit.
A. sativa ssp. sativa var. glaberrima subv. secunda (Malz.) Tab. Mor., loc. cit.
A. sativa ssp. sativa var. pilosa (Koel.) Tab. Mor., op. cit. 598
A. sativa ssp. sativa var. pilosa subv. homomalla (Malz.) Tab. Mor., op. cit. Tab. 1.
A. sativa ssp. sativa var. pilosa subv. pilifera (Malz.) Tab. Mor., op. cit. 599
A. sativa ssp. sativa var. pilosa subv. pilifera (Malz.) Tab. Mor., op. cit. Tab. 1.
A. sativa ssp. sativa var. sinensis (Fisch.) Tab. Mor., loc. cit.
A. sativa ssp. sativa var. sinensis subv. culinaris (Alef.) Tab. Mor., loc. cit.
A. sativa ssp. sativa var. sinensis subv. decortica (Malz.) Tab. Mor., loc. cit.
A. sativa ssp. sativa var. sinensis subv. mongolica (Pissarev ex Vav.) Tab. Mor., loc. cit.
A. sativa ssp. sativa var. sinensis subv. nudata (Malz.) Tab. Mor., loc. cit.

A. sativa ssp. sativa var. subpilosa (Thell.) Tab. Mor., op. cit. 599
A. sativa ssp. sativa var. subpilosa f. cinerea Tab. Mor., op. cit. 600 [162]
A. sativa ssp. sativa var. subpilosa f. subpilifera (Malz.) Tab. Mor., loc. cit.
A. sativa ssp. sativa var. subuniflora (Trab.) Tab. Mor., op. cit. 602
A. sativa ssp. sativa var. subuniflora subv. glabriuscula (Malz.) Tab. Mor., op. cit. Tab. 1
A. sativa ssp. sativa var. subuniflora subv. kasanensis (Vav.) Tab. Mor., loc. cit.
A. sativa ssp. sativa var. subuniflora subv. longipila (Malz.) Tab. Mor., loc. cit.
A. pseudosativa (Thell.) Herter, Revist. Sudamer. Bot. 6:141 (1940)
A. byzantina var. solida subv. secunda (Malz.) Maire et Weiller, Fl. Afr. Nord 2:289 (1953)
A. sativa var. glaberrima (Thell.) Maire et Weiller, op. cit. 282
A. sativa var. glaberrima subv. diffusa (Neilr.) Maire et Weiller, loc. cit.
A. sativa var. subuniflora (Trab.) Maire et Weiller, loc. cit.

Description

Annual plants. Juvenile growth prostrate to erect, erect in side types and naked types, prostrate or erect in sativa types and byzantina types. Color appearance green to glaucous. Culm erect. Relatively tall plants, 50–180 cm high. Ligules acute. Panicle equilateral or flagged. Spikelets of indeterminate length, 1.7–5.0 cm long without the awns, and the longest ones are found in the naked types; each spikelet (Fig. 284) has 1–7 florets. Glumes equal in length or nearly so, 25–30 cm long, with 9–11 nerves; all the florets are nondisarticulating at maturity. Awns absent, rudimentary, or, when present, inserted at about middle of lemma to below sinus of lemma tip. Lemma structure tough, or resembling glumes in naked types; lemma tips (Fig. 285) bidenticulate, and the denticulae often not equal in length and shape; lemmas without macrohairs below awn insertion or very few hairs present around the insertion. Paleas with 1–3 rows of cilia along the edges of the keels, mostly with 2; palea back beset with prickles or glabrous. Lodicules (Figs. 286–289) sativa type with varying sizes and shapes of side lobes, which are sometimes difficult to discern; lodicules never bearing prickles. Epiblast sativa type (Figs. 290, 291) or fatua type (Fig. 292), varying in size, 0.45–0.65 mm wide. Chromosome number $2n=42$. Genome AACCDD.

Distribution

Found in (Fig. 293) Algeria, Argentina, Australia, Austria, Belgium, Canada, Chile, China, Cuba, Czechoslovakia, Denmark, Egypt, Ethiopia, France, Germany, Hungary, India, Iran, Israel, Italy, Japan, Korea, Luxemburg, Netherlands, Portugal, Romania, Spain, Sweden, Switzerland, Taiwan, Tunisia, Turkey, the United Kingdom, the USA, the USSR, and Yugoslavia.

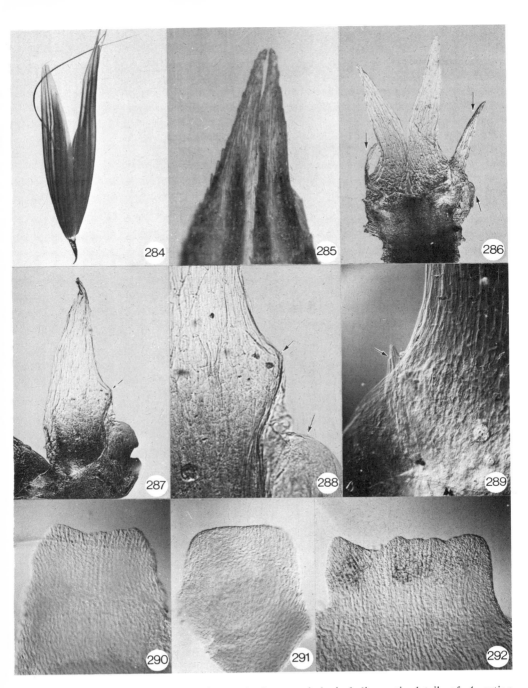

Figs. 284–292. Morphological and micromorphological diagnostic details of *A. sativa*. 284. Regular spikelet with slightly damaged tip of the glumes; the tip is often found this way (× 3). 285. Bidenticulate tip of lemma (× 50). 286. A pair of sativa-type lodicules; note the side lobes (upper arrows) and the membranous wing (lower arrow) (× 50). 287. Sativa-type lodicule with a small side lobe (arrow) (× 50). 288. Magnification of the side lobe (upper arrow) of Fig. 287; note the membranous wing (lower arrow) (× 120). 289. Magnification of the basal part of another lodicule, showing a small side lobe (arrow) (× 120). 290. Typical sativa-type epiblast (× 120). 291. Sativa-type epiblast of an extremely small size (× 120). 292. Fatua-type epiblast also found in *A. sativa* (× 120).

A. sativa

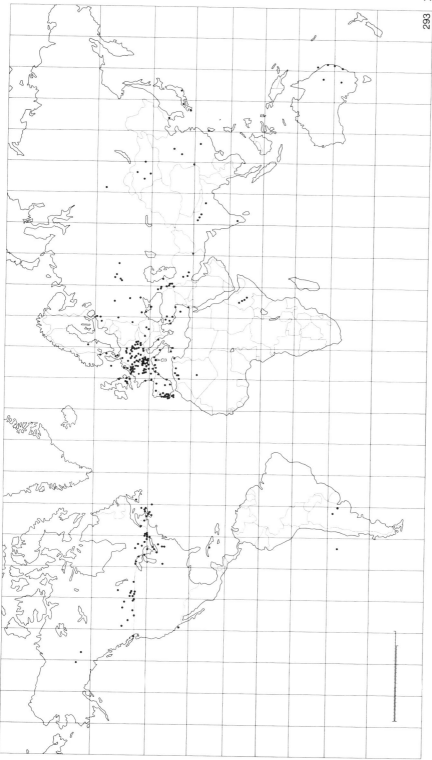

Fig. 293. Distribution map of *A. sativa*. Mercator projection.

Phenology

Flowers from April to November, mostly from June to September in the northern hemisphere; and from October to December in the southern hemisphere, and very rarely to March.

Habitat

Cultivated extensively in the temperate and north-temperate belts of both hemispheres. Also adventitious in cereal fields, road shoulders, and various disturbed places. It requires more humid conditions than wheat and barley, and does not need rich soil to the extent that these two cereals do.

Chorotype

Euro-Siberian, west and central Asiatic Sino-Japanese. It has been recently introduced to Atlantic North American, south Brazilian and the Pampas, East Australian, and South African, also Indian (i.e., the flanks of the Himalayas) regions.

Mode of dispersal

Chiefly by man through such means as cultivation and seed trade, dissemules that escape cultivation often propagate locally. In the naked forms, the caryopsis is the diaspore; otherwise there is no diaspore, that is, the entire spikelet is harvested. In some cases, the spikelets may fracture from the glumes; in fatuoids, the mode of dispersal is identical with that of *A. fatua*. The whole panicle in this species is a man-made diaspore.

Selected specimens

ALGERIA, Alger Kouba ditione urbis, 1879, *Gandoger No. 325* (PRC); ARGENTINA, Buenos Aires Pergamino, 12 November 1925, *Parodi No. 6598* (BAA); AUSTRALIA, N.S.W. Kulnura near Gosford, 5 December 1935, *White No. 10239* (BR); AUSTRIA, Wien, 18 July 1869, *Schur No. 4324* (P); BELGIUM, Anvers, 11 July 1882, *Hennen* (BR); CANADA *Alberta,* Cardston, August 1951, *Gillett* (DAO); CANADA *British Columbia,* Kerrisdale near Vancouver, 13 September 1913, *Davidson* (UBC); CANADA *Manitoba,* Otterburne, 12 July 1954, *Bernard No. 54/149* (MT); CANADA *Newfoundland,* Corner Brook, 17 August 1948, *Rouleau No. 468* (CAN, DAO); CANADA *Nova Scotia,* Inverness Glendyer, rail yard, 12 August 1951, *Smith et al. No. 4974* (ACAD, DAO, and MT); CANADA *Ontario,* Algoma Wama Michipicoten, 13 July 1928, *Hosie et al. No. 1515* (DAO and TRT); CANADA *Prince Edward Island,* Kings County Souris, 5 August 1952, *Erskine No. 1471* (DAO); CANADA *Quebec,* Chambly, 26 July 1906, *Victorin No. 827* (MT); CANADA *Saskatchewan,* Whitewood Remblai route, 4 August 1951, *Boivin and Gillett No. 8485* (DAO and SASK); CANADA *Yukon,* Mayo 1 mi north roadside, 1 August 1949, *Calder and Gillett No. 4089* (DAO); CHILE, Juan Fernandez Island, *Adanson* (PRC); CHINA, Szechwan Chengtu grassy land, 4 May 1938, *Fung No. 1232* (W); CUBA, *Serre,* (BR); CZECHOSLOVAKIA, Bohemia Sloupnice, July 1895, *Fleischer* (PRC); DENMARK, Nyborg, 22 July 1900, *KS No. 132565* (CAN); EGYPT, oestlichen Rande des Deltas, 25 August 1880, *Schweinfurth* (BR); *ETHIOPIA,* Addis Alem 8000 ft, 1964,

Kuckuk No. AB15 (DAO); FRANCE, Nievre Montsanche, 18 August 1883, *Ozanon and Gillet No. 725* (BR, G, MT, and PRC); GERMANY, Frankfurt Schroiebur in Agris, July 1866, *Golenz* (CGE); HUNGARY, (PRC); INDIA, Ganges Inf. Regio Trop., *Thompson and Hooker* (BM and CGE); IRAN, near Tavriza field near road, 1910, *Kudryavtsev No. 145* (WIR); ISRAEL, Jerusalem Monastry of the Cross, 26 May 1926, *Markovitch No. 888B* (WIR); ITALY, Lucca Maulina, 4 June 1860, *Beccari* (FI); JAPAN, Kyushu Kumamoto Minamata, 7 June 1936, *Kaneta* (TI); KOREA, 14 August 1943, *Toh Shim* (TI); LUXEMBURG, Limcule, 1918, *Dolisy (BR);* NETHERLANDS, *Schleicher* (FI); PORTUGAL, Verride in agris, 28 May 1949, *Fernandes and Soussa No. 3155* (BR); ROMANIA, Transilvania, *Schur* (WU); SPAIN, La Sierra de Segura dans le bois, 17 June 1850, *Bourgeau No. 943* (CGE and W); SWEDEN, Goteborg, September 1899, *Stuxberg* (PRC); SWITZERLAND, Arosa Graubunden Bahnhof, 23 July 1915, *Thellung* (BAS); TAIWAN, Taipei, 1961, *Liu* (TI); TUNISIA, Tunis, 12 May 1902, *Cuenod* (G); TURKEY, near Ketchy-Borlu among fields, 29 June 1927 (G); UNITED KINGDOM, Bransford Worcester edge of field, 30 July 1903, *Bickham* (CGE); USA *Illinois,* Urbana Champaign Co., 1 July 1940, *Jones No. 12252* (MT); USA *California,* Los Angeles, 19 August 1910, *Blake No. 776* (BR); USA, *N. Dakota,* Benson Assiniboia, 9 September 1902, *Lunell* (CGE); USA *Pennsylvania,* Pittsburg Allegheny, October 1921, *Jennings No. 296* (MT); USA, *W. Virginia,* Morgantown Agriculture Exp. Station, 1925, *Garber No. 827* (WIR); USSR *Armenia,* Caucasia, *Hohenacker No. 30* (COI); USSR *Azerbaidzhan,* Nakhichevan rocky slope, 1910, *Regel No. 180* (WIR); USSR *Crimea,* Krym, 1902, *Horak* (PRC); USSR *Georgia,* Tiflis, 1924, *Zhukovsky No. 2216* (WIR); USSR, *RSFSR,* Ingria, *Babington No. 778* (CGE); USSR *Mongolia,* Khora mountains along Tola river, 1922, *Pissarev No. 1246* (WIR); USSR *Bushkir,* Ufa, 17 June 1912, *Mordvinkina No. 3414* (WIR); YUGOSLAVIA, Dujeve, July 1899, *Horaz* (PRC).

Identification hints and remarks

Only occasionally can this species be confused with *A. abyssinica* and *A. nuda*. It is confused with the former because of the size and shape of the spikelets, but identification can conclusively be effected by examining the epiblasts. They are brevis type in *A. abyssinica* and sativa type or fatua type in *A. sativa*. The naked forms of *A. sativa* can be confused with *A. nuda* (see "Identification hints" under *A. nuda*). The most serious confusion is between *A. sativa* and *A. fatua* × *sativa* F$_1$ hybrids because the spikelets are often morphologically identical. If so, conclusive identification is effected by inspection of the lodicules (see Baum 1968*b*, 1969*a* and the appropriate keys and descriptions in this work).

Notes

[1] The protolog states:
"5. AVENA calycibus dispermis, seminibus laevibus. *Hort. cliff.* 25. *Hort. ups.* 20. *Mat. med.* 38. *Roy lugdb.* 65.
Avena nigra. *Bauh. pin.* 23.
β Avena alba *Bauh. pin.* 23.
Habitat . . . ."

This name does not pose a serious typification problem. Linnaeus repeated the phrase name of the Hortus cliffortianus, and it is also used in Hortus upsaliensis. The specimen associated with the phrase name is conserved in BM and its identity agrees with *A. sativa* as used by most authors and in its traditional meaning. But the specimen quotes "var. β," that is, *alba*. At Stockholm (S) I was able to examine another syntype, namely, that of the Hortus upsaliensis that bears the following annotation "4 Avena sativa Hort. Ups. 1.20.1.β" and "Herb. Casströmii." It, too, is var. β, and is taxonomically identical.

A specimen with the same identity annotated "sativa 5" by Linnaeus and now conserved in London (LINN), No. 95.6 (Savage 1945, p. 19), although closely linked to the Species Plantarum because of Linnaeus's annotation of the number 5, may be considered as another syntype. Because all the Linnean material examined is of one species, and the collection at BM and S are most likely to have been in Linnaeus' hands, I designate the BM specimen as the lectotype (Fig. 294), and the S and LINN specimens as paralectotypes.

I was unable to examine the Van Royen paralectotype. Furthermore, my searches in Uppsala (UPS) were in vain, for in the Burser herbarium I did not find any specimen pertinent to the Bauhin synonym *Avena nigra* or to var. β. For the latter it is not serious because the lectotype and the paralectotype from S are also its types.

There is another specimen in LINN, No. 95.7 (Savage 1945, p. 19), annotated by Linnaeus "A. ponderosa H.U." but never published, and known only from Jackson (1912). It is also *A. sativa* L.

[2] It is illegitimate because *A. sativa* L. is quoted as synonym.

[3] Schreber's herbarium is at M (Stafleu 1967:434) but I was unable to find the type there. I have designated as lectotype a specimen annotated "Avena orientalis — ex horto hic in agris quoque colitur" by Schreber himself, and now conserved in W. The specimen was sent by Schreber to Wulfen, who annotated it "Avena orientalis à Schrebero." There is a specimen in M that is annotated "Avena orientalis Schreb. spicil. Erl." in handwriting that slightly resembles Schreber's and it is an *A. fatua* × *sativa* F_1 hybrid.

[4] This is the hexaploid naked type, as is clearly evident from Fig. 36. I was unable to find the type in P.

[5] Often *A. heteromalla* Moench, Meth. Pl. 195 (1794) is cited in the literature, but Moench quotes Haller. This is, therefore, a misquote. I was unable to find Haller's type in Paris but the description and figures are clearly *A. sativa*. In addition I found many specimens that are very close in time to Haller or to Moench. The K herb. has a specimen sent to J. Gay by "Batard" in 1818. K has also a specimen that used to belong to Dr. Roth and is labeled *heteromalla* Haller; it has been sent to Dr. D. Turner. The LE herb. has a sheet labeled *"Avena heteromalla.* 1786 H.P." (H.p. stands for Hortus parisiensis.) In S and C are also very old specimens annotated *A. heteromalla*. All the above-mentioned specimens belong here and have flagged-type panicles.

[6] I found three syntypes in P. They are labeled: (i) "dans mon jardin," (ii) "le long des chemins," and (iii) "le 26 juillet dans mon jardin." All belong here.

A. sativa

Fig. 294. Lectotype of *A. sativa*.

A. sativa

[7] I saw a type in P; it belongs here.

[8] This is presumably *A. sativa* (see Note 6), although I have not seen the type. In the text *A. alba* is mentioned (p. 22).

[9] Shrank quotes "A. sativa nigra Bauhin: 23" as syn.; it thus belongs here, although I have not seen the type material.

[10] According to Dr. G. Barioli (personal communication) of Vicenza, Italy, the types of Arduino were destroyed during World War II. According to the description and to Table 1 it belongs here. Sometimes it is wrongly cited as *A. sativa* var. *tartarica* Ard. in Saggi Sc. Acc. Padov. 2:100 (1789).

[11] Gilibert cites *A. sativa* L. in synonymy.

[12] The Thuillier collection of the Fl. Env. Paris, ed. 2, is in G. I have obtained the holotype; it is labeled "Avena novae species/à longumeaux/f en juin" by Thuillier. I have also found two isotypes in BR. All three types belong here.

[13] I was unable to see any related type material so my identification is based on the descriptions only. The epithets *alba* and *nigra* are later homonyms.

[14] I was unable to see the type; it is probably lost. Taborda de Morais (1937:230–231) gives good arguments for putting it here. The citation does not include the word *var.*, but the rank is not in doubt because in var. *sesquialtera* Brotero states "Varietas Av. agrariae."

[15] I have examined in BM herb. an authentic specimen annotated "Avena georgica" by Roemer. It used to belong to Shuttleworth (Murray 1904) who purchased Roemer's herb. This specimen belongs here.

[16] A specimen is in BM herb. labeled "Avena rubra" by Roemer. I presume that it is the holotype; it belongs here.

[17] This is mentioned in the protolog as a synonym of *A. rubra;* therefore, the authentic specimens may be part of the type collection of *A. rubra*. I have examined three such specimens, and they all bear the same handwriting. One is in LE and two are in M. All belong here, except one specimen in M, which is annotated "Avena sativa podolica," but which is *A. sativa* $\times$ *fatua* F_1 hybrid; this specimen is on the same sheet and to the left of specimen No. 6949.

[18] I was unable to find the type; but based on the description only, it belongs here.

[19] I could not detect any original specimens. The closest is a specimen from LE, probably collected by Fisher, which belonged to herb. Ledebour. It belongs here. Fisher's other specimens were collected years after the Roem. et Schult. publication, but they are the same naked hexaploid oat.

[20] This seems to be a new name given by R. and S. instead of *A. georgica* (see Note 15). In herb. M, I found a specimen labeled "Avena sativa L. /Ex sem. Georgiano a cl. Pallas/misso 1779," which may be the specimen ("type") referred to by R. and S. In LE I saw a specimen labeled "A. georgiana hort. mosq. 1809," but it is *A. fatua* × *sativa* F_1 hybrid.

[21] According to the description only, it probably belongs here. No one knows where Gray's types are.

[22] It is illegitimate because it is a later homonym of S.F. Gray (1821). It probably belongs here according to the description.

[23] Metzger was probably the first taxonomist to clarify the confusion between the two naked-type oats. I have not seen the type (and I doubt if there is any), but the description and the figures in the protolog are so clear that the identification is undoubtedly that of *A. sativa* (the naked type).

[24] I have obtained the holotype and an isotype from TUB, and they belong here. Schüz says (loc. cit.) "A. trisperma Autorum"; in W. (herb. Wulfen) I saw an earlier specimen, which was annotated between 1780 and 1790: "a Schrebero Avena trisperma" and this ties in well with Schüz's statement "Autorum."

[25] According to the description, it likely belongs here. I have not seen the type.

[26] It is a later homonym to that of Metzger. I was unable to see the type (which was in B and has been lost), but the clear description in this instance makes identification easy.

[27] It is obviously the *A. sativa* naked type. According to the protolog Link excluded *A. nuda* L. from the synonymy; therefore, that variety is not based on it.

[28] It is superfluous to var. *mutica* S.F. Gray (1821).

[29] It is superfluous to var. *aristata* Schlecht. (1823).

[30] Linnaeus is erroneously quoted, but it is obvious that the epithets are taken from Bauhin's entries in Linnaeus's protolog.

[31] It is a later homonym predated by Schrank (1789) or Haller (1775).

[32] It is part of a multinomial, and therefore illegitimate. All Krause "epithets" here are multinomials or should be understood as such. According to their descriptions they probably belong here.

[33] It is a later homonym predated by S.F. Gray (1821), and it probably belongs here according to the description.

[34] This is a later homonym predated by Reichenb. (1830). I have placed it here on the basis of the description only.

[35] This is a later homonym. According to the description it belongs here.

[36] According to the description it belongs here; I was unable to detect the type, which might still be in HEID.

[37] This is the typical variety and therefore belongs here.

[38] I was unable to obtain the type from NCY, but according to the description it belongs here. The protolog states: "Fleur inférieure enveloppant la fleur supérieure stérile"; this description is similar to what breeders call "bosom oats" in *A. sativa*.

[39] It is a later homonym to Schrank (1789) and belongs here according to the description.

[40] I have not seen the type; it should be in NY (Stafleu 1967:509). The description points to *A. sativa* side oats.

[41] I have examined the holotype in B. It is labeled by Koch "Avena byzantina C. Koch/=A. fatua b. glabrescens Coss./Brussa/K. Koch." It clearly belongs here.

[42] It is probably a naked-type *A. sativa* according to the meaning of the epithets, but this does not preclude the possibility of hybrids.

[43] It is probably a side oat of *A. sativa* according to the meaning of the epithets, but this does not preclude the possibility of hybrids.

[44] These varieties are more likely to belong here than anywhere else. LE has a specimen labeled "A. sativa var. aristata C. Maly," which belongs here; I cannot say if this is the type material of *A. orientalis* var. *aristata* Maly.

[45] It belongs here: it is clearly the side-type oats of *A. sativa* according to description; and furthermore, *A. orientalis* Schreb. is cited as a synonym.

[46] According to the description, it is equivalent to the common-type oats with the diffused panicle, and thus it belongs here.

[47] It is a later homonym to that of Schlecht. (1823), and it very likely belongs here.

[48] I saw an authentic specimen in P, but it is not a type.

[49] I found in P a specimen annotated by Schur, but it is not a type.

[50] According to the description it clearly belongs here. I have not found an authentic specimen.

[51] Provancher specimens are in QPH; I was unable to obtain the type. It is a later homonym of Schrank (1789).

[52] Provancher specimens are in QPH; I was unable to obtain the type. According to the description it is a side oat of *A. sativa*. It is a later homonym of var. *secunda* Wood.

[53] This is a later homonym. It is likely to belong here because it is placed between *A. sativa* and *A. orientalis* in Schur's treatment and is said to be cultivated. Moreover, authentic specimens may be those of *A. fusciflora*, which belongs here (see Note 54).

[54] In P and WU I have seen authentic specimens, which belong here. The P specimen is also annotated *A. orpheosperma* by Schur.

[55] This is a later homonym according to Schlecht. (1823).

[56] According to the description it belongs here.

[57] According to the description it is a naked side oats of *A. sativa*.

[58] It is a later homonym, and it is also a common oat according to the description. Alefeld types were in B and were destroyed during World War II (Stafleu 1967:4).

[59] See Note 58 above for Alefeld types. All these names are put in this list of synonyms on the basis of the protologs alone.

[60] According to the description it belongs here.

[61] It is a later homonym.

[62] It is clearly described as a side oats of *A. sativa*.

[63] This is a later homonym. It could be regarded as the typical variety also. See Note 64 for Ducommun's types.

[64] According to the description it belongs here and also it is a later homonym. Ducommun's specimens are in LAU, but these types were not found in that herbarium.

[65] I found an isotype in P; it was numbered 4324B by Schur and was collected near Vienna on July 18, 1869. It is the naked type of *A. sativa*. The original collection of Schur is supposed to be in LODZ.

[66] According to the text these names are meant to be cultivars of *A. sativa*, that is, these are the latinized cultivar names. These names were quoted by Chase and Niles (1962), but not all were quoted by them, for instance *A. hyemalis* appears on p. 502 of Heuzé's work.

[67] I have obtained from JE the holotype and isotype; they belong here. They are clearly annotated by Hausknecht. The identification of the holotype is conclusive because the annotation indicates a date and details on the site of the collection. Although the annotation of the isotype does not have all these details, it fits closely.

[68] I was unable to see or detect the type, but according to the protolog it belongs here.

[69] Körnicke's specimens used to be in Institut für Landwirtschaftl. Bot. Univ. Bonn; according to Prof. Ulrich (personal communication), they were all lost in World War II. I searched in various other herbaria for type specimens and duplicate types but did not find any. The synonymy has been assessed on the descriptions only. The var. *chinensis* is an illegitimate combination because of earlier *A. sativa* var. *chinensis*.

[70] I tried to locate the type in W, but I was unable to find it. It probably belongs here, but I cannot be certain.

[71] There is certainly no type. There is a picture taken from the Catalog by "C. Platz and Sons." According to the description and picture it might belong here.

[72] It can be positively identified (although I was unable to locate the type) because of the clear description and details given in the figure.

[73] I have obtained the holotype and the isotype from JE. The holotype is duly annotated by Haussknecht. The isotype seems to be a later separation into a duplicate sheet by a curator. The specimens are clearly the naked type of *A. sativa*.

[74] Ascherson and Graebner quoted Krause, but because Krause's epithet is not valid it should be omitted. See also Note 32 above.

[75] It has not been validly published; according to Article 44 of the Code, 1908 is the publication deadline for a new taxon on the basis of a figure and analysis of the essential characters. In P I found the original material, which I designated erroneously lectotype and paralectotype. That 'lectotype' was annotated by Trabut "*Avena algeriensis*/A. sativa pro parte/derivé de l'*A. sterilis*/juin 1910." The two specimens definitely belong here.

[76] I regard the original specimen in P as holotype and a duplicate in W as isotype. Both specimens are duly annotated by Trabut and were collected in June 1908 in Sersou. They obviously belong here.

[77] I have examined the holotype from herb. BAS, and it belongs here. It is labeled "Kiesgrube, Hardan, adventiv 3.7.1911."

[78] It is equivalent to the typical variety because Thellung, among others, quotes "A. byzantina C. Koch sens. stric.," that is, *A. sterilis* ssp. *byzantina* var. *byzantina*. It therefore belongs here.

[79] I have examined the holotype from herb. BAS, and it belongs here. The data on the label reads "Seeaufschüttung beim Belvoir in Zürich II, 23.6.1913."

[80] I was unable to find the type in BAS; it might belong here according to the description and is classified under *A. sativa* by Thellung. I could not detect the types because Thellung did not mention any specific collection for them.

[81] The specimen annotated "Seeaufschüttung beim Belvoir, Zürich II, 23. 6.1913" should be regarded as holotype. I obtained it for examination from BAS and it belongs here.

[82] These epithets, although published as specific epithets, were meant to be published as races or strains of *A. sativa*. Litwinow stated: "The following strains are already established." These epithets, therefore, should not be regarded as specific-name synonyms; they are listed here because they have appeared in the literature. I was unable to examine original material of these three races.

[83] I have examined three possible syntypes from BAS: "Bahnof Aarou 10.8.1917," "Getreidelagerhäusen in Komanshorn 16.7.1917," and "Basel Land: Komport in Neuallschwoil 31.7.1915"; all three were collected by P. Aellen, and they belong here.

[84] I have seen a great number of specimens from Arosa collected by Thellung, which are potential syntypes. There is no problem of identity; all belong here. There is, however, a typification problem because they are not annotated *f. chlorathera* by Thellung, but they fit with the protolog.

[85] Although it could be an F_1 hybrid of *A. sterilis* × *sativa* according to the description, it belongs here. I have examined the holotype from BAS; it is labeled "Zürich 3, Lagerhaus, Giesshübel 19.7.1917" and was collected by Thellung.

[86] It belongs here according to the circumscription given in the protolog and the author's intent. I have, however, not selected a type because no material was available.

[87] According to the protolog it would be an *A. sativa* naked type. I was unable to obtain the type.

[88] *Avena diffusa* is perhaps erroneously regarded by Vavilov as a species, and all these varieties are described under that species. At other places in the same work, some of these varieties are mentioned instead under *A. sativa*. I have listed them here under *A. diffusa* because all are given in one table by Vavilov, and as such they are quoted by Chase and Niles (1962). Vavilov did not effect any combinations of *"A. diffusa"* at the specific level, although he likely implied some. Mordvinkina, one of his students, annotated the types of some of these varieties, such as *A. sativa* ssp. *diffusa* var. *asiatica* Vav. This kind of combination is found in the lectotype, which bears the number 2089; it was collected in 1923 from a plant grown from seeds in plots or in a greenhouse at Leningrad, and is now conserved at WIR. Mordvinkina told me (when I visited Leningrad in 1973) that Vavilov's taxa were described from

panicles brought back by him from expeditions and subsequently from seeds of these panicles, which were propagated locally in Leningrad.

[89] I did not find specimens that antedate the publication of this taxon but I found in WIR a contemporaneous specimen, No. 3414, which I designate as neotype.

[90] I have selected No. 6392 as lectotype; it was collected in 1925 and is now conserved in WIR. It belongs here.

[91] I have designated No. 3396, collected in 1926, as neotype. I found it in herb. WIR, and it belongs here.

[92] I have chosen as lectotype No. 2081 in herb. WIR, which was collected in 1923. I regard No. 3394, collected by Zhukovski in 1924, as a paralectotype. They belong here.

[93] I have designated No. 3416 from WIR as a neotype. It was collected by Stoletova in 1926, and belongs here.

[94] I could not find any specimen that antedated the publication; therefore, I chose No. 2376 from WIR as neotype. It was collected by Stoletova in 1926 and belongs here.

[95] No. 8030, which I found in WIR, is the most likely specimen to be considered as type. It fits the protolog but postdates the publication. Therefore, I designate this specimen as neotype. This specimen, however, has strigosa-type lodicules; this is the only case of this kind that I found among the hexaploids. In all likelihood, however, it is an *A. sativa*. (Definitely, more material should be obtained for study from Mongolia and China because naked oats prevail there and this area seems to be their center of diversity.)

[96] According to the description it is very likely to belong here. I was unable to find in WIR any specimen related in time and content to the protolog.

[97] According to the circumscription it belongs here. Malzew states that all these varieties are cultivated. In the Monograph the subspecies contains a number of varieties and some subvarietal units of which many are taken from Vavilov's varieties of *A. diffusa* (see Note 88). Because the var. *glabra* in Malzew's monograph (p. 303) is mentioned as "frequens," it could be a good choice for the lectotype of this subspecies; the var. *subglabra* is mentioned as "non raro colitur." Of these two, I select specimen No. 829 from WIR as lectotype (see Note 111) of the subspecies, and it belongs here.

[98] Malzew quotes *praegravis* Krause, and makes a combination. Because Krause's name is not valid (see Note 32), the epithet should be attributed to Malzew. In WIR I have found an appropriate specimen, which I chose as lectotype of this subspecies; it is in No. 827, collected by Garber in 1925. It is also the holotype of var. *leiantha* subv. *subeligulata* Malz. (see Note 118).

[99] It is a hypothetical form according to Thellung, who stated that he had not yet found this form. This contravenes with Article 34 of the Code, that is, it is not valid. It belongs here, however, because of its implied identification (of that circumscription) under ssp. *sativa* by its author (see Thellung op. cit. p. 426, line 2).

[100] Thellung stated that this is a new name given for *A. sativa* var. *nigra* Schrank, which belongs here. I was unable to find any of Thellung's authentic material.

[101] I have examined the holotype in BAS, and it belongs here.

[102] I examined the two syntypes from BAS, and they belong here. I designate the Thellung specimen as lectotype and consequently the Aellen specimen becomes a paralectotype.

[103] I have obtained from BAA a spikelet and a photo of the holotype, and it belongs here.

[104] I was unable to find a type in WIR, but the picture in table 51 and the description clearly identify it as an *A. sativa* naked type.

[105] I have examined the holotype from WIR. It is No. 1156 and it belongs here.

[106] I was able to examine the two syntypes from WIR. I designate No. 1155 as lectotype. The paralectotype No. 1154 does not belong here; it is an *A. sativa* × *sterilis* F_1 hybrid.

[107] I have examined the two syntypes from WIR, and have designated No. 148 as lectotype. Both syntypes belong here.

[108] I have seen the holotype in WIR, it is No. 145, and it belongs here.

[109] Specimen No. 167 from WIR is the holotype, and it belongs here. In the protolog Schmidt is mentioned as the collector in 1910, but he probably collected only a panicle or seeds that were subsequently grown, and so the holotype is probably one of those propagations. The holotype was duly numbered by Malzew.

[110] Because I was unable to find a specimen type in WIR, I chose table 44 in Malzew (loc. cit.) as the lectotype. From the description and the figure, it is *A. sativa,* the naked type.

[111] Although no specimens are cited in the protolog, I am selecting "No. 829 R.J. Garber West Virginia Agr. Expt. St. Morgantown W.Va. 1925" from WIR as lectotype. It is duly annotated by Malzew and belongs here.

[112] I decided that they belong here strictly on the basis of their descriptions. No specimen is mentioned in the protolog.

[113] On the basis of the figures (Tab. 43, figs. 1, 2 of Malzew loc. cit.) it belongs here, although it possibly is an F_1 hybrid or a heterozygous fatuoid.

[114] This taxon may be out of place here because the description states that the lemma is beset with hairs below the awn insertion (that is, it might be an *A. sterilis* × *sativa* F_1 hybrid). I was unable to see the specimen cited as the holotype; an examination of it should be conclusive.

[115] I have selected "No. 18, Kusk, Exp. Farm Bagorodskaiya 12.7.1911" from WIR as lectotype, although no specimens are cited in the protolog.

[116] It is based on many synonyms according to the protolog. Some of these are *A. sativa* var. *inermis* Kcke., *A. sativa* var. *chinensis* (Fisch.) Kcke., and *A. sterilis* var. *degenerans* Hausskn., which belong here. Also from Tab. 70 and 71 it is obvious that this is *A. sativa*, the naked type.

[117] Although no specimens are cited in the protolog, I have found in WIR a syntype, which I designate as lectotype; this is No. 831 and was duly annotated by Malzew.

[118] I have found the holotype in WIR; it is No. 827 and it belongs here. In the same sheet there are also a number of isotypes.

[119] No specimens are cited in the protolog. I found in WIR specimens annotated by Malzew, and collected by him in 1912 and 1913; their numbers are 494 and 494(2). I regard these specimens as syntypes, and I chose the former as lectotype. They belong here.

[120] I was unsuccessful in locating the type collection in WIR. In the protolog Malzew refers to Christie specimens, among others, and to Christie (1912); these represent *A. sativa* in all likelihood.

[121] Malzew made the combination *pseudo transiens* (Thell.) Malz., but this is not valid because the basionym is based on a "hypothetical form" by Thellung; therefore, the epithet should be attributed to Malzew only. I was unable to examine in LEP the two syntypes mentioned in the protolog; I saw at K two specimens collected by Bouly de Lesdain on different days in the same year. One of these is *A. fatua*, the other is *A. sativa*. I think, therefore, that the specimen of Bouly de Lesdain mentioned in Malzew's protolog is, very likely, an *A. sativa*.

[122] It probably belongs here according to the description. I could not uncover the type collection; furthermore, no specimen is cited.

[123] One of the subvarieties (i.e., subvar. *culta*) is based on the type of *A. byzantina* (see Note 78 and Malzew p. 400). Thus the type of *A. byzantina* is also included within var. *macrotricha*. Therefore, I regard this taxon as equivalent to *A. sterilis* ssp. *byzantina* var. *byzantina*, and as such, it belongs here.

[124] I have examined the holotype in WIR. Its number is 971a, and it is also mentioned in the protolog; however, it also bears No. 148, which is probably Trabut's number because the original Trabut's label also has that number. The holotype belongs here.

[125] I have seen all the specimens cited for some of the appropriate varieties. All belong here (except No. 888A). They are conserved in WIR and I have selected No. 888b as lectotype.

[126] According to the description it belongs here, although no specimen is cited in the protolog.

[127] The only specimen quoted in the protolog is No. 888b; it is listed under subv. *pilosiuscula* (see Note 128) and I designate it as lectotype for the variety.

[128] I have examined the holotype (i.e., No. 888b) from WIR, and it belongs here.

[129] I was unable to find the type collection in WIR, but from the description it probably belongs here. The only subvar. with a documented specimen (see Note 130) belongs here.

[130] I have seen Thellung's specimen from BAS. I could not find the holotype in WIR. It is quite possible that the BAS specimen is the holotype. It is annotated by Thellung var. *transietissima*. It belongs here, but because the lodicule is slightly stunted it may be a heterozygous fatuoid. Chase and Niles (1962) quoted Thell. ex Malz., which is not quite right because Thellung labeled the specimen *A. ludoviciana* var. *transietissima*. I designate the BAS specimen as lectotype.

[131] Malzew cited the type of *A. sterilis* ssp. *byzantina* f. *pseudosativa* Thell. and the name as synonym. The holotype belongs here (see Note 81). According to present-day Rules this is equal to var. *pseudosativa*.

[132] According to the description it might be *A. sativa*. I was unable to see the type collection in COI.

[133] The drawing on p. 26 shows that it is very likely *A. sativa* with a flagged panicle; however, I did not find the type in WIR.

[134] The drawing on p. 25 shows that it is probably *A. sativa* with a flagged panicle; however, I did not find the type in WIR. This specimen and the preceding one have the same type locality and were discovered at the same time, probably from the same sample.

[135] Often Krause is quoted as author of the basionym, but I have shown that Krause's epithet is not valid (see Note 98). I have found in LE a specimen annotated *A. praegravis* in 1931 by Roshevitz and also subsequently (1933) labeled *A. georgiana* by Nevski. I designate this specimen as lectotype. On the other hand, after the rectification of Krause's invalid epithet, the basionym implied could be that of Malzew. In either case, it belongs here.

[136] It is based on *A. fatua* ssp. *praegravis* prol. *grandiuscula* Malz., and belongs here (see Note 116).

[137] This is presumably based on *A. fatua* ssp. *macrantha* prol. *nudata* Malz., which is cited in synonymy, and it belongs here (see Note 104).

[138] It is based on *A. sterilis* ssp. *byzantina* f. *pseudosativa* Thell., and belongs here (see Note 81).

[139] It is the typical subspecies and is equal to ssp. *sativa* according to present-day Rules.

[140] I have seen the type collection from ELVAS and have selected No. 2384 as lectotype, and it belongs here.

[141] Carvalho e Vasconscellos quotes Krause, but I regard this epithet as a new combination of Malzew (see Note 98), not of Krause.

[142] No type is cited but obviously it might belong here according to the description, although a great number of accessions that we collected in Turkey (Baum et al. 1972) were F_1 hybrids or fatuoids.

[143] I have examined two pertinent collections in WIR and I have selected No. 7730/1 as lectotype. No. 1808, which also belongs here, is a paralectotype.

[144] I was unable to find a type of this taxon in WIR.

[145] I found three syntypes in WIR; all are numbered 1902. I designate as loctotype the collection of 1928; the 1929 and 1930 specimens are paralectotypes. The three types belong here.

[146] I have chosen No. 4641 from WIR as lectotype, and it belongs here. I was unable to find the type of f. *hibernans* and f. *smirnensis*.

[147] According to the description and figure 214 on p. 416, it belongs here; I was unable to find a type in WIR.

[148] From the description it should be an *A. sativa,* but I did not find any material at WIR.

[149] Same as Note 148.

[150] I found two syntypes in WIR, and designated No. 1903 as lectotype; No. 1910 thus becomes a paralectotype. The two belong here.

[151] I have found two syntypes in WIR, which belong here. Both bear the number 1782. I designated the specimen collected on 21 June 1930 as lectotype, and that collected on 23 July 1928 becomes a paralectotype. I was unable to find type material for ff. *aestivalis, rhodiorum* and *cyprica*.

[152] Among the specimens to be considered as syntypes in WIR are Nos. 5441 and 2021; in addition "Avena sterilis selection" is mentioned in the protolog. The latter is probably C.I. 2891 (cultivar Cassel), which I

regard as *A. sativa*. I have, however, selected No. 5441 as lectotype and it belongs here with the paralectotype.

[153] Among various syntypes in WIR I have chosen No. 2389 as lectotype, and it belongs here.

[154] Among various type material in WIR I have designated No. 2462 as lectotype, and I have seen No. 2460, a paralectotype. Both belong here. The protolog mentions var. 'Golden Giant', which also belongs here taxonomically.

[155] I have designated No. 8036 from WIR as lectotype. It is *A. sativa*, the naked type.

[156] I have designated No. 8026 from WIR as lectotype. It is the naked type of *A. sativa*.

[157] I was unable to find specimens that antedate the publication, that is, potential syntypes. In WIR I found a specimen collected in 1950, No. 5545, which I designate as neotype. This specimen fits well with the protolog and is also duly annotated by Mordvinkina.

[158] From the type material in WIR I chose No. 2144 as lectotype. This was collected in 1927 and it belongs here.

[159] I was unable to find type material of this taxon in WIR. Judging from the protolog it belongs here.

[160] I have selected No. 8023, a specimen from WIR, as lectotype. It belongs here and is clearly the naked type, but the lodicules have prickles and they are slightly stunted as some fatuoids are; I have found this feature also in *A. nuda* var. *mongolica* Vav. (see Note 95). Furthermore, the epiblast here is of the septentrionalis type.

[161] In WIR I selected specimen No. 2366 as lectotype, and it belongs here.

[162] It is No. 1986, and it belongs here; I have examined the holotype from COI. The other new forms described by Taborda de Morais under this variety do not belong here.

Additional note

Judging from the description, *A. gracillima* Keng (1936) is probably an F_1 hybrid, or an F_1-like phenotype, which falls between the regular form and the naked form of *A. sativa*. Alternatively, it could be regarded as an aberrant form that frequently results from lack of complete genetic compatibility between parents (Coffman 1964). The name is therefore invalid on the basis of Article 71, or on the basis that it is not a taxon because it is a result of the interaction (hybrid) of two parents within one single taxon. I was unable to obtain the type, but see Note 5 in the Non Satis Notae Section.

(29) A. STERILIS L.

Homotypic synonyms

A. sterilis L., Sp. Pl., ed. 2, 118 (1762) [1]
A. macrocarpa Moench, Meth. Pl. 196 (1794) nom. illegit. [2]
A. fatua var. sterilis (L.) Fiori et Paoletti, Ic. Fl. Ital. 1:29 (1895)
A. sativa ssp. fatua var. sterilis (L.) Fiori, Nuov. Fl. Anal. Ital. 1:109 (1923)

Heterotypic synonyms

A. affinis Bernh. ex Steud., Nom. Bot., ed. 2, 1:171 (1840) pro syn.
A. persica Steud., Syn. Pl. Glum. 1:230 (1854) [3]
A. ludoviciana Dur., Bull. Soc. Linn. Bordeaux 20:41 (1855) [4]
A. sterilis var. minor Coss. et Dur., Expl. Sci. Alger. 2:109 (1855) [5]
A. ludoviciana var. glabrescens Gren. et Godr., Fl. France 3:513 (1855) [6]
A. sterilis ssp. ludoviciana (Dur.) Gillet et Magne, Nouv. Fl. Franç., ed. 3, 532 (1873)
A. sterilis var. turkestanica Regel, Acta Horti Petrop. 7:633 (1880) [7]
A. sterilis var. typica Regel, loc. cit.
A. nutans St. Lager, Rech. Hist. 43 (1884) nom. illegit. [8]
A. sterilis f. albescens Hausskn. ex Heldr., Herb. Graec. Norm. Exsicc. No. 895 (1885) [9]
A. sterilis var. scabriuscula Perez Lara, Anal. Soc. Esp. Hist. Nat. 15:398 (1886) [10]
A. nuda var. sterilis Hort. Wien. Ill. Gart-Zeit. 15:80 (1890) [11]
A. fatua f. straminea Hausskn., Mitt. Thür. Bot. Ver. N.F. 6:37 (1894) [12]
A. sterilis f. leiophylla Hausskn., op. cit. 44 [13]
A. sterilis var. pseudovilis Hausskn., op. cit. 39 [14]
A. sterilis var. solida Hausskn., op. cit. 40 [15]
A. sensitiva Hort. ex Vilmorin's Blumeng., ed. 3, Sieb. et Voss. 1:1205 (1895), pro syn.
A. sterilis var. ludoviciana (Dur.) Husnot, Gram. Franc. Belg. 39 (1899)
A. sterilis var. ludoviciana subv. glabrescens Husnot, loc. cit. [16]
A. sterilis f. biflora Hausskn., Mitt. Thür. Bot. Ges. 13/14:43 (1899) nom. nud. [17]
A. sterilis f. brachyathera Hausskn., loc. cit., nom. nud. [18]
A. sterilis f. breviglumis Hausskn., loc. cit., nom. nud. [18]
A. sterilis f. contracta Hausskn., loc. cit., nom. nud. [18]
A. sterilis f. fusca Hausskn., loc. cit., nom. nud. [18] [19]
A. sterilis f. longiglumis Hausskn., loc. cit., nom. nud. [18]
A. sterilis f. macrathera Hausskn., loc. cit., nom. nud. [20]
*A sterilis f. nigrescens Hausskn., loc. cit., nom. nud. [18]
*A. sterilis f. patula Hausskn., loc. cit., nom. nud. [18]
*A. sterilis f. quadriflora Hausskn., loc. cit., nom. nud. [18]
*A. sterilis f. quinqueflora Hausskn., loc. cit., nom. nud. [18]
*A. sterilis f. secunda Hausskn., loc. cit., nom. nud. [18]

*A. sterilis f. straminea Hausskn., loc. cit., nom. nud. [18]
*A. sterilis f. triflora Hausskn., loc. cit., nom. nud. [18]
*A. sterilis f. vegeta Hausskn., loc. cit., nom. nud. [18]
*A. sterilis f. 7-nervata Hausskn., loc. cit., nom. nud. [18]
*A. sterilis f. 8-nervata Hausskn., loc. cit., nom. nud. [18]
*A. sterilis f. 9-nervata Hausskn., loc. cit., nom. nud. [18]
*A. sterilis f. 10-nervata Hausskn., loc. cit., nom. nud. [18]
*A. sterilis f. 11-nervata Hausskn., loc. cit., nom. nud. [18]
A. abyssinica var. granulata Chiov., Ann. Ist Bot. Roma 8:343 (1908) [21]
A. fatua var. ludoviciana (Dur.) Fiori, Anal. Ital. 1:72 (1908)
A. ludoviciana var. triflora Tourlet, Cat. Pl. Vasc. Indre Loire 568 (1908) [22]
A. turonensis Tourlet, loc. cit., pro syn.
A. ludoviciana var. triflora f. biaristata Tourlet, loc. cit. [22]
A. ludoviciana var. triflora f. triaristata Tourlet, loc. cit. [22]
A. ludoviciana var. vulgaris Tourlet, loc. cit. [23]
A. sterilis var. micrantha Trab., Bull. Agr. Algerie Tunisie 16:354, fig. d (1910)
A. sterilis ssp. macrocarpa (Moench) Briq., Prodr. Fl. Corse 1:105 (1910)
A. sterilis var. calvescens Trab. et Thell., Vierteljahrs. Nat. Ges. Zürich 56:272 (1911) [24]
A. sterilis ssp. ludoviciana var. glabrescens (Gren. et Godr.) Thell. Vierteljahrs. Nat. Ges. Zürich 56:314 (1912) [25]
A. sterilis ssp. ludoviciana var. lasiathera Thell., loc. cit. [26]
A. sterilis ssp. ludoviciana var. psilathera Thell., loc. cit. [27]
A. sterilis ssp. macrocarpa var. calvescens (Trab. et Thell.) Thell., op. cit. 315
A. sterilis ssp. macrocarpa var. pseudovilis (Hausskn.) Thell., loc. cit.
A. sterilis ssp. macrocarpa var. scabriuscula (Perez Lara) Thell., loc. cit.
A. sterilis ssp. macrocarpa var. solida (Hausskn.) Thell., op. cit. 316
A. sterilis var. pilosa Aznav., Magyar Bot. Lapok 12:182 (1913) [28]
A. sterilis ssp. ludoviciana f. subulifera Thell., Rep. Sp. Nov. Fedde 13:53 (1913) [29]
A. sterilis ssp. macrocarpa f. triaristata Thell., op. cit. 52 [30]
A. sterilis var. brevipila Malz., Bull. Ang. Bot. 7:328 (1914) [31]
A. sterilis var. glabriflora Malz., loc. cit. [32]
A. macrocalyx Sennen, Bull. Soc. Bot. France 68:407 (1922), nom. nud. [33]
A. sativa ssp. fatua var. ludoviciana (Dur.) Fiori, Nuov. Fl. Anal. Ital. 1:109 (1923)
A. sterilis ssp. byzantina var. brachytricha Thell., Rec. Trav. Bot. Neerland. 25:432 (1929), nom. illegit. [34]
A. sterilis ssp. byzantina var. macrotricha subv. pseudovilis (Hausskn.) Malz., Monogr. 398 (1930)
A. sterilis ssp. byzantina var. solida (Hausskn.) Malz., op. cit. 396
A. sterilis ssp. ludoviciana var. glabrescens (Gren. et Godr.) Malz., op. cit. 373
A. sterilis ssp. ludoviciana var. glabrescens subv. turkestanica (Regel) Malz., op. cit. 374

A. sterilis ssp. ludoviciana var. glabriflora (Malz.) Malz., op. cit. 376
A. sterilis ssp. ludoviciana var. media Malz., op. cit. 375 [35]
A. sterilis ssp. ludoviciana var. media subv. armeniaca Malz., loc. cit. [36]
A. sterilis ssp. ludoviciana var. typica Malz., op. cit. 365 [37]
A. sterilis ssp. ludoviciana var. typica subv. hibernans Malz., op. cit. 371 [38]
A. sterilis ssp. ludoviciana var. typica subv. lasiathera (Thell.) Malz., op. cit. 367
A. sterilis ssp. ludoviciana var. typica subv. leiophylla (Hausskn.) Malz., op. cit. 370
A. sterilis ssp. ludoviciana var. typica subv. macrantha Malz., op. cit. 372 [39]
A. sterilis ssp. ludoviciana var. typica subv. micrantha Trab. ex Malz., op. cit. 371 [40]
A. sterilis ssp. ludoviciana var. typica subv. psilathera (Thell.) Malz., op. cit. 367
A. sterilis ssp. ludoviciana var. typica subv. subulifera (Thell.) Malz., op. cit. 371
A. sterilis ssp. macrocarpa var. brevipila (Malz.) Malz., op. cit. 393
A. sterilis ssp. macrocarpa var. setosissima Malz., op. cit. 389 [41]
A. sterilis ssp. macrocarpa var. setosissima subv. scabriuscula (Perez Lara) Malz., op. cit. 390
A. sterilis var. oleodens Marq. in Druce, Rep. Bot. Soc. Exch. Club. Brit. Isles 9:286 (1931), nom. nud. [42]
A. sterilis ssp. ludoviciana (Dur.) Trab. in Jah. et Maire, Cat. Pl. Maroc 1:49 (1931), comb. illegit.
A. sterilis var. ludoviciana (Dur.) Vasc., Rev. Agron. 19:20 (1931) comb. illegit.
A. ludoviciana var. macrantha (Malz.) Roshev. in Fedtsch. Fl. Turkm. 1:107 (1932)
A. ludoviciana var. turkestanica (Regel) Roshev., loc. cit.
A. melillensis Sennen et Maur., Cat. Fl. Rif. Or. 129 (1933), pro syn., A. ludoviciana Dur.
A. sterilis var. glabrescens Dur. ex Douin in Bonnier, Fl. Compl. 12:14 (1934) [43]
A. byzantina ssp. byzantina var. brachytricha (Thell.) Tab. Mor., Bol. Soc. Brot. 2, 13: Tab. 1 (1939)
A. byzantina ssp. byzantina var. macrotricha subv. pseudovilis (Hausskn.) Tab. Mor., loc. cit.
A. byzantina ssp. byzantina var. solida (Hausskn.) Tab. Mor., loc. cit.
A. sterilis ssp. ludoviciana var. subpubescens Tab. Mor., op. cit. 582 [44]
A. sterilis ssp. ludoviciana var. typica subv. nodipilosiuscula Tab. Mor., op. cit. 581 [45]
A. sterilis ssp. ludoviciana var. typica subv. scabrimicrantha Tab. Mor., op. cit. 580 [46]
A. sterilis ssp. macrocarpa var. brevipila subv. armeniaca (Malz.) Tab. Mor., op. cit. Tab. 1
A. sterilis ssp. macrocarpa var. setosissima subv. glabrisetigera Tab. Mor., op. cit. 577 [47]
A. sterilis ssp. macrocarpa var. setosissima subv. glabrisetigera f. fusca Tab. Mor., loc. cit. [48]

A. sterilis ssp. macrocarpa var. setosissima subv. glabrisetigera f. nigrescens Tab. Mor., loc. cit. [49]

A. sterilis ssp. macrocarpa var. setosissima subv. macrantha (Malz.) Tab. Mor., op. cit. Tab. 1

A. sterilis ssp. macrocarpa var. setosissima subv. subulatisetigera Tab. Mor., op. cit. 579 [50]

A. sterilis ssp. macrocarpa var. setosissima subv. trichosubulata Tab. Mor., op. cit. 578 [51]

A. solida (Hausskn.) Herter, Revist. Sudamer. Bot. 6:144 (1940)

A. byzantina var. pseudovilis (Hausskn.) Maire et Weiller, Fl. Afr. Nord 2:289 (1953)

A. byzantina var. pseudovilis subv. pseudovilis Maire et Weiller, loc. cit.

A. byzantina var. solida (Hausskn.) Maire et Weiller, op. cit. 288

A. byzantina var. solida subv. eusolida Maire et Weiller, loc. cit. [52]

A. sterilis ssp. trichophylla var. setigera f. mauritiana Maire et Weiller, op. cit. 284 [53]

Description

Annual plants. Juvenile growth prostrate to erect. Color appearance green. Culms erect. Relatively tall plants 50–120 cm high. Ligules acute to truncate with a point. Panicle equilateral to slightly flagged. Spikelets (Fig. 295) indeterminate, 1.7–4.5 cm long without the awns; each spikelet has 2–5 florets. Glumes equal or nearly so, 24–50 cm long, with 9–11 nerves; only the lowermost floret is disarticulating at maturity; the scars (Fig. 297) are oval to round elliptic. Periphery ring confined to ¼–⅓ of scar. Awns inserted at about the lower ⅓ of the lemma or higher but definitely below the middle. Lemma structure tough; lemma tips bidentate to bisubulate (Fig. 296); lemmas densely beset with macrohairs, sometimes very few, or lemmas glabrous. Paleas with 1–3 rows of cilia along the edges of the keels, mostly 2; palea back (Fig. 298) beset with prickles. Lodicules fatua type (Fig. 299), very rarely bearing a few prickles (Figs. 299, 300). Epiblast fatua type (Fig. 301), 0.55–0.65 mm wide. Chromosome number $2n = 42$. Genome AACCDD.

Distribution

Native to (Fig. 302) Afghanistan, Algeria, Azores, Belgium, Canary Islands, Cape Verde, Corsica, Crete, Cyprus, Czechoslovakia, Egypt, Ethiopia, France, Germany, Greece, Iran, Iraq, Israel, Italy, Jordan, Kenya, Lebanon, Libya, Luxemburg, Madeira, Malta, Morocco, Pakistan, Portugal, Saudi Arabia, Sardinia, Sicily, Spain, Switzerland, Syria, Tunisia, Turkey, United Kingdom, USSR (Armenia, Azerbaidzhan, Crimea, Tadzhik, Turkmen, Ukraine, and Uzbek), and Yugoslavia. Introduced in Argentina, Australia, Costa Rica, Ecuador, Mexico, Peru, Uruguay, USA, and the Union of South Africa.

Phenology

In the northern hemisphere it flowers from January to November depending on local conditions and latitude but mainly from April to August; and in the southern hemisphere it flowers in October and November.

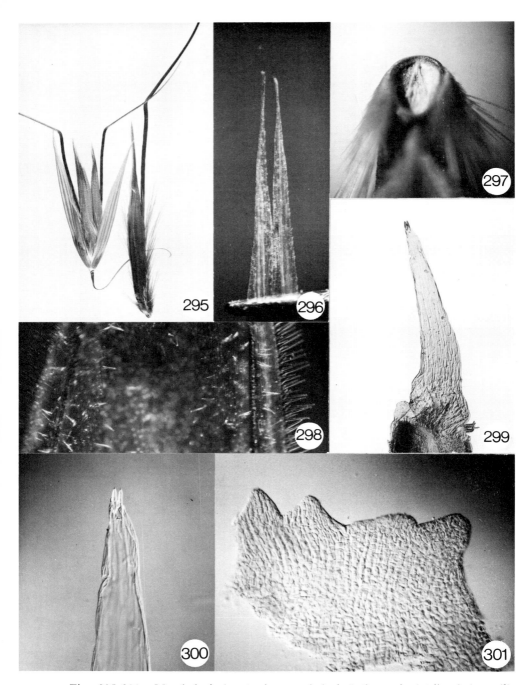

Figs. 295–301. Morphological and micromorphological diagnostic details of *A. sterilis*. 295. Spikelet (left) and the lower floret of a larger spikelet (right); note the variability in size and the macrohairs on the lemma of the floret ($\times$ 2.2). 296. Bisubulate tip of lemma ($\times$ 40). 297. A round elliptic scar; note the periphery ring confined to ⅓ of scar ($\times$ 48). 298. A portion of the back of a palea showing the prickles; note that the prickles are shorter than the cilia along the edge of the keels ($\times$ 76). 299. Fatua-type lodicule with prickles at apex ($\times$ 50). 300. Apex of the lodicule of Fig. 299 showing the prickles in detail ($\times$ 120). 301. Fatua-type epiblast; such large sizes are commonly found in this species ($\times$ 120).

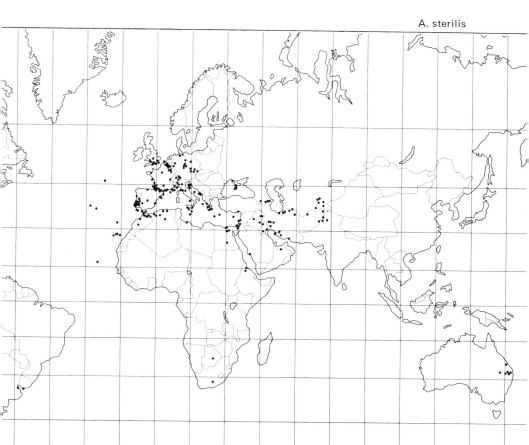

Fig. 302. Distribution map of *A. sterilis*. Mercator projection.

Habitat

This is the species best adapted to a wide range of habitats (Figs. 303–308). Although it populates undisturbed habitats, it is primarily a weedy species. It forms large populations in cultivated fields, preferring deep, fertile soils that are often irrigated. It is found up to 2000 m in elevation on rocky hillsides and slopes, in wadis, on basaltic soil, in forest clearings, on rooftops and balconies, on road shoulders, in olive groves, in vineyards, and in orchards. It infests cultivated fields, especially fields of cereals, and is considered a most noxious weed in the warm temperate belt. It is found also on sandy soil, in oases, and along railroad tracks. It is sometimes found in flooded paddy fields, for example, in Iran. A natural typical habitat would be in a *Quercetum* clearing or in a *Sarcopoterietum,* or similar low shrub Mediterranean association. This species is locally very abundant.

A. sterilis

Figs. 303–308. Six sites of *A. sterilis*. 303. On rooftop in downtown Istanbul, Turkey. 304. Rocky slope above Derband, N Iran. See also Figs. 305–308 next two pages.

Chorotype

Mediterranean and Irano-Turanian, but has been introduced into the southern fringes of the Euro-Siberian region, the various Australian regions, the South African, the Caribbean, the Pampas, and parts of the Andean regions.

Mode of dispersal

The diaspore is the spikelet without the glumes. It is relatively plump and long, and hairy or glabrous. The diaspore has differential germinability; it falls readily at maturity and may be carried short distances by the wind or by ants. It is, however, transported chiefly by man and grazing animals. *Avena sterilis* is harvested with the grain of cultivated cereals and is carried as an impurity, or it may be harvested as hay (e.g., where less-advanced agricultural practices exist) and is transported in this manner to various places.

A. sterilis

Six sites of *A. sterilis*, continued from page 340. Fig. 305. In a lava crater 80 km E of Urfa, SW Anatolia. 306. Contaminating barley field and harvested for hay, near Sanandaj, E Iran.

A. sterilis

Six sites of *A. sterilis*, continued from page 340 and 341. Fig. 307. In a fallow field in Dajhan, Iran. 308. Infesting a cultivated irrigated field near Quasr-Shirin, Iran.

Selected specimens

AFGHANISTAN, Chitial Arandu, 25 April 1958, *Stainton No. 2291* (W); UNION OF S. AFRICA, Löwenschwanz lapides *Ecklon No. 925* (PRC); ALGERIA, Tizi près Mascara, April 1888, *Battandier No. 5688* (BM, MT, and P); ARGENTINA, Buenos Aires Eubaldio, 13 October 1945, *Krapovickas No. 2585* (MT); AUSTRALIA Moreton Sandgate, December 1912, *White* (BRI); AZORES, July 1842, *Watson No. 303* (CGE); BELGIUM, Strail champ d'Avoine, 15 August 1862, *Pire* (BR); CANARY ISLANDS, Gran Canaria west of Las Lagunetas, 14 April 1960, *Lid* (O); CAPE VERDE IS., Santo Antao Is. Rib. Grande, 6 March 1841, *Gray* (P); COSTA RICA, San Jose Terrain inculte, *Pittier No. 1146-18* (BR); CYPRUS, Myrtou Karpass weed in cereals, 1 March 1938, *Syngrassides No. 1757* (G); CZECHOSLOVAKIA, Moravia Trebic, *Krajina* (PCR); ECUADOR, Bolivar Guaranda, 3 August 1939, *Asplund No. 8229* (BR); EGYPT, Mariut Burg el Arab, 14 March 1958, *Täckholm* (G); ETHIOPIA, Eritrea Amasen Monti Lesa, 25 April 1902, *Pappi No. 4901* (FI); FRANCE, Bordeaux rive droite de la Garonne, *Durieu No. 386* (B, BR, M, P, S, UPS, and W); GERMANY, Brandenburg Tamsel, 24 June 1916, *Segel No. 17470* (MT); GREECE, Attica Valle Cephissi Myli, 17 May 1885, *Heldreich No. 895* (B, COI, GB, P, PRC, UPS, W, and WU); IRAN, Radar Dalechi prope pagum, 1842, *Kotschy No. 162* (BM, CGE, FI, G, P, S, UPS, and W); IRAQ, Jarmo, 23 April 1955, *Helbaek No. 1031* (C and W); ISRAEL, Jerusalem Mt. Scopus, 10 April 1931, *Zohary and Jaffe No. 106* (BR, C, COI, LE, M, P, PRC, S, UPS, and W); ITALY, Bordighera olivetis, 28 April 1893 (LE and PRC); JORDAN, Jerash, 1928, *Crowfoot No. 104* (BM); LEBANON, 11 km to Beirut Yamhour, 24 May 1957, *Rechinger No. 13325* (W); LIBYA, Tripolim, 1827, *Dickson* (FI); LUXEMBURG, Victon décombres, September 1909, *Verhulst* (BR); MADEIRA ARCHIPELAGO, Island of Chao, 23 May 1969, *Hansen* (C); MALTA, *Sommerville No. 165* (CGE); MOROCCO, Beni Snassen Berkane steppes, 14 April 1934, *Sennen and Mauricio No. 9604* (BM, G, and P); PAKISTAN, Hazara Hassem plain, *Stewart No. 537893* (CAL); PERU, Lima Huarochiri Matucana dry steppe, 22 May 1940, *Asplund No. 10933* (BR); PORTUGAL, Souselas, May 1931, *Taborda de Morais No. 3473* (COI); SAUDI ARABIA, Taif Highlands, 4 May 1947, *Fitzgerald No. 17062* (G); SICILY, Palermo inter segetes, May, *Todaro No. 1210* (CGE, COI, P, and S); SPAIN, Barcelone Castelldefels Marges, 16 May 1929, *Sennen No. 7072* (COI, G, MA, and MT); SWITZERLAND, Zurich Guterbahnhof, 1 July 1914, *Thellung* (BAS); SYRIA, prope Saida, 24 May 1854, *Gaillardot No. 1877* (BR, LE, P, PRC, and S); TURKEY, Adana, 1925, *Zhukova No. 787* (WIR); UNITED KINGDOM, Vice Herefordshire Rothamsted, 6 July 1945, *Thurston* (BRI and K); URUGUAY, Montevideo, 1838, *Isabelle* (FI); USSR *Armenia*, Alikochak, 1926 *Stoletova No. 915* (WIR); USSR *Azerbaidzhan*, Geokchay-Karasharyan, 30 April 1908, *Woronow No. 598* (WIR); USSR *Crimea*, Yalta Katseneli vineyards, June 1913, *Tsinger No. 520* (WIR); USSR *Tadzhik*, Gissar ridge Kondara, 22 May 1962, *Kamelin No. 546* (LE); USSR *Turkmen*, Aschabad, 5 June 1900, *Sintenis No. 2231* (G); USSR *Ukraine*, Kherson, 15 June 1914, *Paczoski No. 588* (WIR); USSR *Uzbekistan*, Tashkent Chimgan Mts. July 1910, *Suetenko No. 72A* (WIR); YUGOSLAVIA, Ljetanska-Nahiya, May 1903, *Rohlena* (PRC).

Identification hints and remarks

This species cannot generally be confused with others, but there are closely related species that can be misidentified as being *A. sterilis*. These species are *A. atherantha, A. trichophylla,* and *A. maroccana*. The lodicule of *A. sterilis* is fatua type, so this rules out *A. trichophylla*. When the back of the palea is beset with prickles it eliminates from consideration both *A. maroccana* and *A. atherantha* because they have macrohairs instead. Thus the lodicule and back of palea provide enough information for conclusive identification (see also "Identification hints" under *A. atherantha*).

Notes

[1] The protolog states:
"8. AVENA paniculata, calycibus quinquefloris: exterioribus flosculis aristisque basi pilosi, interioribus muticis.
Habitat in Hispania ☉ Alströmer.
Simillimus A. fatuae, ut forte sola varietas, sed triplo omnibus partibus major et cultura constans."

The Linnean herbarium in London (LINN) conserves two sheets annotated by Linnaeus "A. sterilis," namely Nos. 95.11 and 95.12 (Savage 1945, p. 19). It is obvious that Alströmer collected seeds, or plants, or both, of *A. sterilis* and that Linnaeus cultivated them, since "cultura constans" and the collector are mentioned in the protolog. Both sheets bear an "H.U." for Hortus Uppsaliensis. Furthermore, sheet 95.12 bears an "A" for Alströmer, but the specimen is a *Stipa gigantea* Lag. Sheet 95.11 is *A. sterilis* in the traditional sense. I designate this sheet as lectotype (Fig. 309).

The choice of the lectotype is supported by the fact that the original description is in accord with specimen No. 95.11 and not with No. 95.12. Furthermore, the lectotype is conspecific with another closely related specimen, which I found in Stockholm (S) with the annotation "Avena sterilis Hort. Ups." by Dahl who was Linnaeus's secretary in the Uppsala Botanic Garden.

[2] Moench cites *A. sterilis* L. as synonym; therefore, the name is illegitimate according to the rules.

[3] I have seen the type collection in CGE, FI, P, BM, HAL, K, LE, UPS, and G-Boiss; all are Kotschy No. 162. The holotype is in P together with an isotype. The specimens in the other herbaria mentioned are all isotypes and belong here, except one of the two sheets from BM, which is *A. barbata* Pott ex Link.

[4] The protolog lists a number of specific collections, and these are: "Bordeaux, *Durieu;* Coutras, 29 Juin, *Durieu;* Agen, Juin 1852, *Debeaux;* Manzac (Dordogne) Juin 1842, *de Dives;* Blanchardie près de Riberac, *Elly Durieu.*" Additional specimens are implied only. According to Jeanjean et Baudrimont (1961) Actes Soc. Linn. Bordeaux 99:5, the herbarium of Durieu should be in Bordeaux (France). I was unable to see those specimens. Fortunately, I found a collection in P that could be the original; but because its authenticity cannot be confirmed unless we verify the situation in BORD, I regard the P collection as a duplicate of the original. I designate as isolectotype the sheet collected by Durieu at "Bordeaux" in "Juin 1855"; and the P herbarium possesses 3 such isolectotypes. Another isolectotype is at K. The P herb. possesses another sheet with two paralectotypes on it: "Coutras 29 Juin 1854" and "Blanchardie près de Riberac (Dordogne), 11 Oct. 1854 Elly Durieu." All the type collection in P and K belong here. It is quite probable that the exsiccata of Durieu, Nos. 386 and 787, are part of the type collection,

Fig. 309. Lectotype of *A. sterilis*.

[5] I have examined two syntypes of this taxon. In herb. P there are two specimens mounted on the same sheet: (i) *Balansa No. 551,* which I have selected as lectotype, (ii) *Cosson, Aïn-Yagout, Mai 1853,* which should thus be regarded as a paralectotype. In G, I found an isolectotype, which does not belong here because its identity is *A. atherantha.*

[6] Grenier and Godron quote "A. ludoviciana var. glabrescens Dur.," and, therefore, this name is mentioned as "Dur. ex Gren. et Godr." (see for instance Chase and Niles 1962). Durieu described the variability of *A. ludoviciana* on pp. 46 and 47 (1855), but he never published such a name; therefore, this epithet ought to be attributed solely to Grenier and Godron. They quote some specimens mentioned in the protolog of *A. ludoviciana* Dur. and in addition they have "Toulouse," which perhaps could be attributed to their var. *glabrescens.* I was not able to find this specimen, but according to the description it belongs here. I saw a specimen in P annotated var. *glabrescens* by Durieu, and it belongs here. It could be the specimen seen by Grenier and Godron.

[7] The type collection consists of heterogeneous elements, and I examined them all in LE. I have selected *Krause, 20 May 1871,* as lectotype and it belongs here together with the isolectotype. The paralectotype (*Fedtchenko, 27 June 1871*) and an isoparalectotype are *A. fatua,* and the other paralectotype and isoparalectotype (*Fedtchenko, 22 June 1870*) are *A. hybrida.*

[8] It is illegitimate because it is a new name for *A. sterilis* L.: it is a later homonym to *A. nutans* Maratti (1822), and moreover, the protolog says "l'*Avena sterilis* deviendra *A. nutans.*"

[9] According to Lanjouw and Stafleu (1957:266) the original collection of Heldreich is in B. I was able to examine the holotype that belongs here. I saw many isotypes, most of which belong here; they are in herb. JE, LE, UPS, W, WU, P, S, BM, K, COI, GB, and PRC. The isotypes from CGE and a different one from PRC belong to *A. trichophylla.*

[10] The types of var. *maxima* and var. *scabriuscula* are mounted on the same sheet in herb. W. The label agrees with the information in the protolog; moreover, it reads "cum praecedente mixta" (that is, var. *maxima*) and the two labels are identical, except for the names, of course. The protolog mentions also "In Hispanica prope Granada, *Willkomm,*" which I was unable to examine. I have selected the W specimen as lectotype and it belongs here. Paunero (1957:391) mentions that the original examplars of Perez-Lara are conserved in MAF, but I did not see these.

[11] According to the picture, which is taken from the catalog of "Haage and Schmidt," it looks like an *A. sterilis.*

[12] I have examined in JE a type that can be considered as the holotype.

It was collected in 1893 by Bornmueller in Bushir (Iran) and is No. 773; it belongs here.

[13] It is based on *A. persica* Steud. and thus belongs here (see Note 3).

[14] In JE I have examined the holotype and 4 isotypes; I have also seen isotypes in LE, B, and PRC. All the type collection belongs here.

[15] From JE I have obtained the holotype, which is from "hort. Hamburg." and belongs here.

[16] Husnot cites Durieu, as Grenier and Godron did. The specimen in P annotated "var. glabrescens" by Durieu could be the type? In any event the correct name is subv. *glabrescens* Husnot, because the basionym was never published by Durieu (see Note 6 above).

[17] I have examined a syntype in K and it is duly annotated. Similar types from other herbaria are doubtful, but all belong here.

[18] The variability in name of these specimens is detailed, but each taxon is not formally described; therefore, the specimens are *nomina nuda*. Based on these very brief accounts and the fact that they appear under *A. sterilis*, the specimens seem to belong here. In some cases (see Note 17 above) Haussknecht has annotated the specimens, and they can serve as types, although *nomina nuda* do not have types according to the Rules. I recently obtained authentic specimens of f. *7-11-nervata* from JE; they are duly labeled by Haussknecht in pencil. They belong here.

[19] Herb. W has specimens that Haussknecht annotated "*A. sterilis* f. *fuscescens*," which could be taken as types, and they belong here. In BM, BR, K, and LE I found similar specimens. I recently obtained from JE similar 'types,' which all belong here.

[20] The statements on p. 44 can be taken as valid descriptions of this taxon. I found in K and LE possible syntypes labeled "Nauplia," and a few other syntypes in JE labeled "Laurion 1885," "Capri in herbidis 21 April 1898," and "Culta in Winnigan a/Mosel leg. Huter 20 July 1866."

[21] I obtained the holotype from FI; it was collected by Pappi, numbered 4901, and belongs here.

[22] I was unable to see the type material of Tourlet, but the descriptions suggest that these taxa belong here.

[23] It is implied as the typical variety, and therefore, it belongs here because *A. ludoviciana* is a synonym here too.

[24] Herb. BAS has the holotype. The label reads "19 September 1911. Zürich III, Guterbahnhof III 1, vereinzelt," and the specimen belongs here.

[25] Thellung was right (see footnote in Thellung, loc. cit.) when he claimed that Gren. et Godr. were erroneous in citing Durieu (see Note 6 above). Thellung also claimed that Husnot was wrong, but it should be attributed to Husnot because he was the first to use *glabrescens* at the level of subvariety. Although the same combination is sometimes attributed to Malzew (for instance, Chase and Niles 1962), Malzew himself quotes Thellung.

[26] Only one collection can be implied as holotype according to the protolog and I found it in P and W. I selected the W specimen as lectotype because no specific herbarium was mentioned in the protolog. This specimen is "Trabut No. 189 Algerie Tizi près Mascara."

[27] Thellung stated for this variety, in the protolog: "Die häufigere Form," that is, the typical variety by implication. According to the Rules it is equivalent to var. *ludoviciana,* and, therefore, belongs here.

[28] I attempted to obtain the type in RH, but Dr. A. Sükür wrote that the specimen could not be found in the herbarium. According to the protolog it should belong here.

[29] The holotype is in BAS, duly annotated by Thellung. It reads "Lagerhaus Giesshübel, Zürich II, 30 June 1913," and belongs here.

[30] The holotype is in BAS. It is annotated by Thellung and was collected on "2 July 1913 [at] Zürich III, Güterbanhof III 1." It belongs here.

[31] The holotype is conserved in WIR; it is No. 578, which was probably grown in Leningrad from seeds sent to Malzew or Vavilov. The protolog may have been based on these seeds, the panicle, or on specimens grown from these seeds. It belongs here.

[32] The holotype is in WIR. It is a specimen collected by Suetanko in 1910, numbered "72a," and belongs here.

[33] I have examined various specimens labeled *A. macrocalyx* by Sennen, which antedate the publication of the name. Although they were collected in more or less the same area, they have different identities; for instance, the specimen collected "entre le Tenbria et le Turo, 2 May 1919" is *A. trichophylla,* whereas No. 3598 "Massif du Tibidabo, 15 August 1918" is *A. sterilis,* and so also is No. 4222. I have seen these specimens, in BM, K, COI, and MA.

[34] It was published as a hypothetical form and is thus not valid, according to the Rules. It belongs here according to the description. It was first given the rank of variety by Malzew (1930). Thellung published it as "var. vel form."

[35] I have seen the holotype in WIR, and an isotype. Each is labeled No. 352a, and was grown from seeds brought by Nakemov. Its identity is unquestionable.

[36] I have examined the holotype from WIR. It was collected by Stoletova in 1926, and is No. 893. Its identity is *A. sterilis*.

[37] I feel fairly safe in saying that this is the typical variety by implication, and is equivalent to var. *ludoviciana* according to present day Rules; and, therefore, it belongs here.

[38] I was unable to find the syntypes in WIR. From the picture (Tab. 73) and the protolog it seems to belong here. It is of great interest because this taxon seems to be winter-hardy, judging by "planta hibernans" in the protolog.

[39] Very many syntypes are cited in the protolog, and I have examined most of them in WIR. I have selected No. 787 as lectotype, and it belongs here. Some syntypes do not belong here, such as No. 722, which is *A. trichophylla*.

[40] It is based on two syntypes. I was unable to find a specimen in WIR. The closest specimen that I found is in P; it agrees with the protolog but is annotated "A. sterilis microstachya" by Trabut. I am choosing this specimen as lectotype; it also belongs here.

[41] This is the typical variety because one of the subvarieties is "A. sterilis L. sens. strict." (see Malzew, op. cit. p. 389). Because ssp. *macrocarpa* is also the typical subspecies (see Note 2 above), var. *setosissima* is equivalent to *A. sterilis* ssp. *sterilis* var. *sterilis* according to present day Rules.

[42] It is classified under *A. sterilis* by its author, and is, therefore, placed here, reluctantly. I have not seen any specimen related to this name. Furthermore there is no description.

[43] I saw a specimen in P annotated var. *glabrescens* by Durieu but it is of *A. ludoviciana* and is probably the type (see last part of Note 6 above). Chase and Niles (1962) quote Dur. ex Douin erroneously, whereas Douin just quotes Durieu's *glabrescens* as a variety of *A. ludoviciana* but notes a combination with *A. sterilis*, and regards *A. ludoviciana* as a race including that variety. Thus the correct name ought to be as stated in the synonymy; furthermore, there seems to be no doubt of the type.

[44] I have obtained the holotype from COI. It is No. 3855 and it belongs here.

[45] I have examined the holotype from COI. It is No. 3488 and it belongs here.

[46] I have obtained and examined the holotype from COI. It is specimen No. 3489, and it belongs here.

[47] No specimens are mentioned in the protolog. Two forms (f. *fusca* and f. *nigrescens*) are classified under this subvariety, and specimens are cited under them. I regard all these specimens as syntypes and I have selected No. 3638 as lectotype (see Note 48 next below).

A. sterilis

[48] I have examined the holotype from COI and it belongs here. This specimen is No. 3638.

[49] I have examined the holotype, specimen No. 3470, and four paratypes, Nos. 3471, 3472, 3494, and 3646. All are conserved in COI, and belong here.

[50] I have examined the holotype (No. 3476) and two paratypes (Nos. 3475 and 3670) from COI, and they all belong here. The paratypes from LISU I was unable to see.

[51] I have obtained from COI the holotype, No. 3477. It belongs here.

[52] This specimen, because it is the typical subvariety, is equivalent to subv. *solida* according to the Rules, and, therefore belongs here.

[53] I have seen the holotype in MPU and an isotype in BM. The specimens are labeled "Sennen et Mauricio No. 9602" and were collected on the 26 April 1934 in Mazuza, Morocco. They belong here, and the MPU specimen is duly annotated by Maire. A probable paratype "Jalla, Oued Nefifik entre Rabat et Casablanca" also belongs here.

Additional note

Paunero (An. Inst. Bot. Cavan. 15:393.1957) lists *A.* × *ludoviciana* (Dur.) Paunero, as a combination of a hybrid name based on *A. fatua* × *sterilis*. The type of *A. ludoviciana* belongs here (see Note 4), although I must admit that I have not studied many F_1 hybrids between *A. sterilis* and *A. fatua* and certainly I know nothing about hybrid derivatives of such a cross; I doubt if Paunero's views are correct because they are certainly not substantiated in her publication.

(30) A. TRICHOPHYLLA C. KOCH

Homotypic synonyms

A. trichophylla C. Koch, Linnaea 21:393 (1848) [1]
A. fatua var. trichophylla (Koch) Griseb. in Ledeb., Fl. Ross. 4:412 (1853)
A. sterilis ssp. trichophylla (Koch) Malz., Bull. Appl. Bot. Pl. Breed. 20:143 (1929)

Heterotypic synonyms

*A. sterilis var. subulata Batt. et Trab., Fl. Alger. 1:179 (1895) [2]
A. sterilis ssp. macrocarpa var. calviflora Malz., Monogr. 394 (1930) [3]
*A. sterilis ssp. macrocarpa var. setosissima subv. subulata (Batt. et Trab.) Malz., op. cit. 392
A. sterilis ssp. trichophylla var. setigera Malz., op. cit. 381 [4]

A. sterilis ssp. trichophylla var. subcalvescens Malz., op. cit. 383 [5]
A. sterilis ssp. macrocarpa var. calvescens subv. subcalvescens (Malz.) Tab. Mor., Bol. Soc. Brot. 2, 13:Tab. 1 (1939)
A. sterilis ssp. macrocarpa var. setosissima subv. setigera (Malz.) Tab. Mor., loc. cit.

Description

Annual plants. Juvenile growth prostrate to erect. Color appearance green. Culms erect. Relatively tall plants, 30–120 cm high. Ligules acute. Panicle equilateral. Spikelets (Fig. 310) long, 2.0–3.5 cm without the awns; each spikelet has 3–4 florets. Glumes equal in length or nearly so, 25–45 mm long, with 9–11 nerves; only the lowermost floret is disarticulating at maturity; scars (Fig. 313) broad elliptic. Periphery ring only confined to ⅓ of scar. Awns inserted at about lower ⅓ of lemma. Lemma tips (Figs. 311–312) bisubulate, sometimes bidenticulate, the short subule may sometimes have a setula-like structure that is only membranous (Fig. 312); lemmas densely beset with macrohairs below the insertion of the awn, or sometimes only a few hairs present around the insertion only. Paleas with 1–2 rows of cilia along the edges of the keels, rarely 3 rows; palea back beset with prickles or with hairs. Lodicules (Fig. 314) sativa type, often with a very small side lobe, or sometimes similar to the strigosa-type configuration; prickles on the lodicules rarely present; epiblast sativa type to slightly septentrionalis type (Fig. 315), 0.35 mm wide. Chromosome number $2n=42$. Genome AACCDD presumably.

Distribution

Native to (Fig. 316) Afghanistan, Algeria, Canary Islands, Crete, France, Germany, Greece, India, Iran, Iraq, Israel, Italy, Malta, Pakistan, Portugal, Sicily, Spain, Switzerland, Turkey, the United Kingdom, the USSR (Azerbaidzhan, Crimea, Kirghiz, Kazakh, Turkmen, Ukraine), and Yugoslavia. Introduced in Brazil.

Phenology

Flowers from March to July.

Habitat

This species is scattered here and there, rare in the entire area of distribution as well as locally. It grows in fields, arid places, rocky slopes, river banks, and on alluvial soil, steppical land, near hedges and docks. It is often found growing together with *A. sterilis* (with which it is easily confused) in natural habitats and in disturbed habitats.

Chorotype

Irano-Turanian, Mediterranean and Macaronesian, and has been introduced into the southern and western fringes of the Euro-Siberian region and into the Pampas region.

A. trichophylla

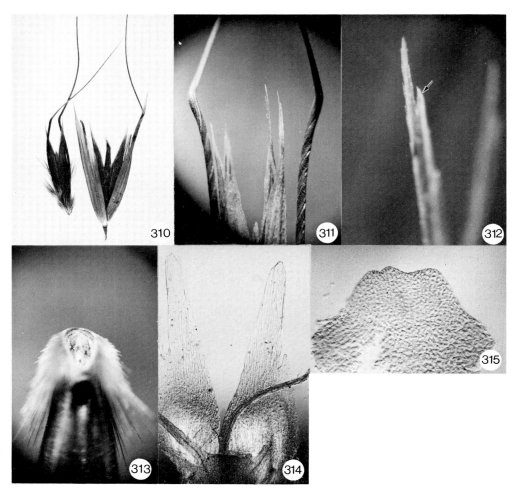

Figs. 310–315. Morphological and micromorphological diagnostic details of *A. trichophylla*. 310. Spikelets: left, without glumes; right, including glumes (× 2). 311. Bisubulate lemma tips (× 16). 312. Detail of a subule with a membranous tooth (arrow) similar to a setula (× 76). 313. Broad elliptic scar; note the shiny periphery ring confined to ⅓ of scar (× 24). 314. A pair of sativa-type lodicules, slightly resembling the strigosa-type configuration (× 50). 315. Sativa-type epiblast that slightly resembles the septentrionalis-type configuration (× 120).

Mode of dispersal

The diaspore is the whole spikelet without the glumes. It is very similar to that of *A. sterilis* in size and shape but the species is not nearly so well adapted to various conditions.

A. trichophylla

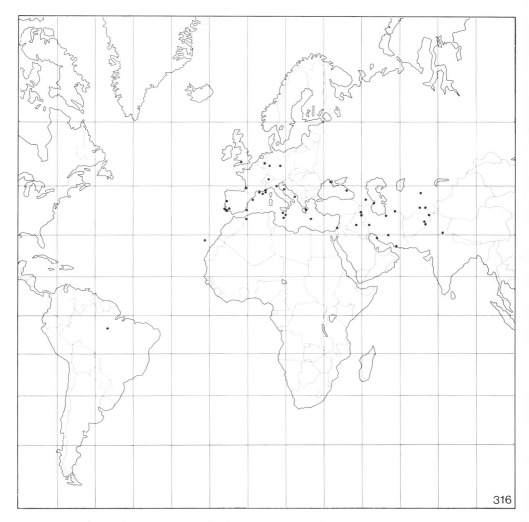

Fig. 316. Distribution map of *A. trichophylla*. Mercator projection.

Selected specimens

AFGHANISTAN, Bada Tschal-Ishkamish, 14 June 1952, *Volk No. 2655* (W); ALGERIA, Oran Union du Sig, 1850, *Durando No. 181* (FI); BRAZIL, *Sellow* (CGE); CANARY ISLANDS, Teneriffe Orotava, 15 March 1855, *Bourgeau* (K); CRETE, La Canee lieux arides, 1883, *Reverchon* (PRC); FRANCE, Villefranche, *Viniklar* (PRC); GERMANY, Saxony Grossfurra, July 1890, *Sterzing* (PRC); GREECE, Attica valle Cephissi Myli 17 May 1885, *Heldreich No. 895* (CGE and PRC); INDIA, Punjab Ludhiana Kalka, 1885, *Drummond No. 21240* (G); IRAN, Shahrud-Bustam Bah-Sar-E-Ceshue, 13 June 1948, *Rechinger No. 5390* (W); IRAQ, Rowanduz gorge deep canyon rock, 15 April 1955, *Helbaek No. 818* (C and W); ISRAEL, Tabor, *No. 18142* (WIR); ITALY, Verona montis Larocea di Garda, May 1908, *Rigo* (PRC); MALTA, near St. George's field, 17 July 1834, *Dikson No. 94* (CGE); PAKISTAN, W-Kurram Parachinar, 11 June 1965, *Rechinger No. 30936* (W); PORTUGAL, Vila Franca Camarao, April 1931, *Tenudo and Brana*

No. 2917 (ELVE); SICILY, Hort. bot. Panormitano, April, *Todaro No. 712* (CGE); SPAIN, Alicante, 1879, *Gandoger* (LY and PRC); SWITZERLAND, Zürich Lagerhaus Giesshube, 21 July 1916, *Thellung* (BAS); TURKEY, Kurdistan Dir Omar Gudrun, 1867, *Haussknecht* (BM); UNITED KINGDOM, Sharpness docks, 10 June 1954, *Townsend No. 73-67-74* (K); USSR *Azerbaidzhan,* Padar Station, 5 June 1927, *Malkov No. 1057* (WIR); USSR *Crimea,* Yalta slopes near Georgievka, 21 May 1912, *Ganeshin No. 517* (WIR); USSR *Kirghiz,* montes meridionales Tian Shan occidentalis prope urbem Taschkent, 7 May 1926, *Popov and Vvedensky* (BR); USSR *Turkmen,* Firyuza pass, 18 June 1924, *No. 18722* (WIR); USSR *Kazakh,* Krgan Kuieka Orpstatkovae ruga, 15 April 1910, *Michelson No. A200* (G); USSR *Ukraine,* Tyahinka, 6 June 1911, *Paczoski No. 1221* (WIR); YUGOSLAVIA, Dalmatia Castelnuovo Boche di Cattaro, 25 April 1898, *Baenitz* (PRC).

Identification hints and remarks

This species is often confused with *A. sterilis,* and it can also be confused with *A. atherantha. Avena trichophylla* is the only species in this sterilis-like group that has a sativa-type lodicule; *A. atherantha* and *A. sterilis* have fatua-type lodicules. The lodicule thus provides a definite answer when there is doubt. *Avena trichophylla,* because the back of the palea in some specimens is beset with macrohairs, may occasionally be confused with *A. maroccana,* but the lodicule of *A. maroccana* is fatua type.

Notes

[1] Although I found the type of *A. byzantina* Koch in B, I was unable to find that of *A. trichophylla* Koch. I have designated the type of *A. sterilis* ssp. *macrocarpa* var. *calviflora* Malz. as the neotype for this species. Based on my interpretation of the protolog of *A. trichophylla,* together with the type locality, I am led to identify the taxon, which I have here circumscribed, as *A. trichophylla.* See Note 3 below for further details on the neotype.

[2] I decided, with doubt, that it belongs here because the protolog states: "glumelle terminée par deux subules fines et longues." This feature is more confined to *A. trichophylla* than to *A. sterilis.* Of course, without a conclusive examination of the type, I can only speculate. I have not detected the type in P, but it may be there.

[3] I have examined the holotype from WIR; it is No. 183. I have also seen 3 isotypes, of which 2 belong here and one may be an *A. sterilis?* The holotype is also the neotype of *A. trichophylla* (Fig. 317).

[4] Because Malzew states (loc. cit.) that this is "Forma typica" (the typical variety), it belongs here. Some of the specimens cited belong here and others are *A. sterilis,* but this is irrelevant to typification.

[5] I have examined the two syntypes from WIR, and have designated specimen No. 1057 as lectotype. It belongs here but the paralectotype No. 1055 belongs to *A. sterilis.*

A. trichophylla

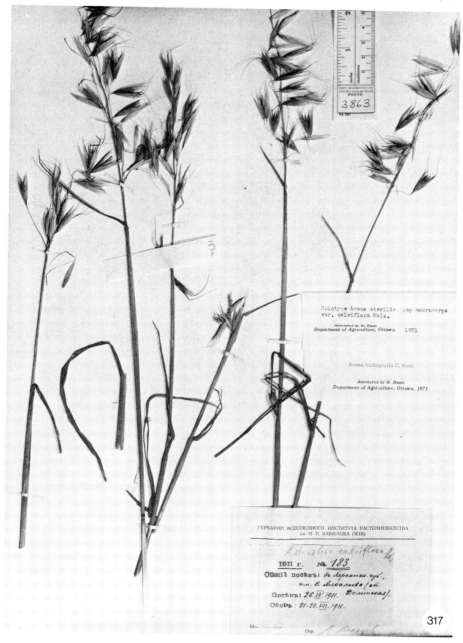

Fig. 317. Neotype of *A. trichophylla*.

(31) A. SATIVA FATUOID

Synonyms

A. sativa var. setosa Kcke. in Kcke. et Werner, Handb. Getreidb. 1:207 (1885) [1]
*A. sativa var. armata Schur, Enum. Pl. Transsilv. 756 (1866) [2]
A. sativa ssp. diffusa var. brunnea subv. setosa (Kcke.) Aschers. et Graebn., Synopsis 235 (1899)
A. fatua var. basifixa Malz., Bull. Appl. Bot. 7:329 (1914) [3]
A. fatua var. pachycarpa Malz., op. cit. 328 [4]
A. fatua ssp. cultiformis Malz., Bull. Appl. Bot. Pl. Breed. 20:142 (1929) [5]
A. fatua ssp. fatua f. basifixa (Malz.) Thell., Rec. Trav. Bot. Neerl. 25:427 (1929)
A. fatua ssp. cultiformis var. leiocarpa Malz., Monogr. 347 (1930) [6]
A. fatua ssp. cultiformis var. leiocarpa subv. aglossicos Malz., loc. cit. [7]
A. fatua ssp. cultiformis var. oligotricha Malz., loc. cit. [8]
A. fatua ssp. cultiformis var. pseudoculta Malz., op. cit. 348 [9]
A. fatua ssp. cultiformis var. pseudoculta subv. americana Malz., op. cit. 349 [10]
A. fatua ssp. cultiformis var. pseudoculta subv. crassiflora Malz., op. cit. 350 [11]
A. fatua ssp. cultiformis var. pseudoculta subv. eucontracta Malz., op. cit. 351 [12]
A. fatua ssp. cultiformis var. pseudoculta subv. pachycarpa (Malz.) Malz., op. cit. 350
A. fatua ssp. cultiformis var. pseudoculta subv. patentissima Malz., op. cit. 348 [13]
A. fatua ssp. cultiformis var. pseudoculta subv. subturgida Malz., op. cit. 351 [14]
A. fatua ssp. fatua var. vilis subv. basifixa (Malz.) Malz., op. cit. 332
A. fatua ssp. meridionalis var. elongata subv. glabrifolia Malz., op. cit. 309 [15]
A. fatua ssp. septentrionalis var. glabella subv. breviflora Malz., op. cit. 299 [16]
A. aemulans Nevski, Act. Univ. Asiae Med. 8b. Bot. 17:5 (1934) [17]
A. cultiformis (Malz.) Malz. in Keller et al., Sorn. Rast. SSSR 1:208 (1934) [18]
A. sativa ssp. nodipilosa var. alboaristata Vasc., Ens. Sem. Melhor. Pl. Bol. 20 Ser. A. 40 (1935) [19]
A. fatua ssp. fatua var. vilis subv. breviflora (Malz.) Tab. Mor., Bol. Soc. Brot. 2, 13:Tab. 1 (1939)
A. fatua ssp. fatua var. vilis subv. glabrifolia (Malz.) Tab. Mor., loc. cit.
A. sterilis ssp. ludoviciana var. glabriflora subv. basifixa (Malz.) Tab. Mor., loc. cit.

Description

Annual plants. Juvenile growth prostrate to erect. Color appearance green to glaucous. Culms erect. Relatively tall plants, 50–180 cm high. Ligules acute. Panicle equilateral to flagged. Spikelets (Fig. 318) short, 15–22 mm long; each spikelet has 2–3 florets, although most have 2. Glumes equal or nearly equal, and similar in appearance to those of *A. sativa* normal plants; all the florets are disarticulating at maturity; scars (Fig. 319) round to round-oval, often slightly heart-shaped. Periphery ring only confined to ⅓–½ of scar. Awns inserted at about middle of lemma. Lemma tips bidenticulate to bisubulate, sometimes almost bilobed; lemmas beset with macrohairs below the insertion of the awn, or only a few hairs around the insertion or macrohairs absent. Paleas with 1–3 rows of cilia along the edges of the keels, often 2 rows only; back of palea beset with prickles or rarely without prickles. Lodicules sativa type but usually stunted or aborted (Figs. 320–323), and never bearing prickles. Epiblast sativa type (Fig. 324) or fatua type (Fig. 325), 0.5–0.6 mm wide. Chromosome number $2n = 42$, 41, or 40. Genome essentially AACCDD.

Distribution

Native to (Fig. 326) Algeria, Argentina, Belgium, Canada, Chile, Czechoslovakia, Egypt, France, Germany, India, Israel, Japan, Luxemburg, Norway, Portugal, Spain, Sweden, Switzerland, the United Kingdom, the USA, and the USSR, but generally to the same areas as *A. sativa*.

Phenology

In the northern hemisphere, it flowers from May to September, and rarely from March. In the southern hemisphere, it flowers in November and December.

Habitat

It is to be expected in the same habitats as *A. sativa*. It is more common in oat fields, but is also found as adventitious plants in various ruderal and segetal habitats.

Mode of dispersal

The diaspores are the florets. The dissemules behave similarly to those of *A. fatua* and are carried as impurities in the harvest by man. This mode of dispersal of *A. sativa* complements that of the cultivated type and it probably was more important before the species was domesticated.

A. sativa fatuoid

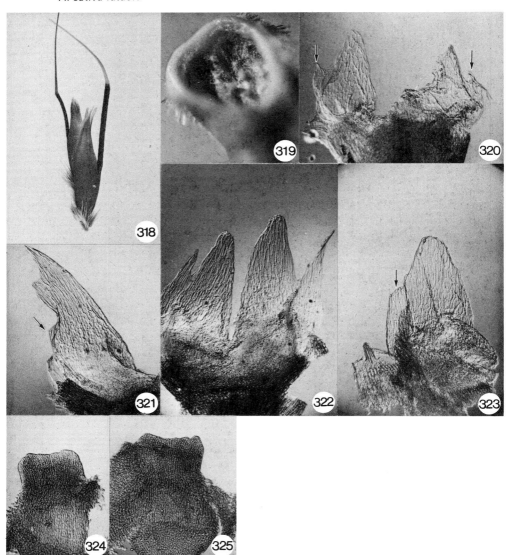

Figs. 318–325. Morphological and micromorphological diagnostic details of *A. sativa* fatuoid. 318. Spikelet without the glumes; note the plumpness similar to that of *A. sativa* (× 4). 319. Scar of 1st floret; note the slightly heart-shaped configuration (× 50). 320. Pair of sativa-type, stunted lodicules; the basic configuration is clearly recognizable. Note the side lobes (arrow) (× 120). 321. Deformed sativa-type lodicule; arrow points to side lobe (× 120). 322. A pair of aborted sativa-type lodicules; the configuration is reminiscent of the strigosa-type lodicule (× 120). 323. Strongly aborted sativa-type lodicule; note the side lobe (arrow) (× 120). 324. Sativa-type epiblast (× 50). 325. Fatua-type epiblast (× 50).

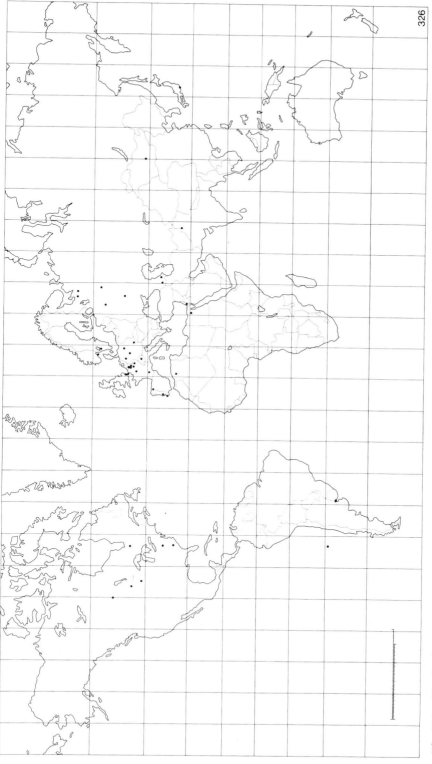

Fig. 326. Distribution map of *A. sativa* fatuoids. Mercator projection.

A. sativa fatuoid

Selected specimens

ALGERIA, 12 km from Tiaret, 10 July 1926, *Morvinkina No. 171* (WIR); ARGENTINA, Buenos Aires Moron, 26 December 1943, *Remiro* (MT); BELGIUM, *Oudemans* (BR); CANADA *Manitoba,* Aweme, 1913, *Criddle No. 539A* (WIR); CANADA *Ontario,* 1912, *Perceival No. 451* (WIR); CHILE, Ilas Juan Fernandez, *Douglas* (CGE); CZECHOSLOVAKIA, Bohemia Litomysl, August 1909, *Obdrzalek* (PRC); EGYPT, près des tombeaux des Mamelouks desert, 28 May 1890, *Burdet No. 503* (G); FRANCE, 1823, *Lejeune* (G); GERMANY, Thuringia pr. Weimar, July 1895, *Haussknecht No. 3387* (COI and G); INDIA, N.W. frontier province Peshawar, 28 March 1935, *Kerstan No. 2171* (K); ISRAEL, Jerusalem Monastery of the Cross, 26 May 1926, *Markovitch No. 888* (WIR); JAPAN, Yokohama environs, *Borel* (G); LUXEMBOURG, Greiveldangees Stoutbredim, *Marchand fils No. 10* (BR); NORWAY, Tönsberg, August 1892, *Lewin No. 132543* (CAN); PORTUGAL, Sezimbra, 1929, *Vasconcellos No. 5106* (ELVAS); SPAIN, Asturia Pravia Vega, 21 July 1927, *Vavilov No. 1141* (WIR); SWEDEN, Vestrogothia circa Lidkoping, *Lindgren No. 89* (CGE); SWITZERLAND, Zürich Glattufer bie Swerzenbach-Fällenden, 8 July 1930, *Koch No. 207678* (CAN); UNITED KINGDOM, Suffolk Nayland, July 1884, *Gray No. 1530* (CGE); USA *North Dakota,* Fargo Halsteds, July 1892, *Bolley No. 193* (BRI); USA *West Virginia,* Morgantown Agric. Exp. Station, 1925, *Garber No. 826* (WIR); USSR *Archangelsk,* Onega Vangudy, 18 August 1910, *Egorov No. 37* (WIR); USSR *Armenia,* Erevan, 1926, *Stoletova No. 905* (WIR); USSR *Azerbaidzhan,* Baku Khynalyk, 27 August 1903, *Razevich* (LE); USSR *Dagestan,* Samur Kurusch, 29 July 1898, *Alexeenko No. 875* (LE); USSR *Mongolia,* grown at Kamennaya Step Exp. Station, Voronezh, 27 July 1923, *Malzew No. 39* (WIR).

Identification hints and remarks

The fatuoids can easily be confused with *A. fatua* and *A. hybrida.* Conclusive identification can be achieved by examining the lodicules and epiblasts. The fatuoids have sativa-type lodicules, which are often malformed or stunted, and the epiblast is fatua type or sativa type; whereas *A. hybrida* has a septentrionalis-type epiblast, and *A. fatua* has a fatua-type lodicule. See also Chapter 11 on fatuoids.

Notes

[1] According to Professor H. Ulrich, Inst. Landwirtsch. Bot. in Bonn, Koernicke's herbarium was there but it was destroyed at the end of World War II (personal communication). Koernicke claims, in his description, that this taxon is the result of a mixture of *A. sativa* and *A. fatua* received from a botanical garden, but that it breeds true to type. This claim and the description, in my judgment, point most likely to *A. sativa* fatuoid or to an F_1 hybrid with *A. sterilis.*

[2] It could belong here, if I judge solely from the protolog, which states "flosculis omnibus medio dorsi aristatis." According to Dr. Riedl (W) the types of Schur are now at LOD, but I was unable to see the type of this taxon.

[3] I have examined the holotype in WIR; it is No. 580, and it belongs here.

[4] The holotype is specimen No. 110 in WIR, and it belongs here.

A. sativa fatuoid

[5] The protolog does not list any specimens, but the Monograph (Malzew 1930) can serve for lectotypification. Malzew designated var. *trichocarpa* as "forma typica" and it would seem logical to choose a lectotype from among the specimens listed under this variety, as I have usually done with Malzew's taxa. I decided, in this particular instance, not to be consistent. As can be seen in the Notes below and in the synonymy, all the varieties and subvarieties belong here except the "typical variety." Even the "typical variety" has one paralectotype (No. 1141), which belongs here (see Note 19 of *A. fatua*). Because the protolog and the description in the monograph are based on all the material, I am selecting a lectotype from among the material of all the varieties. The designated lectotype is specimen No. 538, which is also the lectotype of *A. fatua* ssp. *cultiformis* var. *pseudoculta* subv. *americana* Malz. (See Note 10 below.) All the syntypes are conserved in WIR.

[6] I have selected specimen No. 451 from WIR as lectotype. This specimen and paralectotypes Nos. 592 and 581 belong here.

[7] I have examined the holotype; it is No. 826 and is conserved in WIR. This specimen was collected from a plant grown locally in 1927 from seeds obtained from Garber in 1925, and it belongs here.

[8] The illustration and the description undoubtedly point to this entity. I was unable to find any type specimen in WIR.

[9] I designate the lectotype of subv. *americana* as lectotype of this variety (see Note 10 below).

[10] I have selected No. 538 from WIR as lectotype. This specimen was collected from a plant grown locally in 1914 from seeds or panicle obtained in 1913 from Criddle in Canada. The paralectotypes Nos. 830 and 1026 also belong here.

[11] I have selected No. 450 as lectotype. This specimen and paralectotype No. 541 are conserved in WIR, and both belong here.

[12] I have examined the holotype in WIR, it is No. 539, and it belongs here. It was grown in Leningrad in 1914 from seeds obtained from Criddle (Canada) in 1913.

[13] The holotype is No. 579 from WIR and it belongs here. It was collected in 1913 from a plant grown from seeds obtained in 1912.

[14] I have examined the holotype in WIR. It is No. 539a and it belongs here. Like many other specimens, this one was grown locally in 1914 from seeds obtained in 1913 from Criddle (Canada).

[15] I was unable to see the holotype in WIR, but it is likely to belong here because it is glabrous as in *A. sativa* and because it was collected by Criddle in Canada; this conclusion is also supported by the figure (Tab. 46).

[16] I have examined the holotype in WIR. It is No. 828, and it belongs here.

[17] It is obviously based on *A. fatua* var. *basifixa* Malz. (see Note 3 for the identity) according to the protolog. The name may even have been taken from a sentence in Malzew (op. cit. 333) "...spicularum aemulat."

[18] The account of the grasses in Keller was written by Roshevitz; however, Malzew was a member of the editorial board of Vol. 1 Therefore, Malzew may have had a part in the authorship of the account on *Avena,* and thus the combination should be attributed to Malzew as stated and implied, and not to Roshevitz.

[19] I have obtained the type collection from Oeiras (Portugal). It is No. 5106 and contains 3 specimens. The lectotype and one paralectotype belong here; the other paralectotype is an *A. sativa*.

(32) A. SATIVA HETEROZYGOUS FATUOID

Synonyms

    A. fatua ssp. fatua f. glabricalla Thell., Ber. Schweiz. Bot. Ges. 30/31:78 (1922) [1]
  *A. fatua ssp. praegravis var. microtricha Malz., Monogr. 355 (1930) [2]
    A. fatua ssp. sativa var. pilosa subv. glabricalla (Thell.) Malz., op. cit. 337
    A. sterilis ssp. nodipubescens var. longiseta subv. diathera Malz., op. cit. 386 [3]
    A. sterilis ssp. nodipubescens var. longiseta subv. laevigata Malz., op. cit. 385 [4]
  * × A. mutata Sampaio, Bol. Soc. Brot. 2, 7:118 (1931) [5]
    A. sativa ssp. diffusa var. rossica Petrop., Bull. Appl. Bot. Pl. Breed. Suppl. 45:29 (1931) [6]
    A. byzantina var. rubra Mordv. in Wulff, Fl. Cult. Pl. 2:415 (1936) [7]
    A. byzantina ssp. byzantina var. macrotricha subv. diathera (Malz.) Tab. Mor., Bol. Soc. Brot. 2, 13:Tab. 1 (1939)
    A. byzantina ssp. byzantina var. macrotricha subv. laevigata (Malz.) Tab. Mor., loc. cit.
    A. sativa ssp. praegravis var. microtricha (Malz.) Tab. Mor., loc. cit.
    A. sativa ssp. sativa var. glabricalla (Thell.) Tab. Mor., loc. cit.
    A. sativa ssp. sativa var. subpilosa subv. heteroclita Tab. Mor., op. cit. 601 [8]

Description

Same as *A. sativa* fatuoid except that: the awns are similar to those in *A. sativa,* or only those of the first florets are well developed; the florets are not disarticulating at maturity, although they may fracture without leaving a scar (Fig. 327); a few macrohairs may be present on the lemmas around the insertion of the awn, or no macrohairs below the insertion; paleas have 2–3 rows of cilia along the edges of the keels.

A. sativa heterozygous fatuoid

Fig. 327. Spikelet of *A. sativa* heterozygous fatuoid without the glumes; note the place of fracture at the base of the spikelet ($\times$ 5).

Distribution

Native to Australia, Canada, Germany, Switzerland, USA, USSR, but generally to the same areas as *A. sativa*.

Phenology

Same as that of *A. sativa*.

Habitat

Grows sporadically in the areas where *A. sativa* is found.

Mode of dispersal

Identical with that of *A. sativa* and dispersed similarly. See Chapter 11 on fatuoids.

Notes

[1] According to the description it belongs here. I was unable to see the type.

[2] According to the excellent figure it is likely to belong here, although it might also be an *A. sativa*. No specimens are cited and I have not found any specimens in WIR that might belong to the type collection.

[3] Same remark as Note 2 above.

[4] Although no specimens are cited in the protolog, I found in WIR the holotype, labeled "Typus" and two isotypes. These specimens are No. 888; they belong here.

[5] According to the protolog it could belong here or it might be an *A. sativa* × *sterilis* F_1 hybrid. I was unable to see the type, which might be in COI.

[6] It typically belongs here according to the description and the figure (on p. 30), but I did not find the type.

[7] I have selected specimen No. 1717 as lectotype; it also bears No. 4745. The Isolectotype is Nos. 1717 and 4742. Both specimens are in WIR and belong here.

[8] I have obtained from COI the holotype, specimen No. 3997, and two paratypes Nos. 3842 and 3996. All three belong here.

(33) A. SATIVA × FATUA F_1 HYBRID

Synonyms

    A. sativa var. arduennensis Lej. et Court., Comp. Fl. Belg. 1:73 (1828) [1]
*A. sativa var. arduennensis Bluff et Nees, Comp. Fl. Germ. ed. 2, 1:149 (1836) [2]
    A. sativa var. arduennensis Schur, Oesterr. Bot. Zeitschr. 10:70 (1860) [3]
    A. sativa var. atrocarpa Schur, Enum. Pl. Transsilv. 756 (1866) [4]
    A. fatua var. transiens Hausskn., Mitt. Geogr. Ges. Thür. 3:238 (1885) [5]
    A. glabrata Haussm. in Harz. Landw. Samenkunde 2:1317 (1885) [6]
    A. sativa × fatua Aschers. et Graebn., Synopsis 2:242 (1899)
    A. fatua ssp. fatua var. transiens subv. unilateralis Thell., Repert. Sp. Nov. 13:54 (1913)
    A. fatua ssp. fatua f. transiens (Hausskn.) Thell., Rec. Trav. Bot. Neerl. 25:427 (1929)
    A. fatua ssp. sativa var. pilosa subv. unilateralis (Thell.) Malz., Monogr. 338 (1930)

A. sativa ssp. praegravis var. cinerascens Vascon., Ens. Sem. Melhor. Pl. Bol. 20 Ser. A, 40 (1935) [7]

A. sativa var. cinerea f. vacceorum Mordv. in Wulff, Fl. Cult. Pl. 2:380 (1936) [8]

×A. fatua × sativa var. transiens (Hausskn.) Tab. Mor., Bol. Soc. Brot. 2, 11:69 (1936)

A. sativa ssp. sativa var. pilosa subv. unilateralis (Thell.) Tab. Mor., Bol. Soc. Brot. 2, 13:Tab. 1 (1939)

A. sativa ssp. sativa var. transiens (Hausskn.) Tab. Mor., op. cit. 603

Description

Same as *A. sativa* except: lodicules fatua type; lemmas sometimes with a few hairs around the insertion of the awn, or if no awn present the hairs may occur at about the middle of the lemma; often a weak awn is present, at least on the first floret. Figure 328 shows a spikelet of such a hybrid.

Fig. 328. Spikelet of *A. sativa* × *fatua* F_1 hybrid including the glumes; note the similarity in appearance to a spikelet of *A. sativa*, and note the weak awn (× 3).

A. sativa × fatua F₁ hybrid

Distribution

Native to (Fig. 329) Austria, Belgium, Bulgaria, Canada, France, Germany, Japan, Portugal, Spain, Switzerland, the United Kingdom, the USSR, and Yugoslavia, but to be expected wherever both *A. sativa* and *A. fatua* are growing.

Phenology

Flowers from April to September.

Habitat

Found among cultivated oats or in various ruderal, segetal, and disturbed places. It is to be expected in the area of overlap of the two parents; the frequency of occurrence in samples may be up to 12%; see reports for Canada by Baum (1968*a*).

Mode of dispersal

Identical with that of *A. sativa* and dispersed by similar means, that is, passes as the normal crop because of the identical shape and appearance of the spikelets.

Selected specimens

AUSTRIA, Steiermark cultivated fields at Marburg, 30 July 1902, *Heyek* (GB); BELGIUM, Anvers Hoboken champ, 16 June 1916, *Hennen* (BR); CANADA *Alberta*, Lethbridge 10 km W champs, 18 August 1960, *Boivin and Gillett No. 8962* (WIN); CANADA *British Columbia*, Kicking Horse River near railway, 13 August 1890, *Macoun No. 30097* (CAN); CANADA *Manitoba*, Souris, 10 July 1951, *Dore and Boivin No. 3176* (DAO); CANADA *New Brunswick*, Bristol 27 km N of Woodstock, 24 August 1955, *Scoggan No. 13089* (ACAD); CANADA *Nova Scotia*, Cumberland Isle au Haute, 10 August 1953, *Schofield No. 3942* (ACAD); CANADA *Ontario*, Algoma Pancake Bay, 12 July 1935, *Taylor et al. No. 971* (CAN and TRT); CANADA *Quebec*, Montmorency Chateau-Richer 22 August 1963, *Rousseau No. 63-1304* (QFA); FRANCE, Rhone Arnas, 10 May 1920, *Gandoger No. 3585* (LY and MT); GERMANY, Thuringia Weimar arvis neglectis, July 1895, *Hausknecht* (PRC); JAPAN, Honshu Okayama Funaho-Mura, 19 May 1914, *Morikawa* (TI); PORTUGAL, Azambrya Farfalae, April 1926. *Machado and Passos No. 647* (ELVAS); SPAIN, Basque Province, 1928, *Mordvinkina No. 2391* (WIR); SWITZERLAND, Granges Wallis Rebgelanden, July 1902, *Wolf No. 333* (PRC); UNITED KINGDOM, Cambridge Hardwick Woodfield, 16 July 1954, *Walters* (CGE); USSR, Mosqua, 1809, (LE); YUGOSLAVIA, Montenegro Podgorica, May 1903, *Rohlena* (PRC).

A. sativa × fatua F₁ hybrid

Fig. 329. Distribution map of *A. sativa* × *fatua* F₁ hybrids. Mercator projection.

Notes

[1] It should not be confused with *A. strigosa* var. *arduennensis* by the same author, which was probably described as var. β (see p. 72) "glumella atro-fusca. Colitur copiose in Arduenna . . ." (see Note 3 of *A. strigosa*). I found a type of *A. sativa* var. *arduennensis* in P; it is annotated by Lejeune and belongs here.

[2] I have seen the protolog, and I can only guess that the authors meant the same as above. It is with doubt, therefore, that I have placed it here; I was unable to find a type.

[3] Same arguments and conclusion as in Note 2.

[4] Riedl (personal communication) said that the types of Schur are in LOD. I was unable to see the type collection of this variety. According to the description it should belong here, but it could also be an *A. sativa*.

[5] I have obtained the holotype from JE. It is duly annotated by Hausknecht, it was collected in August 1883, and it clearly belongs here.

[6] This is a name given for *A. sativa* × *fatua*, and, therefore, belongs here. I was unable to find the voucher to back this up.

[7] I have examined the type obtained from Oeiras (Portugal). It is No. 8806 and clearly belongs here.

[8] I have selected No. 2391 in WIR as lectotype, and it belongs here. It is a specimen collected from plants grown locally from the original seeds or panicle; and it bears an additional number, 4699/1.

(34) A. SATIVA × STERILIS F_1 HYBRID

Synonyms

    A. sativa var. semiaristata Schur, Oesterr. Bot. Zeitschr. 10:70 (1860) [1]
    A. sativa var. subaristata Schur, Enum. Pl. Transsilv. 756 (1866) [2]
    A. barbata ssp. segetalis Nyman, Consp. Fl. Eur. 810 (1882) [3]
    A. segetalis Bianca ex Nyman, loc. cit. pro syn. [4]
    A. syriaca Boiss. et Ball. ex Boiss., Fl. Orient. 5:542 (1884) pro syn. A. sterilis L. [5]
    A. sterilis f. aprica Hausskn., Mitt. Thür. Bot. Ver. N.F. 6:38 (1894) [6]
    A. sterilis var denudata Hausskn., op. cit. 40 [7]
  *A. sativa var. modigenita Kcke., Arch. Biontologie 2:436 (1908) [8]
  *A. sativa var. quadriflora Kcke., loc. cit. [8]
    A. sativa var. submontana Kcke., loc. cit. [8]
    A. sativa var. biaristata Hack. ex Trab., Comp. Rend. Acad. Sci. (Paris) 149:228 (1909), nom. illegit. [9]
    A. sterilis ssp. byzantina var. culta f. denudata (Hausskn.) Thell., Vierteljahrs. Nat. Ges. Zürich 56:319 (1912)

A. sterilis var. segetalis Trab., 4e Conf. Intern. Genet. Paris 1911:336 (1913) [10]

A. sterilis ssp. macrocarpa f. segetalis (Trab.) Thell., Rep. Sp. Nov. Fedde 13:52 (1913)

A. sterilis ssp. byzantina var. induta Thell., Rec. Trav. Bot. Neerland. 25:431 (1929) [11]

A. fatua ssp. praegravis var. polytricha Malz., Monogr. 354 (1930) [12]

A. sterilis ssp. byzantina var. macrotricha subv. biaristata (Hack. ex Trab.) Malz., op. cit. 399

A. sterilis ssp. byzantina var. solida subv. induta (Thell.) Malz., op. cit. 398

A. sterilis ssp. byzantina var. solida subv. segetalis (Nyman) Malz., op. cit. 397

A. sterilis ssp. macrocarpa var. setosissima subv. aprica (Hausskn.) Malz., op. cit. 392

A. sterilis ssp. nodipubescens var. longiseta Malz., op. cit. 385 [13]

A. sterilis ssp. nodipubescens var. longiseta subv. asperata Malz., loc. cit. [14]

A. sativa ssp. diffusa var. transiens f. pseudopolytricha Petrop., Bull. Appl. Bot. Pl. Breed. Suppl. 45:28 (1931) [15]

A. haussknechtii Nevski, Acta Univ. Asiae Med. 8b. Bot. 17:7 (1934) [16]

A. byzantina var. pilosorubida Vasc., Ens. Sem. Melhor. Pl. Bol. 20 Ser. A, 40 (1935) [17]

A. sativa ssp. grandiglumis Vasc., op. cit. 39 [18]

× A. fatua × sativa var. major Tab. Mor., Bol. Soc. Brot. 2, 11:69 (1936) [19]

A. sativa var. haussknechtii Vav. et Mordv. in Wulff, Fl. Cult. Pl. 2:383 (1936) [20]

A. byzantina ssp. byzantina var. denutata (Hausskn.) Tab. Mor., Bol. Soc. Brot. 2, 13:Tab. 1 (1939)

A. byzantina ssp. byzantina var. induta (Thell.) Tab. Mor., loc. cit.

A. byzantina ssp. byzantina var. induta subv. segetalis (Nyman) Tab. Mor., loc. cit.

A. byzantina ssp. byzantina var. macrotricha subv. asperata (Malz.) Tab. Mor., loc. cit.

A. sativa ssp. praegravis var. major Tab. Mor., op. cit. 605 [21]

A. sativa ssp. praegravis var. polytricha (Malz.) Tab. Mor., op. cit. Tab. 1

A. sativa ssp. sativa var. brachytricha subv. subcalosa Tab. Mor., op. cit. 604 [22]

A. sativa ssp. sativa var. subpilosa subv. gigantea Tab. Mor., op. cit. 601 [23]

A. sativa ssp. sativa var. subpilosa subv. subtransiens Tab. Mor., loc. cit. [24]

A. sativa ssp. sativa var. subpilosa f. pallida Tab. Mor., op. cit. 600 [25]

A. byzantina var. pseudovilis subv. biaristata (Hackel ex Trabut) Maire et Weiller, Fl. Afr. Nord 2:289 (1953)

A. byzantina var. solida subv. induta (Thell.) Maire et Weiller, op. cit. 288

A. sativa × sterilis F₁ hybrid

Description

Same as *A. sativa* except: lodicule fatua type; epiblast only fatua type; the lowermost floret (Fig. 330) usually with a well-developed awn, that is, with a twisted column, a bend, and a subule, or the lowermost floret with a weak awn; the uppermost floret awnless (Fig. 330); only the lowermost floret with macrohairs on the lemma below the awn insertion or glabrous but most have only a few hairs around the insertion; spikelet (Fig. 330) elongated, resembling *A. sterilis* or *A. byzantina*.

Distribution

Native to (Fig. 331) Algeria, Belgium, Corsica, France, Germany, India, Iran, Israel, Luxemburg, Portugal, Sicily, Spain, Switzerland, Syria, Turkey, the USA, and the USSR, but to be expected in the area of overlap between *A. sativa* and *A. sterilis*.

Phenology

Flowers from April to September.

Fig. 330. Spikelets of *A. sativa* × *sterilis* F₁ hybrid: two with glumes and one without glumes (to the right); note the elongated shape, the well-developed awns on the first floret, the lack of awns on the upper florets, and as can be seen on the spikelet without the glumes, the macrohairs on the lower floret (× 2).

A. sativa × sterilis F₁ hybrid

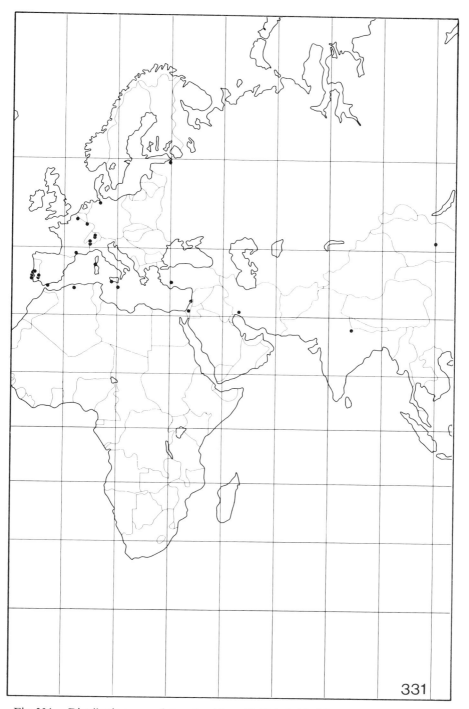

Fig. 331. Distribution map of *A. sativa* × *sterilis* F₁ hybrids. Mercator projection.

A. sativa × sterilis F$_1$ hybrid

Habitat

Found mostly among cultivated oats or in ruderal and segetal places, where the two parents coexist, and to be expected in the whole area of overlap of these two.

Mode of dispersal

Identical with that of *A. sativa,* but often recognized as an impurity in the crop and screened by man.

Selected specimens

ALGERIA, Algers Trois Palmiers, *Trabut No. 1148* (W); BELGIUM, Bruxelles Leuze Champs, July 1887, *Henry* (BR); CORSICA, Entre Ajaccio et Campo di Loro, 20 April 1911, *Thellung* (G); FRANCE, Peyreleau, 5 July 1961, *Berghen* (BR); GERMANY, Hamburg Bot. Garden, August 1889, *Haussknecht* (JE); INDIA, Lucknow, (BR); IRAN, Khuzestan 25 km NE Haft-Kel hills, 3 May 1960, *Bent and Wright No. 03-401* (W); ISRAEL, Jerusalem Monastery of the Cross, 26 May 1926, *Markovitch No. 888A* (WIR); LUXEMBOURG, Berchiruez Villus-La-Loue, 1920, *Verhulst* (BR); PORTUGAL, Cantanhede, June 1934, *Vasconcellos No. 8805* (ELVAS); SICILY, Siracusa Avola, *Todaro No. 712* (BR, COI, and FI); SPAIN, Sierra de Segura dans les bois, 17 June 1850, *Bourgeau No. 943* (FI and G); SWITZERLAND, Zürich Guterbahnhof, 26 September 1917, *Thellung* (BAS); SYRIA, Coelesyria Chtaura, 11 June 1882, *Blanche* (G); TURKEY, near Dinar among fields, 29 June 1927, *Zhukovsky No. 1154* (WIR); USSR *Leningrad,* 10 August 1911, *Shevelov No. 1190* (WIR); USSR *Mongolia,* grown at Kamennaya Step Exp. Station Voronezh, 1 August 1923, *Malzew No. 45* (WIR).

Notes

[1] Schur types should be in LOD (Riedl, personal communication) but I was unable to find them. I was also unable to find material from the type collection in other herbaria. On the basis of the protolog it very likely belongs here.

[2] Same remark as Note 1 above, except that it could be *A. sativa* as well.

[3] This is the correct name for what is known as *A. segetalis* Bianca ex Nyman (Index Kewensis, Chase and Niles 1962, etc.). In Nyman it is given as subspecies because an asterisk appears before the name (see Introduction of Nyman's work and see last sentence of Note 10 below). The name is based on Todaro's collection No. 712 of which I saw many specimens in various herbaria. Nyman's herbarium is in S (Stafleu 1967: 341) and I have examined the holotype. I have also examined isotypes from K, LE (2 specimens), P, BR, CGE, and COI (2 specimens). All belong here except for the CGE specimen, which is *A. trichophylla*.

[4] This is given as a synonym of *A. barbata* ssp. *segetalis* by Nyman, and is based on the Exsiccatum published by Todaro (see Note 3 above). It is a name suggested by Bianca to Todaro and validated or taken by Nyman.

[5] It belongs here; I have seen the 'holotype' in G-Boiss. and examined an 'isotype' in G. The label reads: "Dans les cultures à Ord Sebbagh 14.3.1853, *Blanche* 286." It obviously has been published as a synonym of *A. sterilis* and is therefore, not valid, but its 'typification' is important for the next taxon (see Note 6 below).

[6] For this taxon Hausknecht, among others, wrote in the protolog "Solche Formen wurden nach von Blanche bei Tripolis in Syrien gesammelten Exemplaren von Boiss. fruher als *A. syriaca* bezeichnet." This is based on Blanche No. 286 (see Note 5 above).

[7] From JE I obtained the holotype, which Hausknecht had annotated "Avena sterilis L. v. solida denudata straminea glabra." It obviously belongs here.

[8] According to Prof. Ullrich, Inst. Landw. Bot. Univ. Bonn, the types of Koernicke had been there but were destroyed in 1945 by bombardments. These taxa were put in this synonymy on the basis of their descriptions only; var. *submontana,* more clearly than the other two, belongs here.

[9] It is a later homonym to Alefeld (1866). I have examined an isotype in W. It is specimen No. 1148; it was collected by Trabut at "Trois palmiers (Alger)" but it belongs to Hackel's herbarium. On one of the labels Trabut wrote "A. sativa × sterilis?" The holotype is in P and also belongs here.

[10] It is based on the same type material of *A. barbata* ssp. *segetalis* Nym. except that the holotype is in P (see Note 3 for details on the type). Trabut stated in the protolog "rapporté à tort par Nyman à l'*A. barbata,*" which supports my interpretation of Nyman's rank of that taxon (see Note 3).

[11] I have examined one specimen from BAS, duly annotated by Thellung and collected by him in "Zurich 4 Guterbahnhof IV 1, 26.9.1917." I designate this specimen as lectotype, and it belongs here. I have also examined a paralectotype, which I obtained from BAA, but it does not belong here; it is actually the holotype of *A. byzantina* var. *induta* Thell. (see Note 103 of *A. sativa*). Is it possible that Thellung was confused and considered the two species as one? Or could it be that he changed his mind and one year later made the transfer from *A. sterilis* to *A. byzantina*? This situation should be sorted out if possible.

[12] I have examined all the syntypes from WIR, and all belong here. I designate No. 418 as lectotype; thus Nos. 417 and 1100 become paralectotypes.

[13] No specimens are cited in the protolog, but three subvarieties are classified under this variety and several specimens, which can be regarded as syntypes, are cited for each subvariety. I have designated the holotype of subv. *asperata* as lectotype of this variety (see Note 14 below).

[14] I have seen the holotype in WIR, it is No. 88a, and it belongs here. On

[15] From the short description and from the figure it is likely to belong here, although I was unable to find the type.

[16] This is based on *A. sterilis* var. *denudata* Hausskn. (Note 7 above) and therefore belongs here.

[17] I have obtained the holotype from Oeiras (Portugal); it is No. 8000 and belongs here.

[18] I am choosing the specimens listed by Vasconcellos under var. *cinerea* as syntypes for the subspecies. For var. *hirsuta* I have chosen a lectotype (see Note 140 under *A. sativa*). The two specimens, No. 2406, belong here and I designated one as lectotype and the other becomes an isolectotype. The paralectotype No. 8993 is an *A. sativa*.

[19] I have not seen the type, but according to the description it should belong here.

[20] It seems, at least in part, to be based on *A. fatua* var. *transiens* Hausskn. Furthermore, the picture is very revealing and points to this identity. I was unable to find type material at WIR.

[21] I was unable to see the holotype from COI, but there is a photograph of the type in the publication. The photograph and the description point to this identity, but with some doubt of course.

[22] Although in the text on p. 604 it is not called subv. *subcalosa,* this name appears on fig. 2 of table 3 and in Table 1 at the end of the publication. It is based on Hausknecht's var. *transiens* (Mitt. Geogr. Ges. 3:238, fig. 4 (1884)) and therefore, belongs here.

[23] I have examined the holotype that I obtained from COI. It is No. 2970 and it belongs here.

[24] I have examined the holotype from COI. It is No. 3766 and it belongs here.

[25] I have examined the holotype, specimen No. 3490, and it belongs here. Two paratypes are conserved in COI: No. 3491 is a heterozygous fatuoid and No. 3478 is an *A. sativa*.

23. Non Satis Notae

A. azo-carti R. Phil., Anal. Univ. Chile 94:29 (1896) [1]
A. barbata f. glabra Cavara, Bull. Orto Bot. Univ. Napoli 9:43 (1928) [2]

A. barbata var. triflora f. subglabra Merino, Fl. Descr. Illustr. Galicia 3:300 (1909) [1]

A. benghalensis Schwaegr. ex Steud., Nom. Bot. ed. 2, 1:171 (1840) [3]

A. euhybrida Hausskn. ex Thell., Mitt. Naturw. Ges. Winterthur. 12:128 (1918) [4]

A. fatua var. intermedia Merino. Fl. Descr. Illustr. Galicia 3:301 (1909), nom. illegit. [1]

A. fatua var. mollis Keng, Sinensia 11:411 (1940) [5]

A. fatua ssp. fatua f. pseudobasifixa Thell., Rec. Trav. Bot. Neerland. 25:427 (1928), nom. illegit. [6]

A. fatua ssp. fatua f. pseudopilibarbis Thell., op. cit. 426, nom. illegit. [6]

A. fatua f. pseudotransiens Thell., Repert. Sp. Nov. Fedde 13:55 (1913), nom. illegit. [6]

A. magellanica Beauv., Ess. Agrost. 153 (1812), nom. nud.

A. mutica Krocker, Fl. Siles. Suppl. 4:187 (1823)

A. polonica Schwaegr. ex Schmalh., Fl. Centr. et S. Russia 2:619 (1897), nom. nud. [7]

A. purpurea Gueldenst. in Pallas, Reisen Russland Cauc. 1:189 (1787), nom. nud. [8]

A. sativa var. uniflora S.F. Gray, Nat. Arr. Brit. Pl. 2:130 (1821), nom. illegit.

A. sativa var. uniflora Stokes, Bot. Mat. Med. 1:158 (1812)

A. strigosa ssp. strigosa prol. brevis var. hepatica Mordv. in Wulff, Fl. Cult. Pl. 2:430 (1936) [9]

Notes

I was unable to locate or obtain types of the specimens listed here. Furthermore, I was unable to judge the synonymy on the basis of the descriptions alone.

[1] The types of R. Phil and Merino should be in SGO.

[2] The type of Cavara should be in NAP.

[3] Steudel's types are in P and I found many of them, but I failed to locate the type of this taxon even after many searches in various genera in the general herbarium. The type is probably misfiled or lost.

[4] I did not find the type among all Thellung's specimens in BAS and Z, and have no clue to its location.

[5] Keng types should be in N.

[6] These taxa have no types because these are 'hypothetical forms'.

[7] I tried to find authentic specimens in LE but did not find any.

[8] I looked in BM, LE, and M for authentic specimens but in vain. I thought that they might possibly have been in BM because Pallas' specimens are there, or in LE or M where Ledebour's specimens are.

[9] I was unable to find in WIR a specimen that antedates the publication and could not find a good candidate specimen for Neotype.

24. Names wrongly cited

A. brevis Breb., Fl. Norm. ed. 3. 349 (1859) [1]
A. clauda var. eriantha C. Mueller in Walp., Ann. Bot. 6:1000 (1861) [2]
A. hirsuta (Roth) Mathieu, Fl. Gen. Belg. 1:606 (1853) [3]
A. pilosa Aucher ex Dur., Revue Bot. Duchartre 1:360 (1845), pro syn. A. clauda Dur. [4]
A. pseudofatua Schur ex Hausskn., Mitt. Thür. Bot. Ver. N.F. 6:37 (1894) [5]
A. sterilis Del. ex Boiss., Fl. Or. 5:543 (1884), pro syn. A. fatua L. [6]

Notes

These names were erroneously cited by Chase and Niles (1962) or Index Kewensis or both.

[1] According to the context Breberisson must mean Schreber because he is using other conventional names; the omission of the author does not automatically attribute the name to Breberisson.

[2] This name is quoted correctly in Walpers.

[3] Mathieu used to put all the authorities in parentheses, and so he did not mean to make a combination, which would be unnecessary in this case.

[4] This is a misidentification quoted by Durieu. It was never intended to be published as a name.

[5] This is an erroneous citation because it was never effected by Haussknecht. It is cited under *A. fatua* as a name published by Schur in 1866. I found two such authentic specimens in P annotated by Schur but after 1866, and these are *A. fatua*.

[6] This is an identification, not a name, published by Delile.

Part 3. Digest

Digest

This digest contains all the specific epithets that were once described under, or combined with, the genus *Avena*. It contains, in addition, all the specific and infraspecific epithets of *Avena* proper as recognized in this book, and as such can also serve as an index.

For various reasons, many of the species that were once described as *Avena* no longer belong to this genus. As can be seen from the summary (Table 17), from the time of Linnaeus to Roemer and Schultes a number of genera were segregated from *Avena*. Many more autonomous genera were recognized because of increased knowledge and information on this group of plants. It is imperative, therefore, to provide this list of epithets to document the information relevant to *Avena* and allied genera, which are recognized today.

I also found it necessary and of some use to document here all the species names described under *Avenastrum, Helictotrichon,* and *Avenochloa,* not only because many of these names were earlier attached to *Avena,* but also because under various classification schemes these genera are strongly interlinked.

Information about *Avena* species has often been published under various other names (e.g., about distribution, ecology, uses, diseases, chemical composition, and folklore) and it is hoped that this digest will help in retrieving all this information so that it can be properly interpreted and used according to contemporaneous concepts.

All the names, which are retained in *Avena* in this digest, are obviously synonyms of the correct names of the taxa recognized in this work. Therefore this digest can also serve as an index to synonyms.

Abbreviations:
    A. = Avena
    Av. = Avenula
    Aven. = Avenastrum
    Aveno. = Avenochloa
    D. = Danthoniastrum
    H. = Helictrotrichon

Avena abietorum Ohwi, Acta Phytotax. Geobot. 2:162 (1933)
    = A. abietetorum Ohwi
    = H. abietetorum (Ohwi) Ohwi
Avena abietetorum Ohwi, Jour. Jap. Bot. 17:441 (1941)
    = H. abietetorum Ohwi,
Avena abyssinica Hochst. in Schimper Iter Abyss. Sect. 3, Unit. Iter No. 1877 (1846)

TABLE 17. SUMMARY OF NEW GENERA.

New genera and transfers of epithets from *Avena* to these genera from the time of Linnaeus (1753) until Besser (1827).

Linnaeus, 1753, 1756, 1762, 1767	Persoon 1805	Palisot de Beauvois 1812, 1817	Besser ex R. et S. 1827
Avena elatior	*Avena elatior*	*Arrhenatherum elatior*	————
Avena fatua	*Avena fatua*	*Avena fatua*	*Avena fatua*
Avena flavescens	*Trisetum pratense*	*Trisetum flavescens*	*Acrospelion flavescens*
Avena fragilis	*Avena fragilis*	*Gaudinia fragilis*	————
Avena loeflingiana	*Trisetum hispanicum*	*Trisetum "loeflingi"*	————
Avena nuda	*Avena nuda*	*Avena nuda*	*Avena nuda*
Avena pennsylvanica	*Avena "pensylvanica"*	*Trisetum pennsylvanicum*	————
Avena pratensis	*Avena pratensis*	*Avena pratensis*	*Helictotrichon pratensis*
Avena sativa	*Avena sativa*	*Avena sativa*	*Avena sativa*
Avena sesquitertia	*Avena sesquitertia*	*Trisetum sesquitertium*	————
Avena sibirica	*Stipa sibirica*	*Avena sibirica*	————
Avena spicata	*Avena glumosa*	*Danthonia spicata*	————
Avena sterilis	*Avena sterilis*	*Avena sterilis*	*Avena sterilis*
Avena stipiformis	*Avena stipiformis*	*Chaetaria stipiformis*	————

Avena abyssinica var. baldratiana Cif., Atti Ist. Bot. Univ. Pavia Ser. 5.2:152 (1944)
    = ? A. abyssinica Hochst.
Avena abyssinica var. chiovendae Mordv. in Wulff, Kult. Fl. USSR 2:421 (1936)
    = A. abyssinica Hochst.
Avena abyssinica var. granulata Chiov., Ann. Ist. Bot Roma 8:343 (1908)
    = A. sterilis L.
Avena abyssinica var. neoschimperi Cif., Atti Ist. Bot Univ. Pavia Ser. 5,2:152 (1944)
    = ? A. abyssinica Hochst.
Avena abyssinica f. glaberrima Chiov., Ann. Ist. Bot. Roma 8:343 (1907)
    = A. abyssinica Hochst.
Avena adsurgens Schur, Enum. Pl. Transsilv. 162 (1866)
    = Heuf. praeusta (Reichb.) Schur
    = H. alpinum (Smith) Henrard
    = Aven. adsurgens Schur
    = Aven. adsurgens (Schur) Javorka
Avena adzharica Alboff, Prodr. Fl. Colch. 257 (1895), or Acta Horti Tiflis 1, Suppl. 1:257 (1895)
    = Aven. adzharicum (Alboff) Roschev.
    = H. adzharicum (Alboff) Henrard
    = Aveno. adzharica (Alboff) Holub
Avena aemulans Nevski, Acta Univ. Asiae Med. 8b. Bot. 17:5 (1934)
    = A. sativa fatuoid
Avena aenea Hook. f., Fl. Brit. India 7:279 (1896)
    = Trisetum aeneum (Hook. f.) R.R. Stewart, Brittonia 5:431 (1945)
Avena affinis Bernhardi ex Steudel, Nomencl. ed. 2, 1:171, 173 (1840)
    = A. sterilis L.
Avena agraria Brot., Fl. Lusit. 1:105 (1804)
    = A. hispanica Ard.
Avena agraria var. mutica Brot., Fl. Lusit. 1:106 (1804)
    = A. sativa L.
Avena agraria var. sesquialtera Brot., Fl. Lusit. 1:106 (1804)
    = A. hispanica Ard.
Avena agropyroides Boiss., Diagn. Pl. Orient. Nov. 2 (13):50 (1854)
    = Aven. agropyroides (Boiss.) Halacsy
    = H. agropyroides (Boiss.) Henrard
    = Aveno. agropyroides (Boiss.) Holub
Avena agrostidea Fries, Nov. Fl. Suec. Mant. 3:3 (1842)
    = Trisetum subalpestre (Hartm.) Neum., Sver. Fl. 755 (1901)
Avena agrostoides Griseb., Spicil. Fl. Rumel. 2:454 (1845)
    = Trisetum myrianthum (Bertol.) Parl., Fl. Ital. 1:270 (1848)
Avena airoides Koel., Descr. Gram. 298 (1802)
    = Trisetum spicatum (L.) Richt., Pl. Eur. 1:59 (1890)
Avena airoides Pourret ex Willd., Ges. Naturf. Freund. Berlin Mag. 2:290 (1808)
    = Agrostis pourretii Willd., loc. cit.
Avena alba Vahl, Symb. Bot. 2:24 (1791)
    = Arrhenatherum erianthum Boiss. et Reut.
Avena alba ssp. abyssinica (Hochst.) Löve and Löve, Bot. Not. 114:50 (1961)
    = A. abyssinica Hochst.

Avena alba var. barbata (Pott ex Link) Maire et Weiller, Fl. Afr. Nord 2:275 (1953)
 = A. barbata Pott ex Link
Avena alba var. barbata subv. fallax Maire et Weiller, in Maire Cat. Pl. Afr. Nord No. 2857 (1939)
 = A. barbata Pott ex Link
Avena alba var. barbata f. fallax (Maire et Weiller) Maire et Weiller, Fl. Afr. Nord 2:276 (1953)
 = A. barbata Pott ex Link
Avena alba var. barbata f. genuina (Aschers. et Graeb.) Maire et Weiller, Fl. Afr. Nord 2:276 (1953)
 = A. barbata Pott ex Link
Avena alba var. barbata f. triflora Maire et Weiller, Bull. Soc. Sci. Nat. Phys. Maroc 37:145 (1957)
 = A. barbata Pott ex Link
Avena alba var. hirtula (Lang.) Emberger et Maire, Bull. Soc. Sci. Nat. Phys. Maroc 37:145 (1957)
 = A. hirtula Lag.
Avena alba var. hirtula subv. aristulata (Malz.) Maire et Weiller, Fl. Afr. Nord 2:277 (1953)
 = A. wiestii Steud.
Avena alba var. hirtula subv. glabrifolia (Malz.) Maire et Weiller, Fl. Afr. Nord 2:277 (1953)
 = A. wiestii Steud.
Avena alba var. hirtula subv. minor (Lange) Emberger et Maire, Bull. Soc. Sci. Nat. Phys. Maroc 37:145 (1957)
 = A. hirtula Lag.
Avena alba var. wiestii (Steud.) Maire et Weiller, Fl. Afr. Nord 2:277 (1953)
 = A. wiestii Steud.
Avena albicans Lestib., Bot. Belg. 2:35 (1827)
 = ? A. nuda L.
Avena albinervis Boiss., Voy. Bot. Espagne 2:656 (1844)
 = Aven. albinerve (Boiss.) Vierh.
 = H. albinerve (Boiss.) Henrard
 = Aveno. albinervis (Boiss.) Holub
Avena algeriensis Trab. Bull. Agr. Algerie Tunisie 16:354 (1910), and Bull. Soc. Hist. Nat. Afr. Nord 2:151 (1910)
 = A. sativa L.
Avena almeriensis Gandog., Bull. Soc. Bot. France 52:443 (1905)
 = A. barbata Pott ex Link
Avena alopecuros Roth, Catalecta Bot. 3:20 (1806)
 = ? Trisetum alopecuros (Roth) Roem. et Schult., Syst. Veg. 2:660 (1817)
Avena alpestris Host, Icon. Gram. Austr. 3:27 (1805)
 = Trisetum alpestre (Host.) P. Beauv., Agrostogr. 88 (1812)
Avena alpina (L.) Trin, Fund. Agrost. 157 (1820)
 = Aira alpina L., Sp. Pl. ed. 1:65 (1753)
Avena alpina Latour., Chlor. Lugd. 3 (1785)
 = ? H. versicolor (Vill.) Pilger
Avena alpina Honck. Vollst. Syst. Verz. Aller. Gew. Deutschl. 302 (1782)
 = ? Deschampsia caespitosa (L.) Beauv., Agrost. 160 (1812)

Avena alpina J.E. Smith, Trans. Linn. Soc. 10:335 (1811)
 = H. alpinum (Smith) Henrard
 = Aveno. alpina (Smith) Holub
Avena alta Cavan. ex Roem. et Schult., Syst. Veg. 2:691 (1817)
 = A. hispanica Ard. ex Saggi
Avena altaica Stephan ex Roshev., Not. Syst. Herb. Hort. Bot. Petrop. 3:85 (1922)
 = Trisetum altaicum Roshev. in Komarov, Fl. USSR 2:254 (1934)
Avena altior Hitchc., Proc. Biol. Soc. Washington 43:96 (1930)
 = H. altius (Hitchc.) Ohwi
Avena ambigua Schönh., Fl. Thür. 517 (1850)
 = A. strigosa Schreb.
Avena americana Scribn., U.S. Dept. Agr. Div. Agrost. Bull. 7:183 (1897)
 = H. hookeri (Scribn.) Henrard
Avena amethystina DC., Fl. Franç. 3:37 (1805)
 = H. pubescens (Huds.) Pilger
Avena anathera Nabelek, Publ. Fac. Sci. Univ. Masaryk, No. 111:12 (1929)
 = Anatherum tauricolum Nabelek, loc. cit.
Avena andropogoides Steud., Flora 12:486 (1929) sphlam. andropogonoides
 = Triraphis rehmanni Hack., Bull. Herb. Boiss. 3:388 (1895)
Avena anglica Hort. ex Roem. et Schultes, Syst. Veg. 2:669 (1817)
 = A. sativa L.
Avena anisopogon Raspail, Ann. Sci. Nat. 1.5:439 (1825)
 = Anisopogon avenaceus R. Br., Prodr. 176 (1810)
Avena antarctica (R.Br.) Roem. et Schult., Syst. Veg. 2:676 (1817)
 = ? Trisetum antarcticum (Forst.) Trin., Mem. Acad. Petersb. 6, 1:61 (1831)
Avena antarctica Thunb., Prodr. Pl. Cap. 22 (1794)
 = Aven. antarcticum (Thunb.) Stapf
 = H. capense Schweickert and H. leoninum (Steud.) Schweick.
Avena arduensis Lej. ex Steud., Nom. Bot. ed. 2, 1:171 (1840)
 = A. strigosa Schreb.
Avena arenaria (Labill.) Spreng., Syst. Veg. 1:333 (1825)
 = Trisetum lineare (Forssk.) Boiss. Diagn. 1, 13:49 (1854)
Avena argaea Boiss., Fl. Orient. 5:546 (1884)
 = H. argaeum (Boiss.) Parsa
 = Aveno. argaea (Boiss.) Holub
Avena argentea Willd., Enum. Pl. Hort. Berol. 125 (1809)
 = Trisetum argenteum (Willd.) Roem. et Schult., Syst. Veg. 2:665 (1817)
Avena argentoideum Schur, Oesterr. Bot. Zeitschr. 10:74 (1860)
 = Trisetum pratense Pers., Syn. 1:97 (1805)
Avena ariguensis Steud., Syn. Pl. Glum. 1:233 (1854)
 = Danthonia sp.
Avena aristelliformis Senn. ex St.-Yves, Candollea 4:456 (1931)
 = H. pratense (L.) Pilger
Avena aristiodoides Thunb., Prodr. Pl. Cap. 22 (1794)
 = Pentaschistis aristidoides (Thunb.) Stapf
Avena armeniaca Schischk., Bericht. Tomsk. Staats Univ. 81: 418 (1929)
 = Aven. armeniacum (Schischk.) Roshev.
 = H. armeniacum (Schischk.) Grossh.
 = Aveno. armeniaca (Schischk.) Holub

Avena arundinacea Delile, Fl. Egypte 171 (1812)
    = Asthenatherum forsskahlii (Vahl) Nevski, Act. Univ. As. Med. Ser. 8b Bot. Fasc. 17:8 (1934)
Avena arvensis Salisb., Prodr. Stirp. 23 (1796)
    = Bromus arvensis L., Sp. Pl. 77 (1753)?
Avena aspera Munro ex Thwaites, Enum. Pl. Zeyl. 372 (1864)
    = Aven. asperum (Munro ex Thwaites) Vierh.
    = H. virescens (Nees ex Steud.) Henrard
    = H. asperum (Munro ex Thwait.) Bor

Avena atherantha Presl, Cyp. Gram. Sicul. 30 (1820)

Avena atropurpurea (Wahlb.) Link, Hort. Berol. 1:119 (1827)
    = Deschampsia atropurpurea (Wahl.) Scheele, Flora 27:56 (1844)
Avena australis Parl., Fl. Ital. 1:285 (1848)
    = Aven. australe (Parl.) Halácsy
    = H. australe (Parl.) Holub
    = H. compressum (Heuff.) Potztal
Avena azo-carti R. Phil., Anal. Univ. Chile 94:29 (1896)
    = Non satis notae
Avena barbata Brot., Fl. Lusit. 1:108 (1804)
    = A. barbata Pott ex Link

Avena barbata Pott ex Link, Jour. Bot. Schrader 2:315 (1799)

Avena barbata ssp. barbata (Pott ex Link) Tab. Mor., Bol. Soc. Brot. II. 13:617 (1939)
    = A. barbata Pott ex Link
Avena barbata ssp. barbata var. subtypica (Malz.) Tab. Mor., op. cit. 621.
    = ? A. wiestii Steud.
Avena barbata ssp. barbata var. subtypica subv. triflorisubtypica Tab. Mor., op. cit. 622.
    = A. lusitanica (Tab. Mor.) Baum
Avena barbata ssp. barbata var. typica (Malz.) Tab. Mor., op. cit. 617.
    = A. barbata Pott ex Link
Avena barbata ssp. barbata var. typica subv. genuina (Willk.) Tab. Mor., op. cit. 618.
    = A. lusitanica (Tab. Mor.) Baum
Avena barbata ssp. barbata var. typica subv. glabritriflora Tab. Mor., op. cit. 621.
    = A. barbata Pott ex Link
Avena barbata ssp. barbata var. typica subv. hirsuta (Moench) Tab. Mor., op. cit. 619.
    = A. matritensis Baum
Avena barbata ssp. barbata var. typica subv. triflora (Willk.) Tab. Mor., op. cit. 620.
    = A. hirtula Lag.
Avena barbata ssp. hirtula (Lag.) Tab. Mor., op. cit. 622.
    = A. hirtula Lag.
Avena barbata ssp. hirtula var. calva Tab. Mor., op. cit. 626.
    = A. barbata Pott ex Link
Avena barbata ssp. hirtula var. caspica (Hausskn.) Tab. Mor., op. cit. Tab. 1.
    = A. barbata Pott ex Link
Avena barbata ssp. hirtula var. deserticola (Malz.) Tab. Mor., op. cit. Tab. 1.
    = A. wiestii Steud.

Avena barbata ssp. hirtula var. malzevii Tab. Mor., op. cit. 622.
= A. hirtula Lag.
Avena barbata ssp. hirtula var. malzevii subv. lusitanica Tab. Mor., op. cit. 624.
= A. lusitanica (Tab. Mor.) Baum
Avena barbata ssp. hirtula var. malzevii subv. minor (Lange) Tab. Mor., op. cit. 625.
= A. hirtula Lag.
Avena barbata ssp. hirtula var. malzevii subv. pseudostrigosa (Malz.) Tab. Mor., op. cit. 623.
= A. matritensis Baum
Avena barbata ssp. hirtula var. malzevii subv. subaristulata Tab. Mor., op. cit. 625.
= A. lusitanica (Tab. Mor.) Baum
Avena barbata ssp. hirtula var. malzevii subv. trifloriaristulata Tab. Mor., op. cit. 623.
= A. barbata Pott ex Link
Avena barbata ssp. hirtula var. subcalva Tab. Mor., op. cit. 625.
= A. lusitanica (Tab. Mor.) Baum
Avena barbata ssp. longiglumis (Dur.) Lindb., Act. Soc. Sci. Fenn. N. Ser. B. 1:12 (1932).
= A. longiglumis Durieu
Avena barbata ssp. segetalis Nyman, Consp. Fl. Eur. 810 (1882).
= A. sativa × sterilis F_1 hybrid
Avena barbata ssp. vaviloviana (Malz.) Tab. Mor., Bol. Soc. Brot. II. 13:626 (1939)
= A. vaviloviana (Malz.) Mordv.
Avena barbata ssp. vaviloviana var. glabra (Hausskn.) Tab. Mor., op. cit. Tab. 1.
= A. abyssinica × vaviloviana F_1 hybrid
Avena barbata ssp. vaviloviana var. intercedens (Thell.) Tab. Mor., op. cit. Tab. 1.
= ? A. abyssinica × vaviloviana F_1 hybrid
Avena barbata ssp. vaviloviana var. pilosiuscula (Thell.) Tab. Mor., op. cit. 627.
= A. vaviloviana (Malz.) Mordv.
Avena barbata ssp. vaviloviana var. pseudoabyssinica (Thell.) Tab. Mor., op. cit. Tab. 1.
= A. abyssinica Hochst.
Avena barbata ssp. wiestii (Steud.) Mansf., Kulturpfl. Beih. 2:479 (1959)
= A. wiestii Steud.
Avena barbata var. abbreviata Hausskn., Mitt. Thür. Bot. Ver. N.F. 13/14:51 (1899)
= A. abyssinica Hochst.
Avena barbata var. atherantha (Presl) Grossh., Trudy Bot. Inst. Azerbaidzh. Fil. Akad. Nauk SSSR 8:207 (1939)
= A. atherantha Presl
Avena barbata var. caspica Hausskn., Mitt. Thür. Bot. Ver. N.F. 6:41 (1894)
= A. barbata Pott ex Link
Avena barbata var. eubarbata Maire, in Jahand. et Maire, Cat. Pl. Maroc 1:50 (1931)
= A. barbata Pott ex Link

Avena barbata var. fuscescens Batt. et Trab., Fl. Alger. Monoc. 62 (1884)
   = A. hirtula Lag.
Avena barbata var. genuina Willk. in Willk. et Lange, Prodr. Fl. Hisp. 1:68 (1862)
   = A. lusitanica (Tab. Mor.) Baum
Avena barbata var. hirtula (Lag.) Perez-Lara, Anal. Soc. Españ. Hist. Nat. 15:398 (1886)
   = A. hirtula Lag.
Avena barbata var. hoppeana (Scheele) Richt., Pl. Eur. 1:62 (1890)
   = ? A. barbata Pott ex Link
Avena barbata var. minor Lange, Naturhist. For. Kjøbenhaven Vid. Medd. II. 1:39 (1860)
   = A. hirtula Lag.
Avena barbata var. sallentiana (Pau) Jahand. et Maire, Cat. Pl. Maroc 1:50 (1931)
   = A. barbata Pott ex Link
Avena barbata var. saxatilis Lojac., Fl. Sicul. 3:302 (1909)
   = A. matritensis Baum
Avena barbata var. solida Hausskn., Mitt. Thür. Bot. Ver. N.F. 6:41 (1894)
   = A. barbata Pott ex Link
Avena barbata var. triflora Willkomm. in Willkomm. et Lange, Prodr. Fl. Hisp. 1:68 (1862)
   = A. hirtula Lag.
Avena barbata var. triflora Trab., 4^{em} Conf. Genet. Paris: 433, f.9 (1913)
   = A. barbata Pott ex Link
Avena barbata var. triflora f. subglabra Merino, Fl. Descr. Illustr. Galicia 3:300 (1909)
   = Non satis notae
Avena barbata var. wiestii (Steud.) Hausskn., Mitt. Thür. Bot. Ver. N.F. 6:45 (1894).
   = A. wiestii Steud.
Avena barbata f. glabra Cavara, Bull. Orto Bot. Univ. Napoli 9:43 (1928)
   = Non satis notae
Avena baregensis (Laff. et Miegev.) Nyman, Consp. Fl. Eur. 813 (1882)
   = Trisetum agrostideum Fries, Mant. 3:180 (1842?)
Avena baumgartenii Steud., Syn. Pl. Glum. 1:233 (1854)
   = ? Helictotrichon sp.
Avena beguinotiana Pampan., Archivio Bot. Forli 12:18 (1936)
   = A. ventricosa Bal. ex Coss.
Avena bellardi Colla, Herb. Pedem. 6:34 (1836)
   = ? Helictotrichon sp.
Avena benghalensis Schwaegr. ex Steud., Nom. Bot. ed. 2:171 (1840)
   = Non satis notae
Avena besseri Griseb. in Ledeb., Flor. Ross. 4:415 (1853)
   = Aven. besseri (Griseb.) Koczwara
   = H. besseri (Griseb.) Janchen
   = Arrhenatherum desertorum (Less.) Potztal, Bot. Jahrb. 75:329 (1951)
Avena bifida (Thunb.) Beauv., Ess. Agrost. 89, 153, 155 (1812)
   = Trisetum bifidum (Thunb.) Ohwi, Fl. Japan 149 (1965)
Avena bipartita Link, Hort. Berol. 1:113 (1827)
   = Danthonia semiannularis R. Brown, Prodr. 177 (1810)

Avena blavii Aschers. et Janka in Janka, Termesz. Füzet. 1:99 (1877)
= Aven. blavii (Aschers. et Janka) Beck
= H. blavii (Aschers. et Janka) Hubb.
= Aveno. blavii (Aschers. et Janka) Holub
Avena bolivaris Sennen, Butll. Inst. Catal. Hist. Nat. 18:176 (1918)
= A. matritensis Baum
Avena borbonia Steud., Syn. Pl. Glum. 1:234 (1854)
= ? Helictotrichon sp.
Avena bornmuelleri Domin, Repert. Sp. Nov. Fedde 2:31 (1906)
= Trisetum bornmuelleri Domin
Avena breviaristata Barratte ex Batt. et Trab., Fl. Algerie Monoc. 2:184 (1895)
= H. breviaristatum (Barr. ex Batt. et Trab.) Henrard
= Aveno. breviaristata (Barr. ex Batt. et Trab.) Holub
Avena brevifolia Schrad. ex Spreng., Syst. Veg. 1:334 (1825)
= ? Trisetum rigidum (M.B.) Roem. et Schult., Syst. 2:662 (1817)
Avena brevifolia Host, Gram. Austr. 283 (1805)
= Trisetum distichophyllum (Vill.) P. Beauv., Agrostogr. 88 (1812)
Avena brevis Breb., Fl. Norm. ed. 3:349 (1859)
= A. See Names Wrongly Cited
Avena brevis Roth, Bot. Abh. Beob. 42 (1787)
Avena brevis var. uniflora (Parl.) Drouet, Mem. Soc. Acad. Dept. Aube Troyes 30:206 (1866)
= A. brevis Roth
Avena bromoides Gouan, Hort. Monsp. 52 (1762)
= Aven. bromoides (Gouan) Vierh.
= H. bromoides (Gouan) Hubbard
= Aveno. bromoides (Gouan) Holub
= Arrhenatherum bromoides (Gouan) Sampaio, An. Fac. Ci. Porto 17:45 (1931)
Avena brownei (R.Br.) Spreng., Syst. Veg. 1:336 (1825)
= Danthonia pallida R. Brown, Prodr. 177 (1810)
Avena bruhnsiana Gruner, Bull. Soc. Nat. Moscou 40:458 (1867)
= A. ventricosa Bal. ex Coss.
Avena bulbosa Willd., Ges. Naturf. Fr. Berlin Neue Schr. 2:116 (1799)
= Arrhenatherum elatius (L.) J.S. et C.B. Presl, Fl. Cechica 17 (1819)
Avena burnoufii (Req.) Nyman, Syll. 413 (1854)
= Trisetum pratense Pers., Syn. 1:97 (1805)
Avena byzantina C. Koch, Linnaea 21:392 (1848)
= A. sativa L.
Avena byzantina var. anopla Mordv. in Wulff, Fl. Cult. Pl. 2:415 (1936)
= A. sativa L.
Avena byzantina var. anopla subv. anatolica Mordv. loc. cit.
= A. sativa L.
Avena byzantina var. cinnamomea Mordv. in Wulff, op. cit. 416
= ? A. sativa L.
Avena byzantina var. cremea Mordv. in Wulff, op. cit. 415
= A. sativa L.
Avena byzantina var. culta (Thell.) Mordv. op. cit. 413
= A. sativa L.
Avena byzantina var. culta f. Zhukovskii Mordv. in Wulff, op. cit. 415
= A. sativa L.

Avena byzantina var. graeca Mordv. in Wulff, op. cit. 417
    = ? A. sativa L.
Avena byzantina var. incana Mordv. in Wulff, op. cit. 416
    = ? A. sativa L.
Avena byzantina var. induta Thell. in Parodi, Rev. Fac. Agron. Vet. Univ. Buenos Aires 7:165 (1930)
    = A. sativa L.
Avena byzantina var. maroccana Mordv. in Wulff, Fl. Cult. Pl. 2:416 (1936)
    = A. sativa L.
Avena byzantina var. nigra Mordv., loc. cit.
    = A. sativa L.
Avena byzantina var. pilosorubida Vasc., Ens. Sem. Melhor. Pl. Bol. 20 Ser. A:40 (1935)
    = A. sativa $\times$ sterilis F_1 hybrid
Avena byzantina var. pseudovilis (Hausskn.) Maire et Weiller, Fl. Afr. Nord 2:289 (1953)
    = A. sterilis L.
Avena byzantina var. pseudovilis subv. biaristata (Hackel ex Trab.) Maire et Weiller, loc. cit.
    = A. sativa $\times$ sterilis F_1 hybrid
Avena byzantina var. pseudovilis subv. pseudovilis (Hausskn.) Maire et Weiller, loc. cit.
    = A. sterilis L.
Avena byzantina var. rubida (Krause) Vasc., Ens. Sem. Melhor. Pl. Bol. 20:40 (1935)
    = A. sativa L.
Avena byzantina var. rubra Mordv. in Wulff, Fl. Cult. Pl. 2:415 (1936)
    = A. sativa heterozygous fatuoid
Avena byzantina var. solida (Hausskn.) Maire et Weiller, Fl. Afr. Nord 2:288 (1953)
    = A. sterilis L.
Avena byzantina var. solida subv. eusolida Maire et Weiller, loc. cit.
    = A. sterilis L.
Avena byzantina var. solida subv. induta (Thell.) Maire et Weiller, loc. cit.
    = A. sativa $\times$ sterilis F_1 hybrid
Avena byzantina var. solida subv. secunda (Malz.) Maire et Weiller, op. cit. 289.
    = A. sativa L.
Avena byzantina var. solida f. praecox Mordv., op. cit. 413
    = A. sativa L.
Avena byzantina var. ursina Mordv. in Wulff, Fl. Cult. Pl. 2:417 (1936)
    = A. sativa L.
Avena byzantina ssp. byzantina (Koch) Tab. Mor., Bol. Soc. Brot. II. 13:Tab. 1 (1939)
    = A. sativa L.
Avena byzantina ssp. byzantina var. brachytricha (Thell.) Tab. Mor., loc. cit.
    = see under A. sterilis L.
Avena byzantina ssp. byzantina var. denudata (Hausskn.) Tab. Mor., loc. cit.
    = A. sativa $\times$ sterilis F_1 hybrid
Avena byzantina ssp. byzantina var. hypatricha (Thell.) Tab. Mor., loc. cit.
    = A. sativa L.

Avena byzantina ssp. byzantina var. induta (Thell.) Tab. Mor., loc. cit.
= A. sativa × sterilis F_1 hybrid

Avena byzantina ssp. byzantina var. induta subv. pilosiuscula (Malz.) Tab. Mor., loc. cit.
= A. sativa L.

Avena byzantina ssp. byzantina var. induta subv. secunda (Malz.) Tab. Mor., loc. cit.
= A. sativa L.

Avena byzantina ssp. byzantina var. induta subv. segetalis (Nyman) Tab. Mor., loc. cit.
= A. sativa × sterilis F_1 hybrid

Avena byzantina ssp. byzantina var. macrotricha (Malz.) Tab. Mor., loc. cit.
= A. sativa L.

Avena byzantina ssp. byzantina var. macrotricha subv. asperata (Malz.) Tab. Mor., loc. cit.
= A. sativa × sterilis F_1 hybrid

Avena byzantina ssp. byzantina var. macrotricha subv. biaristata (Hack. ex Trabut) Tab. Mor., loc. cit.
= A. sativa L.

Avena byzantina ssp. byzantina var. macrotricha subv. culta (Thell.) Tab. Mor., loc. cit.
= A. sativa L.

Avena byzantina ssp. byzantina var. macrotricha subv. diathera (Malz.) Tab. Mor., loc. cit.
= A. sativa L. or A. sativa heterozygous fatuoid

Avena byzantina ssp. byzantina var. macrotricha subv. hypomelanathera (Thell.) Tab. Mor., loc. cit.
= A. sativa L.

Avena byzantina ssp. byzantina var. macrotricha subv. laevigata (Malz.) Tab. Mor., loc. cit.
= A. sativa heterozygous fatuoid

Avena byzantina ssp. byzantina var. macrotricha subv. pseudovilis (Hausskn.) Tab. Mor., loc. cit.
= A. sterilis L.

Avena byzantina ssp. byzantina var. macrotricha subv. solidissima (Thell.) Tab. Mor., loc. cit.
= A. sativa L.

Avena byzantina ssp. byzantina var. macrotricha subv. subculta (Malz.) Tab. Mor., loc. cit.
= A. sativa L.

Avena byzantina ssp. byzantina var. solida (Hausskn.) Tab. Mor., loc. cit.
= A. sterilis L.

Avena byzantina ssp. byzantina var. solida subv. solidiflora (Malz.) Tab. Mor., loc. cit.
= A. sativa L.

Avena byzantina ssp. pseudosativa (Thell.) Tab. Mor., Bol. Soc. Brot. II, 13:610 (1939)
= A. sativa L.

Avena byzantina ssp. pseudosativa var. thellungiana (Malz.) Tab. Mor., op. cit. 611.
= A. sativa L.

Avena byzantina ssp. pseudosativa var. transietissima (Thell.) Tab. Mor., loc. cit.
= A. sativa L.
Avena caespitosa (L.) Kuntze, Taschenfl. Leipzig 45 (1867)
= ? Aira caespitosa L., Sp. Pl. 64 (1753)
= ? Deschampsia caespitosa (L.) Beauv., Agrost. 91 (1812)
Avena caffra Stapf., Kew Bull. Misc. Inf. 1897:293 (1897)
= Aven. caffrum (Stapf) Stapf
= H. longifolium (Nees) Schweickerdt
Avena calicina Vill., Fl. Delph. 10 (1785)
= See A. calycina Vill.
Avena callosa Turcz. ex Griseb. in Ledeb., Fl. Ross. 4:416 (1853)
= Schizachne purpurascens (Torr.) Swallen, Wash. Acad. Sci. Jour. 18:204 (1928)
Avena calycina Vill., Hist. Pl. Dauphine 1:315 (1786)
= Danthonia calycina (Vill.) Rchb., Ic. 1:44 Tab. 103 fig. 1713, 1714 (1834)
Avena calycina Poiret in Lam., Tabl. Encycl. 1:200 (1791) et Suppl. 1:540 (1811)
= Danthonia calycina (Poiret) Rocm. et Schult. Syst. Veg. 2:691 (1817)
Avena canariensis Nees ex Steud., Nom. Bot. ed. 2:171 (1840)
= Pentameris macrantha (Schrad.) Nees, Linnaea 7:316 (1832)
Avena canariensis Baum, Rajhathy et Sampson, Can. J. Bot. 51:759 (1973)
Avena candoliei Nyman, Consp. Fl. Eur. 812 (1882)
= T. flavescens (L.) P. Beauv. Agrostogr. 88 (1812)
Avena canescens (L.) Weber ex Wigg., Prim. Fl. Hols. 9 (1780)
= Corynephorus canescens (L.) Beauv. Agrost. 90 (1812)
Avena cantabrica Lag. in Herrera, Agr. 1:141 (1818)
= H. filifolium (Lag.) Henrard
Avena capensis Burm. f., Fl. Ind. 3 (1768)
= ? Themeda quadrivalvis (L.) O. Ktze., Rev. Gen. Pl. 2:794 (1891)
Avena capensis L. f., Suppl. Pl. 112 (1781)
= Aristida capensis (L. f.) Schult. in Thunb. Fl. Capensis ed. Schult. 105 (1823)
Avena capensis Steud., Flora 12:481 (1829)
= Pentaschistis aspera (Thunb.) Stapf in Thiselton-Dyer, Fl. Capensis 7:501 (1899)
Avena capillacea Beauv., Ess. Agrost. 89, 153 (1812)
= Aira caryophyllacea L., Sp. Pl. 66 (1753)
Avena capillaris (Host) Mert. et Koch, Deutschl. Fl. ed. 3, 1:573 (1823)
= ? Aira capillaris Host, Gram. Austr. 4:20 (1809)
= ? Aira elegans Willd. ex Gaudin, Agrost. Helv. 1:130 (1811)
Avena carmeli Boiss., Diagn. Pl. Orient. Nov. 2 (13):50 (1854)
= Parapholis incurva (L.) C.E. Hubb., Blumea, Suppl. 3:14 (1946)
Avena caroliniana Walter, Fl. Carol. 81 (1788)
= ? Trisetum pennsylvanica (L.) Beauv. ex Roem. et Schult., Syst. Veg. 2:658 (1817)
Avena carpatica Host, Icon. Gram. Austr. 4:18 (1809)
= Trisetum carpaticum (Host) Roem. et Schult., Syst. Veg. 2:663 (1817)
Avena caryophyllea Sibth. et Smith, Fl. Graec. Prodr. 1:67 (1806)
= H. cycladum (Rech. et Scheff.) Rech.

Avena caryophyllea (L.) Weber ex Wigg., Prim. Fl. Hols. 10 (1780)
 = Aira caryophyllea L., Sp. Pl. 66 (1753)
Avena cavanillesii (Trin.) Bluff Nees et Schauer, Comp. Fl. Germ. ed. 2, 1:143 (1836)
 = Trisetum cavanillesii Trin., Mem. Acad. Petersb. Ser. 6, 1:63 (1830)
 = Lophochloa cavanillesii (Trin.) Bor, Grasses India Pak. 445 (1960)
Avena cavanillesii Hort. ex Roem. et Schult., Syst. Veg. 2:691 (1817)
 = A. hispanica Ard. ex Saggi
Avena cavanillesii (Trin.) Koch, Syn. Fl. Germ. Helv. 1:797 (1837)
 = Trisetum cavanillesii Trin., Act. Petrop. 1830:63 (1830)
 = Lophochloa cavanillesii (Trin.) Bor, Grasses India Pak. 445 (1960)
Avena cavanillesii Lag., Varied. Cienc. 2(4):39 (1805)
 = Macrochloa arenaria Kunth, Enum 1:279 (1833)
Avena cernua Kunth, Rev. Gram. 1: Suppl. 26 (1830)
 = Trisetum cernuum Trin., Mem. Acad. St. Petersb. Ser. VI, 1:61 (1830)
Avena chinensis Fisch. ex Roem. et Schult., Syst. Veg. 2:669 (1817)
 = A. sativa L.
Avena chinensis Metzger, Eur. Cerealien 53-54 (1824)
 = A. sativa L.
Avena chinensis Link, Handb. Gewächs. 1:43 (1829)
 = A. sativa L.
Avena chlorantha Link, Linnaea 17:401 (1843)
 = Trisetum aureum (Ten.) Ten., Fl. Nap. 2:378 (1820)
Avena ciliaris Kit. ex Schult., Oesterr. Fl. ed. 2, 1:268 (1814)
 = Trisetum carpaticum (Host) Roem. et Schult., Syst. Veg. 2:663 (1817)
Avena cinerea Hort. ex Roem. et Schult., Syst. Veg. 2:669 (1817)
 = A. sativa L.
Avena clarkei Hook. f., Fl. Brit. Ind. 7:278 (1896)
 = Trisetum clarkei (Hook. f.) R. R. Stewart, Brittonia 5:431 (1945)
Avena clauda Durieu, Rev. Bot. Duchartre 1:360 (1845)
Avena clauda var. eriantha Bal. ex Coss. et Dur., Bull. Soc. Bot. France 1:15 (1854)
 = A. clauda Dur.
Avena clauda var. solida Hausskn., Mitt. Thür. Bot. Ver. N.F. 6:43 (1894)
 = A. eriantha Dur.
Avena clauda subv. eriantha (Bal. ex Coss. et Dur.) Malz., Monogr. 232 (1930)
 = A. clauda Dur.
Avena clauda subv. leiantha Malz., Monogr. 232 (1930)
 = A. clauda Dur.
Avena coarctata Desf., Cat. Pl. Paris ed. 3:22 (1829)
 = Gaudinia geminiflora
Avena colorata Steud., Flora 12:481 (1829)
 = Pentaschistis colorata (Steud.) Stapf in Thiselton-Dyer Fl. Capensis 7:491 (1899)
Avena compacta Boiss. et Heldr. in Boiss. Diagn. Pl. Orient. Nov. 1 (7):122 (1846)
 = Aven. compactum (Boiss. et Heldr.) Halácsy
 = H. compactum (Boiss. et Heldr.) Henrard
 = Arrhenatherum neumayerianum (Vis.) Potztal, Bot. Jahrb. 75:330 (1951)

= D. compactum (Boiss. et Heldr.) Holub
Avena compressa Heuff., Flora 18:244 (1835)
    = Aven. compressum (Heuff.) Vierh.
    = H. compressum (Heuff.) Henrard
    = Aveno. compressa (Heuff.) Holub
Avena condensata Link, Enum. Hort. Berol. 1:82 (1821)
    = Trisetum aureum. (Ten.) Ten., Fl. Nap. 2:378 (1820)
Avena convoluta Presl, Cyp. Gram. Sicul. 31 (1820)
    = Aven. convolutum (Presl) Halács.
    = H. convolutum (Presl) Henrard
    = Arrhenatherum filifolium (Lag.) Potztal, Bot. Jahrb. 75:329 (1951)
Avena coquimbensis Roem. et Schult., Syst. Veg. 2:677 (1817)
    = Hierochlöe utriculata Kunth
Avena coronensis Schur, Enum. Pl. Transsilv. 763 (1866)
    = Aven. convolutum (Presl) Halács.
Avena corymbosa Nyman, Consp. Fl. Eur. 814 (1882)
    = Aira tenorei Guss., Fl. Sic. Prodr. 1:62 (1827)
Avena crassifolia Font-Quer, Butll. Inst. Catal. Hist. Nat. 20:189 (1920)
    = ? Helictotrichon sp.
Avena cristata Roem. et Schult., Syst. Veg. 2:758 (1817)
    = Agropyron cristatum (L.) Gaertn., Nov. Comm. Petrop. 14:540 (1770)
Avena cultiformis (Malz.) Malz. in Keller et al. eds., Sorn. Rast. SSSR 1:208 (1934)
    = A. sativa fatuoid
Avena cupaniana (Guss.) Nyman, Syll. 414 (1854)
    = Aira cupaniana Guss., Fl. Sicul. Syn. 1:145 (1842)
Avena cuspidata Willd. ex Spreng., Syst. Veg. 1:253 (1825)
    = Calamagrostis cuspidata
Avena daenensis Boiss., Diagn. Pl. Orient. Nov. 1:7:123 (1846)
    = Trisetum rigidum (M.B.) Roem. et Schult. Syst. Veg. 2:662 (1817)
Avena dahurica Kom. ex Roshev. in Komarov, Fl. URSS 2:275 (1934)
    = Aven. dahuricum (Kom.) Roshev.
    = H. dahuricum (Kom.) Kitagawa
    = Aveno. dahurica (Kom.) Holub
Avena damascena Rajhathy et Baum, Can. J. Genet. Cytol. 14:646 (1972)
Avena decora Janka in Termesz. Fuzetek 8:28 (1884)
    = Aven. decorum (Janka) Vierh.
    = H. decorum (Janka) Henrard
    = Arrhenatherum decorum (Janka) Potztal, Bot. Jahrb. 75:329 (1951)
Avena delavayi Hack., Oesterr. Bot. Zeitschr. 52:189 (1902)
    = H. delavayi (Hack.) Henrard
Avena depauperata Sieber ex Trin., Mem. Acad. Petersb. 6. Math. Phys. Nat. 1:68 (1830)
    = Danthonia provincialis DC. in Lam. et DC., Fl. Fr. 3:33 (1805)
Avena desertorum Less., Linnaea 9:208 (1834)
    = H. desertorum (Less.) Pilg.
    = Aven. desertorum (Less.) Podper.
    = Arrhenatherum desertorum (Less.) Potztal, Bot. Jahrb. 75:329 (1951)
Avena deusta Ball, Jour. Linn. Soc. London 16:719 (1878)
    = A. barbata Pott ex Link

Avena deyeuxioides H.B.K., Nova Gen. Sp. Pl. 1:147 (1815)
 = Trisetum deyeuxioides (H.B.K.) Kunth, Rev. Gram. 102 (1829)
Avena diffusa (Neilr.) Aschers. et Graebn., Syn. Mitteleur. Fl. 2:234 (1899)
 = A. sativa L. (See also note in "special cases" in Materials and Methods p. 20)
Avena diffusa var. asiatica Vav., Bull. Appl. Bot. Pl. Breed. 16:94 (1926)
 = A. sativa L.
Avena diffusa var. bashkirorum Vav., loc. cit.
 = A. sativa L.
Avena diffusa var. iranica Vav., loc. cit.
 = A. sativa L.
Avena diffusa var. kasanensis Vav., loc. cit.
 = A. sativa L.
Avena diffusa var. persica Vav., loc. cit.
 = A. sativa L.
Avena diffusa var. segetalis Vav., loc. cit.
 = A. sativa L.
Avena diffusa var. volgensis Vav., loc. cit.
 = A. sativa L.
Avena dispermis Mill., Gard. Dict. ed. 8 Avena No. 1 (1768)
 = A. sativa L.
Avena distans Schur, Oesterr. Bot. Zeitschr. 20:22 (1870)
 = A. sativa L.
Avena disticha Lam., Encycl. 1:333 (1783)
 = Trisetum rigidum (M.B.) Roem. et Schult.
Avena distichophylla Vill., Prosp. 16 (1779)
 = Trisetum distichophyllum (Vill.) P. Beauv., Agrostogr. 88 (1812)
Avena dodii Stapf in Thiselton-Dyer, Fl. Capensis 7:475 (1899)
 = Aven. dodii Stapf
 = H. dodii (Stapf) Schweickerdt
 = Arrhenatherum dodii (Stapf) Potztal, Bot. Jahrb. 75:328 (1951)
Avena dubia Leers, Fl. Herborn. 41 (1775)
 = Ventenata dubia (Leers) F. Schultz, Pollich. 20-21:273 (1863)
Avena dufourei (Boiss. et Reut.) Nyman, Consp. Fl. Eur. 813 (1882)
 = Trisetum dufourei Boiss. et Reut., Pug. 122 (1852)
Avena dura (L.) Salisb., Prodr. Stirp. 22 (1796)
 = Festuca duriuscula L., Sp. Pl. 74 (1753)
Avena editissima Pampanini, Bull. Soc. Bot. Ital. 1915:29 (1915)
 = Danthonia cachemeriana Jaub et Spach var. minor Hook. f.
Avena elata Salisb., Prodr. Stirp. 23 (1796)
 = Arrhenatherum elatius (L.) J.S. et C.B. Presl, Fl. Cech. 17 (1819)
Avena elata Forssk., Fl. Aegypt. Arab. 19 (1775)
 = ? H. bromoides (Gouan) C.E. Hubbard
Avena elatior L., Sp. Pl. 79 (1753)
 = Arrhenatherum elatius (L.) J.S. et C.B. Presl, Fl. Cech. 17 (1819)
Avena elegans (Willd.) Aschers., Verh. Bot. Ver. Brand. 3/4:279 (1862)
 = Aira capillaris Host, Gram. Austr. 4:20 (1809)
Avena elephantina Spreng. ex Steud., Nom. Bot. ed. 2, 1:172 (1840)
 = Triraphis rehmanni Hack., Bull. Herb. Boiss. 3:388 (1895)
Avena elephantina Thunb., Prodr. Fl. Capensis 23 (1794)
 = Danthonia elephantina (Thunb.) Nees, Fl. Afr. Austr. 334 (1841)

Avena elongata H. B. et K., Gen. et Sp. 1:148 (1815)
  = Trisetum spicatum (L.) Richt., Pl. Eur. 1:59 (1890)
Avena eriantha (Boiss.) Hack. Oesterr. Bot. Zeitschr. 27:122 (1877)
  = Arrhenatherum erianthum Boiss. et Reut., Pug. 121 (1852)
Avena eriantha Dur., Rev. Bot. Duchartre 1:360 (1845)
Avena eriantha var. acuminata Coss. et Dur., Bull. Soc. Bot. France 1:14 (1854)
  = A. clauda × eriantha F_1 hybrid
Avena euhybrida Hausskn. ex Thell., Mitt. Naturw. Ges. Winterthur. 12:128 (1918)
  = name for A. fatua × sterilis F_1 hybrid
Avena fallax Roem. et Schult., Syst. Veg. 2:692 (1817)
  = Arrhenatherum montanum (Vill.) Potztal, Bot. Jahrb. 75:329 (1951)
Avena fallax Tenore, Fl. Napol. 3:961 (1829)
  = Arrhenatherum filifolium (Lag.) Potztal, Bot. Jahrb. 75:329 (1951)
Avena fatua L., Sp. Pl. 80 (1753)
Avena fatua var. abbreviata Hausskn., Mitt. Geogr. Ges. (Thür.) Jena 3:239 (1885)
  = A. fatua L.
Avena fatua var. ambigua (Schönh.) Hausskn., op. cit. 237
  = A. strigosa Schreb.
Avena fatua var. barbata (Pott ex Link) Fiori et Paoletti, Icon. Fl. Ital. 1:29 (1855)
  = A. barbata Pott ex Link
Avena fatua var. basifixa Malz., Bull. Appl. Bot. 7:329 (1914)
  = A. sativa fatuoid
Avena fatua var. contracta Hausskn., op. cit. 239
  = A. hybrida Peterm.
Avena fatua var. flavescens Schur, Oesterr. Bot. Zeitschr. 10:71 (1860)
  = A. fatua L.
Avena fatua var. genuina Ducom., Taschenb. Schweiz. Bot. 863 (1869)
  = A. fatua L.
Avena fatua var. glabra Ducom., Taschenb. Schweiz. Bot. 863 (1869)
  = A. hybrida Peterm.
Avena fatua var. glabrata Peterm., Fl. Bienitz 13 (1841)
  = A. fatua L.
Avena fatua var. glabrata Stapf in Rendle, Jour. Linn. Soc. Bot. 36:401 (1904)
  = ? A. fatua L. or ? A. hybrida Peterm.
Avena fatua var. glabrescens Coss. Bull. Soc. Bot. France 1:15 (1854)
  = A. fatua L.
Avena fatua var. glabrescens Schur, Oesterr. Bot. Zeitschr. 10:71 (1860)
  = ? A. fatua L.
Avena fatua var. glabrescens Hausskn., Mitt. Geogr. Ges. Thür. 3:237 (1885)
  = A. fatua L.
Avena fatua var. glabrescens Tourlet, Cat. Pl. Vasc. Indre et Loire 568 (1908)
  = A. hybrida Peterm.
Avena fatua var. grandiflora Scheele, Flora 27:57 (1844)
  = ? A. fatua L.
Avena fatua var. hirsuta Neilr., Fl. Nieder. Oesterr. 59 (1859)
  = A. fatua L.

Avena fatua var. hybrida (Peterm.) Aschers., Fl. Brandenb. 1:828 (1864)
= A. hybrida Peterm.
Avena fatua var. intermedia (Lestib.) Lej. et Court., Comp. Fl. Belg. 1:71 (1828)
= ? A. hybrida Peterm. or ? A. sativa fatuoid
Avena fatua var. intermedia (Lindgr.) Hartm., Handb. Skand. Fl. ed. 4:30 (1843)
= A. hybrida Peterm.
Avena fatua var. intermedia (Lindgr.) Ducom., Taschenb. Schweiz. Bot. 863 (1869)
= A. hybrida Peterm.
Avena fatua var. intermedia (Lindgr.) Husnot, Gram. Fr. Belg. 39 (1896)
= A. hybrida Peterm.
Avena fatua var. intermedia (Lindgr.) Vasc., Rev. Agron. 19:19 (1931)
= A. hybrida Peterm.
Avena fatua var. intermedia Merino, Fl. Descr. Illustr. Galicia 3:301 (1909)
= Non satis notae
Avena fatua var. ludoviciana (Dur.) Fiori, Anal. Ital. 1:72 (1908)
= A. sterilis L.
Avena fatua var. mollis Keng, Sinensia 11:411 (1940)
= Non satis notae
Avena fatua var. nigrescens Hausskn., Mitt. Geogr. Ges. Thür. 3:237 (1885)
= A. fatua L.
Avena fatua var. nigrescens f. albescens Hausskn. loc. cit.
= A. fatua L.
Avena fatua var. nigrescens f. cinerascens Hausskn. loc. cit.
= A. fatua L.
Avena fatua var. nigrescens Schur, Oesterr. Bot. Zeitschr. 10:71 (1860)
= A. fatua L.
Avena fatua var. pachycarpa Malz., Bull. Appl. Bot. Pl. Breed. 7:328 (1914)
= A. sativa fatuoid
Avena fatua var. pilosa Syme ex Wolley-Dod, Fl. Sussex 507 (1937)
= probably A. fatua L.
Avena fatua var. pilosissima S. F. Gray, Nat. Arr. Brit. Pl. 2:131 (1821)
= ? probably A. fatua L.
Avena fatua var. pilosissima subv. scabrida (Malz.) Cabrera, Rev. Invest. Agricolas 11:378 (1957)
= A. fatua L.
Avena fatua var. sativa (L.) Hausskn., Mitt. Geogr. Ges. Thür. 3:238 (1885)
= A. sativa L.
Avena fatua var. sativa f. secunda Hausskn., op. cit. 239
= A. sativa L.
Avena fatua var. sterilis Fiori et Paoletti, Ic. Fl. Ital. 1:29 (1895)
= A. sterilis L.
Avena fatua var. subuniflora Trab., Bull. Agr. Alg. Tunis. Maroc 16:360 (1910)
= A. sativa L.
Avena fatua var. transiens Hausskn., Mitt. Geogr. Ges. Thür. 3:238 (1885)
= A. fatua $\times$ sativa F_1 hybrid
Avena fatua var. trichophylla (Koch) Griseb. in Ledeb., Fl. Ross. 4:412 (1853)
= A. trichophylla C. Koch

Avena fatua var. vilis (Wallr.) Hausskn., Mitt. Thür. Bot. Ver. N.F. 6:39 and 45 (1894)
  = ? A. hybrida Peterm. or ? A. fatua L.
Avena fatua var. vulgaris Tourlet, Cat. Pl. Vasc. Indre Loire 568 (1908)
  = A. fatua L.
Avena fatua f. deserticola Hausskn., Mitt. Thür. Bot. Ges. 13/14:46 (1899)
  = A. occidentalis Dur.
Avena fatua f. pseudotransiens Thell., Rep. Sp. Nov. Fedde 13:55 (1913)
  = name not validly published (hypothetical form)
Avena fatua f. straminea Hausskn., op. cit. 37
  = A. sterilis L.
Avena fatua ssp. cultiformis Malz., Bull. Appl. Bot. Pl. Breed. 20:142 (1929)
  = A. sativa fatuoid
Avena fatua ssp. cultiformis var. leiocarpa Malz., Monogr. 347 (1930)
  = A. sativa fatuoid
Avena fatua ssp. cultiformis var. leiocarpa subv. aglossicos Malz., loc. cit.
  = A. sativa fatuoid
Avena fatua ssp. cultiformis var. oligotricha Malz., loc. cit.
  = ? A. sativa fatuoid
Avena fatua ssp. cultiformis var. pseudoculta Malz., op. cit. 348
  = A. sativa fatuoid
Avena fatua ssp. cultiformis var. pseudoculta subv. americana Malz., op. cit. 349
  = A. sativa fatuoid
Avena fatua ssp. cultiformis var. pseudoculta subv. crassiflora Malz., op. cit. 350
  = A. sativa fatuoid
Avena fatua ssp. cultiformis var. pseudoculta subv. eucontracta Malz., op. cit. 351
  = A. sativa fatuoid
Avena fatua ssp. cultiformis var. pseudoculta subv. pachycarpa (Malz.) Malz., op. cit. 350
  = A. sativa fatuoid
Avena fatua ssp. cultiformis var. pseudoculta subv. patentissima Malz., op. cit. 348
  = A. sativa fatuoid
Avena fatua ssp. cultiformis var. pseudoculta subv. subturgida Malz., op. cit. 351
  = A. sativa fatuoid
Avena fatua ssp. cultiformis var. trichocarpa Malz., op. cit. 346
  = A. fatua L.
Avena fatua ssp. eufatua Hyland., Nord. Karlvax. 1:286 (1953)
  = A. fatua L.
Avena fatua ssp. fatua (L.) Thell., Vierteljahrs. Nat. Ges. Zürich 56:319 (1912)
  = A. fatua L.
Avena fatua ssp. fatua var. glabrata (Peterm.) Malz., Monogr. 320 (1930)
  = A. fatua L.
Avena fatua ssp. fatua var. glabrata subv. flocculosa Malz., op. cit. 322
  = A. fatua L.
Avena fatua ssp. fatua var. hybrida (Peterm.) Thell., Vierteljahrs. Nat. Ges. Zürich 56:322 (1912)

= A. hybrida Peterm.
Avena fatua ssp. fatua var. hybrida subv. petermanni Thell., op. cit. 324
= A. hybrida Peterm.
Avena fatua ssp. fatua var. intermedia (Lestib.) Lej. et Court., Comp. fl. Belg. 1:71 (1828)
= A. hybrida Peterm.
Avena fatua ssp. fatua var. intermedia subv. longispiculata (Malz.) Tab. Mor., Bol. Soc. Brot. Ser. 2, 13:Tab. 1 (1939)
= A. hybrida Peterm.
Avena fatua ssp. fatua var. intermedia subv. minima Tab. Mor., op. cit. 593
= A. fatua L.
Avena fatua ssp. fatua var. intermedia subv. sparsepilosa (Malz.) Tab. Mor., op. cit. Tab. 1
= A. hybrida Peterm.
Avena fatua ssp. fatua var. intermixta Thell., Vierteljahrs. Nat. Ges. Zürich 56:325 (1912)
= A. sativa L.
Avena fatua ssp. fatua var. pilibarbis Thell., Rep. Sp. Nov. Fedde 13:54 (1913)
= ? A. barbata Pott ex Link
Avena fatua ssp. fatua var. pilosissima (S.F. Gray) Malz., Monogr. 316 (1930)
= A. fatua L.
Avena fatua ssp. fatua var. pilosissima subv. biflora Tab. Mor., Bol. Soc. Brot. Ser. 2, 13:591 (1939)
= A. fatua L.
Avena fatua ssp. fatua var. pilosissima subv. biflora f. cinerascens (Hausskn.) Tab. Mor., op. cit. Tab. 1
= A. fatua L.
Avena fatua ssp. fatua var. pilosissima subv. deserticola (Hausskn.) Malz., Monogr. 320 (1930)
= A. occidentalis Dur.
Avena fatua ssp. fatua var. pilosissima subv. parva Tab. Mor., Bol. Soc. Brot. Ser. 2, 13:592 (1939)
= A. occidentalis Dur.
Avena fatua ssp. fatua var. pilosissima subv. pilibarbis (Thell.) Malz., Monogr. 317 (1930)
= ? A. barbata Pott ex Link
Avena fatua ssp. fatua var. pilosissima subv. puberula (Malz.) Tab. Mor., Bol. Soc. Brot. Ser. 2, 13:Tab. 1 (1939)
= ? possibly A. occidentalis Dur.
Avena fatua ssp. fatua var. pilosissima subv. scabrida Malz. op. cit. 318.
= A. fatua L.
Avena fatua ssp. fatua var. pilosissima subv. scabrida f. albescens (Hausskn.) Tab. Mor., Bol. Soc. Brot. Ser. 2, 13:590 (1939)
= A. fatua L.
Avena fatua ssp. fatua var. pilosissima subv. scabrida f. nigrescens (Hausskn.) Tab. Mor., op. cit. 591
= A. fatua L.
Avena fatua ssp. fatua var. pilosissima subv. scabriuscula (Malz.) Tab. Mor., op. cit. Tab. 1
= A. hybrida Peterm.

Avena fatua ssp. fatua var. pilosissima subv. valdepilosa (Malz.) Tab. Mor., op. cit. 591
    = A. hybrida Peterm.
Avena fatua ssp. fatua var. pilosissima subv. villosa (Malz.) Tab. Mor., op. cit. Tab. 1
    = ? A. occidentalis Dur.
Avena fatua ssp. fatua var. transiens subv. unilateralis Thell. Rep. Sp. Nov. 13:54 (1913)
    = A. sativa × fatua F_1 hybrid
Avena fatua ssp. fatua var. vilis (Wallr.) Malz., Monogr. 326 (1930)
    = ? A. hybrida Peterm. or ? A. fatua L.
Avena fatua ssp. fatua var. vilis subv. basifixa (Malz.) Malz., op. cit. 332
    = A. sativa fatuoid
Avena fatua ssp. fatua var. vilis subv. breviflora (Malz.) Tab. Mor., op. cit. Tab. 1
    = A. sativa fatuoid
Avena fatua ssp. fatua var. vilis subv. glabella (Malz.) Tab. Mor., loc. cit.
    = A. hybrida Peterm.
Avena fatua ssp. fatua var. vilis subv. glabella (Malz.) Tab. Mor., loc. cit.
    = A. hybrida Peterm.
Avena fatua ssp. fatua var. vilis subv. glabrifolia (Malz.) Tab. Mor., loc. cit.
    = ? A. sativa fatuoid
Avena fatua ssp. fatua f. basifixa (Malz.) Thell., Rec. Trav. Bot. Neerl. 25:427 (1929)
    = A. sativa fatuoid
Avena fatua ssp. fatua f. glabrata (Peterm.) Thell., op. cit. 426
    = A. fatua L.
Avena fatua ssp. fatua f. glabricalla Thell., Ber. Schweiz. Bot. Ges. 30/31:78 (1922)
    = ? A. sativa heterozygous fatuoid
Avena fatua ssp. fatua f. hybrida (Peterm.) Thell., Rec. Trav. Bot. Neerl. 25:426 (1929)
    = A. hybrida Peterm.
Avena fatua ssp. fatua f. intermedia (Lestib.) Thell., op. cit. 426
    = ? A. hybrida Peterm. or ? A. sativa fatuoid
Avena fatua ssp. fatua f. pilibarbis (Thell.) Thell., op. cit. 426
    = ? A. barbata Pott ex Link
Avena fatua ssp. fatua f. pilosissima (S.F. Gray) Thell., op. cit. 426
    = probably A. fatua L.
Avena fatua ssp. fatua f. pseudobasifixa Thell., op. cit. 427
    = not validly published, given as hypothetical form
Avena fatua ssp. fatua f. pseudopilibarbis Thell., op. cit. 426
    = not validly published, given as hypothetical form
Avena fatua ssp. fatua f. pseudotransiens Thell., op. cit. 427
    = A. fatua L.
Avena fatua ssp. fatua f. transiens (Hausskn.) Thell., op. cit. 427
    = A. sativa × fatua F_1 hybrid
Avena fatua ssp. macrantha (Hack.) Malz., Bull. Appl. Bot. Pl. Breed. 20:141 (1929)
    = A. sativa L.
Avena fatua ssp. macrantha var. asiatica (Vav.) Malz., Monogr. 312 (1930)
    = A. sativa L.

Avena fatua ssp. macrantha var. asiatica subv. iranica (Vav.) Malz., op. cit. 313
 = A. sativa L.
Avena fatua ssp. macrantha var. calva Malz., loc. cit.
 = A. sativa L.
Avena fatua ssp. macrantha var. longipila Malz., op. cit. 312
 = A. sativa L.
Avena fatua ssp. macrantha var. pilifera Malz., op. cit. 311
 = A. sativa L.
Avena fatua ssp. macrantha var. pilifera subv. homomalla Malz., op. cit. 312
 = A. sativa L.
Avena fatua ssp. macrantha var. pilifera subv. subpilifera Malz., op. cit. 311
 = A. sativa L.
Avena fatua ssp. macrantha prol. nudata Malz., op. cit. 313
 = A. sativa L.
Avena fatua ssp. meridionalis Malz., Bull. Appl. Bot. Pl. Breed. 20:140 (1929)
 = A. hybrida Peterm.
Avena fatua ssp. meridionalis var. elongata Malz., op. cit. 309
 = A. hybrida Peterm.
Avena fatua ssp. meridionalis var. elongata subv. glabrifolia Malz., op. cit. 309
 = ? A. sativa fatuoid
Avena fatua ssp. meridionalis var. grandis Malz., Monogr. 306 (1930)
 = A. hybrida Peterm.
Avena fatua ssp. meridionalis var. grandis subv. puberula Malz., op. cit. 306
 = ? possibly A. occidentalis Dur.
Avena fatua ssp. meridionalis var. grandis subv. scabriuscula Malz., loc. cit. 306
 = A. hybrida Peterm.
Avena fatua ssp. meridionalis var. grandis subv. villosa Malz., op. cit. 307
 = ? A. occidentalis Dur.
Avena fatua ssp. meridionalis var. longiflora Malz., op. cit. 308
 = A. hybrida Peterm.
Avena fatua ssp. meridionalis var. longispiculata Malz., op. cit. 308
 = ? A. hybrida Peterm.
Avena fatua ssp. nodipilosa Malz., Bull. Appl. Bot. Pl. Breed. 20:140 (1929)
 = A. sativa L.
Avena fatua ssp. nodipilosa var. glabra Malz., Monogr. 303 (1930)
 = A. sativa L.
Avena fatua ssp. nodipilosa var. glabra subv. pilosiuscula (Vav.) Malz., loc. cit.
 = A. sativa L.
Avena fatua ssp. nodipilosa var. glabra subv. pseudoligulata Malz., loc. cit.
 = A. sativa L.
Avena fatua ssp. nodipilosa var. glabra subv. secunda Malz. loc. cit.
 = A. sativa L.
Avena fatua ssp. nodipilosa var. glabriuscula Malz., op. cit. 301
 = A. sativa L.
Avena fatua ssp. nodipilosa var. glabriuscula subv. kasanensis (Vav.) Malz., op. cit. 302
 = A. sativa L.

Avena fatua ssp. nodipilosa var. piligera Malz., op. cit. 301
= A. sativa L.
Avena fatua ssp. nodipilosa var. subglabra Malz., op. cit. 302
= A. sativa L.
Avena fatua ssp. nodipilosa var. subglabra subv. speltiformis (Vav.) Malz., loc. cit.
= A. sativa L.
Avena fatua ssp. nodipilosa prol. decortica Malz., op. cit. 303
= A. sativa L.
Avena fatua ssp. nodipilosa prol. decortica subv. mongolica (Pissarev ex Vav.) Malz., op. cit. 304
= A. sativa L.
Avena fatua ssp. nuda (L.) Thell., Vierteljahrs. Naturf. Ges. Zürich 56:328 (1912)
= A. nuda L.
Avena fatua ssp. praegravis (Krause) Malz., Bull. Appl. Bot. Pl. Breed. 20:142 (1929)
= A. sativa L., see Note [98]
Avena fatua ssp. praegravis Malz., loc. cit.
Avena fatua ssp. praegravis var. leiantha Malz., Monogr. 355 (1930)
= A. sativa L.
Avena fatua ssp. praegravis var. leiantha subv. subeligulata Malz., op. cit. 356
= A. sativa L.
Avena fatua ssp. praegravis var. leiantha subv. turgida (Erikss.) Malz., op. cit. 355
= A. sativa L.
Avena fatua ssp. praegravis var. macrotricha Malz., op. cit. 354
= A. sativa L.
Avena fatua ssp. praegravis var. macrotricha subv. arundinacea (Schur) Malz., op. cit. 355
= A. sativa L.
Avena fatua ssp. praegravis var. macrotricha subv. norvegica Malz., op. cit. 354
= A. sativa L.
Avena fatua ssp. praegravis var. macrotricha Malz., op. cit. 355
= ? A. sativa L. or A. sativa heterozygous fatuoid
Avena fatua ssp. praegravis var. polytricha Malz., op. cit. 354
= A. sativa $\times$ sterilis F_1 hybrid
Avena fatua ssp. praegravis prol. grandiuscula Malz., op. cit. 357
= A. sativa L.
Avena fatua ssp. praegravis prol. grandiuscula subv. affinis (Koern.) Malz., op. cit. 357
= A. sativa L.
Avena fatua ssp. sativa (L.) Thell., Vierteljahrs. Naturf. Ges. Zürich 56:325 (1912)
= A. sativa L.
Avena fatua ssp. sativa var. brachytricha (Thell.) Malz., Monogr. 338 (1930)
= A. sativa L.
Avena fatua ssp. sativa var. brachytricha subv. pseudotransiens Malz., op. cit. 339
= A. sativa L.

Avena fatua ssp. sativa var. brachytricha subv. spelticola Malz., op. cit. 339
= ? A. sativa L.
Avena fatua ssp. sativa var. contracta (Neilr.) Thell., Vierteljahrs. Naturf. Ges. Zürich 56:326 (1912)
= A. sativa L.
Avena fatua ssp. sativa var. diffusa (Neilr.) Thell., loc. cit.
= A. sativa L.
Avena fatua ssp. sativa var. glaberrima (Thell.) Malz., Monogr. 340 (1930)
= A. sativa L.
Avena fatua ssp. sativa var. glaberrima subv. contracta (Neilr.) Malz., op. cit. 341
= A. sativa L.
Avena fatua ssp. sativa var. glaberrima subv. culinaris (Alefeld) Malz., op. cit. 344
= A. sativa L.
Avena fatua ssp. sativa var. glaberrima subv. eligulata (Vav.) Malz., op. cit. 342
= A. sativa L.
Avena fatua ssp. sativa var. pilosa (Koel.) Malz., op. cit. 336
= Probably A. sativa L.
Avena fatua ssp. sativa var. pilosa subv. glabricalla (Thell.) Malz., op. cit. 337
= A. sativa heterozygous fatuoid
Avena fatua ssp. sativa var. pilosa subv. subpilosa (Thell.) Malz., op. cit. 337
= A. sativa L.
Avena fatua ssp. sativa var. pilosa subv. unilateralis (Thell.) Malz., op. cit. 338
= A. sativa $\times$ fatua F_1 hybrid
Avena fatua ssp. sativa var. subuniflora (Trabut) Thell., Vierteljahrs. Naturf. Ges. Zürich 56:327 (1912)
= A. sativa L.
Avena fatua ssp. sativa f. brachytricha Thell., Rep. Sp. Nov. Fedde 13:55 (1913)
= A. sativa L.
Avena fatua ssp. sativa f. chlorathera (Thell.) Thell., Rec. Trav. Bot. Neerl. 25:428 (1929)
= A. sativa L.
Avena fatua ssp. sativa f. glaberrima Thell., Rep. Sp. Nov. Fedde 13:54 (1913)
= A. sativa L.
Avena fatua ssp. sativa f. macrathera Thell., loc. cit.
= A. sativa L.
Avena fatua ssp. sativa f. pseudosubpilosa Thell., Rec. Trav. Bot. Neerl. 25:428 (1929)
= Not validly published: given as hypothetical form.
Avena fatua ssp. sativa f. pseudosubuniflora Thell., Rep. Sp. Nov. Fedde 13:55 (1913)
= A. sativa L.
Avena fatua ssp. sativa f. setulosa Thell., op. cit. 55
= A. sativa L.
Avena fatua ssp. sativa f. subdecidua Thell., Rec. Trav. Bot. Neerl. 25:427 (1929)
= A. sativa L.

Avena fatua ssp. sativa f. subpilosa (Thell.) Thell., op. cit. 428
: = A. sativa L.
Avena fatua ssp. sativa f. subuniflora (Trabut) Thell., Rep. Sp. Nov. Fedde 13:55 (1913)
: = A. sativa L.
Avena fatua ssp. septentrionalis (Malz.) Malz., Bull. Appl. Bot. Pl. Breed. 20:140 (1929)
: = A. hybrida Peterm.
Avena fatua ssp. septentrionalis var. glabella Malz., Monogr. 298 (1930)
: = A. hybrida Peterm.
Avena fatua ssp. septentrionalis var. glabella subv. breviflora Malz., op. cit. 299
: = A. sativa fatuoid
Avena fatua ssp. septentrionalis var. glabripaleata Malz., op. cit. 295
: = A. hybrida Peterm.
Avena fatua ssp. septentrionalis var. sparsepilosa Malz., op. cit. 297
: = A. hybrida Peterm.
Avena fatua ssp. septentrionalis var. valdepilosa Malz., op. cit. 294
: = A. hybrida Peterm.
Avena fedtschenkoi Hack., Acta Horti Petrop. 26:55 (1906)
: = Aven. fedtschenkoi (Hack.) Roshev.
: = H. fedtschenkoi (Hack.) Henrard
Avena fertilis All., Auct. Fl. Pedem. 45 (1789)
: = Ventenata dubia (Leers) F. Schultz, Pollich. 20, 21:273 (1863)
Avena festucaeformis Hochst. ex A. Rich., Flora 38:275 (1855)
: = H. elongatum (Hochst. ex A. Rich.) C.E. Hubbard
: = Arrhenatherum elongatum (Hochst. ex A. Rich.) Potztal, Bot. Jahrb. 75:328 (1951)
Avena festucoides (Desf.) Raspail, Ann. Sci. Nat. Ser. 1, 5:439 (1825)
: = Ampelodesmos
Avena filifolia Lag., Elench. Plant. 4 (1816)
: = Aven. filifolium (Lag.) Fritsch
: = H. filifolium (Lag.) Henrard
: = Arrhenatherum filifolium (Lag.) Potztal, Bot. Jahrb. 75:329 (1951)
Avena filiformis G. Forst., Fl. Ins. Austr. Prodr. 9 (1786)
: = Calamagrostis filiformis (Forst.) Pilger, Det. Kgl. Danske Vid. Selsk. 3, 2:9 (1921)
: = Deyeuxia forsteri (Roem. et Schult.) Kunth, Enum. 1:244 (1833)
: = Agrostis avenacea Gmel., Syst. Nat. 2:171 (1791)
Avena flaccida Hack. ex Hook. f., Fl. Brit. Ind. 7:280 (1896)
: = ? Festuca gigantea Vill. see Bor, Grasses Burma Ceylon India Pakistan 448 (1960)
Avena flava Hort. ex Roem. et Schult., Syst. Veg. 2:669 (1817)
: = A. sativa L.
Avena flavescens Bout. ex Willk. et Lange, Prodr. Fl. Hisp. 1:72 (1861)
: = Trisetum
Avena flavescens L., Sp. Pl. 80 (1753)
: = Trisetum flavescens (L.) Beauv., Ess. Agrost. 88 (1812)
Avena flexuosa (L.) Mert. and Koch, in Roehl. Deutschl. Fl. ed. 3, 1:570 (1823)
: = Deschampsia flexuosa (L.) Trin., Acad. St. Petersb. Mem. Ser. 6 Sci. Nat. 4 (2):9 (1836)

Avena flexuosa Schrank, Denkschr. Baier. Bot. Ges. Regensburg 1:7 (1818)
   = Deschampsia flexuosa (L.) Trin., Mem. Acad. St. Petersb. Ser. 6, 4 (2):9 (1836)
Avena forsskalii Vahl, Symb. Bot. 2:25 (1791)
   = Asthenatherum forsskalii (Vahl) Nevski, Acta Univ. Asiae Med. Ser. 8 b, Bot. Fasc. 17:8 (1934)
Avena forsteri Kunth, Rev. Gram. 1:104 (1829)
   = Deyeuxia forsteri (Roem. et Schult.) Kunth, Enum. 1:244 (1833)
Avena fragilis L., Sp. Pl. 80 (1753)
   = Gaudinia fragilis (L.) Beauv., Agrostogr. 164 (1812)
Avena freita Ortega ex Sprengel, Bot. Gart. Halle 14 (1800)
   = A. hispanica Ard.
Avena fusca Ard. ex Saggi, Accad. Padov. 2:ab. 4 (1789)
   = A. hispanica Ard.
Avena fusca Kit. ex Schult., Fl. Austr. ed. 2, 1:268 (1814)
   = Trisetum carpathicum (Host) Roem. et Schult., Syst. Veg. 2:663 (1817)
Avena fusca Schur, Enum. Pl. Transsilv. 756 (1866)
   = A. sativa L.
Avena fuscoflora Schur, loc. cit.
   = A. sativa L.
Avena gallecica (Willk. and Lange) Nyman, Syll. Suppl. 71 (1865)
   = Trisetum
Avena geminiflora Kunth, Rev. Gram. 1:103 (1829); 2:299 (1830)
   = ? Gaudinia sp.
Avena georgiana Roem. et Schult., Syst. Veg. 2:669 (1817)
   = A. sativa L.
Avena georgica Zuccagni, Obs. Bot. Cent. 1:14, No. 31 (1806)
   = ? A. sativa L.
Avena georgica Boiss., Fl. Orient. 5:540 (1884)
   = Gaudinopsis macra (M.B.) Eig. Rep. Sp. Nov. Fedde 26:77 (1929)
Avena gigantea (L.) Salisbury, Prodr. Stirp. 23 (1796)
   = Festuca gigantea (L.) Vill., Hist. Pl. Dauph. 2:110 (1787)
Avena glabra C. Koch, Linnaea 19:5 (1847)
   = H. pubescens (Huds.) ? Pilger
Avena glabrata Brot. ex Steud., Nom. Bot. ed. 2, 1:172 (1840)
   = Aira pulchella (Beauv.) Link, Hort. Berol. 1:130 (1827)
Avena glabrata Haussm. in Harz, Samenkunde 2:1318 (1885)
   = A. sativa $\times$ fatua F_1 hybrid
Avena glabrescens (Marq.) Herter, Revista Sudamer. Bot. 6:141 (1940)
   = A. strigosa Schreb.
Avena glacialis Boiss., Elench. 87 (1838)
   = Trisetum glaciale Boiss., Elench. 87 (1838)
Avena glauca La Peyer. ex Roem. et Schult., Syst. Veg. 2:672 (1817)
   = H. versicolor (Vill.) Pilger
Avena glomerata Steud., Flora 12:483 (1829)
   = Pentaschistis curviflora (Schrad.) Stapf in Thiselton-Dyer, Fl. Capensis 7:492 (1899)
Avena glumosa Michx., Fl. Bor. Amer. 1:72 (1803)
   = Danthonia spicata (L.) Beauv. ex Roem. et Schult., Syst. Veg. 2:690 (1817)

Avena gmelini (Trin.) Nyman, Syll. Suppl. 71 (1865)
 = Trisetum gmelini Trin., Bull. Soc. Acad. Petersb. 1:66 (1836) Acta Horti Petrop. 6, 2:14 (1835)
Avena gonzaloi Senn. ex St.-Yves, Candollea 4:453 (1931)
 = H. gonzaloi (Senn.) Potztal
 = Aveno. gonzaloi (St.-Yves) Holub
Avena gracilis Moris, Stirp. Sard. Elench. Fasc. 1:50 (1827)
 = Trisetum morisii Trin. ex Steud. Syn. Pl. Gram. 1:225 (1855)
Avena gracillima Keng, Bull. Fan. Mem. Inst. Biol. 7:36 (1936)
 = probably A. sativa L.
Avena grandis Nevski, Acta Univ. Asiae Med. 8b. Bot. 17:6 (1934)
 = A. sativa L.
Avena hackelii Henriq., Bol. Soc. Brot. 20:87 (1905)
 = H. hackelii (Henriq.) Henrard
 = Aven. hackelii (Henriq.) Vierh.
 = Aveno. hackelii (Henriq.) Holub
 = Arrhenatherum hackelii (Henriq.) Sampaio, An. Fac. Sci. Porto 17:45 (1931)
Avena haussknechtii Nevski, Acta Univ. Asiae Med. 8b. Bot. 17:7 (1934)
 = A. sativa $\times$ sterilis F_1 hybrid
Avena heldreichii Parl., Fl. Palerm. 1:111 (1845)
 = H. filifolium (Lag.) Henrard
Avena heteromalla Haller, Nov. Comm. Soc. Sci. Göttingen 6: fig. 32, 33 (1775)
 = A. sativa L.
Avena heteromalla Moench, Meth. Pl. 195 (1794)
 = A. sativa L.
Avena hexantha Steud., Flora 12:487 (1829)
 = Danthonia stricta Schrad. in Schult., Mant. 2:383 (1824)
Avena hideoi Honda, Bot. Mag. Tokyo 40:435 (1926)
 = H. hideoi (Honda) Ohwi
Avena hirsuta Moench, Meth. Pl. Suppl. 64 (1802)
 = ? A. matritensis Baum
Avena hirsuta Roth, Catalecta Bot. 3:19 (1806)
 = ? A. matritensis Baum
Avena hirsuta Hornem., Hort. Hafn. 1:102 (1813)
 = A. barbata Pott ex Link
Avena hirsuta De Moor, Traité Gram. 165 (1854)
 = ? A. barbata Pott ex Link
Avena hirsuta var. humilis Nees, Nov. Act. Acad. Caes. Leop. Carol. 19, Suppl. 1:158 (1843)
 = A. barbata Pott ex Link
Avena hirsuta var. racemosa Lojac., Natural. Sicil. 4:138 (1885)
 = A. barbata Pott ex Link
Avena hirsuta var. sallentiana (Pau) Senn. et Mauriç., Cat. Fl. Fl. Rif. Or. 129 (1933)
 = ? A. barbata Pott ex Link
Avena hirta Schrad., Goett. Anz. Ges. Wiss. 3:2075 (1821)
 = H. hirtulum (Steud.) Schweickerdt
Avena hirtifolia Boiss., Diagn. Pl. Orient. Nov. Ser. 2, 3:128 (1859)
 = H. pubescens (Huds.) Pilger
Avena hirtula Lag., Gen. et Sp. Nov. 4 (1816)

Avena hispanica Ard. ex Saggi, Accad. Padov. 2:112 (1789)
Avena hispanica Hort. ex Roem. et Schult., Syst. Veg. 2:691 (1817)
 = A. hispanica Ard. ex Saggi
Avena hispanica Lange, Naturhist. For. Kjobenhavn Meddel. 2. 1:41 (1860)
 = Arrhenatherum erianthum Boiss. et Reut., Pug. 121 (1852)
Avena hispida L. f., Suppl. Pl. 111 (1781)
 = Tristachya leucothrix Trin. ex Nees, Agrost. Bras. 460 (1829)
Avena hispida (Lange) Nyman, Syll. Suppl. 71 (1865)
 = Trisetum hispidum Lange, Pug. 42 (1860)
Avena hookeri Scribn. in Hackel, True grasses 123 (1890)
 = H. hookeri (Scribn.) Henrard
 = Aveno. hookeri (Scribn.) Holub
Avena hoppeana Scheele, Flora 27:57 (1844)
 = A. barbata ? Pott ex Link
Avena hostii Boiss. et Reut., Pugill. 121 (1852)
 = Arrhenatherum parlatorei (Woods) Potztal, Bot. Jahrb. 75:329 (1951)
 = Av. hostii (Boiss. et Reut.) Dum.
Avena hugeninii De Not. ex Steud., Syn. Pl. Gram. 1:425 (1855)
 = ? H. pubescens (Huds.) Pilger
Avena hungarica Lucé, Topograph. Nachricht. Insel Oesel 20 (1823)
 = ? A. sativa L.
Avena hybrida Peterm. Fl. Bienitz 13 (1841)

Avena hybrida Peterm. ex Reichenb., Fl. Saxon. 17 (1842)
Avena hydrophila F. Muell. ex Hook. f., Fl. Tasm. 2:121 (1858)
 = Amphibromus neesii Steud., Syn. Pl. Gram. 328 (1855)
Avena insubrica (Aschers. et Graebn.) Dalla Torre et Harms, Fl. Tirol. 6 (2):195 (1906)
 = Aven. insubricum (Aschers. et Graebn.) Fritsch
 = ? H. pubescens (Huds.) Pilger
Avena intermedia Bellardi ex Colla, Herb. Pedem. 6:35 (1836)
 = H. setacea (Vill.) Henr.
Avena intermedia Guss., Fl. Sic. Prodr. Suppl. 16 (1832)
 = Aira tenorei Guss., Fl. Sic. Prodr. 1.62 (1827)
Avena intermedia Lestib., Bot. Belg. 3:36 (1827)
 = ? A. hybrida Peterm. or A. sativa fatuoid
Avena intermedia Lindgr., Bot. Not. 1841:151 (1841)
 = A. hybrida Peterm. or A. sativa fatuoid
Avena involucrata Schrad., Gött. Anz. Ges. Wiss. 3:2075 (1821)
 = Chaetobromus schraderi Stapf in Thiselton-Dyer, Fl. Capensis 7:538 (1899)
Avena involuta Presl ex Steud., Nom. Bot. ed. 2, 1:172 (1840)
 = H. sempervirens (Vill.) Pilger
Avena jahandiezii R. Litard. ex Jahand. et Maire, Bull. Soc. Hist. Nat. Afr. Nord 16:67 (1925)
 = H. jahandiezii (Lit.) Potztal
 = Aveno. jahandiezii (Lit.) Holub

Avena japonica Steud., Syn. Pl. Gram. 1:231 (1854)
 = A. hybrida Peterm.
Avena junghuhnii Buese in Miquel, Pl. Jungh. 345 (1854)
 = H. junghuhnii (Buese) Henrard

Avena koenigii Spreng., Syst. Veg. 1:336 (1825)
    = Neyraudia arundinacea (L.) Henr., Meded. Herb. Leid. No. 58:8 (1929)

Avena kotschyi (Boiss.) Steud., Syn. Pl. Gram. 235 (1854)
    = Arrhenatherum kotschyi Boiss., Diagn. Pl. Or. Nov. Ser. 1, 7:122 (1846)

Avena krylovii Pavlov, Animadvers. Syst. Herb. Univ. Tomsk No. 5–6:1 (1933)
    = Aven. krylovii Pavlov
    = H. krylovii (Pavlov) Henrard

Avena lachnantha (Hochst. ex A. Rich.) Hook. f., Jour. Linn. Soc. Bot. 7:227 (1864)
    = Aven. lachnanthum (Hochst. ex A. Rich.) Vierh.
    = Aven. lachnanthum (Hochst. ex A. Rich.) Pilger
    = Aven. lachnanthum (Hochst. ex A. Rich.) Hubbard
    = Arrhenatherum lachnanthum (Hochst. ex A. Rich.) Potztal, Bot. Jahrb. 75:328 (1951)

Avena laconica (Boiss. et Orph.) Nyman, Syll. Suppl. 71 (1865)
    = Trisetum laconicum Boiss. et Orph. in Boiss., Diagn. Pl. Or. Nov. Ser. 2, 4:129 (1859)

Avena laeta Salisb., Prodr. Stirp. 22 (1796)
    = Brachypodium pinnatum (L.) Beauv., Agrost. 155 (1812)

Avena laevigata Schur, Oesterr. Bot. Zeitschr. 10:72 (1860)
    = Aven. laevigatum (Schur) Domin
    = H. laevigatum (Schur) Potztal
    = H. pubescens (Huds.) Pilger

Avena laevis Hackel, Oesterr. Bot. Zeitschr. 27:46 (1877)
    = Aven. laeve (Hackel) Vierh.
    = H. laeve (Hackel) Potztal

Avena lagascae Sennen, Plantes d'Espagne 1926, No. 5980 (Exsiccata)
    = A. hirtula lag.

Avena lanata (L.) Cav., Descr. Pl. 308 (1802)
    = Holcus lanatus L., Sp. Pl. 1048 (1753)

Avena lanata (L.) Koel., Descr. Gram. 303 (1802)
    = Holcus lanatus L., Sp. Pl. 1048 (1753)

Avena lanata Schrad., Goett. Anz. Ges. Wiss. 3:2075 (1821)
    = Danthonia lanata (Schrad.) Schrad. in Schult., Mant. 2:386 (1824)

Avena lanuginosa Gilib., Exercit. Phyt. 2:539 (1792)
    = A. fatua L.

Avena lasiantha Link, Linnaea 9:135 (1834)
    = Bromus intermedius Guss., Prodr. Fl. Sic. 1:114 (1827)

Avena latifolia Kit. ex Host, Icon. Gram. Austr. 4:19 (1809)
    = H. planiculme (Schrad.) Pilger

Avena leiantha Keng, Bull. Fan Mem. Inst. Biol. 7:35 (1936)
    = H. leianthum (Keng) Ohwi

Avena lejocolea Gola, Mem. Accad. Sci. Torino Ser. 2, 62:61 (1912)
    = H. setaceum (Vill.) Henrard

Avena lendigera (L.) Salisb., Prodr. Stirp. 23 (1796)
    = Gastridium ventricosum (Gouan) Schinz et Thell., Vierteljahrs. Naturf. Ges. Zürich 58:39 (1913)

Avena leonina Steud., Flora 12:484 (1829)
    = H. leoninum (Steud.) Schweickerdt

= Arrhenatherum leoninum (Steud.) Potztal, Bot. Jahrb. 75:328 (1951)
Avena leptostachys Hook. f., Fl. Antarct. 378 (1847)
= ? Trisetum sp.
Avena letourneuxii Trab., Bull. Soc. Bot. France 36:411 (1890)
= H. letourneuxii (Trab.) Henrard
= Aveno. letourneuxii (Trab.) Holub
Avena levis Hack., Oesterr. Bot. Zeitschr. 27:122 (1877)
= H. pratense (L.) Pilger
Avena lodunensis Delastr., Cat. Hort. Pict. 1835
= ? H. sulcatum (J. Gay) Henr.
Avena loefflingiana Geuns ex Roem. et Schult., Mant. 2:370 (1824)
= Aira praecox L. Sp. Pl. 65 (1753)
Avena loeflingiana L., Sp. Pl. 79 (1753)
= Trisetum loeflingianum (L.) Beauv., Agrost. 88 (1812)
Avena longa Stapf, Kew Bull. Misc. Inf. 1897:292 (1897)
= Aven. longum (Stapf) Stapf
= Arrhenatherum longum (Stapf) Potztal, Bot. Jahrb. 75:328 (1951)
= H. longum (Stapf) Schweickerdt
Avena longepedicellata Senn. ex St.-Yves, Candollea 4:448 (1931)
= H. pratense (L.) Pilger
Avena longepilosa Sennen, Diagn. Nouv. Pl. Espagne et Maroc 1928–35:41 (1936)
= see A. longipilosa Senn. ex St.-Yves
Avena longespiculata Senn. ex St.-Yves, Candollea 4:448 (1931)
= H. pratense (L.) Pilger
Avena longifolia Thore, Prom. Golfe Gascogne 92 (1810)
= Arrhenatherum thorei (Duby) Desv., Cat. Dord. 153 (1840)
= Aven. longifolium (Thore) G. Sampaio
= Pseudarrhenatherum longifolium (Thore) Rouy, Bull. Soc. Bot. France 68:401 (1921)
= Arrhenatherum longifolium (Thore) Dulac
Avena longifolia Req. et DC. ex Duby, Bot. Gall. 1:514 (1828)
= H. pratense (L.) Pilger
Avena longifolia (R.Br.) Spreng. Syst. Veg. 1:336 (1825)
= Danthonia longifolia R.Br., Prodr. 176 (1810)
Avena longiglumis Durieu in Duchartre Rev. Bot. 1:359 (1846)
Avena longiglumis var. genuina Maire et Weiller, Fl. Afr. Nord 2:272 (1953)
= A. longiglumis Dur.
Avena longiglumis var. genuina subv. australis Maire, Fl. Afr. Nord 2:272 (1953)
= A. longiglumis Dur.
Avena longiglumis var. tripolitana Maire et Weiller in Maire, Fl. Afr. Nord 2:272 (1953)
= A. longiglumis Dur.
Avena longiglumis subv. glabrifolia Malz., Monogr. 240 (1930)
= A. longiglumis Dur.
Avena longiglumis subv. pubifolia Malz., op. cit. 239
= A. longiglumis Dur.
Avena longipilosa Senn. ex St-Yves, Candollea 4:448 (1931)
= H. pratense (L.) Pilger
Avena lucida Bertol. Fl. Ital. 1:701 (1834)
= H. pubescens (Huds.) Pilger

Avena ludoviciana Durieu, Bull. Soc. Linn. Bordeaux 20:41 (1855)
    = A. sterilis L.
Avena ludoviciana var. glabrescens Gren. et Godr., Fl. France 3:513 (1855)
    = A. sterilis L.
Avena ludoviciana var. macrantha (Malz.) Roshev, in Fedtsch. Fl. Turkm. 1:107 (1932)
    = A. sterilis L.
Avena ludoviciana var. transietissima Thell. ex Malz., Monogr. 378 (1930)
    = A. sativa L.
Avena ludoviciana var. triflora Tourlet, Cat. Pl. Vasc. Indre et Loire 568 (1908)
    = A. sterilis L.
Avena ludoviciana var. triflora f. biaristata Tourlet, loc. cit.
    = A. sterilis L.
Avena ludoviciana var. triflora f. triaristata Tourlet, loc. cit.
    = A. sterilis L.
Avena ludoviciana var. turkestanica (Regel) Roshev. in Fedtsch. Fl. Turkm. 1:107 (1932)
    = A. sterilis L.
Avena ludoviciana var. vulgaris Tourlet, Cat. Pl. Vasc. Indre et Loire 568 (1908)
    = A. sterilis L.
Avena lupulina L.f., Suppl. 113 (1781)
    = Danthonia lupulina (L.f.) Roem. et Schult., Syst. Veg. 2:690 (1817)
Avena lusitanica (Tab. Mor.) Baum, in this work
Avena lutea L.f., Suppl. 112 (1781)
    = Trisetum luteum Pers.
Avena macilenta (DC.) Guss., Fl. Sic. Prodr. Suppl. 1:29 (1832)
    = Avellinia michelii (Savi) Parl., Pl. Nov. 59 (1842)
Avena macra Stev. ex M.B., Fl. Taur.-Cauc. 1:77 (1808)
    = Gaudinopsis macra (Stev. ex M.B.) Eig, Rep. Sp. Nov. Fedde 26:77 (1929)
Avena macrantha (Hack.) Malz. in Keller et al., Sorn. Rast. U.R.S.S. 1:206 (1934) (Feb.)
    = ? A. sativa L.
Avena macrantha (Hack.) Nevski, Acta Univ. Asiae Med. Ser. 8b, Bot. Fasc. 17:6 (1934) (April)
    = ? A. sativa L.
Avena macrocalycina Steud., Flora 12:482 (1829)
    = Danthonia macrantha Schrad. in Schult., Mant. 2:385 (1824)
Avena macrocalyx Sennen, Bull. Soc. Bot. France 68:407 (1922)
    = A. sterilis L.
Avena macrocarpa Moench, Meth. Pl. 196 (1794)
    = A. sterilis L.
Avena macrostachya Balansa ex Coss. et Dur., Bull. Soc. Bot. France 1:318 (1855)
Avena magellanica Beauv., Ess. Agrost. 153 (1812)
    = see Non satis notae
Avena magna Murphy et Terrell, Science 159:103-104 (1968)
    = A. maroccana Gdgr.
Avena malabarica Heyne ex Hook. f., Fl. Brit. Ind. 7:69 (1896)
    = Jausenella griffithiana (C. Muell.) Bor. Kew Bull. 1955:98 (1955)

Avena mandoniana Coss. et Bal., Bull. Soc. Bot. France 15:185 (1868)
 = A. brevis Roth
Avena marginata Lowe, Trans. Cambridge Phil. Soc. 6:529 (1838)
 = H. pratense (L.) Pilger
Avena maroccana Gandog., Bull. Soc. Bot. France 55:658 (1908)
Avena maxima Presl. Fl. Sicul. 44 (1826)
 = A. atherantha Presl
Avena mediolanensis Bal. et De Not. ex Comolli Fl. Comense 1:147 (1836)
 = Trisetum myrianthum (Bertol.) Parl., Fl. Ital. 1:270 (1848)
Avena melillensis Senn. et Mauricio, Cat. Fl. Rif. Or. 129 (1933)
 = A. sterilis L.
Avena meridionalis (Malz.) Roshev. in Fedtsch., Fl. Turkom. 1:105 (1932)
 = A. hybrida Peterm.
Avena meridionalis var. grandis (Malz.) Roshev. in Fedtsch., Fl. Turkom. 1:106 (1932)
 = A. hybrida Peterm.
Avena micans Hook. f., Fl. Brit. Ind. 7:279 (1896)
 = Trisetum micans (Hook. f.) Bor, Grasses 448 (1960)
Avena michelii (Savi) Guss., Fl. Sic. Syn. 151 (1843)
 = Avellinia michelii (Savi) Parl., Pl. Nov. 59 (1842)
Avena micrantha Scribn., U.S. Dept. Agric. Div. Agrost. Circ. 19:3 (1900)
 = Triniochloa micrantha (Scribn.) Hitch., Contr. U.S. Nat. Herb. 17:304 (1913)
Avena miranda Sennen, Pl. Espagne No. 6709 (1928)
 = H. pratense (L.) Pilger
Avena mollis (L.) Salisb., Prodr. Stirp. 23 (1796)
 = Bromus mollis L., Sp. Pl. ed. 2, 1:112 (1762)
Avena mollis (L.) Koel., Descr. Gram. 300 (1802)
 = Holcus mollis L., Syst. Nat. ed. 10, 2:1305 (1759)
Avena mollis Michx., Fl. Bor. Amer. 1:72 (1803)
 = Trisetum spicatum (L.) Richt., Pl. Eur. 1:59 (1890)
Avena mongolica Roshev., Bull. Jard. Bot. Princ. U.R.S.S. 27:96 (1928)
 = H. mongolicum (Roshev.) Henrard
Avena montana (L.) Weber ex Wigg., Prim. Fl. Holsat. 9 (1780)
 = Deschampsia flexuosa (L.) Trin., Acad. St. Petersb. Mem. 6, Sci. Nat. 2:9 (1836)
Avena montana Brot., Fl. Lusit. 1:109 (1804)
 = Pseudarrhenatherum longifolium (Thore) Rouy, Bull. Soc. Bot. France 68:401 (1921)
Avena montana Roem. et Schult., Syst. Veg. 2:674 (1817)
 = H. pratense (L.) Pilger
Avena montana Vill., Pl. Dauph. 1:286 (1786)
 = Aven. montanum (Vill.) Vierh.
 = H. montanum (Vill.) Henrard
 = Arrhenatherum montanum (Vill.) Potztal, Bot. Jahrb. 75:329 (1951)
 = Av. montana (Vill.) Dum.
Avena montevidensis Hack., Oesterr. Bot. Zeitschr. 52:188 (1902)
 = Amphibromus quadridentulus (Doell) Swallen, Am. J. Bot. 18:414 (1931)
Avena mortoniana Scribn., Bot. Gaz. 21:133 (1896)
 = H. mortonianum (Scribn.) Henrard

Avena multiculmis (Dum.) Nyman, Syll. Suppl. Fl. Eur. 71 (1865)
    = Aira caryophyllea L., Sp. Pl. 66 (1753)
Avena muralis Salisb., Prodr. Stirp. 22 (1796)
    = Vulpia myuros (L.) Gmel.
Avena muralis Steud. ex Lechl., Berb. Amer. Austr. 52 (1857)
    = ? Trisetum trinii
Avena muricata Spreng., Neue Entdeck. 1:247 (1820)
    = Pentaschistis aspera (Thunb.) Stapf in Thisleton-Dyer, Fl. Capensis 7:501 (1899)
Avena muriculata Stapf, Kew Bull. Misc. Inf. 1897:291 (1897)
    = H. elongatum (Hochst. ex A. Rich.) C.E. Hubbard
    = Arrhenatherum elongatum (Hochst.) Potztal, Bot. Jahrb. 75:328 (1951)
Avena murphyi Ladiz., Israel Jour. Bot. 20:24 (1971)
Avena mutata Sampaio Bol. Soc. Brot. 2, 7:118 (1931)
    = ? A. sativa heterozygous fatuoid
Avena mutica Krocker, Suppl. Fl. Siles. 4:187 (1823)
    = Non satis notae
Avena myriantha Bertol., Fl. Ital. 1:722 (1834)
    = Trisetum myrianthum (Bertol.) Parl., Fl. Ital. 1:270 (1848)
Avena mysorensis Spreng., Syst. Veg. 1:337 (1825)
    = Tripogon bromoides Roem. et Schult., Syst. Veg. 2:600 (1817)
Avena nana Kunze ex Steud., Nom. Bot. ed. 2, 1:172 (1840)
    = Aira praecox L., Sp. Pl. 65 (1753)
Avena neesii Hook. f., Jour. Linn. Soc. Bot. 7:229 (1864)
    = Aven. rigidulum Pilger
    = H. rigidulum (Pilger) Hubbard
    = Arrhenatherum rigidulum (Pilger) Potztal, Bot. Jahrb. 75:329 (1951)
Avena neglecta Savi, Pl. Pis. 1:132 (1798)
    = Trisetum paniceum (Lam.) Pers., Syn. 1:97 (1805)
    = Rostraria neglecta (Savi) Holub, Fol. Geobot. Phytotax. 9:271 (1974)
Avena nemoralis (Huds.) Salisb., Prodr. Stirp. 23 (1796)
    = Bromus asper Murr., Prodr. Stirp. Gott. 42 (1770)
Avena nervosa Lam., Tabl. Encycl. 1:201 (1791)
    = A. strigosa Schreb.
Avena nervosa R.Br., Prodr. Fl. Nov. Holl. 1:178 (1810)
    = Amphibromus neesii Steud., Syn. Pl. Glum. 1:328 (1854)
    = Aven. nervosum Vierh.
Avena neumayeriana Visiani, Fl. Dalm. 3:339 (1850-1852)
    = H. neumayerianum (Vis.) Henrard
    = Arrhenatherum neumayerianum (Vis.) Potztal, Bot. Jahrb. 75:330 (1951)
    = D. compactum (Boiss. et Heldr.) Holub
Avena newtonii Stapf, Kew Bull. Misc. Inf. 1897:291 (1897)
    = H. newtonii (Stapf) C.E. Hubbard
Avena nigra Wallr., Linnaea 14:544 (1840)
    = ? A. fatua L.
Avena nigra Heuzé, Pl. Alimentaires 1:505 (1873)
    = A. sativa L.
Avena nitens L. ex Jacks., Ind. Linn. Herb. 42 (1912)
    = ? Arrhenatherum elatius (L.) J.S. et C.B. Presl
Avena nitida Desf., Fl. Atlant. 1:102 (1798)

    = Trisetum nitidum (Desf.) Pers., Syn. 1:97 (1805)
Avena nodipilosa (Malz.) Malz., Sorn. Rast. U.R.S.S. 1:205 (1934)
    = A. sativa L.
Avena nodosa Cullum, It. Scan. 251
    = Arrhenatherum elatius (L.) J.S. et C.B. Presl
Avena nodosa J. Walker, Econ. Hist. Hebrides 2:23 (1808)
    = Arrhenatherum elatior J.S. et C.B. Presl
Avena noeana (Boiss.) Nyman, Syll. Suppl. 71 (1865)
    = Trisetum aureum Tenore, Fl. Nap. 2:278 (1820)
Avena notarisii Parl., Fl. Ital. 1:279 (1848)
    = H. sempervirens (Vill.) Pilger
Avena nuda L., Demonstr. Pl. 3 (1753)
Avena nuda var. biflora Haller, Nov. Comm. Soc. Sci. Göttingen 6:Tab. 6, fig. 35 (1775)
    = probably A. nuda L.
Avena nuda var. chinensis Fish. ex Roem. et Schult., Syst. Veg. 2:669 (1817)
    = A. sativa L.
Avena nuda var. chinensis Kunth, Rev. Gram. 1:103 (1829)
    = A. sativa L.
Avena nuda var. elegantissima Hort., Wien. Illustr. Gart.-Zeit. 11:79 (1886)
    = ? A. sativa L.
Avena nuda var. mongolica Vav., Bull. Appl. Bot. Pl. Breed. 16:47, 175 (1926)
    = A. sativa L.
Avena nuda var. multiflora Haller, Nov. Comm. Soc. Sci. Göttingen 6: Tab. 6, fig. 36 (1775)
    = probably A. sativa L.
Avena nuda var. quadriflora Opiz, Seznam Rostl. České 20 (1852)
    = ? A. sativa L.
Avena nuda var. sterilis Hort., Wien. Illustr. Gart.-Zeit. 15:80 (1890)
    = A. sterilis L.
Avena nuda var. triflora Haller, Nov. Comm. Soc. Sci. Göttingen 6:Tab. 6, fig. 34 (1775)
    = ? A. nuda L.
Avena nuda ssp. wiestii (Steud.) Löve and Löve, Bot. Not. 114:50 (1961)
    = A. wiestii Steud.
Avena nudibrevis Vavilov, Bull. Appl. Bot. Pl. Breed. 16:48, 176 (1926)
    = A. nuda L.
Avena nutans Marat., Fl. Rom. 1:77 (1822)
    = ? Helictotrichon sp.
Avena nutans St. Lager, Rech. Hist. 43 (1884)
    = A. sterilis L.
Avena nutkaensis J. Presl, Rel. Haenk. 1:254 (1830)
    = Trisetum cernuum Trin., Acad. St. Petersb. Mem. 6. Math. Phys. Nat. 1:61 (1830)
Avena occidentalis Durieu, Cat. Graines Bordeaux 25 (1865)
Avena odorata (L.) Koel., Descr. Gram. 299 (1802)
    = Hierochloe odorata (L.) Beauv., Ess. Agrost. 62 (1812)
Avena oligostachya Munro, Jour. Linn. Soc. Bot. 19:193 (1882)
    = H. oligostachyum (Munro) Holub
    = Arrhenatherum oligostachyum (Munro) Potztal, Willdenovia 4(3):400 (1968)

= Duthiea oligostachya (Munro) Stapf, Hook. Icon. Pl. Tab. 2474 (1896)

Avena opulenta Lojac., Fl. Sic. 3:304 (1909)
= ? H. pratensis (L.) Pilger

Avena orientalis Schreb., Spicil. Fl. Lips. 52 (1771)
= A. sativa L.

Avena orientalis var. aristata Maly in Opiz. Seznam Rostl. České 20 (1852)
= ? A. sativa L.

Avena orientalis var. mutica S.F. Gray, Nat. Arr. Brit. Pl. 2:130 (1821)
= ? A. sativa L.

Avena orientalis var. mutica Spenner ex Schuebler et Mertens, Fl. Wurtemb. 71 (1834)
= ? A. sativa L.

Avena orientalis var. mutica Peterm., Fl. Lips. 105 (1838)
= ? A. sativa L.

Avena orientalis var. mutica Maly in Opiz, loc. cit.
= ? A. sativa L.

Avena orientalis var. mutica Ducomm., Taschenb. Schweiz. Bot. 863 (1869)
= A. sativa L.

Avena orientalis var. nuda Link, Hort. Berol. 2:254 (1833)
= A. sativa L.

Avena orientalis var. turgida Eriks., Bot. Centralbl. 38:787 (1889)
= A. sativa L.

Avena ovata (Cav.) Spreng., Syst. Veg. 1:335 (1825)
= Trisetum ovatum (Cav.) Pers., Syn. Pl. 1:98 (1805)

Avena ovina (L.) Salisb., Prodr. Stirp. 22 (1796)
= Festuca ovina L., Sp. Pl. 73 (1753)

Avena palaestina (Boiss.) Steud., Syn. Pl. Glum. 1:425 (1855)
= Arrhenatherum palestinum Boiss., Diagn. Pl. Or. Nov. Ser. 1, 13:51 (1854)

Avena pallens Link, Jour. Bot. Schrader 2:314 (1799)
= Aven. pallens (Link) Sampaio
= Pseudoarrhenatherum pallens (Link) Holub, Taxon 15:167 (1966)

Avena pallida Salisb., Prodr. Stirp. 24 (1796)
= Holcus lanatus L., Sp. Pl. 1048 (1753)

Avena pallida Thunb., Prodr. Pl. Cap. 22 (1794)
= Pentaschisitis thunbergii (Kunth) Stapf in Thiselton-Dyer, Fl. Capensis 7:507 (1899)

Avena paludosa (Roth) Heynh., Nom. 1:107 (1840)
= Deschampsia caespitosa (L.) Beauv., Ess. Agrost. 91 (1812)

Avena palustris Michx., Fl. Bor. Amer. 1:72 (1803)
= Trisetum pennsylvanicum (L.) Beauv. ex Roem. et Schult., Syst. Veg. 2:658 (1817)

Avena panicea Lam., Tabl. Encycl. 1:202 (1791)
= Trisetum neglectum Roem. et Schult., Syst. Veg. 2:660 (1817)

Avena panicea Link, Jour. Bot. Schrader 2:314 (1799)
= Trisetum neglectum Roem. et Schult., Syst. Veg. 2:660 (1817) or ? Polypogon

Avena panicea Pourret ex Willd., Ges. Naturf. Freund. Berlin Mag. 2:290 (1808)
= ? Trisetum neglectum Roem. et Schult., Syst. Veg. 2:660 (1817)

Avena panicea Hort. ex Roem. et Schult., Syst. Veg. 1:660 (1817)
= ? Koeleria, ? Lophochloa
Avena panormitana Lojac., Pl. Sic. 3:303 (1909)
= Arrhenatherum bromoides (L.) Sampaio, An. Fac. Sc. Porto 17:45 (1931)
= H. bromoides (Gouan) Hubbard
Avena papillosa Schrad., Gött. Anz. Ges. Wiss. 3:2075 (1821)
= ? Pentaschistis thunbergii Stapf in Thiselton-Dyer, Fl. Capensis 7:507 (1899)
Avena papillosa Steud., Flora 12:434 (1829)
= Pentaschistis subulifolia Stapf in Thiselton-Dyer, Fl. Capensis 7:499 (1899)
Avena paradensis Kit. ex Kanitz, Verh. Zool. Bot. Ges. Wien 13:543 (1863)
= Helictotrichon sp.
Avena paradoxa Willd. ex Kunth, Enum. Pl. 1:118 (1833)
= Aristida dichotoma Michx., Fl. Bor. Amer. 1:41 (1803)
Avena parlatorii Woods, The Tour. Fl. 405 (1850)
= Arrhenatherum parlatorei (Woods) Potztal, Bot. Jahrb. 75:329 (1951)
= Aven. parlatorei (Woods) Beck.
= H. parlatorei (Woods) Pilger
Avena parviflora Desf., Fl. Atlant. 1:103 (1798)
= Trisetum parviflorum (Desf.) Pers., Syn. 1:97 (1805)
= Rostraria parviflora (Desf.) Holub, Fol. Geobot. Phytotax. 9:271 (1974)
Avena patens St. Lager in Cariot, Etude des Fleurs ed. 8, 2:921 (1889)
= A. fatua L.
Avena patulipes Nyman, Syll. Suppl. 71 (1865)
= ? Aira caryophyllea L., Sp. Pl. 66 (1753)
Avena pauciflora Auct. ex Roem. et Schult., Syst. Veg. 2:660 (1817)
= Trisetum paniceum (Lam.) Pers., Syn. 1:97 (1805)
Avena paupercula Phil., Linnaea 29:94 (1858)
= ? not Avena
Avena pendula Gilib., Exerc. Phyt. 2:539 (1792)
= A. sativa L.
Avena penicillata Willd. ex Steud., Nom. Bot. ed. 2, 1:172 (1840)
= Trisetum rigidum (M.B.) Roem. et Schult., Syst. Veg. 2:662 (1817)
Avena pennsylvanica L., Sp. Pl. 79 (1753)
= Trisetum pennsylvanicum (L.) Beauv. ex Roem. et Schult., Syst. Veg. 2:658 (1817)
Avena pennsylvanica Forssk., Fl. Aegypt.-Arab. 23 (1775)
= Asthenatherum forsskalii (Vahl) Nevski, Acta Univ. Asiae Med. Ser. 8b, Bot. 17:8 (1934)
Avena pennsylvanica Muhl., Descr. Gram. 185 (1817)
= Trisetum pennsylvanicum (L.) Beauv. ex Roem. et Schult., Syst. Veg. 2:658 (1817)
Avena persarum Nevski, Acta Univ. Asiae Med. 8b, Bot. 17:6 (1934)
= A. sativa L.
Avena persica Steud., Syn. Pl. Glum. 1:230 (1854)
= A. sterilis L.
Avena peruviana Hort. ex Willd., Ges. Naturf. Freund. Berlin Mag. 2:290 (1808)
= ? Trisetum paniceum (Lam.) Pers., Syn. 1:97 (1805)

Avena phleoides D'Urv., Mem. Soc. Linn. Paris 4:601 (1826)
= ? Trisetum subspicatum Beauv., Ess. Agrost. 88 (1812)
Avena pilosa Scopoli, Fl. Carn. ed. 2, 1:86 (1772)
= A. fatua L.
Avena pilosa (Roem. et Schult.) M.B., Fl. Taur. Cauc. 3:84 (1819)
= A. eriantha Dur.
Avena pilosa Presl, Rel. Haenk. 1:253 (1830)
= ? Trisetum prelii (Kunth) E. Desv. in C. Gay, Fl. Chil. 6:347 (1854)
Avena pilosa var. glabriflora Grun., Bull. Soc. Nat. Moscou 40:459 (1867)
= A. eriantha Dur.
Avena pilosa var. pubiflora Grun., loc. cit.
= A. eriantha Dur.
Avena pilosa subv. glabriflora (Grun.) Malz., Monogr. 236 (1930)
= A. eriantha Dur.
Avena pilosa subv. pubiflora (Grun.) Malz., op. cit. 235
= A. eriantha Dur.
Avena planiculmis J.E. Smith in Sowerby, Engl. Bot. 30:Tab. 2141 (1810)
= H. alpinum (Smith) Henrard
Avena planiculmis Schrad., Fl. Germ. 1:381 (1806)
= Aven. planiculme (Schrad.) Opiz
= H. planiculme (Schrad.) Besser ex Rechb.
= Aveno. planiculmis (Schrad.) Holub
Avena planifolia St. Lager in Cariot, Etude des Fleurs ed. 8, 922 (1889)
= A. montana Vill.
Avena plesiantha (Jord.) Nyman, Syll. Suppl. 71 (1865)
= Aira caryophyllea L., Sp. Pl. 66 (1753)
Avena podolica Pascal. ex Zuccagni in Roem., Coll. Bot. 126 (1809)
= A. sativa L.
Avena podolica Blocki ex Koczw., Oesterr. Bot. Zeitschr. 75:239 (1926)
= ? Helictotrichon sp.
Avena polonica Schwaegr. ex Schmalh., Fl. Centr. et S. Russia 2:619 (1897)
= Non satis notae
Avena polyneura Hook. f., Fl. Brit. India 7:277 (1896)
= H. polyneurum (Hook. f.) Henrard
Avena ponderosa L. ex Jacks., Ind. Linn. Herb. in Suppl. Proceed. Linn. Soc. London 124th Session 42 (1912)
= A. sativa L.
Avena praecocioides Litw., Bull. Appl. Bot. 8:564 (1915)
= A. sativa L.
Avena praecoqua Litw., loc. cit.
= A. sativa L.
Avena praecox (L.) Beauv., Ess. Agrost. 89 (1812)
= Aira praecox L., Sp. Pl. 65 (1753)
Avena praegravis Roshev. in Komarov, Fl. U.R.S.S. 2:268 (1934)
= A. sativa L.
Avena praegravis var. arundinacea (Schur) Roshev., loc. cit.
= A. sativa L.
Avena praegravis var. macrotricha (Malz.) Roshev., loc. cit.
= A. sativa L.
Avena praeusta Reichenb., Fl. Germ. Excurs. 140 (1830)
= H. alpinum (Smith) Henrard
Avena praeusta Hausskn. in Bornm., Bull. Herb. Boiss. 4:16 (1896)

= H. laevigatum (Schur) Potztal
Avena pratensis L., Sp. Pl. 80 (1753)
= Aven. pratense (L.) Opiz
= Av. pratensis (L.) Dum.
= H. pratense (L.) Pilger
= Arrhenatherum pratense (L.) Sampaio, An. Fac. Sci. Porto 7:45 (1931)
= Aveno. pratensis (L.) Holub
Avena precatoria Thuill., Fl. Env. Paris ed. 2, 58 (1799)
= Arrhenatherum elatius (L.) J.S. et C.B. Presl, Fl. Čechica 17 (1819)
Avena preslei Kunth, Enum. Pl. 1:304 (1833)
= ? Trisetum preslei (Kunth) E. Desv. in C. Gay, Fl. Chil. 6:347 (1854)
Avena pressia Opiz, Seznam Rostl. České 20 (1852)
= ? A. strigosa Schreb.
Avena prostrata Ladizinsky, Israel Jour. Bot. 20:297 (1971)
= A. hirtula Lag.
Avena provincialis (Jord.) Nyman, Syll. Suppl. 71 (1865)
= Aira provincialis Jord., Pug. 142 (1852)
Avena provincialis (Lam. et DC.) Poiret, Encyc. Suppl. 1:541 (1811)
= Danthonia provincialis Lam. et DC., Fl. Fr. 3:33 (1805)
Avena pruinosa Trab. ex Hackel et Trab., Bull. Soc. Bot. France 36:411 (1890)
= H. pruinosum (Trab. ex Hack. et Trab.) Henrard
= Aveno. pruinosa (Trab. ex Hack. et Trab.) Holub
Avena pseudolucida Schur, Oesterr. Bot. Zeitschr. 10:72 (1860)
= H. pubescens (Huds.) Pilger
Avena pseudosativa (Thell.) Herter, Revista Sudamer. Bot. 6:141 (1940)
= A. sativa L.
Avena pseudoviolacea Kern. in Dalla Torre, Atl. Alpenfl. 228 (1882)
= Aven. pseudoviolaceum (Kern.) Fritsch
= ? H. alpinum (Smith) Henrard
Avena puberula Guss. in Tenore, Prodr. Fl. Neapol. App. 5 (1823)
= Avellinia michelii (Savi) Parl., Pl. Nov. 59 (1842)
Avena pubescens Huds., Fl. Angl. 42 (1762)
= Aven. pubescens (Huds.) Opiz
= Heuf. pubescens (Huds.) Schur
= Arrhenatherum pubescens (Huds.) Sampaio, An. Fac. Sc. Porto 17:45 (1931)
= H. pubescens (Huds.) Pilger
= Av. pubescens (Huds.) Dum.
= Aveno. pubescens (Huds.) Holub
Avena pulchella Beauv., Ess. Agrost. 89 (1812)
= ? Aira lendigera Lag., Gen. Sp. 38 (1816)
Avena pumila Desf., Fl. Atlant. 1:103 (1798)
= Trisetum pumilum (Desf.) Kunth, Enum. 1:297 (1833)
= Lophochloa pumila (Desf.) Bor. Grass. Burma Ceyl. Ind. Pak. 445 (1960)
Avena pungens Senn., Pl. Esp. No. 6276 (1927)
= H. pratense (L.) Pilger
Avena purpurascens DC., Cat. Hort. Monsp. 82 (1813)
= Trisetum flavescens (L.) Beauv., Ess. Agrost. 88 (1812)

Avena purpurea L.f., Suppl. Pl. 112 (1781)
  = Karroochloa purpurea (L.f.) Conert et Türpe, Senck. Biol. 50:303 (1969)

Avena purpurea Gueldenst. in Pallas, Reisen Russland Cauc. 1:189 (1787)
  = Non satis notae

Avena pusilla Web. ex Wigg., Prim. Fl. Hols. 9 (1780)
  = Aira praecox L., Sp. Pl. 65 (1753)

Avena pygmaea (Spreng.) Reichenb., Icon. Fl. Germ. Helv. ed. 2, 19 f. 195 (1850)
  = ? Trisetum hispanicum Pers., Syn. 1:97 (1805)

Avena quadridentula Doell in Mart., Fl. Bras. 2 (3):100 (1878)
  = Amphibromus quadridentulus (Doell) Swallen, Am. Jour. Bot. 18:414 (1931)

Avena quadriseta Labill., Nov. Holl. Pl. 1:25 (1804)
  = Deyeuxia quadriseta (Labill.) Benth., Fl. Austr. 7:581 (1878)

Avena quinqueseta Steud., Flora 12:485 (1829)
  = Aven. quinquesetum (Steud.) Stapf
  = H. quinquesetum (Steud.) Schweickerdt
  = Arrhenatherum quinquesetum (Steud.) Potztal, Bot. Jahrb. 75:328 (1951)

Avena racemosa Thuill., Fl. Env. Paris ed. 2, 59 (1799)
  = A. sativa L.

Avena redolens Pers., Syn. Pl. 1:100 (1805)
  = ? Hierochloa redolens (Pers.) Roem. et Schult. 2:514 (1817)

Avena requienii Mutel, Fl. Franç. 4:62 (1837)
  = H. requienii (Mutel) Henrard
  = Aveno. requienii (Mutel) Holub

Avena riabushinskii Kom., Repert. Sp. Nov. Fedde 13:86 (1914)
  = ? H. pratense

Avena rigida M.B., Fl. Taur. Cauc. 1:77 (1808)
  = Trisetum rigidum (M.B.) Roem. et Schult., Syst. Veg. 2:662 (1817)

Avena rigida Steud., Flora 12:482 (1829)
  = Danthonia rigida (Steud.) Steud., Syn. Pl. Glum. 1:243 (1854)

Avena rotae De Not. ex Parl., Fl. Ital. 1:264 (1848)
  = ? Trisetum argenteum (Willd.) Roem. et Schult., Syst. Veg. 2:665 (1817)

Avena rothii Stapf, Kew Bull. 1897:292 (1897)
  = Arrhenatherum lachnanthum (Hochst.) Potztal, Bot. Jahrb. 75:328 (1951)
  H. lachnanthum (Hochst. ex A. Rich.) C.E. Hubbard

Avena roylei (Hook. f.) Keng, Bull. Fan Mem. Inst. Biol. 7:36 (1936)
  = ? H. asperum (Munro ex Thwait.) Bor

Avena rubra Zuccagni Obs. Bot. Cent. 1:14, No. 30 (1806) et in Roem. Collect. Bot. 126 (1809)
  = A. sativa L.

Avena rufescens Pančic, Addit. Fl. Serb. 238 (1884)
  = ? Helictotrichon sp.

Avena rupestris J.F. Gmel., Syst. Nat. 194 (1791)
  = Agrostis alpina Scop., Fl. Carn. 1:60 (1772)

Avena ruprechtii Griseb. in Ledeb., Fl. Ross. 4:418 (1853)
  = Trisetum sibiricum Rupr., Beitr. Pflanzk. Russ. Reich. 2:65 (1845)

Avena sallentiana Pau Bol. Soc. Aragon Cienc. Nat. 1918:133 (1918)
= ? A. barbata Pott ex Link
Avena sarracenorum Gandog., Bull. Soc. Bot. France 60:402 (1913)
= Aven. sarracenorum (Gandog.) C.E. Hubbard et Sandwith
= H. sarracenorum (Gandog.) Holub
Avena sativa L., Sp. Pl. 79 (1753)
Avena sativa f. chlorathera Thell., Ber. Schweiz. Bot. Ges. 26/29:172 (1920)
= A. sativa L.
Avena sativa f. pilosiuscula Vav., Bull. Appl. Bot. 16(2):91, 210 (1926)
= A. sativa L.
Avena sativa f. subpilosa Thell., loc. cit.
= A. sativa L.
Avena sativa var. abyssinica (Hochst.) Kcke. in Kcke. et Werner, Handb. Getreidb. 1:216 (1885)
= A. abyssinica Hochst.
Avena sativa var. abyssinica (Hochst.) Engler, Abhandl. Preuss. Akad. Wiss. 2:129 (1891)
= A. abyssinica Hochst.
Avena sativa var. affinis Kcke., op. cit. 208
= A. sativa L.
Avena sativa var. alba Haller, Nov. Comm. Soc. Sc. Göttingen 6:Tab. 4, fig. 29-30 (1775)
= A. sativa L.
Avena sativa var. alba Koel., Descr. Gram. 290 (1802)
= ? A. sativa L.
Avena sativa var. arduennensis Lej. et Court., Comp. Fl. Belg. 1:73 (1828)
= A. sativa × fatua F_1 hybrid
Avena sativa var. arduennensis Bluff et Nees, Comp. Fl. Germ. ed. 2, 1:149 (1836)
= A. sativa × fatua F_1 hybrid
Avena sativa var. arduennensis Schur, Oesterr. Bot. Zeitschr. 10:70 (1860)
= A. sativa × fatua F_1 hybrid
Avena sativa var. aristata Alefeld, Landwirth. Flora 321 (1866)
= A. sativa L.
Avena sativa var. aristata Ducomm., Taschenb. Schweiz. Bot. 863 (1869)
= ? A. sativa L.
Avena sativa var. aristata Krause, Getr. 7:13 (1840)
= ? A. sativa L.
Avena sativa var. aristata Schlecht., Fl. Berolin. 1:51 (1823)
= ? A. sativa L.
Avena sativa var. aristata Schuebl. et Martens, Fl. Wurtemberg 71 (1834)
= A. sativa L.
Avena sativa var. aristata f. alba Schuebl. et Martens, loc. cit.
= A. sativa L.
Avena sativa var. aristata f. nigra Schuebl. et Martens, loc. cit.
= A. sativa L.
Avena sativa var. aristata Schur, Oesterr. Bot. Zeitschr. 10:70 (1860)
= ? A. sativa L.
Avena sativa var. armata Schur, Enum. Pl. Transsilv. 756 (1866)
= ? A. sativa fatuoid
Avena sativa var. arundinacea Schur, loc. cit.
= A. sativa L.

Avena sativa var. atrocarpa Schur, loc. cit.
  = ? A. sativa or A. sativa × fatua F_1 hybrid
Avena sativa var. aurea Koern. in Koern. et Werner, Handb. Getreidb. 1:210 (1885)
  = A. sativa L.
Avena sativa var. bashkirorum (Vavilov) Nevski, Act. Univ. Asiae Med. 8b, Bot. 17:5 (1934)
  = A. sativa L.
Avena sativa var. biaristata Alefeld, Landw. Fl. 322 (1866)
  = ? A. nuda L.
Avena sativa var. biaristata Hackel ex Trab., Compt. Rend. Acad. Sci. Paris 149:228 (1909)
  = A. sativa × sterilis F_1 hybrid
Avena sativa var. braunii Kcke. in Kcke. et Werner, Handb. Getreidb. 1:216 (1885)
  = A. abyssinica Hochst.
Avena sativa var. brevis (Roth) Fiori et Paoletti, Icon. Fl. Ital. 1:29 (1895)
  = A. brevis Roth
Avena sativa var. brevis (Roth) Kcke. in Kcke. et Werner, Handb. Getreidb. 1:213 (1885)
  = A. brevis Roth
Avena sativa var. brevistrigosa Kcke., Syst. Uebers. Cereal. 17 (1873)
  = A. hispanica Ard.
Avena sativa var. brunnea Kcke. in Kcke. et Werner, Handb. Getreidb. 1:207 (1885)
  = A. sativa L.
Avena sativa var. chinensis Doell, Rhein. Fl. 99 (1843)
  = A. sativa L.
Avena sativa var. chinensis (Metzger) Aschers., Fl. Brand. 1:827 (1864)
  = A. sativa L.
Avena sativa var. chinensis (Fisch.) Kcke., op. cit. 208
  = A. sativa L.
Avena sativa var. chinensis (Fisch.) Vilm., Blumengartn. ed. 3, Sieb. et Voss 1:1205 (1896)
  = A. sativa L.
Avena sativa var. cinerea Kcke. in Kcke. et Werner, Handb. Getreidb. 1:207 (1885)
  = A. sativa L.
Avena sativa var. cinerea f. scabra Mordv. in Wulff, Kulturn. Flora SSSR 2:380 (1936)
  = A. sativa L.
Avena sativa var. cinerea f. vacceorum Mordv. loc. cit.
  = A. sativa × fatua F_1 hybrid
Avena sativa var. contracta Neil., Fl. Nieder-Oesterr. 58 (1859)
  = A. sativa L.
Avena sativa var. culinaris Alefeld, Landw. Fl. 322 (1866)
  = A. sativa L.
Avena sativa var. diffusa Neilreich, loc. cit.
  = A. sativa L.
Avena sativa var. eligulata Vav. ex Mordv. in Wulff, Kulturn. Flora SSSR 2:384 (1936)
  = A. sativa L.

Avena sativa var. flava Kcke. in Kcke. et Werner, Handb. Getreidb. 1:207 (1885)
= A. sativa L.
Avena sativa var. fusca Alefeld, Landw. Fl. 321 (1866)
= A. sativa L.
Avena sativa var. fusciflora Schur, Oesterr. Bot. Zeitschr. 10:71 (1860)
= A. sativa L.
Avena sativa var. genuina Godr., Fl. Lorr. 3:150 (1844)
= A. sativa L.
Avena sativa var. glaberrima (Thell.) Maire et Weiller, Fl. Afr. Nord 2:282 (1953)
= A. sativa L.
Avena sativa var. glaberrima subv. diffusa (Neilr.) Maire et Weiller, loc. cit.
= A. sativa L.
Avena sativa var. grisea Kcke. in Kcke. et Werner, Handb. Getreidb. 1:210 (1885)
= A. sativa L.
Avena sativa var. gymnocarpa Kcke., op. cit. 208
= A. sativa L.
Avena sativa var. haussknechtii Vav. et Mordv. in Wulff, Fl. Cult. Pl. 2:383 (1936)
= A. sativa $\times$ sterilis F_1 hybrid
Avena sativa var. heteromalla Haller, Nov. Comm. Soc. Sci. Göttingen 6:18 Fig. 32, 33 (1776)
Avena sativa var. heteromorpha Opiz, Sezna Rostl. České 20 (1852)
Avena sativa var. hildebrantii Kcke. in Kcke. et Werner, Handb. Getreidb. 1:216 (1885)
= A. abyssinica Hochst.
Avena sativa var. homomalla (Malz.) Mordv. in Wulff, op. cit. 385
= A. sativa L.
Avena sativa var. hyemalis Alefeld, Landw. Fl. 321 (1866)
= A. sativa L.
Avena sativa var. inermis Kcke. in Kcke. et Werner, Handb. Getreidb. 1:217 (1885)
= A. sativa L.
Avena sativa var. inermis f. elegantissima Mordv. et Rodina in Wulff, op. cit. 386
= A. sativa L.
Avena sativa var. inermis f. tardiflora Mordv. et Rodina, loc. cit.
= A. sativa L.
Avena sativa var. involvens Godr., Fl. Lorr. 3:150 (1844)
= A. sativa L.
Avena sativa var. krausei Kcke. in Kcke. et Werner, Handb. Getreidb. 1:210 (1885)
= A. sativa L.
Avena sativa var. krausei f. baydarica Mordv. in Wulff, Kulturn. Flora SSSR 2:379 (1936)
= A. sativa L.
Avena sativa var. krausei f. citrina Mordv., loc. cit.
= A. sativa L.
Avena sativa var. leucocarpa Schur, Enum. Pl. Transsilv. 756 (1866)
= ? A. sativa L.

Avena sativa var. leucosperma Sweet, Hort. Brit. 452 (1826)
= ? A. sativa L.
Avena sativa var. ligulata Vav. ex Mordv. in Wulff, Kulturn. Flora SSSR 2:384 (1936)
= A. sativa L.
Avena sativa var. lusitanica Sampaio, Bol. Soc. Brot. Ser. 2, 7:117 (1931)
= ? A. sativa L.
Avena sativa var. macrantha Hack., Bot. Jahrb. Engler 6:244 (1885)
= ? A. sativa L.
Avena sativa var. maculata Mordv. in Wulff, Kulturn. Flora SSSR 2:387 (1936)
= A. sativa L.
Avena sativa var. melanosperma Sweet, Hort. Brit. 452 (1827)
= ? A. sativa L.
Avena sativa var. melanosperma Reichenb., Fl. Germ. 1:52 (1830)
= A. hispanica Ard.
Avena sativa var. melanosperma Peterm., Fl. Bienitz 13 (1841)
= A. sativa L.
Avena sativa var. modigenita Kcke., Arch. Biontologie 2:436 (1908)
= ? A. sativa × sterilis F_1 hybrid
Avena sativa var. montana Alefeld, Landw. Fl. 320 (1866)
= A. sativa L.
Avena sativa var. mutica Alefeld, loc. cit.
= A. sativa L.
Avena sativa var. mutica S.F. Gray, Nat. Arr. Brit. Pl. 2:130 (1821)
= A. sativa L.
Avena sativa var. mutica Peterm., Fl. Bienitz 13 (1841)
= A. sativa L.
Avena sativa var. mutica Lestib., Botan. Belg. 2:35 (1827)
= A. sativa L.
Avena sativa var. mutica Schlecht., Fl. Berolin. 1:52 (1823)
= A. sativa L.
Avena sativa var. mutica Schuebler et Mertens, Fl. Wurtenberg 71 (1834)
= A. sativa L.
Avena sativa var. mutica Maly in Opiz, Seznam Rostl. České 20 (1852)
= A. sativa L.
Avena sativa var. mutica Schur, Oesterr. Bot. Zeitschr. 10:70 (1860)
= A. sativa L.
Avena sativa var. mutica Stokes, Bot. Mat. Med. 1:158 (1812)
= A. sativa L.
Avena sativa var. mutica f. gigantea Mordv. in Wulff, Kurturn. Flora SSSR 2:376 (1936)
= A. sativa L.
Avena sativa var. niger Alefeld, Landw. Fl. 321 (1866)
= A. sativa L.
Avena sativa var. nigra S.F. Gray, Nat. Arr. Brit. Pl. 2:130 (1821)
= A. sativa L.
Avena sativa var. nigra Haller, Nov. Comm. Soc. Sci. Göttingen 6:Tab. 6, fig. 31 (1775)
= A. sativa L.
Avena sativa var. nigra Koel., Descr. Gram. 290 (1802)
= ? A. sativa L.

Avena sativa var. nigra Krause, Getr. Heft 7:15 (1835-37)
= ? A. sativa L.

Avena sativa var. nigra Provancher, Fl. Canad. 689 (1862)
= ? A. sativa L.

Avena sativa var. nigra Schrank, Baier, Fl. 1:375 (1789)
= A. sativa L.

Avena sativa var. nigra Schur, Oesterr. Bot. Zeitschr. 10:70 (1860)
= A. sativa L.

Avena sativa var. nigra Wood, Class-book ed. 2, 610 (1847)
= A. sativa L.

Avena sativa var. nigra Opiz, Seznam Rostl. České 20 (1852)
= ? A. sativa L.

Avena sativa var. nigra subv. glabra Koel., Descr. Gram. 290 (1802)
= A. sativa L.

Avena sativa var. nigra subv. pilosa Koel, loc. cit.
= probably A. sativa L.

Avena sativa var. normalis Opiz, Seznam Rostl. České 20 (1852)
= ? A. sativa L.

Avena sativa var. nuda (L.) Kcke. in Kcke. et Werner, Handb. Getreidb. 1:218 (1885)
= A. nuda L.

Avena sativa var. nuda (L.) Schmalh., Fl. Central and S. Russia 2:618 (1897)
= A. nuda L.

Avena sativa var. obtusata Alefeld, Landw. Fl. 321 (1866)
= A. sativa L.

Avena sativa var. orientalis (Schreb.) Hook. f., Fl. Brit. India 7:275 (1896) (April)
= A. sativa L.

Avena sativa var. orientalis (Schreb.) Vilm. Blumengartn. ed. 3 Silb. et Voss 1:1205 (1896)
= A. sativa L.

Avena sativa var. orientalis (Schreb.) Schmalh. Fl. Central and S. Russia 2:617 (1897)
= A. sativa L.

Avena sativa var. praecox Metzger, Europ. Cereal. 71 (1824)
= A. sativa L.

Avena sativa var. praecox Opiz, Seznam Rostl. České 20 (1852)
= ? A. sativa L.

Avena sativa var. praegravis Krause, Abbild. Beschr. Heft 7:7 (1837)
= ? A. sativa L.

Avena sativa var. pugnax Alefeld, Landw. Fl. 321 (1866)
= A. sativa L.

Avena sativa var. quadriflora Kcke., Arch. Biontologie 2:436 (1908)
= ? A. sativa × sterilis F_1 hybrid

Avena sativa var. rubida Krause, Getr. Heft 7:13 (1837)
= A. sativa L.

Avena sativa var. rufa Schur, Oesterr. Bot. Zeitschr. 10:70 (1860)
= A. sativa L.

Avena sativa var. schimperi Kcke. in Kcke. et Werner, Handb. Getreib. 1:216 (1885)
= A. abyssinica Hochst.

Avena sativa var. secunda Provancher, Fl. Canada. 2:689 (1862)
= A. sativa L.
Avena sativa var. secunda Wood, Class-book ed. 2, 610 (1847)
= A. sativa L.
Avena sativa var. segetalis (Vav.) Nevski, Act. Univ. Asiae Med. 8b. Bot. 17:5 (1934)
= A. sativa L.
Avena sativa var. semiaristata Schur, Oesterr. Bot. Zeitschr. 10:70 (1860)
= ? A. sativa × sterilis F_1 hybrid
Avena sativa var. sericea Hook. f., Fl. Brit. Ind. 7:275 (1896)
= A. fatua L.
Avena sativa var. setosa Kcke. in Kcke. et Werner, Handb. Getreidb. 1:211 (1885)
= ? A. sativa fatuoid or A. sativa × sterilis F_1 hybrid
Avena sativa var. strigosa (Schreb.) Fiori et Paoletti, Icon. Fl. Ital. 1:29 (1895)
= A. strigosa Schreb.
Avena sativa var. strigosa (Schreb.) Kcke. in Kcke. et Werner, Handb. Getreidb. 1:213 (1885)
= A. strigosa Schreb.
Avena sativa var. subaristata Schur, Enum. Pl. Transsilv. 756 (1866)
= A. sativa × sterilis F_1 hybrid
Avena sativa var. submontana Kcke., Arch. Biontologie 2:436 (1908)
= A. sativa × sterilis F_1 hybrid
Avena sativa var. submutica Opiz, Seznam Rostl. České 20 (1852)
= ? A. sativa L.
Avena sativa var. subuniflora (Trab.) Thell., Vierteljahrs. Naturf. Ges. Zürich 56:227 (1911)
= A. sativa L.
Avena sativa var. subuniflora (Trab.) Maire et Weiller, Fl. Afr. Nord 2:282 (1953)
= A. sativa L.
Avena sativa var. trisperma (Schuebler) Koch in Peterm., Fl. Bienitz 13 (1841)
= A. sativa L.
Avena sativa var. trisperma (Schuebler) Aschers., Fl. Brand. 1:827 (1864)
= A. sativa L.
Avena sativa var. trisperma (Schuebler et Mert.) Ducom., Taschenb. Schweiz. 863 (1869)
= ? A. sativa × fatua F_1 hybrid
Avena sativa var. tristis Alefeld, Landw. Fl. 322 (1866)
= A. sativa L.
Avena sativa var. turgida Krause, Abbild. Beschr. Heft 7:7-8 (1837)
= A. sativa L.
Avena sativa var. uniflora S.F. Gray, Nat. Arr. Brit. Pl. 2:130 (1821)
= Non satis notae
Avena sativa var. uniflora Stokes, Bot. Mat. Med. 1:158 (1812)
= Non satis notae
Avena sativa var. variegata Haller, Nov. Comm. Soc. Sci. Göttingen 6:Tab. 6, fig. 30 (1775)
= A. sativa L.

Avena sativa var. volgensis (Vav.) Nevski, Acta Univ. Asiae Med. 8 b, Bot. 17:5 (1934)
   = A. sativa L.
Avena sativa var. vulgaris Alefeld, Landw. Fl. 320 (1866)
   = A. sativa L.
Avena sativa ssp. autumnalis Marquand, Welsh Pl. Br. St. Bull. Ser. C, 2:8 (1922)
   = A. sativa L.
Avena sativa ssp. brevis (Roth) Aschers. et Graebn., Synopsis 2:237 (1899)
   = A. brevis Roth
Avena sativa ssp. diffusa (Neilr.) Aschers. et Graebn., op. cit. 234
   = A. sativa L.
Avena sativa ssp. diffusa (Neilr.) Hütonen, Suonen Kasvis 198 (1933)
   = A. sativa L.
Avena sativa ssp. diffusa var. aristata (Krause) Aschers. et Graebn., Synopsis 2:234 (1899)
   = ? A. sativa L.
Avena sativa ssp. diffusa var. aristata subv. trisperma (Schuebler) Aschers. et Graebn., loc. cit.
   = ? A. sativa L.
Avena sativa ssp. diffusa var. aurea (Koern.) Aschers. et Graebn., op. cit. 235
   = A. sativa L.
Avena sativa ssp. diffusa var. aurea subv. krausei (Koern.) Aschers. et Graebn., op. cit. 235
   = A. sativa L.
Avena sativa ssp. diffusa var. brunnea (Koern.) Aschers. et Graebn., op. cit. 235
   = A. sativa L.
Avena sativa ssp. diffusa var. brunnea subv. montana (Alefeld) Aschers. et Graebn., op. cit. 235
   = A. sativa L.
Avena sativa ssp. diffusa var. brunnea subv. setosa (Koern.) Aschers. et Graebn., op. cit. 235
   = ? A. sativa fatuoid or A. sativa × sterilis F_1 hybrid
Avena sativa ssp. diffusa var. grisea (Koern.) Aschers. et Graebn., op. cit. 235
   = A. sativa L.
Avena sativa ssp. diffusa var. grisea subv. cinerea (Koern.) Aschers. et Graebn., op. cit. 235
   = A. sativa L.
Avena sativa ssp. diffusa var. mutica (Alefeld) Aschers. et Graebn., op. cit. 234
   = A. sativa L.
Avena sativa ssp. diffusa var. mutica subv. praegravis (Langeth.) Aschers. et Graebn., op. cit. 234
   = A. sativa L.
Avena sativa ssp. diffusa var. nigra (Krause) Aschers. et Graebn., op. cit. 235
   = ? A. sativa L.
Avena sativa ssp. diffusa var. rossica Petrop., Bull. Appl. Bot. Pl. Breed. Suppl. 45:29 (1931)
   = ? A. sativa heterozygous fatuoid
Avena sativa ssp. diffusa var. transiens f. pseudopolytricha Petrop., op. cit. 28
   = ? A. sativa × sterilis F_1 hybrid

Avena sativa ssp. diffusa var. rubida (Krause) Aschers. et Graebn., op. cit. 235
   = A. sativa L.
Avena sativa ssp. fatua (L.) Fiori, Nuov. Fl. Anal. Ital. 1:109 (1923)
Avena sativa ssp. fatua var. barbata (Pott ex Link) Fiori, loc. cit.
   = A. barbata Pott ex Link
Avena sativa ssp. fatua var. fatua (L.) Fiori, loc. cit.
   = A. fatua L.
Avena sativa ssp. fatua var. ludoviciana (Dur.) Fiori, loc. cit.
   = A. sterilis L.
Avena sativa ssp. fatua var. sterilis (L.) Fiori, loc. cit.
   = A. sterilis L.
Avena sativa ssp. genuina Vascon., Ens. Sem. Melhor. Pl. Bol. 20 Ser. A:40 (1935)
   = A. sativa L.
Avena sativa ssp. genuina var. brunnea (Kcke.) Vascon., loc. cit.
   = A. sativa L.
Avena sativa ssp. genuina var. mutica (Alefeld) Vascon., loc. cit.
   = A. sativa L.
Avena sativa ssp. grandiglumis Vascon., Ens. Sem. Melhor. Pl. Bol. 20 Ser. A:39 (1935)
   = A. sativa × sterilis F_1 hybrid
Avena sativa ssp. grandiglumis var. cinerea (Kcke.) Vascon., loc. cit.
   = A. sativa L.
Avena sativa ssp. grandiglumis var. hirsuta Vascon., Ens. Sem. Melhor. Pl. Bol. 20 Ser. A:39 (1935)
   = A. sativa L.
Avena sativa ssp. nodipilosa (Malz.) Vascon., Revista Agron. 19:19 (1931)
   = A. sativa L.
Avena sativa ssp. nodipilosa var. alboaristata Vascon., Ens. Sem. Melhor. Pl. Bol. 20 Ser. A:40 (1935)
   = A. sativa fatuoid
Avena sativa ssp. nuda (L.) Gillet et Magne, Nouv. Fl. France ed. 3:532 (1873)
   = A. nuda L.
Avena sativa ssp. nuda (L.) Aschers. et Graebn., Synopsis 2:237 (1899)
   = A. nuda L.
Avena sativa ssp. nuda var. affinis (Kcke.) Aschers. et Graebn., op. cit. 238
   = A. sativa L.
Avena sativa ssp. nuda var. biaristata (Alefeld) Aschers. et Graebn., op. cit. 238
   = A. nuda L.
Avena sativa ssp. nuda var. chinensis (Fisch.) Aschers. et Graebn., op. cit. 238
   = A. sativa L.
Avena sativa ssp. nuda var. gymnocarpa (Kcke.) Aschers. et Graebn., op. cit. 238
   = A. sativa L.
Avena sativa ssp. nuda var. inermis (Kcke.) Aschers. et Graebn., op. cit. 238
   = A. sativa L.
Avena sativa ssp. orientalis (Schreb.) Aschers. et Graebn., op. cit. 235
   = A. sativa L.
Avena sativa ssp. orientalis var. armata Petrop., Bull. Appl. Bot. Pl. Breed. Suppl. 45:25 (1931)

= A. sativa L.
Avena sativa ssp. orientalis var. borealis Petrop., loc. cit.
= A. sativa L.
Avena sativa ssp. orientalis var. flava (Kcke.) Aschers. et Graebn., Synopsis 2:236 (1899)
= A. sativa L.
Avena sativa ssp. orientalis var. tartarica (Ard.) Aschers. et Graebn. op. cit. 235
= A. sativa L.
Avena sativa ssp. orientalis var. tristis (Alefeld) Aschers. et Graeb., op. cit. 236
= A. sativa L.
Avena sativa ssp. praegravis (Langeth.) Vascon., Ens. Sem. Melhor. Pl. Bol. 20 Ser. A:40 (1935)
= A. sativa L.
Avena sativa ssp. praegravis var. cinerescens Vascon., loc. cit.
= A. sativa × fatua F_1 hybrid
Avena sativa ssp. praegravis var. grandiuscula (Malz.) Tab. Mor., Bol. Soc. Brot. Ser. 2, 13:Tab. 1 (1939)
= A. sativa L.
Avena sativa ssp. praegravis var. grandiuscula subv. affinis (Kcke.) Tab. Mor., loc. cit.
= A. sativa L.
Avena sativa ssp. praegravis var. leiantha (Malz.) Tab. Mor., op. cit. 606
= A. sativa L.
Avena sativa ssp. praegravis var. leiantha subv. subeligulata (Malz.) Tab. Mor., op. cit. Tab. 1
= A. sativa L.
Avena sativa ssp. praegravis var. leiantha subv. turgida (Erikss.) Tab. Mor., op. cit. Tab 1
= A. sativa L.
Avena sativa ssp. praegravis var. macrotricha (Malz.) Tab. Mor., op. cit. 604
= A. sativa L.
Avena sativa ssp. praegravis var. macrotricha subv. arundinacea (Schur.) Tab. Mor., op. cit. Tab. 1
= A. sativa L.
Avena sativa ssp. praegravis var. major Tab. Mor., op. cit. 605
= A. sativa × sterilis F_1 hybrid
Avena sativa ssp. praegravis var. microtricha (Malz.) Tab. Mor., op. cit. Tab. 1
= A. sativa heterozygous fatuoid
Avena sativa ssp. praegravis var. norvegica (Malz.) Tab. Mor., op. cit. Tab. 1
= A. sativa L.
Avena sativa ssp. praegravis var. polytricha (Malz.) Tab. Mor., op. cit. Tab. 1
= A. sativa × sterilis F_1 hybrid
Avena sativa ssp. sativa (L.) Fiori, Nuov. Fl. Anal. Ital. 1:108 (1923)
= A. sativa L.
Avena sativa ssp. sativa (L.) Tab. Mor., op. cit. 598
= A. sativa L.
Avena sativa ssp. sativa var. brachytricha (Thell.) Tab. Mor., op. cit. Tab. 1
= A. sativa L.

Avena sativa ssp. sativa var. brachytricha subv. asiatica (Vav.) Tab. Mor., op. cit. Tab. 1
= A. sativa L.
Avena sativa ssp. sativa var. brachytricha subv. iranica (Vav.) Tab. Mor., op. cit. Tab. 1
= A. sativa L.
Avena sativa ssp. sativa var. brachytricha subv. spelticola (Malz.) Tab. Mor., op. cit. Tab. 1
= ? A. sativa L.
Avena sativa ssp. sativa var. brachytricha subv. speltiformis (Vav.) Tab. Mor., op. cit. Tab. 1
= A. sativa L.
Avena sativa ssp. sativa var. brachytricha subv. subcallosa Tab. Mor., op. cit. 604
= A. sativa $\times$ sterilis F_1 hybrid
Avena sativa ssp. sativa var. brachytricha subv. subglabra (Malz.) Tab. Mor., op. cit. Tab. 1
= A. sativa L.
Avena sativa ssp. sativa var. brevis (Roth) Fiori, Nuov. Fl. Anal. Ital. 1:105 (1923)
= A. brevis Roth
Avena sativa ssp. sativa var. diffusa (Neilr.) Fiori, Nuov. Fl. Anal. Ital. 1:109 (1923)
= A. sativa L.
Avena sativa ssp. sativa var. glaberrima (Thell.) Tab. Mor., op. cit. Tab. 1
= A. sativa L.
Avena sativa ssp. sativa var. glaberrima subv. calva (Malz.) Tab. Mor., op. cit. Tab. 1
= A. sativa L.
Avena sativa ssp. sativa var. glaberrima subv. contracta (Neilr.) Tab. Mor., op. cit. Tab. 1
= A. sativa L.
Avena sativa ssp. sativa var. glaberrima subv. eligulata (Vav. ex Mordv.) Tab. Mor., op. cit. Tab. 1
= A. sativa L.
Avena sativa ssp. sativa var. glaberrima subv. glabra (Malz.) Tab. Mor., op. cit. Tab. 1
= A. sativa L.
Avena sativa ssp. sativa var. glaberrima subv. pilosiuscula (Vav.) Tab. Mor., op. cit. Tab. 1
= A. sativa L.
Avena sativa ssp. sativa var. glaberrima subv. pseudoligulata (Malz.) Tab. Mor., op. cit. Tab. 1
= A. sativa L.
Avena sativa ssp. sativa var. glaberrima subv. secunda (Malz.) Tab. Mor., op. cit. Tab. 1
= A. sativa L.
Avena sativa ssp. sativa var. glabricalla (Thell.) Tab. Mor., op. cit. Tab. 1
= A. sativa heterozygous fatuoid
Avena sativa ssp. sativa var. nuda (L.) Fiori, Nuov. Fl. Anal. Ital. 1:109 (1923)
= A. nuda L.

Avena sativa ssp. sativa var. orientalis (Schreb.) Fiori, loc. cit.
    = A. sativa L.
Avena sativa ssp. sativa var. pilosa (Koel.) Tab. Mor., op. cit. 598
    = ? A. sativa L.
Avena sativa ssp. sativa var. pilosa subv. homomalla (Malz.) Tab. Mor., op. cit. Tab. 1
    = A. sativa L.
Avena sativa ssp. sativa var. pilosa subv. pilifera (Malz.) Tab. Mor., op. cit. 599
    = A. sativa L.
Avena sativa ssp. sativa var. pilosa subv. pilifera (Malz.) Tab. Mor., op. cit. Tab. 1
    = A. sativa L.
Avena sativa ssp. sativa var. pilosa subv. unilateralis (Thell.) Tab. Mor., op. cit. Tab. 1
    = A. sativa $\times$ fatua F_1 hybrid
Avena sativa ssp. sativa var. sinensis (Fisch.) Tab. Mor., op. cit. Tab. 1
    = A. sativa L.
Avena sativa ssp. sativa var. sinensis subv. culinaris (Alefeld) Tab. Mor., op. cit. Tab. 1
    = A. sativa L.
Avena sativa ssp. sativa var. sinensis subv. decortica (Malz.) Tab. Mor., op. cit. Tab. 1
    = A. sativa L.
Avena sativa ssp. sativa var. sinensis subv. mongolica (Pissarev) Tab. Mor., op. cit. Tab. 1
    = A. sativa L.
Avena sativa ssp. sativa var. sinensis subv. nudata (Malz.) Tab. Mor., op. cit. Tab. 1
    = A. sativa L.
Avena sativa ssp. sativa var. strigosa (Schreb.) Fiori, Nuov. Fl. Anal. Ital. 1:109 (1923)
    = A. strigosa Schreb.
Avena sativa ssp. sativa var. subpilosa (Thell.) Tab. Mor., op. cit. 599
    = A. sativa L.
Avena sativa ssp. sativa var. subpilosa subv. gigantea Tab. Mor., op. cit. 601
    = A. sativa $\times$ sterilis F_1 hybrid
Avena sativa ssp. sativa var. subpilosa subv. heteroclita Tab. Mor., op. cit. 601
    = A. sativa heterozygous fatuoid
Avena sativa ssp. sativa var. subpilosa subv. subtransiens Tab. Mor., op. cit. 601
    = A. sativa $\times$ sterilis F_1 hybrid
Avena sativa ssp. sativa var. subpilosa f. cinerea Tab. Mor., op. cit. 600
    = A. sativa L.
Avena sativa ssp. sativa var. subpilosa f. pallida Tab. Mor., op. cit. 600
    = A. sativa $\times$ sterilis F_1 hybrid
Avena sativa ssp. sativa var. subpilosa f. subpilifera (Malz.) Tab. Mor., op. cit. 600
    = A. sativa L.
Avena sativa ssp. sativa var. subuniflora (Trab.) Tab. Mor., op. cit. 602
    = A. sativa L.

Avena sativa ssp. sativa var. subuniflora subv. glabriuscula (Malz.) Tab. Mor., op. cit. Tab. 1
   = A. sativa L.
Avena sativa ssp. sativa var. subuniflora subv. kasanensis (Vav.) Tab. Mor., op. cit. Tab. 1
   = A. sativa L.
Avena sativa ssp. sativa var. subuniflora subv. longipila (Malz.) Tab. Mor., op. cit. Tab. 1
   = A. sativa L.
Avena sativa ssp. sativa var. transiens (Hausskn.) Tab. Mor., op. cit. 603
   = A. sativa $\times$ fatua F_1 hybrid
Avena sativa ssp. strigosa (Schreb.) Aschers. Aschers. et Graebn., Synopsis 2:236 (1899)
   = A. strigosa Schreb.
Avena scabriuscula Lag., Varied. Cienc. 4:212 (1805)
   = Trisetum scabriusculum (Lag.) Coss., Pl. Crit. Esp. 129 (1849)
Avena scabrivalvis Trin., Mem. Acad. St. Petersb. 6. Sci. Nat. 2:28 (1836)
   = Amphibromus scabrivalvis (Trin.) Swallen, Am. J. Bot. 18:413 (1931)
Avena schelliana Hack. in Korsh., Acta Horti Petrop. 12:419 (1892)
   = Aven. schellianum (Hack.) Podp.
   = H. schellianum (Hack.) Kitagawa
   = Aveno. schelliana (Hack.) Holub
Avena scheuchzeri All., Fl. Pedem. 2:255 (1785)
   = H. versicolor (Vill.) Pilger
   = Av. scheuchzeri (All.) Dum.
   = Heuf. scheuchzeri (All.) Schur
Avena secalina (L.) Salisb., Prodr. Stirp. 22 (1796)
   = Bromus secalinus L., Sp. Pl. 76 (1753)
Avena secunda Salisb., loc. cit.
   = Festuca elatior L., Sp. Pl. 75 (1753)
Avena sedenensis Clar. ex Lam. et DC., Fl. Française ed. 3, 3:719 (1805)
   Arrhenatherum montanum (Vill.) Potztal, Bot. Jahrb. 75:329 (1951)
   = H. sedenensis (DC.) Holub
Avena segetalis Bianca ex Nym., Consp. Fl. Eur. 810 (1882)
   = see A. barbata subsp. segetalis Nyman
Avena sempervirens Besser, Enum. Pl. Volh. 6 (1822)
   = Aven. desertorum (Less.) Podp.
   = H. desertorum (Less.) Pilger
   = Arrhenatherum desertorum (Less.) Potztal, Bot. Jahrb. 75:329 (1951)
Avena sempervirens Vill., Prosp. 17 (1799)
   = Aven. sempervirens (Vill.) Vierh.
   = H. sempervirens (Vill.) Pilger
   = Av. sempervirens (Vill.) Dum.
   = Heuf. sempervirens (Vill.) Schur
Avena sensitiva Hort. ex Vilm., Blumengartn. ed. 3, 1:1205 (1895)
   = A. sterilis L.
Avena septentrionalis Malz., Bull. Angew. Bot. 6:915 (1913)
   = A. hybrida Peterm.
Avena serrulatiglumis Sennen et Mauricio in Sennen, Diagn. Nouv. Pl. Espagne et Maroc 1928-35:248 (1936)
   = A. hirtula Lag.

Avena sesquiflora (Trin.) Griseb. in Ledeb., Fl. Ross. 4:419 (1853)
 = Calamagrostis sesquiflora (Trin.) Tzvel. in Tolmatchev, Fl. Arct. URSS 2:74 (1964)
Avena sesquitertia Hort. ex Steud., Nom. Bot. ed. 1, 95 (1821)
 = A. barbata Pott ex Link
Avena sesquitertia L., Mant. Pl. 1:34 (1767)
 = Aven. sesquitertium (L.) Fritsch
 = ? H. pubescens (Huds.) Pilger or H. pratense (L.) Pilger
Avena setacea Baumg. ex Schur, Enum. Pl. Transsilv. 753 (1866)
 = ? Deschampsia flexuosa (L.) Trin., Mem. Acad. St. Petersb. Ser. 6, 2:9 (1836)
Avena setacea Muhl. ex Trin., Mem. Acad. St. Petersb. Ser. 6, 1:87 (1830)
 = Aristida dichotoma Michx., Fl. Bor. Amer. 1:41 (1803)
Avena setacea Vill., Prosp. 16 (1779)
 = Aven. setaceum (Vill.) Vierh.
 = Av. setacea (Vill.) Dum.
 = H. setaceum (Vill.) Henrard
 = Arrhenatherum setaceum (Vill.) Potztal, Bot. Jahrb. 75:330 (1951)
 = Heuf. setacea (Vill.) Schur
Avena setifolia Brot., Fl. Lusit. 108 (1804)
 = Arrhenatherum pallens Link, Hort. Berol. 1:124 (1821)
 = ? Avena setigera Harz., Land w. Samenkunde 2:1326 (1885)
Avena 6-flora Larrañaga, Escritos D.A. Larrañaga (Pub. Inst. Hist. Geogr. Urug.) 2:49 (1923)
 = probably A. sativa L.
Avena shatilowiana Litwinow, Bull. Appl. Bot. 8:564 (1915)
 = A. sativa L.
Avena sibirica L., Sp. Pl. 79 (1753)
 = Stipa sibirica (L.) Lam., Tableau Encycl. 1:158 (1791)
Avena sibthorpii Nyman, Syll. 412 (1854)
 = ? H. compressum (Heuff.) Potztal
Avena sicula Spreng., Syst. Veg. 1:335 (1825)
 = Trisetum aureum (Ten.) Ten., Fl. Nap. 2:378 (1820)
Avena sikkimensis Hook. f., Fl. Brit. Ind. 7:280 (1896)
 = Trisetum flavescens (L.) Beauv., Ess. Agrost. 88 (1812)
Avena smithii Porter ex A. Gray, Man. Bot. ed. 5, 640 (1867)
 = Melica smithii (Porter ex A. Gray) Vasey, Bull. Torrey Bot. Club 15:294 (1888)
Avena solida (Hausskn.) Herter, Revist. Sudamer. Bot. 6:144 (1940)
 = A. sterilis L.
Avena spicaeformis Beauv., Ess. Agrost. 154 (1812)
 = ? Danthonia spicata (L.) Beauv. ex Roem. et Schult., Syst. Veg. 2:690 (1817)
Avena spicata (L.) Fedtsch., Acta Horti Petrop. 28:76 (1908)
 = Trisetum spicatum (L.) Richt., Pl. Eur. 1:59 (1908)
Avena spicata Gilib., Exerc. Phyt. 2:540 (1792)
 = H. pratense (L.) Pilger
Avena spicata L., Sp. Pl. 80 (1753)
 = Danthonia spicata (L.) Beauv. ex Roem. et Schult., Syst. Veg. 2:690 (1817)
Avena splendens Auct. ex Steud., Nom. Bot. ed. 2, 1:173 (1840)
 = H. pratense (L.) Pilger

Avena splendens Boiss., Elench. 88 (1838)
  = H. albinerve (Boiss.) Henrard
Avena splendens Bellardi ex Colla, Herb. Pedem. 6:34 (1836)
  = see A. bellardi Colla
Avena splendens Guss., Fl. Sic. Prodr. 1:126 (1827)
  = Trisetum flavescens (L.) Beauv., Ess. Agrost. 88 (1812)
Avena squarrosa Schrank, Denkschr. Baier. Bot. Ges. Regensburg 1(2):7 (1818)
  = ? Trisetum spicatum (L.) Richt., Pl. Eur. 1:59 (1890)
Avena sterilis L., Sp. Pl. ed. 2, 118 (1762)
Avena sterilis (L.) Salisb., Prodr. Stirp. 22 (1796)
  = Bromus sterilis L.
Avena sterilis f. albescens Hausskn. ex Heldr. Herb. Graec. Norm. Exsicc. No. 895 (1885)
  = A. sterilis L.
Avena sterilis f. aprica Hausskn., Mitt. Thür. Bot. Ver. N.F. 6:38 (1894)
  = A. sativa × sterilis F_1 hybrid
Avena sterilis f. brachyathera Hausskn., Mitt. Thür. Bot. Ges. 13/14:43 (1899)
  = ? A. sterilis L.
Avena sterilis f. biflora Hausskn., loc. cit.
  = A. sterilis L.
Avena sterilis f. breviglumis Hausskn., loc. cit.
  = ? A. sterilis L.
Avena sterilis f. contracta Hausskn., loc. cit.
  = ? A. sterilis L.
Avena sterilis f. fusca Hausskn., loc. cit.
  = A. sterilis L.
Avena sterilis f. leiophylla Hausskn., Mitt. Thür. Bot. Ver. N.F. 6:44 (1894)
  = A. sterilis L.
Avena sterilis f. longiglumis Hausskn., Mitt. Thür. Bot. Ges. 13/14:43 (1899)
  = ? A. sterilis L.
Avena sterilis f. macrathera Hausskn., loc. cit.
  = ? A. sterilis L.
Avena sterilis f. nigrescens Hausskn., loc. cit.
  = ? A. sterilis L.
Avena sterilis f. parallela Hausskn., Mitt. Thür. Bot. Ver. 3:240 (1884)
  = A. sativa L.
Avena sterilis f. patula Hausskn., Mitt. Thür. Bot. Ges. 13/14:43 (1899)
  = ? A. sterilis L.
Avena sterilis f. quadriflora Hausskn., loc. cit.
  = ? A. sterilis L.
Avena sterilis f. quinqueflora Hausskn., loc. cit.
  = ? A. sterilis L.
Avena sterilis f. secunda Hausskn., loc. cit.
  = ? A. sterilis L.
Avena sterilis f. solidissima Thell., Naturw. Wochenchr. N.F. 17(32):455 (1918)
  = A. sativa L.
Avena sterilis f. straminea Hausskn., loc. cit.
  = ? A. sterilis L.
Avena sterilis f. triflora Hausskn., loc. cit.

= ? A. sterilis L.
Avena sterilis f. vegeta Hausskn., loc. cit.
= ? A. sterilis L.
Avena sterilis f. 7-nervata Hausskn., loc. cit.
= ? A. sterilis L.
Avena sterilis f. 8-nervata Hausskn., loc. cit.
= ? A. sterilis L.
Avena sterilis f. 9-nervata Hausskn., loc. cit.
= ? A. sterilis L.
Avena sterilis f. 10-nervata Hausskn., loc. cit.
= ? A. sterilis L.
Avena sterilis f. 11-nervata Hausskn., loc. cit.
= ? A. sterilis L.
Avena sterilis var. algeriensis (Trab.) Trab., Jour. Hered. 5:77 (1914)
= A. sativa L.
Avena sterilis var. brevipila Malz., Bull. Ang. Bot. 7:328 (1914)
= A. sterilis L.
Avena sterilis var. calvescens Trabut et Thell., Vierteljahrs. Nat. Ges. Zürich 56:272 (1911)
= A. sterilis L.
Avena sterilis var. degenerans Hausskn., Mitt. Thür. Bot. Ver. N.F. 6:40 (1894)
= A. sativa L.
Avena sterilis var. denudata Hausskn., loc. cit.
= A. sterilis × sativa F_1 hybrid
Avena sterilis var. glabrescens Douin, in Bonnier Fl. Compl. 12:14 (1934)
= A. sterilis L.
Avena sterilis var. glabriflora Malz., Bull. Appl. Bot. 7:328 (1914)
= A. sterilis L.
Avena sterilis var. ludoviciana (Dur.) Husnot, Gram. Franç., Belg. 39 (1899)
= A. sterilis L.
Avena sterilis var. ludoviciana subv. glabrescens Husnot, loc. cit.
= A. sterilis L.
Avena sterilis var. maxima Perez-Lara, Anal. Soc. Espan. Hist. Nat. 15:398 (1886)
= A. longiglumis Dur.
= A. atherantha Presl
Avena sterilis var. micrantha Trab., Bull. Agr. Algerie Tunisie 16:354 fig. d (1910)
= ? A. sterilis L.
Avena sterilis var. minor Coss. et Dur., Expl. Alger. 2:109 (1855)
= A. sterilis L.
Avena sterilis var. oleodens Marq. in Druce, Rep. Bot. Soc. Exch. Club Brit. Isles 9:286 (1931)
= ? A. sterilis L.
Avena sterilis var. pilosa Aznav., Magyar. Bot. Lapok 12:182 (1913)
= A. sterilis L.
Avena sterilis var. pseudovilis Hausskn., Mitt. Thür. Bot. Ver. N.F. 6:39 (1894)
= A. sterilis L.

Avena sterilis var. scabriuscula Perez-Lara, Anal. Soc. Esp. Hist. Nat. 15:398 (1886)
  = A. sterilis L.
Avena sterilis var. segetalis Trab., 4e Conf. Inter. Genet. Paris 1911:336 (1913)
  = A. sativa × sterilis F_1 hybrid
Avena sterilis var. solida Hausskn., Mitt. Thür. Bot. Ver. N.F. 6:40 (1894)
  = A. sterilis L.
Avena sterilis var. subulata Batt. et Trab., Fl. Alger. 1:179 (1895)
  = A. sterilis L. or A. trichophylla C. Koch
Avena sterilis var. turkestanica Regel, Acta Horti Petrop. 7(1):633 (1880)
  = A. sterilis L.
Avena sterilis var. typica Regel, loc. cit.
  = A. sterilis L.
Avena sterilis ssp. barbata (Pott ex Link) Gillett et Magne Nouv. Fl. France ed. 3:532 (1873)
Avena sterilis ssp. byzantina (Koch) Thell., Vierteljahrs. Nat. Ges. Zürich 56:316 (1912)
  = A. sativa L.
Avena sterilis ssp. byzantina var. biaristata (Hack. ex Trab.) Thell., loc. cit.
  = A. sativa × sterilis F_1 hybrid
Avena sterilis ssp. byzantina var. brachytricha Thell., Rec. Trav. Bot. Neerland. 25:432 (1929)
  = a hypothetical form, therefore illegitimate; see A. sterilis L.
Avena sterilis ssp. byzantina var. culta Thell., Vierteljahrs. Nat. Ges. Zürich 56:317 (1912)
  = A. sativa L.
Avena sterilis ssp. byzantina var. culta f. denudata (Hausskn.) Thell., op. cit. 319
  = A. sterilis × sativa F_1 hybrid
Avena sterilis ssp. byzantina var. hypatricha Thell., Rec. Trav. Bot. Neerland. 25:432 (1929)
  = A. sativa L.
Avena sterilis ssp. byzantina var. hypomelanathera Thell., op. cit. 431
  = A. sativa L.
Avena sterilis ssp. byzantina var. induta Thell., op. cit. 431
  = A. sativa × sterilis F_1 hybrid
Avena sterilis ssp. byzantina var. macrotricha Malz., Monogr. 398 (1930)
  = A. sativa L.
Avena sterilis ssp. byzantina var. macrotricha subv. biaristata (Hack. ex Trab.) Malz., op. cit. 399
  = A. sativa × sterilis F_1 hybrid
Avena sterilis ssp. byzantina var. macrotricha subv. culta (Thell.) Malz., op. cit. 400
  = A. sativa L.
Avena sterilis ssp. byzantina var. macrotricha subv. hypomelanathera (Thell.) Malz., op. cit. 399
  = A. sativa L.
Avena sterilis ssp. byzantina var. macrotricha subv. pseudovilis (Hausskn.) Malz., op. cit. 398
  = A. sterilis L.
Avena sterilis ssp. byzantina var. macrotricha subv. solidissima (Thell.) Malz.,

op. cit. 401
  = A. sativa L.
Avena sterilis ssp. byzantina var. pseudosativa (Thell.) Thell., Rec. Trav. Bot. Neerland. 25:433 (1929)
  = A. sativa L.
Avena sterilis ssp. byzantina var. solida (Hausskn.) Malz., Monogr. 396 (1930)
  = A. sterilis L.
Avena sterilis ssp. byzantina var. solida subv. induta (Thell.) Malz., op. cit. 398
  = A. sativa × sterilis F_1 hybrid
Avena sterilis ssp. byzantina var. solida subv. secunda Malz., op. cit. 398
  = A. sativa L.
Avena sterilis ssp. byzantina var. solida subv. segetalis (Nym.) Malz., op. cit. 397
  = A. sativa × sterilis F_1 hybrid
Avena sterilis ssp. byzantina f. pseudosativa Thell., Repert. Sp. Nov. Fedde 13:53 (1913)
  = A. sativa L.
Avena sterilis ssp. byzantina f. solidissima Thell., Naturw. Wochenschr. N.F. 17(32):455 (1918)
  = A. sativa L.; see A. sterilis f. solidissima Thell.
Avena sterilis ssp. ludoviciana (Dur.) Gillet et Magne, Nouv. Fl. Franc. ed. 3, 532 (1873)
  = A. sterilis L.
Avena sterilis ssp. ludoviciana (Dur.) Trab. in Jah. et Maire, Cat. Pl. Maroc 1:49 (1931)
  = A. sterilis L.
Avena sterilis ssp. ludoviciana var. glabrescens (Gren. et Godr.) Thell. Vierteljahrs. Ges. Zürich 56:314 (1912)
  = A. sterilis L.
Avena sterilis ssp. ludoviciana var. glabrescens (Gren. et Godr.) Malz., Monogr. 373 (1930)
  = A. sterilis L.
Avena sterilis ssp. ludoviciana var. glabrescens subv. turkestanica (Regel) Malz., Monogr. 374 (1930)
  = A. sterilis L.
Avena sterilis ssp. ludoviciana var. glabriflora (Malz.) Malz., op. cit. 376
  = A. sterilis L.
Avena sterilis ssp. ludoviciana var. glabriflora subv. basifixa (Malz.) Tab. Mor., Bol. Soc. Brot. Ser. 2, 13:Tab. 1 (1939)
  = A. sativa fatuoid
Avena sterilis ssp. ludoviciana var. lasiathera Thell., Vierteljahrs. Ges. Zürich 56:314 (1912)
  = A. sterilis L.
Avena sterilis ssp. ludoviciana var. media Malz., Monogr. 375 (1930)
  = A. sterilis L.
Avena sterilis ssp. ludoviciana var. media subv. armeniaca Malz., op. cit. 375
  = A. sterilis L.
Avena sterilis ssp. ludoviciana var. psilathera Thell., Vierteljahrs. Nat. Ges. Zürich 56:314 (1912)
  = A. sterilis L.

Avena sterilis ssp. ludoviciana var. subpubescens Tab. Mor., Bol. Soc. Brot. Ser. 2, 13:582 (1939)
   = A. sterilis L.
Avena sterilis ssp. ludoviciana var. typica Malz., Monogr. 365 (1930)
   = A. sterilis L.
Avena sterilis ssp. ludoviciana var. typica subv. hibernans Malz., op. cit. 371
   = ? A. sterilis L.
Avena sterilis ssp. ludoviciana var. typica subv. lasiathera (Thell.) Malz., op. cit. 367
   = A. sterilis L.
Avena sterilis ssp. ludoviciana var. typica subv. leiophylla (Hausskn.) Malz., op. cit. 370
   = A. sterilis L.
Avena sterilis ssp. ludoviciana var. typica subv. macrantha Malz., op. cit. 372
   = A. sterilis L.
Avena sterilis ssp. ludoviciana var. typica subv. micrantha Trab. ex Malz., op. cit. 371
   = A. sterilis L.
Avena sterilis ssp. ludoviciana var. typica subv. nodipilosiuscula Tab. Mor., Bol. Soc. Brot. Ser. 2, 13:581 (1939)
   = A. sterilis L.
Avena sterilis ssp. ludoviciana var. typica subv. psilathera (Thell.) Malz., Monogr. 367 (1930)
   = A. sterilis L.
Avena sterilis ssp. ludoviciana var. typica subv. scabrimicrantha Tab. Mor., Bol. Soc. Brot. Ser. 2, 13:580 (1939)
   = A. sterilis L.
Avena sterilis ssp. ludoviciana var. typica subv. subulifera (Thell.) Malz., Monogr. 371 (1930)
   = A. sterilis L.
Avena sterilis ssp. ludoviciana f. subulifera Thell., Repert. Sp. Nov. Fedde 13:53 (1913)
   = A. sterilis L.
Avena sterilis ssp. macrocarpa (Moench) Briq., Prodr. Fl. Corse 1:105 (1910)
   = A. sterilis L.
Avena sterilis ssp. macrocarpa var. brevipila (Malz.) Malz., Monogr. 393 (1930)
   = A. sterilis L.
Avena sterilis ssp. macrocarpa var. brevipila subv. armeniaca (Malz.) Tab. Mor., Bol. Soc. Brot. Ser. 2, 13:Tab. 1 (1939)
   = A. sterilis L.
Avena sterilis ssp. macrocarpa var. calvescens (Trab. et Thell.) Thell., Vierteljahrs. Nat. Ges. Zürich 56:315 (1912)
   = A. sterilis L.
Avena sterilis ssp. macrocarpa var. calvescens subv. subcalvescens (Malz.) Tab. Mor., Bol. Soc. Brot. Ser. 2, 13:Tab. 1 (1939)
   = A. trichophylla C. Koch
Avena sterilis ssp. macrocarpa var. calviflora Malz., Monogr. 394 (1930)
   = A. trichophylla C. Koch
Avena sterilis ssp. macrocarpa var. maxima (Perez-Lara) Thell., Vierteljahrs. Nat. Ges. Zürich 56:315 (1912)
   = A. atherantha Presl

Avena sterilis ssp. macrocarpa var. pseudovilis (Hausskn.) Thell., op. cit. 315
   = A. sterilis L.
Avena sterilis ssp. macrocarpa var. scabriuscula (Perez-Lara) Thell., op. cit. 315
   = A. sterilis L.
Avena sterilis ssp. macrocarpa var. setosissima Malz., Monogr. 389 (1930)
   = A. sterilis L.
Avena sterilis ssp. macrocarpa var. setosissima subv. aprica (Hausskn.) Malz., op. cit. 392
   = A. sativa × sterilis F_1 hybrid
Avena sterilis ssp. macrocarpa var. setosissima subv. glabrisetigera Tab. Mor., Bol. Soc. Brot. Ser. 2, 13:577 (1939)
   = A. sterilis L.
Avena sterilis ssp. macrocarpa var. setosissima subv. glabrisetigera f. fusca Tab. Mor., loc. cit.
   = A. sterilis L.
Avena sterilis ssp. macrocarpa var. setosissima subv. glabrisetigera f. nigrescens Tab. Mor., loc. cit.
   = A. sterilis L.
Avena sterilis ssp. macrocarpa var. setosissima subv. hirsutimaxima Tab. Mor., op. cit. 576
   = A. atherantha Presl
Avena sterilis ssp. macrocarpa var. setosissima subv. macrantha (Malz.) Tab. Mor., op. cit. Tab. 1
   = A. sterilis L.
Avena sterilis ssp. macrocarpa var. setosissima subv. maxima (Perez-Lara) Malz., Monogr. 389 (1930)
   = A. atherantha Presl
Avena sterilis ssp. macrocarpa var. setosissima subv. scabriuscula (Perez-Lara) Malz., op. cit. 390
   = A. sterilis L.
Avena sterilis ssp. macrocarpa var. setosissima subv. setigera (Malz.) Tab. Mor., Bot. Soc. Brot. Ser. 2, 13:Tab. 1 (1939)
   = A. trichophylla C. Koch
Avena sterilis ssp. macrocarpa var. setosissima subv. subulata (Batt. et Trab.) Malz., Monogr. 392 (1930)
   = A. trichophylla C. Koch
Avena sterilis ssp. macrocarpa var. setosissima subv. subulatisetigera Tab. Mor., Bol. Soc. Brot. Ser. 2, 13:579 (1939)
   = A. sterilis L.
Avena sterilis ssp. macrocarpa var. setosissima subv. trichomaxima Tab. Mor., op. cit. 575
   = A. atherantha Presl
Avena sterilis ssp. macrocarpa var. setosissima subv. trichosubulata Tab. Mor., op. cit. 578
   = A. sterilis L.
Avena sterilis ssp. macrocarpa var. solida (Hausskn.) Thell., Vierteljahrs. Nat. Ges. Zürich 56:316 (1912)
   = A. sterilis L.
Avena sterilis ssp. macrocarpa f. segetalis (Trab.) Thell., Rep. Sp. Nov. Fedde 13:52 (1913)
   = A. sativa × sterilis F_1 hybrid

Avena sterilis ssp. macrocarpa f. triaristata Thell., loc. cit.
   = A. sterilis L.
Avena sterilis ssp. nodipubescens Malz., Monogr. 383 (1930)
   = A. sativa L.
Avena sterilis ssp. nodipubescens var. longiseta Malz., op. cit. 385
   = A. sativa × sterilis F_1 hybrid
Avena sterilis ssp. nodipubescens var. longiseta subv. asperata Malz., loc. cit.
   = A. sativa × sterilis F_1 hybrid
Avena sterilis ssp. nodipubescens var. longiseta subv. diathera Malz., op. cit. 386
   = A. sativa heterozygous fatuoid
Avena sterilis ssp. nodipubescens var. longiseta subv. laevigata Malz., op. cit. 385
   = A. sativa heterozygous fatuoid
Avena sterilis ssp. nodipubescens var. longiseta subv. subculta Malz., op. cit. 386
   = A. sativa L.
Avena sterilis ssp. nodipubescens var. solidiflora Malz., op. cit. 384
   = A. sativa L.
Avena sterilis ssp. nodipubescens var. solidiflora subv. pilosiuscula Malz., op. cit. 385
   = A. sativa L.
Avena sterilis ssp. pseudosativa (Thell.) Malz., op. cit. 376
   = A. sativa L.
Avena sterilis ssp. pseudosativa var. subsolida Malz., op. cit. 378
   = ? A. sativa L.
Avena sterilis ssp. pseudosativa var. subsolida subv. transietissima Malz., op. cit. 378
   = A. sativa L.
Avena sterilis ssp. pseudosativa var. thellungiana Malz., op. cit. 378
   = A. sativa L.
Avena sterilis ssp. trichophylla (C. Koch) Malz., Bull. Appl. Bot. Pl. Breed. 20:143 (1929)
   = A. trichophylla C. Koch
Avena sterilis ssp. trichophylla var. setigera Malz., Monogr. 381 (1930)
   = A. trichophylla C. Koch
Avena sterilis ssp. trichophylla var. setigera f. mauritiana Maire et Weiller, Fl. Afr. Nord 2:284 (1953)
   = A. sterilis L.
Avena sterilis ssp. trichophylla var. subcalvescens Malz. Monogr. 383 (1930)
   = A. trichophylla C. Koch
Avena stipaeformis L., Mant. Pl. 34 (1767)
   = ? Pentaschistis, ? Danthonia, ? Chaetobromus
Avena stipoides (H.B.K.) Scribn., U.S. Dept. Agric. Div. Agrost. Circ. 19:4 (1900)
   = Triniochloa stipoides (H.B.K.) Hitchc., Contr. U.S. Nat. Herb. 17:303 (1913)
Avena stipoides Willd. ex Steud., Nom. Bot. ed. 2, 2:146 (1841)
   = Piptochaetium fimbriatum (H.B.K.) Hitchc., Wash. Acad. Sci. Jour. 23:453 (1933)
Avena striata Lam., Encycl. 1:332 (1783)
   = H. sempervirens (Vill.) Pilger

Avena striata Michx., Fl. Bor. Amer. 1:73 (1803)
 = Schizachne purpurascens (Torr.) Swallen, Wash. Acad. Sci. Jour. 18:204 (1928)
Avena stricta Host, Fl. Austr. 1:127 (1827)
 = Danthonia calycina (Vill.) Rchb., Ic. 1:44 Tab. 103 fig. 1713, 1714 (1834)
Avena strigosa Vogler, Schediasma Bot. 22 (1776)
 = Ventenata dubia (Leers) F. Schultz, Pollich. 20/21:273 (1863)
Avena strigosa Schreb., Spic. Fl. Lips. 52 (1771)
Avena strigosa var. abbreviata Hausskn., Mitt. Thür. Bot. Ver. N.F. 6:44 (1894)
 = A. brevis Roth
Avena strigosa var. abyssinica (Hochst.) Hausskn., op. cit. 45
 = A. abyssinica Hochst.
Avena strigosa var. agraria (Brot.) Sampaio, Bol. Soc. Brot. Ser. 2, 7:116 (1931)
 = A. hispanica Ard.
Avena strigosa var. brevis (Roth) Hausskn., op. cit. 45
 = A. brevis Roth
Avena strigosa var. elatior Kunth, Rev. Gram. 1:103 (1829)
 = A. hispanica Ard.
Avena strigosa var. nuda (L.) Hausskn., op. cit. 45
 = A. nuda L.
Avena strigosa var. sesquialtera (Brot.) Hack., Cat. Gram. Portug. 19 (1880)
 = A. hispanica Ard.
Avena strigosa var. tricholepis Holmb., Bot. Not. 1926:182 (1926)
 = A. strigosa Schreb.
Avena strigosa var. uniflora Hack., Oesterr. Bot. Zeitschr. 27:125 (1877)
 = A. hispanica Ard.
Avena strigosa ssp. abyssinica (Hochst.) Thell., Vierteljahrs. Nat. Ges. Zürich 56:335 (1912)
 = A. abyssinica Hochst.
Avena strigosa ssp. abyssinica var. glaberrima (Chiov.) Thell., op. cit. 336
 = A. abyssinica Hochst.
Avena strigosa ssp. abyssinica var. glaberrima Malz., Monogr. 285 (1930)
 = A. abyssinica Hochst.
Avena strigosa ssp. abyssinica var. glaberrima (Chiov.) Tab. Mor., Bol. Soc. Brot. 2, 12:263 (1937)
 = A. abyssinica Hochst.
Avena strigosa ssp. abyssinica var. pilosiuscula Thell., op. cit. 336
 = A. vaviloviana (Malz.) Mordv.
Avena strigosa ssp. abyssinica var. solidiflora (Thell.) Malz., Monogr. 284 (1930)
 = A. lusitanica (Tab. Mor.) Baum
Avena strigosa ssp. abyssinica var. subglaberrima Malz., op. cit. 285
 = A. abyssinica Hochst.
Avena strigosa ssp. agraria (Brot.) Tab. Mor., Bol. Soc. Brot. Ser. 2, 12:240 (1937)
 = A. hispanica Ard.
Avena strigosa ssp. agraria var. agrisubpilosa Tab. Mor., Bol. Soc. Brot. Ser. 2, 13:640 (1939)
 = A. hispanica Ard.

Avena strigosa ssp. agraria subv. sesquialtera (Brot.) Tab. Mor., Bol. Soc. Brot. Ser. 2, 12:241 (1937)
= A. hispanica Ard.
Avena strigosa ssp. agraria subv. sesquialtera f. albobrevis Vasc. ex Tab. Mor., op cit. 242
= ? A. hispanica Ard.
Avena strigosa ssp. agraria subv. sesquialtera f. nigrescens Vasc. ex Tab. Mor., op. cit. 242
= ? A. hispanica Ard.
Avena strigosa ssp. agraria subv. subbrevis (Malz.) Tab. Mor., op. cit. 241
= A. hispanica Ard.
Avena strigosa ssp. agraria subv. subbrevis f. albula Tab. Mor., op. cit. 241
= A. hispanica Ard.
Avena strigosa ssp. agraria subv. subbrevis f. obscura Tab. Mor., op. cit. 241
= A. hispanica Ard.
Avena strigosa ssp. agraria var. totiglabra Tab. Mor., Bol. Soc. Brot. Ser. 2, 13:640 (1939)
= A. hispanica Ard.
Avena strigosa ssp. agraria var. totiglabra subv. sesquialtera (Brot.) Tab. Mor., op. cit. 641
= A. hispanica Ard.
Avena strigosa ssp. agraria var. totiglabra subv. sesquialtera f. albobrevis (Vasc.) Tab. Mor., op. cit. 642
= ? A. hispanica Ard.
Avena strigosa ssp. agraria var. totiglabra subv. sesquialtera f. nigrescens (Vasc.) Tab. Mor., op. cit. 642
= ? A. hispanica Ard.
Avena strigosa ssp. agraria var. totiglabra subv. subbrevis (Malz.) Tab. Mor., op. cit. 640
= A. hispanica Ard.
Avena strigosa ssp. agraria var. totiglabra subv. subbrevis f. albula (Tab. Mor.) Tab. Mor., op. cit. 641
= A. hispanica Ard.
Avena strigosa ssp. agraria var. totiglabra subv. subbrevis f. obscura (Tab. Mor.) Tab. Mor., op. cit. 640
= A. hispanica Ard.
Avena strigosa ssp. barbata (Pott ex Link) Thell., Vierteljahrs. Nat. Ges. Zürich 56:330 (1912)
= A. barbata Pott ex Link
Avena strigosa ssp. barbata var. solida (Hausskn.) Thell., op. cit. 331
= A. barbata Pott ex Link
Avena strigosa ssp. barbata var. subtypica Malz., Monogr. 275 (1930)
= ? A. wiestii Steud.
Avena strigosa ssp. barbata var. typica Malz., op. cit. 270
= A. barbata Pott ex Link
Avena strigosa ssp. barbata var. typica subv. atherantha (Presl) Malz., op. cit. 274
= A. atherantha Presl
Avena strigosa ssp. barbata var. typica subv. genuina (Willk.) Malz., op. cit. 272
= A. lusitanica (Tab. Mor.) Baum
Avena strigosa ssp. barbata var. typica subv. triflora Malz., op. cit. 271

= A. barbata Pott ex Link
Avena strigosa ssp. brevis (Roth) Hayek, Repert. Sp. Nov. Fedde Beih. 30:321 (1933)
 = A. brevis Roth
Avena strigosa ssp. brevis (Roth) Husnot, Gram. 2:38 (1897)
 = A. brevis Roth
Avena strigosa ssp. brevis var. albobrevis Vasc., Ens. Sem. Melhor. Pl. Bol. No. 20 Ser. A:39 (1935)
 = A. hispanica Ard.
Avena strigosa ssp. brevis var. glabrata (Malz.) Tab. Mor., Bol. Soc. Brot. Ser. 2, 12:244 (1937)
 = ? A. strigosa Schreb.
Avena strigosa ssp. brevis var. glabrata subv. turgida (Vav. ex Malz.) Tab. Mor., loc. cit.
 = A. brevis Roth
Avena strigosa ssp. brevis var. nigrescens Vasc., Ens. Sem. Melhor. Pl. Bot. No. 20 Ser. A:39 (1935)
 = A. hispanica Ard.
Avena strigosa ssp. brevis var. semiglabra (Malz.) Tab. Mor., Bol. Soc. Brot. Ser. 2, 12:244 (1937)
 = A. brevis Roth
Avena strigosa ssp. brevis var. trichophora (Malz.) Tab. Mor., loc. cit.
 = A. brevis Roth
Avena strigosa ssp. glabrescens Marquand, Rep. Bot. Soc. Exch. Club Brit. Isles 6:324 (1922)
 = A. strigosa Schreb.
Avena strigosa ssp. glabrescens var. albida Marquand, op. cit. 324
 = A. strigosa Schreb.
Avena strigosa ssp. glabrescens var. cambrica Marquand, op. cit. 324
 = A. strigosa Schreb.
Avena strigosa ssp. hirtula (Lag.) Malz., Monogr. 247 (1930)
 = A. hirtula Lag.
Avena strigosa ssp. hirtula subv. aristulata Malz., op. cit. 252
 = A. wiestii Steud.
Avena strigosa ssp. hirtula subv. glabrifolia Malz., op. cit. 251
 = A. wiestii Steud.
Avena strigosa ssp. hirtula subv. minor (Lange) Malz., op. cit. 249
 = A. hirtula Lag.
Avena strigosa ssp. hirtula subv. pseudostrigosa Malz., op. cit. 251
 = A. matritensis Baum
Avena strigosa ssp. mandoniana Tab. Mor., Bol. Soc. Brot. Ser. 2, 12:245 (1937)
 = A. brevis Roth
Avena strigosa ssp. mandoniana subv. açoreana Tab. Mor., loc. cit.
 = A. brevis Roth
Avena strigosa ssp. orcadensis Marquand, Rep. Bot. Soc. Exch. Club Brit. Isles 6:324 (1922)
 = A. strigosa Schreb.
Avena strigosa ssp. orcadensis var. flava Marquand, op. cit. 325
 = A. strigosa Schreb.
Avena strigosa ssp. orcadensis var. intermedia Marquand, op. cit. 325
 = A. strigosa Schreb.

Avena strigosa ssp. orcadensis var. nigra Marquand, op. cit. 325
= A. strigosa Schreb.
Avena strigosa ssp. pilosa Marquand, op. cit. 323
= A. strigosa Schreb.
Avena strigosa ssp. pilosa var. alba Marquand, op. cit. 324
= A. strigosa Schreb.
Avena strigosa ssp. pilosa var. fusca Marquand, op. cit. 324
= A. strigosa Schreb.
Avena strigosa ssp. strigosa (Schreb.) Thell., Vierteljahrs. Nat. Ges. Zürich 56:331 (1912)
= A. strigosa Schreb.
Avena strigosa ssp. strigosa prol. brevis (Roth) Thell., op. cit. 332
= A. brevis Roth
Avena strigosa ssp. strigosa prol. brevis var. candida Mordv. in Wulff, Fl. Cult. Pl. 2:429 (1936)
= A. brevis Roth
Avena strigosa ssp. strigosa prol. brevis var. candida subv. euuniflora Mordv., op. cit. 430
= A. brevis Roth
Avena strigosa ssp. strigosa prol. brevis var. candida subv. turgida (Vav.) Mordv., op. cit. 429
= A. brevis Roth
Avena strigosa ssp. strigosa prol. brevis var. glabrata Malz., Monogr. 265 (1930)
= ? A. strigosa Schreb.
Avena strigosa ssp. strigosa prol. brevis var. glabrata subv. secunda Mordv., Tr. Sb. Gen. Selek. Semen. 3:367 (1929)
= A. hispanica Ard.
Avena strigosa ssp. strigosa prol. brevis var. glabrata subv. turgida Vav. in Malz., op. cit. 265
= A. brevis Roth
Avena strigosa ssp. strigosa prol. brevis var. hepatica Mordv. in Wulff, Fl. Cult. Pl. 2:430 (1936)
= Non satis notae
Avena strigosa ssp. strigosa prol. brevis var. nigricans Mordv. in Wulff, op. cit. 430
= A. hispanica Ard.
Avena strigosa ssp. strigosa prol. brevis var. secunda (Mordv.) Mordv. in Wulff, op. cit. 431
= A. hispanica Ard.
Avena strigosa ssp. strigosa prol. brevis var. semiglabra Malz., Monogr. 265 (1930)
= A. brevis Roth
Avena strigosa ssp. strigosa prol. brevis var. tephrea Mordv. in Wulff, Fl. Cult. Pl. 2:430 (1936)
= A. strigosa Schreb.
Avena strigosa ssp. strigosa prol. brevis var. tephrea subv. epruinosa Mordv. in Wulff, loc. cit.
= ? A. strigosa Schreb.
Avena strigosa ssp. strigosa prol. brevis var. tephrea subv. longistrigs Mordv. in Wulff, op. cit. 430
= ? A. strigosa Schreb.

Avena strigosa ssp. strigosa prol. brevis var. tephrea subv. rachipubescens Mordv. in Wulff, op. cit. 430
   = A. strigosa Schreb.
Avena strigosa ssp. strigosa prol. brevis var. trichophora Malz., Monogr. 264 (1930)
   = A. brevis Roth
Avena strigosa ssp. strigosa prol. nuda (L.) Malz., op. cit. 266
   = A. nuda L.
Avena strigosa ssp. strigosa var. alba (Marq.) Mordv. in Wulff, op. cit. 425
   = A. strigosa Schreb.
Avena strigosa ssp. strigosa var. albida (Marq.) Mordv. in Wulff, op. cit. 425
   = A. strigosa Schreb.
Avena strigosa ssp. strigosa var. fusca (Marq.) Mordv. in Wulff, op. cit. 425
   = A. strigosa Schreb.
Avena strigosa ssp. strigosa var. gilva Mordv. in Wulff, Fl. Cult. Pl. 2:428 (1936)
   = A. strigosa Schreb.
Avena strigosa ssp. strigosa var. glabrescens (Marq.) Thell., Rec. Trav. Bot. Neerland 25:435 (1929)
   = A. strigosa Schreb.
Avena strigosa ssp. strigosa var. glabrescens subv. elatior (Roem. et Schult.) Malz., op. cit. 260
   = A. hispanica Ard.
Avena strigosa ssp. strigosa var. glabrescens subv. sesquialtera (Brot.) Malz., op. cit. 260
   = A. hispanica Ard.
Avena strigosa ssp. strigosa var. glabrescens subv. subbrevis Malz., op. cit. 261
   = A. hispanica Ard.
Avena strigosa ssp. strigosa var. glabrescens subv. uniflora (Parl.) Malz., op. cit. 262
   = A. brevis Roth
Avena strigosa ssp. strigosa var. glabrescens subv. unilateralis Malz., op. cit. 260
   = A. strigosa Schreb.
Avena strigosa ssp. strigosa var. glabrescens subv. unispermica Tab. Mor., Bol. Soc. Brot. Ser. 2, 12:239 (1937)
   = A. hispanica Ard.
Avena strigosa ssp. strigosa var. glabrescens subv. unispermica f. lucida Tab. Mor., op. cit. 240
   = A. hispanica Ard.
Avena strigosa ssp. strigosa var. glabrescens subv. unispermica f. nigra Tab. Mor., op. cit. 240
   = A. hispanica Ard.
Avena strigosa ssp. strigosa var. glabrescens subv. unispermica f. nigella Tab. Mor., Bol. Soc. Brot. Ser. 2, 13:637 (1939)
   = A. hispanica Ard.
Avena strigosa ssp. strigosa var. glabrescens f. albida (Marq.) Tab. Mor., Bol. Soc., Brot. Ser. 2, 12:239 (1937)
   = A. strigosa Schreb.
Avena strigosa ssp. var. glabrescens f. cambrica (Marq.) Tab. Mor., loc. cit.
   = A. strigosa Schreb.

Avena strigosa ssp. strigosa var. intermedia (Marq.) Mordv. in Wulff, Fl. Cult. Pl. 2:425 (1936)
: = A. strigosa Schreb.
Avena strigosa ssp. strigosa var. kewensis Vav. in Wulff, Fl. Cult. Pl. 2:426 (1936)
: = A. hispanica Ard.
Avena strigosa ssp. strigosa var. melanocarpa Mordv. in Wulff, op. cit. 428
: = ? A. hispanica Ard.
Avena strigosa ssp. strigosa var. nigra (Marq.) Mordv. in Wulff, op. cit. 425
: = A. strigosa Schreb.
Avena strigosa ssp. strigosa var. nuda (L.) Tab. Mor., Bol. Soc. Brot. Ser. 2, 13:639 (1939)
: = A. nuda L.
Avena strigosa ssp. strigosa var. orcadensis (Marq.) Thell., Rec. Trav. Bot. Neerland. 25:435 (1929)
: = A. strigosa Schreb.
Avena strigosa ssp. strigosa var. solida (Hausskn.) Malz., Monogr. 256 (1930)
: = A. barbata Pott ex Link
Avena strigosa ssp. strigosa var. solida subv. tricholepis (Holmb.) Malz., op. cit. 256
: = A. strigosa Schreb.
Avena strigosa ssp. strigosa var. subpilosa Malz., op. cit. 257
: = A. strigosa Schreb.
Avena strigosa ssp. strigosa var. subpilosa subv. orcadensis (Marq.) Malz., op. cit. 257
: = A. strigosa Schreb.
Avena strigosa ssp. strigosa var. subpilosa subv. tricholepis (Holmb.) Tab. Mor., Bol. Soc. Brot. Ser. 2, 13:Tab. 1 (1939)
: = A. strigosa Schreb.
Avena strigosa ssp. strigosa var. tricholepis (Holmb.) Thell., Rec. Trav. Bot. Neerland. 25:434 (1929)
: = A. strigosa Schreb.
Avena strigosa ssp. strigosa var. typica Vav. ex Mordv. in Wulff, Fl. Cult. Pl. 2:427 (1936)
: = A. strigosa Schreb.
Avena strigosa ssp. strigosa var. typica subv. elatior f. divaricata Mordv., loc. cit.
: = A. hispanica Ard.
Avena strigosa ssp. strigosa var. typica subv. sesquialtera (Brot.) Mordv., loc. cit.
: = A. hispanica Ard.
Avena strigosa ssp. strigosa var. unilateralis (Malz.) Mordv. in Wulff, Fl. Cult. Pl. 2:428 (1936)
: = A. strigosa Schreb.
Avena strigosa ssp. vaviloviana Malz., Bull. Appl. Bot. Pl. Breed. 20:138 (1929)
: = A. vaviloviana (Malz.) Mordv.
Avena strigosa ssp. vaviloviana var. glabra (Hausskn.) Malz., op. cit. 280
: = A. abyssinica × vaviloviana F_1 hybrid
Avena strigosa ssp. vaviloviana var. intercedens (Thell.) Malz., op. cit. 282
: = A. abyssinica × vaviloviana F_1 hybrid
Avena strigosa ssp. vaviloviana var. pilosiuscula (Thell.) Malz., op. cit. 281

= A. vaviloviana (Malz.) Mordv.
Avena strigosa ssp. vaviloviana var. pseudoabyssinica (Thell.) Malz., op. cit. 280
　　= A. abyssinica Hochst.
Avena strigosa ssp. wiestii (Steud.) Thell., Vierteljahrs. Nat. Ges. Zürich 56:333 (1912)
　　= A. wiestii Steud.
Avena strigosa ssp. wiestii var. glabra (Hausskn.) Thell., op. cit. 334
　　= A. abyssinica × vaviloviana F_1 hybrid
Avena strigosa ssp. wiestii var. intercedens Thell., op. cit. 334
　　= ? A. abyssinica × vaviloviana F_1 hybrid
Avena strigosa ssp. wiestii var. pseudoabyssinica Thell., op. cit. 334
　　= A. abyssinica Hochst.
Avena strigosa ssp. wiestii var. solidiflora Thell., op. cit. 335
　　= A. lusitanica (Tab. Mor.) Baum
Avena strigosa ssp. wiestii subv. caspica (Hausskn.) Malz., Monogr. 278 (1930)
　　= A. barbata Pott ex Link
Avena strigosa ssp. wiestii subv. deserticola Malz., op. cit. 277
　　= A. wiestii Steud.
Avena subalpestris Hartm., Handb. ed. 3, 4:29 (1843)
　　= ? Trisetum agrostideum (Laest.) Fries, Mant. 3:180 (1842)
Avena subcylindrica Ehrenb. ex Boiss., Fl. Orient. 5:536 (1884)
　　= Trisetum glumaceum Boiss., loc. cit.
Avena subspicata Clairv., Man. Herbor. 17 (1811)
　　= Trisetum spicatum (L.) Richt., Pl. Eur. 1:59 (1890)
Avena subulata Lam., Tabl. Encycl. 1:201 (1791)
　　= Arrhenatherum setaceum (Vill.) Potztal, Bot. Jahrb. 75:330 (1951)
　　= ? H. setaceum (Vill.) Henrard
Avena subvillosa Schur, Enum. Pl. Trassilv. 763 (1866)
　　= ? H. convolutum (Presl) Henrard
Avena suffusca Hitchc., Proc. Biol. Soc. Washington 43:95 (1930)
　　= H. suffuscum (Hitchc.) Ohwi
Avena sulcata J. Gay ex Delastre, Fl. Anal. Descr. Dept. Vienne 477, Tab. 4 (1842)
　　= H. sulcatum (J. Gay) Henrard
　　= Aven. sulcatum (J. Gay) Vierh.
　　= Aven. sulcatum (J. Gay) Hubbard and Sandw.
　　= Av. sulcata (J. Gay) Dum.
　　= Aveno. sulcata (J. Gay) Holub
Avena sylvatica Salisb., Prodr. Stirp. 24 (1796)
　　= Holcus mollis L., Syst. Nat. ed. 10, 2:1305 (1759)
Avena symphicarpa Trin. ex Steud., Nom. Bot. ed. 2, 1:173 (1840)
　　= H. hirtulum (Steud.) Schweickerdt
Avena syriaca Boiss. et Bal. ex Boiss., Fl. Orient. 5:542 (1884)
　　= A. sativa × sterilis F_1 hybrid
Avena tartarica Ard. ex Saggi Accad. Padov. 2:101 Tab. 1 (1789)
　　= A. sativa L.
Avena taygetana Steud., Syn. Pl. Glum. 1:233 (1854)
　　= H. filifolium (Lag.) Henrard
Avena tenera (Kit.) Nyman, Syll. 414 (1854)
　　= ? Aira tenera Kit. in Schult., Oestr. Fl. ed. 2, 1:199 (1814)

Avena tenorii Nyman, Consp. Fl. Eur. 814 (1882)
 = Aira pulchella (Beauv.) Link, Hort. Berol. 1:130 (1827)
Avena tentoensis Honda, Rep. First Sci. Exped. Manchoukuo, Bot. 4(2):7 (1935)
 = Aven. tentoense (Honda) Kitagawa
 = H. tentoense (Honda) Kitagawa
Avena tenuiflora Bertol. ex Reichenb., Icon. Fl. Germ. Helv. 1:41 n. 1688 pl. 97 (1834)
 = ? Trisetum sp.
Avena tenuis Moench, Meth. Pl. 195 (1794)
 = Ventenata dubia (Leers) F. Schultz, Pollich. 20, 21:273 (1863)
Avena thellungii Nevski, Acta Univ. Asiae Med. 8b. Bot. 17:6 (1934)
 = A. sativa L.
Avena thorei Duby in DC., Gall. 1:512 (1828)
 = Aven. longifolium (Thore) G. Sampaio
 = Pseudarrhenatherum longifolium (Thore) Rouy, Bull. Soc. Bot. France 68:401 (1921)
 = Arrhenatherum thorei (Duby) Desv., Cat. Dord. 153 (1840)
Avena tibestica Bruneau de Miré et Quézel, Bull. Soc. Bot. France 106:135 (1958)
 = H. tibesticum (Bruneau de Miré et Quézel) Holub
Avena tibetica Roshev., Bull. Jard. Bot. Princ. U.R.S.S. 27:98 (1928)
 = H. tibeticum (Roshev.) P. C. Keng
Avena tolucensis H.B. et K., Nov. Gen. et Sp. 1:148 (1815)
 = Trisetum toluccense (H.B. et K.) Kunth, Rev. Gram. 1:101 (1829)
 = Trisetum spicatum (L.) Richt., Pl. Eur. 1:59 (1890)
Avena torreyi Nash in Britt. et Brown, Illustr. Fl. ed. 2, 1:219 (1913)
 = Schizachne purpurascens (Torr.) Swallen, Wash. Acad. Sci. Jour. 18:204 (1928)
Avena trabutiana Thellung, Rep. Sp. Nov. Fedde 13:53 (1913)
 = A. sativa L.
Avena triaristata Vill., Hist. Pl. Dauph. 2:148 (1787)
 = ? Ventenata dubia (Leers) F. Schultz, Pollich. 20, 21: 273 (1863)
Avena trichophylla C. Koch, Linnaea 21:393 (1848)
Avena trichopodia Presl, Rel. Haenk. 1:254 (1830)
 = Trisetum deyeuxioides (H.B.K.) Kunth, Rev. Gram. 102 (1829)
Avena triseta Thunb., Prodr. Fl. Capensis 22 (1794)
 = Pentaschistis triseta (Thunb.) Stapf in Thiselton-Dyer, Fl. Capensis 7:495 (1899)
Avena trisperma Auct. ex Roem. et Schult., Syst. Veg. 2:669 (1817)
 = A. sativa L.
Avena trisperma Auct. ex Schuebl., Diss. Inaug. Bot. 8 (1825)
 = A. sativa L.
Avena triticoides (Lindl. ex Mitchell) Jackson, Index Kew. 254 (1895)
 = Astrebla triticoides (Lindl. ex Mitch.) Muell. ex Bentham, Fl. Austr. 7:602 (1878)
Avena truncata Dulac, Fl. Hautes-Pyr. 83 (1867)
 = H. bromoides (Gouan) Hubb.
Avena tuberosa Gilib., Exerc. Phyt. 2:538 (1792)
 = Arrhenatherum elatius (L.) Presl, Fl. Cech. 17 (1819)
Avena turgidula Stapf, Kew Bull. Misc. Inf. 1897:293 (1897)
 = Aven. turgidulum (Stapf) Stapf

   = H. turgidulum (Stapf) Schweickerdt
   = Arrhenatherum turgidulum (Stapf) Potztal, Bot. Jahrb. 75:328 (1951)
Avena turonensis Toulet, Cat. Pl. Vasc. Indre et Loire 568 (1908)
   = A. sterilis L.
Avena uniflora Parlat., Pl. Nov. 84 (1842)
   = A. brevis Roth
Avena unilateralis Brouss. ex Roem. et Schult., Syst. Veg. 2:669 (1817)
   = A. sativa L.
Avena valesiaca Nyman, Syll. 414 (1854)
   = ? Trisetum gaudinianum Boiss., Voy. 2:652 (1844)
   = Lophochloa cavanillesii (Trin.) Bor
Avena varia (Haenke) Raspail, Ann. Sci. Nat. 5:439 (1825)
   = Festuca varia Haenke in Jacq., Collect. Bot. 2:94 (1788)
Avena varia Schur, Verh. Siebenb. Ver. Naturw. 4:85 (1853)
   = Trisetum carpaticum (Host) Roem. et Schult., Syst. Veg. 2:663 (1817)
Avena vasconica Sennen ex St.-Yves, Candollea 4:456 (1931)
   = H. pratense (L.) Pilger
Avena vaviloviana (Malz.) Mordv. in Wulff, Fl. Cult. Pl. 2:422 (1936)
Avena vaviloviana var. glabra (Hausskn.) C. E. Hubbard in Hill, Fl. Trop. Afr. 10:119 (1937)
   = A. abyssinica $\times$ vaviloviana F_1 hybrid
Avena vaviloviana var. intercedens (Thell.) C. E. Hubbard in Hill, op. cit. 120
   = A. abyssinica $\times$ vaviloviana F_1 hybrid
Avena vaviloviana var. pilosiuscula (Thell.) C. E. Hubbard in Hill, op. cit. 119
   = A. vaviloviana (Malz.) Mordv.
Avena vaviloviana var. pseudabyssinica (Thell.) C. E. Hubbard in Hill, op. cit. 119
   = A. abyssinica Hochst.
Avena velutina (Boiss.) Nyman, Syll. 413 (1854)
   = Trisetum velutinum Boiss., Voy. 2:653 (1844)
Avena ventricosa Bal. ex Coss. Bull. Soc. Bot. France 1:14 (1854)
Avena ventricosa ssp. bruhnsiana (Gruner) Malz., Monogr. 242 (1930)
   = A. ventricosa Bal. ex Coss.
Avena ventricosa ssp. ventricosa (Bal. ex Coss.) Malz., op. cit. 241
   = A. ventricosa Bal. ex Coss.
Avena verna Heuzé, Pl. Alim. 1:504 (1873)
   = A. sativa L.
Avena versicolor Vill., Hist. Pl. Dauphiné 142 (1787)
   = Aven. versicolor (Vill.) Fritsch
   = H. versicolor (Vill.) Pilger
   = Aveno. versicolor (Vill.) Holub
Avena vilis Wallr., Linnaea 14:543 (1840)
   = ? A. hybrida Peterm. or ? A. fatua L.
Avena villosa Bertol., Excerpta de Re Herbaria 6 (1820)
   = Trisetum villosum (Bert.) Schult., Mant. 2:368 (1824)
Avena villosa Lag. ex Roem. et Schult., Syst. Veg. 2:670 (1817)
   = A. matritensis Baum
Avena virescens (Regel) Regel, Acta Horti Petrop. 7:635 (1881)
   = Trisetum fedtschenkoi Henrard, Blumea 3:425 (1940)
   = ? Trisetum spicatum (L.) Richt., Pl. Eur. 1:59 (1890)
Avena viridis H.B.K., Nov. Gen. Sp. 1:147 (1815)
   = Trisetum viride (H.B.K.) Kunth, Rev. Gram. 101 (1829)

Avena volgensis (Vav.) Nevski, Acta Univ. Asiae Med. 8b, Bot. 17:5 (1934)
= A. sativa L.

Avena wiestii Steud., Syn. Pl. Glum. 1:231 (1854)

Avena wiestii var. glabra Hausskn., Mitt. Thür., Bot. Ver. N.F. 13/14:49 (1899)
= A. abyssinica × vaviloviana F_1 hybrid

Avena wiestii var. glabra Nabelek, Publ. Fac. Sci. Univ. Masaryk (Brno) No. 111:10 (1929)
= A. barbata Pott ex Link

Avena wiestii var. pseudoabyssinica Thell., Vierteljahrs. Nat. Ges. Zürich 56:307 (1912)
= A. abyssinica Hochst.

Avena wiestii var. solida Hausskn., Mitt. Thür., Bot. Ver. N.F. 6:42 (1894)
= A. lusitanica (Tab. Mor.) Baum

Avena wilhelmsii (Ledeb.) Spreng., Syst. Veg. 1:333 (1825)
= ? Trisetum rigidum (M.B.) Roem. et Schult., Syst. Veg. 2:662 (1817)

Avenastrum adsurgens Schur, Prodan Fl. Determ. Descr. Romania 11 (1923)
= H. alpinum (Smith) Henrard

Avenastrum adsurgens (Schur) Javorka, Magyar Fl. 1:81 (1924)
= H. alpinum (Smith) Henrard

Avenastrum adzharicum (Alboff) Roshev in Komarov Fl. URSS 2:274 (1934)
= H. adzharicum (Alboff) Henrard
= Aveno. adzharica (Alboff) Holub

Avenastrum aetolicum K.H. Rechinger, Beih. Bot. Centralbl. Abt. B, 54:680 (1936)
= H. aetolicum (K.H. Rech.) Holub
= Aveno. aetolica (K.H. Rech.) Holub

Avenastrum agropyroides (Boiss.) Halácsy, Consp. Fl. Graec. 3:370 (1904)
= H. agropyroides (Boiss.) Henrard
= Aveno. agropyroides (Boiss.) Holub

Avenastrum albinerve (Boiss.) C. E. Hubbard et Sandwith, Kew Bull. 1928:154 (1928)
= H. albinerve (Boiss.) Henrard
= Aveno. albinervis (Boiss.) Holub

Avenastrum albinerve (Boiss.) Vierh., Verh. Ges. Deutsch. Naturf. Leipzig 85:672 (1914)
= H. albinerve (Boiss.) Henrard
= Aveno. albinervis (Boiss.) Holub

Avenastrum alpinum (Smith) Fritsch, Excursfl. Oesterr. 53 (1897)
= H. alpinum (Smith) Henrard
= Aveno. alpina (Smith) Holub

Avenastrum antarcticum (Thunb.) Stapf in Thiselton-Dyer, Fl. Capensis 7:476 (1899)
= H. capense Schweickerdt and H. leoninum (Steud.) Schweickerdt

Avenastrum armeniacum (Schisk.) Roshev. in Komarov Fl. URSS 2:274 (1934)
= H. armeniacum (Schisk.) Henrard
= Aveno. armeniaca (Schisk.) Holub

Avenastrum asiaticum Roshev. Bull. Jard. Bot. Acad. Sci. URSS 30:772 (1932)
= H. asiaticum (Roshev.) Henrard
= H. versicolor (Vill.) Pilger prol. caucasica Holub

  = Aveno. asiatica (Roshev.) Holub
Avenastrum asperum (Munro ex Thwaites) Vierh., Verh. Ges. Deutsch. Naturf. Leipzig 85:672 (1914)
  = H. virescens (Nees) Henrard
  = H. asperum (Munro) Bor
Avenastrum ausserdorferi (Aschers. et Graebn.) Fritsch, Excursfl. Oesterr. ed. 2:58 (1909)
  = H. alpinum (Smith) Henrard
Avenastrum australe (Parl.) Halácsy, Consp. Fl. Graec. 3:370 (1904)
  = H. australe (Parl.) Holub
  = H. compressum (Heuff.) Potztal
Avenastrum avenoides Stapf in Salisb., Ind. Kew. Suppl. 11:109 (1953)
  = H. avenoides (Stapf) Camus
Avenastrum besseri (Griseb.) Koczwara in Dostal, Květena ČSR ed. 2:2021 (1950)
  = H. besseri (Griseb.) Janchen
Avenastrum blavii (Aschers. et Janka) Beck, Ann. Naturhist. Hofmus. Wien 5:561 (1890)
  = H. blavii (Aschers. et Janka) Henrard
  = Aveno. blavii (Aschers. et Janka) Holub
Avenastrum bromoides (Gouan) Vierh., Verh. Ges. Deutsch. Naturf. Leipzig 85:672 (1914)
  = H. bromoides (Gouan) Henrard
  = Aveno. bromoides (Gouan) Holub
Avenastrum bromoides (Gouan) C. E. Hubbard et Sandwith, Kew. Bull. Misc. Inf. 1928:154 (1828)
  = H. bromoides (Gouan) Henrard
Avenastrum caffrum (Stapf) Stapf in Thiselton-Dyer, Fl. Capensis 7:477 (1899)
  = H. longifolium (Nees) Schweickerdt
Avenastrum caryophylleum (L.) Jess., Deutschl. Gräs. 118 (1863)
  = Aira caryophyllea L., Sp. Pl. 66 (1753)
Avenastrum compactum (Boiss. et Heldr.) Halácsy, Consp. Fl. Graec. 3:370 (1904)
  = H. compactum (Boiss. et Heldr.) Henrard
  = Arrhenatherum neumayerianum (Vis.) Potztal. Bot. Jahrb. 75:330 (1951)
  = D. compactum (Boiss. et Heldr.) Holub
Avenastrum compressum (Heuff.) Vierh., Verh. Ges. Deutsch. Naturf. Leipzig 85:672 (1914)
  = H. compressum (Heuff.) Henrard
  = Aveno. compressa (Heuff.) Holub
Avenastrum compressum (Heuff.) Deg. in Jav. Magyar Fl. 1:81 (1924)
  = H. compressum (Heuff.) Henrard
  = Aveno. compressa (Heuff.) Holub
Avenastrum conjugens (Hack.) Gayer, Neue Beitr. Fl. Komitates Vas. III, 1:7–8 (1932)
  = H. conjugens (Hack.) Wilder
Avenastrum convolutum (Presl) Halácsy, Consp. Fl. Graec. 3:369 (1904)
  = H. convolutum (Presl) Henrard
  = Arrhenatherum filifolium (Lag.) Potztal, Bot. Jahrb. 75:329 (1951)

Avenastrum cycladum Rech. et Scheff., Magyar Bot. Lapok 33:20 (1934)
= H. cycladum (Rech. et Scheff.) Rech.
= Aveno. cycladum (Rech. et Scheff.) Holub
Avenastrum dahuricum (Kom.) Roshev. in Kom., Fl. URSS 2:275 (1934)
= H. dahuricum (Kom.) Henrard
= Aveno. dahurica (Kom.) Holub
Avenastrum decorum (Janka) Vierh., Verh. Ges. Deutsch. Naturf. Leipzig 85:672 (1914)
= H. decorum (Janka) Henrard
= Arrhenatherum decorum (Janka) Potztal, Bot. Jahrb. 75:329 (1951)
Avenastrum desertorum (Less.) Podpera, Bot. Jahrb. Engler 34: Beibl. 76:9 (1904)
= H. desertorum (Less.) Pilger
= Arrhenatherum desetorum (Less.) Potztal, Bot. Jahrb. 75:329 (1951)
Avenastrum dodii Stapf in Thiselton-Dyer, Fl. Capensis 7:475 (1899)
= H. dodii (Stapf) Schweickerdt
= Arrhenatherum dodii (Stapf) Potztal, Bot. Jahrb. 75:328 (1951)
Avenastrum dregeanum (Steud.) Stapf, in Thiselton-Dyer, Fl. Capensis 7:473 (1899)
= H. namaquense Schweickerdt
Avenastrum elatius (L.) Jess., Deutschl. Gräser und Getreidearten 53 f. 97. 52 (1863)
= Arrhenatherum elatius (L.) Presl, Fl. Cech. 17 (1819)
Avenastrum elongatum (Hochst.) Pilger, Notzibl. Bot. Gart. Berlin 9:518 (1926)
= H. elongatum (Hochst.) Hubbard
= Arrhenatherum elongatum (Hochst.) Potztal, Bot. Jahrb. 75:328 (1951)
Avenastrum fedtschenkoi (Hack.) Roshev. in Komarov, Fl. U.R.S.S. 2:280 (1934)
= H. fedtschenkoi (Hack.) Henrard
Avenastrum filifolium (Lag.) Fritsch, Excurs fl. Oesterr. 54 (1897)
= H. filifolium (Lag.) Henrard
= Arrhenatherum filifolium (Lag.) Potztal, Bot. Jahrb. 75:329 (1951)
Avenastrum flabellatum Peter, Rep. Sp. Nov. Fedde Beih. 40: Anhang 98 (1930)
= ? Diplachne
Avenastrum flavescens Jess., Deutschl. Gräs. 21, 54, 119, 215 (1863)
= Trisetum
Avenastrum hackelii (Henriq.) Vierh., Verh. Ges. Deutsch. Naturf. Leipzig 85:672 (1914)
= H. hackelii (Henriq.) Henrard
= Aveno. hackelii (Henriq.) Holub
Avenastrum hissaricum Roshev., Bull. Jard. Bot. Acad. Sci. U.R.S.S. 30:772 (1932)
= H. hissaricum (Roshev.) Henrard
Avenastrum humbertii A. Camus, Bull. Soc. Bot. France 78:9 (1931)
= H. humbertii (A. Camus) Henrard
Avenastrum insubricum (Aschers. et Graebn.) Fritsch, Exkursionfl. ed. 3:688 (1922)
= ? H. pubescens (Huds.) Pilger

Avenastrum krylovii Pavlov, Animadvers. Syst. Herb. Univ. Tomsk No. 5–6:1 (1933)
 = H. krylovii (Pavlov) Henrard
Avenastrum lachnanthum (Hochst. ex A. Rich.) Vierh., Verh. Ges. Deutsch. Naturf. Leipzig 85:672 (1914)
 = H. lachnanthum (Hochst. ex A. Rich.) Hubbard
 • = Arrhenatherum lachnanthum (Hochst. ex A. Rich.) Potztal, Bot. Jahrb. 75:328 (1951)
Avenastrum lachnanthum (Hochst. ex A. Rich.) Pilger, Notizbl. Bot. Gart. Berlin 9:521 (1926)
 = see above: Aven. lachanthum (Hochst. ex A. Rich.) Vierh.
Avenastrum laeve (Hackel) Vierh., loc. cit.
 = H. laeve (Hackel) Potztal
Avenastrum laevigatum (Schur) Domin, Preslia 13–15:40 (1935)
 = H. laevigatus (Schur) Potztal
Avenastrum longifolium (Thore) Sampaio, Bol. Soc. Brot. Ser. 2, 7:119 (1931)
 = Pseudarrhenatherum longifolium (Thore) Rouy, Bull. Soc. Bot. France 68:401 (1921)
Avenastrum longum (Stapf) Stapf in Thiselton-Dyer, Fl. Capensis 7:473 (1899)
 = H. longum (Stapf) Schweickerdt
 = Arrhenatherum longum (Stapf) Potztal, Bot. Jahrb. 75:328 (1951)
Avenastrum majus Pilger, Notizbl. Bot. Gart. Berlin 9:519 (1926)
 = Arrhenatherum milanjianum (Rendle) Potztal, Bot. Jahrb. 75:329 (1951)
 = H. milanjianum (Rendle) Hubbard
Avenastrum mannii Pilger, Notizbl. Bot. Gart. Berlin 9:520 (1926)
 = H. mannii (Pilger) C. E. Hubbard
 = Arrhenatherum mannii (Pilger) Potztal, Bot. Jahrb. 75:329 (1951)
Avenastrum mongolicum (Roshev.) Roshev. in Komarov, Fl. U.R.S.S. 2:280 (1934)
 = H. mongolicum (Roshev.) Henrard
Avenastrum montanum (Vill.) Vierh., Verh. Ges. Deutsch. Naturf. Leipzig 85:672 (1914)
 = H. montanum (Vill.) Henrard
 = Av. montana (Vill.) Dum.
 = Arrhenatherum montanum (Vill.) Potztal, Bot. Jahrb. 75:329 (1951)
Avenastrum nervosum Vierh., Verh. Ges. Deutsch. Naturf. Leipzig 85:672 (1914)
 = Amphibromus neesii Steud., Syn. Pl. Glum. 1:328 (1854)
Avenastrum neumayerianum (Vis.) Beck, Wiss. Mitt. Bosn. Herzegow. 9:434 (1904)
 = H. neumayerianum (Vis.) Henrard
 = Arrhenatherum neumayerianum (Vis.) Potztal, Bot. Jahrb. 75:330 (1951)
 = D. compactum (Boiss. et Helder.) Holub
Avenastrum pallens (Link) Sampaio, Bol. Soc. Brot. Ser. 2, 7:119 (1931)
 = Pseudoarrhenatherum pallens (Link) Holub, Taxon 15:167 (1966)
Avenastrum parlatorii (Woods) Beck, Fl. Nieder.-Oesterr. 1:73 (1890)
 = Arrhenatherum parlatorei (Woods) Potztal, Bot. Jahrb. 75:329 (1951)
 = H. parlatorei (Woods) Pilger

Avenastrum planiculme (Schrad.) Opiz, Sezn. Rostl. České 20 (1852)
 = H. planiculme (Schrad.) Besser ex Rechb.
 = Aveno. planiculmis (Schrad.) Holub
Avenastrum praecox Jess., Deutschl. Gräs. 218 (1863)
 = Aira praecox L., Sp. Pl. 65 (1753)
Avenastrum pratense (L.) Opiz, Seznam Rostl. České 20 (1852)
 = Av. pratense (L.) Dum.
 = H. pratense (L.) Pilger
 = Arrhenatherum pratense (L.) Sampaio, An. Fac. Sci. Porto 7:45 (1931)
 = Aveno. pratensis (L.) Holub
Avenastrum pseudoviolaceum (Kern.) Fritsch, Excursfl. Oesterr. 53 (1897)
 = ? H. alpinum (Smith) Henrard
Avenastrum pubescens (Huds.) Opiz, Seznam Rostl. České 20 (1852)
 = Heuf. pubescens (Huds.) Schur
 = Arrhenatherum pubescens (Huds.) Sampaio, An. Fac. Sci. Porto 17:45 (1931)
 = H. pubescens (Huds.) Pilger
 = Av. pubescens (Huds.) Dum.
 = Aveno. pubescens (Huds.) Holub
Avenastrum quinquenerve Steut and Rattray, Proc. Rhodesia Sci. Ass. 32:42 (1933)
 = H. elongatum (Hochst.) C. E. Hubbard
 = Arrhenatherum elongatum (Hochst.) Potztal, Bot. Jahrb. 75:328 (1951)
Avenastrum quinquesetum (Steud.) Stapf in Thiselton-Dyer, Fl. Capensis 7:474 (1899)
 = H. quinquesetum (Steud.) Schweickerdt
 = Arrhenatherum quinquesetum (Steud.) Potztal, Bot. Jahrb. 75:328 (1951)
Avenastrum rigidulum Pilger, Notizbl. Bot. Gart. Berlin 9:519 (1926)
 = Arrhenatherum rigidulum (Pilger) Potztal, Bot. Jahrb. 75:329 (1951)
 = H. rigidulum (Pilger) C. E. Hubbard
Avenastrum sarracenorum (Gandog.) C. E. Hubbard et Sandwith, Kew Bull. 1928:154 (1928)
 = H. sarracenorum (Gandog.) Holub
Avenastrum schellianum (Hack.) Podp., Publ. Fac. Sci. Univ. Masaryk 27:54 (1923)
 = H. schellianum (Hack.) Kitagawa
 = Aveno. schelliana (Hack.) Holub
Avenastrum sempervirens (Vill.) Vierh., Verh. Ges. Deutsch. Naturf. Leipzig 85:672 (1914)
 = H. sempervirens (Vill.) Pilger
Avenastrum sesquitertium (L.) Fritsch, Excursfl. Oesterr. 53 (1897)
 = ? H. pubescens (Huds.) Pilger
Avenastrum setaceum (Vill.) Vierh., Verh. Ges. Deutsch. Naturf. Leipzig 85:672 (1914)
 = Av. setacea (Vill.) Dum.
 = H. setaceum (Vill.) Henrard
 = Arrhenatherum setaceum (Vill.) Potztal, Bot. Jahrb. 75:330 (1951)
Avenastrum subdecurrens Degen ex Podpera, Acta Soc. Sc. Nat. Morav. 2 (Fasc. 10):594 (1926)

 = H. pratense (L.) Pilger
Avenastrum sulcatum (J. Gay) Vierh., Verh. Ges. Deutsch. Naturf. Leipzig 85:672 (1914)
 = H. sulcatum (J. Gay) Henrard
 = Av. sulcata (J. Gay) Dum.
 = Aveno. sulcata (J. Gay) Holub
Avenastrum sulcatum (J. Gay) Hubbard and Sandw., Kew Bull. Misc. Inf. 1928:155 (1928)
 = see above: Aven. sulcatum (J. Gay) Vierh.
Avenastrum tentoense (Honda) Kitagawa, Rep. Inst. Sc. Res. Manchoukuo 2:282 (1938)
 = H. tentoense (Honda) Kitagawa
Avenastrum tianschanicum Roshev., Bull. Jard. Bot. Acad. Sci. U.R.S.S. 30:773 (1932)
 = H. tianschanicum (Roshev.) Henrard
Avenastrum trisetoides Kitagawa, Rep. Inst. Sc. Res. Manchoukuo 2:282 (1938)
 = H. trisetoides (Kitagawa) Kitagawa
Avenastrum turgidulum (Stapf) Stapf in Thiselton-Dyer, Fl. Capensis 7:474 (1899)
 = Arrhenatherum turgidulum (Stapf) Potztal, Bot. Jahrb. 75:328 (1951)
 = H. turgidulum (Stapf) Schweickerdt
Avenastrum umbrosum (Hochst. et Steudel) Pilger, Notizbl. Bot. Gart. Berlin 9:521 (1926)
 = H. umbrosum (Hochst. ex Steudel) C. E. Hubbard
 = Arrhenatherum umbrosum (Hochst. ex Steud.) Potztal, Bot. Jahrb. 75:328 (1951)
Avenastrum versicolor (Vill.) Fritsch, Excursfl. Oesterr. 53 (1897)
 = H. versicolor (Vill.) Pilger
 = Aveno. versicolor (Vill.) Holub
Avenastrum vierhapperi Adamovic in Rad. Jugosl. Akad. 195:93 (1913)
 = ? Avenochloa
Avenula hostii (Boiss. et Reut.) Dum., Bull. Soc. Roy. Bot. Belg. 7:68 (1868)
 = Arrhenatherum parlatorei (Woods) Potztal, Bot. Jahrb. 75:329 (1951)
Avenula montana (Vill.) Dum., Bull. Soc. Roy. Bot. Belg. 7:68 (1868)
 = H. montanum (Vill.) Henrard
 = Arrhenatherum montanum (Vill.) Potztal, Bot. Jahrb. 75:329 (1951)
Avenula pratensis (L.) Dum., Bull. Soc. Roy. Bot. Belg. 7:68 (1868)
 = Aven. pratense (L.) Opiz
 = H. pratense (L.) Pilger
 = Arrhenatherum pratense (L.) Sampaio, An. Fac. Sci. Porto 7:45 (1931)
Avenula pubescens (Huds.) Dum., Bull. Soc. Roy. Bot. Belg. 7:68 (1868)
 = Aven. pubescens (Huds.) Opiz
 = Heuf. pubescens (Huds.) Schur
 = Arrhenatherum pubescens (Huds.) Sampaio, An. Fac. Sci. Porto 17:45 (1931)
 = H. pubescens (Huds.) Pilger
Avenula scheuchzeri (All.) Dum., Bull. Soc. Roy. Bot. Belg. 7:68 (1868)
 = H. versicolor (Vill.) Pilger

Avenula sempervirens (Vill.) Dum., Bull. Soc. Roy. Bot. Belg. 7:68 (1868)
: = Aven. sempervirens (Vill.) Vierh.
: = H. sempervirens (Vill.) Pilger

Avenula setacea (Vill.) Dum., Bull. Soc. Roy. Bot. Belg. 7:68 (1868)
: = Aven. setaceum (Vill.) Vierh.
: = H. setaceum (Vill.) Henrard
: = Arrhenatherum setaceum (Vill.) Potztal, Bot. Jahrb. 75:330 (1951)

Avenula sulcata (J. Gay) Dum., Bull. Soc. Roy. Bot. Belg. 7:68 (1868)
: = H. sulcatum (J. Gay) Henrard
: = Aven. sulcatum (J. Gay) Vierh.

Avenochloa adzharica (Alboff) Holub, Acta Hort. Bot. Prag. 1962:83 (1962)
Avenochloa aetolica (K.H. Rechinger) Holub, loc. cit.
Avenochloa agropyroides (Boiss.) Holub, loc. cit.
Avenochloa albinervis (Boiss.) Holub, loc. cit.
Avenochloa alpina (Smith) Holub, loc. cit.
Avenochloa agraea (Boiss.) Holub, loc. cit.
Avenochloa armeniaca (Schischk.) Holub, loc. cit.
Avenochloa asiatica (Roshev.) Holub, loc. cit.
Avenochloa blavii (Aschers. et Janka) Holub, loc. cit.
Avenochloa breviaristata (Barratte) Holub, loc. cit.
Avenochloa bromoides (Gouan) Holub, loc. cit.
Avenochloa cincinnata (Ten.) Holub, loc. cit.
Avenochloa compressa (Heuffel) Holub, loc. cit.
Avenochloa cycladum (K.H. Rechinger et Scheffer) Holub, loc. cit.
Avenochloa dahurica (Komar.) Holub, loc. cit.
Avenochloa gonzaloi (St.-Yves) Holub, loc. cit.
Avenochloa hackelii (Henriq.) Holub, loc. cit.
Avenochloa hookeri (Scribn.) Holub, loc. cit.
Avenochloa jahandiezii (Litard.) Holub, loc. cit.
Avenochloa letourneuxii (Trab.) Holub, loc. cit.
Avenochloa levis (Hackel) Holub, loc. cit.
Avenochloa planiculmis (Schrad.) Holub, loc. cit.
Avenochloa pratensis (L.) Holub, loc. cit.
Avenochloa pruinosa (Hackel et Trab.) Holub, loc. cit.
Avenochloa pubescens (Huds.) Holub, loc. cit.
Avenochloa requienii (Mutel) Holub, loc. cit.
Avenochloa schelliana (Hackel) Holub, loc. cit.
Avenochloa sulcata (J. Gay) Holub, loc. cit.
Avenochloa taurica (Prokudin) Holub, loc. cit.
Avenochloa versicolor (Vill.) Holub, loc. cit.

Danthonia strigosa (Schreb.) Beauv., Agrostogr. 92 (1812)
: = A. strigosa Schreb.

Danthoniastrum compactum (Boiss. et Heldr.) Holub, Folia Geobot. Phytotox. 5:436 (1970)

Helictotrichon abietetorum (Ohwi) Ohwi, Acta Phytotax. Geobot. 6:151 (1937)

Helictotrichon adzaricum (Alboff) Grossh., Trudy Bot. Inst. Azerbaidzh. Fil. Akad. Nauk S.S.S.R. 8:214 (1939)
: = Aveno. adzharica (Alboff) Holub

Helictotrichon aetolicum (K.H. Rechinger) Holub, Preslia 31:50 (1959)
: = Aveno. aetolica (K.H. Rechinger) Holub

Helictotrichon agropyroides (Boiss.) Henr., Blumea 3:430 (1940)

= Aveno. agropyroides (Boiss.) Holub
Helictotrichon albinerve (Boiss.) Henr., op. cit. 429
   = Aveno. albinervis (Boiss.) Holub
Helictotrichon alpinum (Smith) Henr., op. cit. 431
   = Aveno. alpinum (Smith) Holub
Helictotrichon altaicum Tzvel., Pl. As. Orient. 4:101 (1968)
Helictotrichon altius (Hitchc.) Ohwi, Jour. Jap. Bot. 17:440 (1941)
Helictotrichon angustum C.E. Hubbard, Kew Bull. 1936:330 (1936)
Helictotrichon argaeum (Boiss.) Parsa, Fl. Iran 5:641 (1951)
   = Aveno. argaea (Boiss.) Holub
Helictotrichon armeniacum (Schischk.) Grossh., Trudy Bot. Inst. Azerbaidzh. Fil. Akad. Nauk S.S.S.R. 8:214 (1939)
   = Aveno. armeniaca (Schischk.) Holub
Helictotrichon asiaticum (Roshev.) Grossh., op. cit. 215
   = Aveno. asiatica (Roshev.) Holub
Helictotrichon asperum (Munro ex Thwait.) Bor, Ind. For. Rec. N.S.I. Bot. 68 (1938)
Helictotrichon australe (Parlat.) Holub in Nemec et al., Opiz et Pflanzentaxonom. 126 (1958)
Helictotrichon avenoides (Stapf) Camus, Rev. Bot. Appl. 27:276 (1947)
Helictotrichon barbatum (Nees) Schweickerdt, Bothalia 3:190 (1937)
Helictotrichon besseri (Griseb.) Janchen, Phyton 5:64 (1953)
Helictotrichon blavii (Aschers. et Janka) C.E. Hubbard, Kew Bull. 1939:101 (1939)
   = Aveno. blavii (Aschers. et Janka) Holub
Helictotrichon breviaristatum (Barratte ex Batt. et Trab.) Henrard, Blumea 3:430 (1940)
   = Aveno. breviaristata (Barratte ex Batt. et Trab.) Holub
Helictotrichon bromoides (Gouan) C.E. Hubbard, Kew Bull. 1939:101 (1939)
   = Aveno. bromoides (Gouan) Holub
Helictotrichon bulbosum (Hitchc.) Parodi, Revist. Argentina Agron. 16(4):211 (1949)
Helictotrichon burmanicum Bor, Kew Bull. 1951:445 (1952)
Helictotrichon capense Schweickerdt, Bothalia 3:193 (1937)
Helictotrichon cartilagineum C.E. Hubbard, Kew Bull. 1936:331 (1936)
Helictotrichon compactum (Boiss. et Heldr.) Henr., Blumea 3:340 (1940)
Helictotrichon compressum (Heuff.) Henr., op. cit. 429
   = Aveno. compressa (Heuff.) Holub
Helictotrichon conjugens (Hack.) Widder, Neue Beitr. Fl. Komit. Vas (Eisenburg) 3, 1:7-8 (1932)
Helictotrichon convolutum (Presl) Henrard, Blumea 3:430 (1940)
Helictotrichon cycladum (K.H. Rechinger et Scheff.) K.H. Rechinger, Denkschr. Akad. Wiss. Math.-Naturw. Wien 105:791 (1943)
   = Aveno. cycladum (K.H. Rechinger et Scheff.) Holub
Helictotrichon dahuricum (Komarov) Kitagawa, Rep. Inst. Sc. Research Manchoukuo 3, App. 1:77 (1939)
   = Aveno. dahurica (Komarov) Holub

Helictotrichon decorum (Janka) Henr., Blumea 3:430 (1940)
Helictotrichon delavayi (Hack.) Henr., op. cit. 427
Helictotrichon desertorum (Less.) Pilger, Repert. Sp. Nov. 45:7 (1938)
Helictotrichon dodii (Stapf) Schweickerdt, Bothalia 3:197 (1937)

Helictotrichon elongatum (Hochst. ex A. Rich.) C.E. Hubbard, Kew Bull. 1936:335 (1936)
Helictotrichon fedtschenkoi (Hack.) Henr., Blumea 3:429 (1940)
Helictotrichon filifolium (Lag.) Henr., op. cit. 430
Helictotrichon friesiorum (Pilger) C.E. Hubbard, Kew Bull. 1936:333 (1936)
Helictotrichon galpinii Schweickerdt, Bothalia 3:192 (1937)
Helictotrichon gonzaloi (Senn.) Potztal, Bot. Jahrb. Engler 75:331 (1951)
    = Aveno. gonzaloi (Senn.) Holub
Helictotrichon hackelii (Henriq.) Henrard, Blumea 3:430 (1940)
    = Aveno. hackelii (Henriq.) Holub
Helictotrichon hideoi (Honda) Ohwi, Acta Phytotax. et Geobot. Kyoto 6:292 (1937)
Helictotrichon hirtulum (Steud.) Schweickerdt, Bothalia 3:193 (1937)
Helictotrichon hissaricum (Roshev.) Henr., Blumea 3:431 (1940)
Helictotrichon hookeri (Scribn.) Henr., op. cit. 429
    = Aveno. hookeri (Scribn.) Holub
Helictotrichon humbertii (Camus) Henr., op. cit. 429
Helictotrichon jahandiezii (Litard.) Potztal, Bot. Jahrb. Engler 75:330 (1951)
    = Aveno. jahandiezii (Litard.) Holub
Helictotrichon junghuhnii (Buese) Henr., Blumea 3:425 (1940)
Helictotrichon krylovii (Pavl.) Henr., op. cit. 431
Helictotrichon lachnanthum (Hochst. ex A. Rich.) C.E. Hubbard, Kew Bull. 1936:335 (1936)
Helictotrichon laeve (Hack.) Potztal, Bot. Jahrb. Engler 75:331 (1951)
Helictotrichon laevigatum (Schur) Potztal, op. cit. 330
Helictotrichon leianthum (Keng) Ohwi, Jour. Jap. Bot. 17:440 (1941)
Helictotrichon leoninum (Steud.) Schweickerdt, Bothalia 3:191 (1937)
Helictotrichon letourneuxi (Trab.) Henr., Blumea 3:430 (1940)
    = Aveno. letourneuxii (Trab.) Holub
Helictotrichon longifolium (Nees) Schweickerdt, Bothalia 3:195 (1937)
Helictotrichon longum (Stapf) Schweickerdt, op. cit. 189
Helictotrichon macrostachyum (Bal. ex Coss. et Dur.) Henr., Blumea 3:430 (1940)
    = A. macrostachya Bal. ex Coss. et Dur.
Helictotrichon maitlandii C.E. Hubb. ex Hutch. in Hutch. et Dalziel, Fl. West. Trop. Afr. 2:528 (1926)
Helictotrichon mannii (Pilger) C.E. Hubb. ex Hutch., op. cit. 528
Helictotrichon milanjianum (Rendle) C.E. Hubbard, Kew Bull. 1936:334 (1936)
Helictotrichon mongolicum (Roshev.) Henr., Blumea 3:431 (1940)
Helictotrichon montanum (Vill.) Henr., op. cit. 430
Helictotrichon mortonianum (Scribn.) Henrard, op. cit. 429
Helictotrichon namaquense Schweickerdt, Bothalia 3:189 (1937)
Helictotrichon natalense (Stapf) Schweickerdt, op. cit. 194
Helictotrichon neumayerianum (Vis.) Henr., Blumea 3:430 (1940)
Helictotrichon newtonii (Stapf) C.E. Hubbard, Kew Bull. 1936:334 (1936)
Helictotrichon oligostachyum (Munro) Holub, Preslia 31:50 (1959)
Helictotrichon parlatorei (Woods) Pilger, Repert. Sp. Nov. 45:7 (1938)
Helictrotrichon parviflorum (Hook. f.) Bor, Kew Bull. 1951:445 (1952)
Helictotrichon petzense Melzer, Oesterr. Bot. Zeitschr. 114:307 (1967)
Helictotrichon phaneroneuron C.E. Hubbard, Kew Bull. 1936:332 (1936)

Helictotrichon planiculme (Schrad.) Pilger, Repert. Sp. Nov. Fedde 45:6 (1938)

Helictotrichon planiculme (Schrad.) Besser ex Rchb., Fl. Germ. Excurs. 140, 6 (1830)
= Aveno. planiculmis (Schrad.) Holub

Helictotrichon polyneurum (Hook. f.) Henr., Blumea 3:425 (1940)

Helictotrichon pratense (L.) Pilger, Repert. Sp. Nov. Fedde 45:6 (1938)
= Aveno. pratensis (L.) Holub

Helictotrichon pruinosum (Hack. ex Trab.) Henr., Blumea 3:430 (1940)
= Aveno. pruinosa (Hack. ex Trab.) Holub

Helictotrichon pubescens (Huds.) Pilger, Repert. Sp. Nov. 45:6 (1938)
= Aveno. pubescens (Huds.) Holub

Helictotrichon quinquesetum (Steud.) Schweickerdt, Bothalia 3:188 (1937)

Helictotrichon requienii (Mutel) Henr., Blumea 3:430 (1940)
= Aveno. requienii (Mutel) Holub

Helictotrichon rigidulum (Pilger) C.E. Hubbard, Kew Bull. 1936:335 (1936)

Helictotrichon roylei (Hook. f.) Keng, Claves Gen. et Spec. Gram. Sinic. 200 (1957)

Helictotrichon sarracenorum (Gandoger) Holub in Nemec et al., Opiz. et Pflanzentaxon. 124 (1958)

Helictotrichon schellianum (Hack.) Kitagawa, Rep. Inst. Sc. Research Manchoukuo 3, App. 1:78 (1939)
= Aveno. schelliana (Hack.) Holub

Helictotrichon schmidii (Hook. f.) Henr. Blumea 3:427 (1940)

Helictotrichon sedenensis (DC.) Holub, Folia Geobot. Phytotox. 5:436 (1970)

Helictotrichon sempervirens (Vill.) Pilger, Repert. Sp. Nov. Fedde 45:7 (1938)

Helictotrichon setaceum (Vill.) Henr., Blumea 3:430 (1940)

Helictotrichon suffuscum (Hitchc.) Ohwi, Jour. Jap. Bot. 17:440 (1941)

Helictotrichon sulcatum (J. Gay) Henr., Blumea 3:430 (1940)
= Aveno. sulcata (J. Gay) Holub

Helictotrichon sumatrense Ohwi, Bull. Tokyo Sci. Mus. No. 18:7 (1947)

Helictotrichon tauricum Prokudin, Trudy. Nauch.-Issl. Inst. Biol. Charkov. Univ. 13:198 (1950)
= Aveno. taurica (Prokudin) Holub

Helictotrichon tentoense (Honda) Kitagawa, Rep. Inst. Sc. Research Manchoukuo 2:282 (1938)

Helictotrichon thomasii C.E. Hubbard, Kew Bull. 1936:500 (1936)

Helictotrichon tianschanicum (Roshev.) Henr., Blumea 3:429 (1940)

Helictotrichon tibesticum (de Miré et Quezel) Holub, Acta Univ. Carol. Biol. 1962:155 (1962)

Helictotrichon tibeticum (Roshev.) Keng f. in Keng, Claves Gen. et Spec. Gramin. Sinic. 200 (1957)

Helictotrichon trisetoides Kitagawa, Rep. Inst. Sc. Research Manchoukuo 2:282 (1938)

Helictotrichon turgidulum (Stapf) Schweickerdt, Bothalia 3:196 (1937)

Helictotrichon umbrosum (Hochst. ex Steud.) C.E. Hubbard, Kew Bull. 1936:334 (1936)

Helictotrichon versicolor (Vill.) Pilger, Repert. Sp. Nov. Fedde 45:7 (1938)
= Aveno. versicolor (Vill.) Holub

Helictotrichon virescens (Nees ex Steud.) Henr., Blumea 3:425 (1940)

Heuffelia australis (Parl.) Fourr., Ann. Soc. Linn. Lyon, N.S. 17:183 (1869)
= H. australe (Parl.) Holub
Heuffelia bromoides (Gouan) Schur, Enum. Pl. Transs. 762 (1866)
= H. bromoides (Gouan) Hubbard
= Aveno. bromoides (Gouan) Holub
Heuffelia compressa (Heuff.) Schur, loc. cit.
= H. compressum (Heuff.) Henrard
= Aveno. compressa (Heuff.) Holub
Heuffelia convoluta (Presl) Schur, op. cit. 763
= H. convolutum (Presl) Henrard
Heuffelia laevigata (Schur) Schur, op. cit. 761
= H. laevigatum (Schur) Potztal
Heuffelia lucida (Bert.) Schur, op. cit. 761
= ? Aveno. pubescens (Huds.) Holub
Heuffelia montana (Vill.) Fourr., Ann. Soc. Linn. Lyon, N.S. 17:183 (1869)
= H. montanum (Vill.) Henrard
Heuffelia planiculmis (Schrad.) Schur, Enum. Pl. Transs. 762 (1866)
= H. planiculme (Schrad.) Pilger
= Aveno. planiculmis (Schrad.) Holub
Heuffelia praeusta (Rchb.) Schur, op. cit. 762
= H. alpinum (Smith) Henrard
Heuffelia pratensis (L.) Schur, op. cit. 762
= H. pratense (L.) Pilger
= Aveno. pratensis (L.) Holub
Heuffelia pubescens (Huds.) Schur, op. cit. 760
= H. pubescens (Huds.) Pilger
= Aveno. pubescens (L.) Holub
Heuffelia scheuchzeri (All.) Schur, op. cit. 763
= H. versicolor (Vill.) Pilger
Heuffelia sempervirens (Vill.) Schur, op. cit. 763
= H. sempervirens (Vill.) Pilger
Heuffelia setacea (Vill.) Schur, op. cit. 763
= H. setaceum (Vill.) Henrard
Preissia strigosa (Schreb.) Opiz, Seznam Rostl. České 79 (1852)
= A. strigosa Schreb.

Literature Cited

This list provides bibliographic details for all literature cited, except for the nomenclatural literature citations, which are given in the individual synonymies of Part 2.

Anderson, E. 1952. Plants, man and life. Little Brown, Boston. 245 pp.
Anderson, T. S., and Bertelsen, F. 1972. Scanning electron microscope studies of pollen of cereals and other grasses. Grana 12:79-86.
Ascherson, P., and Graebner, P. 1899. Synopsis der Mitteleuropäischen Flora pp. 233, 234.
Baker, E. A. and Parsons, E. 1971. Scanning electron microscopy of plant cuticles. J. Microsc. (Oxf.) 94:39-49.

Bartcher, R. L. 1966. FORTRAN IV program for estimation of cladistic relationships using the IBM 7040. Computer Contribution No 6, State Geol. Surv., Univ. of Kans.

Bauhin, C. 1671. Pinax theatri botanici. Ludovici Regis, Basel. Reprint of the 1623 issue. p. 23.

Baum, B. R. 1968a. Delimitation of the genus *Avena*. Can. J. Bot. 46:121-132.

Baum, B. R. 1968b. On some relationships between *A. sativa* and *A. fatua* (Gramineae) as studied from Canadian material. Can. J. Bot. 46:1013-1024.

Baum B. R. 1969a. The role of the lodicule and epiblast in determining natural hybrids of *Avena sativa* × *fatua* in cultivated oats. Can. J. Bot. 47:85-91.

Baum, B. R. 1969b. The use of lodicule type in assessing the origin of *Avena* fatuoids. Can. J. Bot. 47:931-944.

Baum, B. R. 1970. The problem of classifying cultivars with special emphasis on oat (*Avena*) cultivars. Can. J. Bot. 48:1373-1381.

Baum, B. R., and Thompson, B. K. 1970. Registers with pedigree charts for cultivars: their importance, their contents, and their preparation by computer. Taxon 19:762-768.

Baum, B. R. 1971a. Additional taxonomic studies on *Avena* fatuoids: some morphological attributes seen using the scanning electron microscope. Can. J. Bot. 49:647-649.

Baum, B. R. 1971b. Organophyletic trends in several micromorphological floral traits in the hexaploid cultivated oats *(Avena)*. Evolution 25:235-241.

Baum, B. R. 1972. *Avena septentrionalis,* and the semispecies concept. Can. J. Bot. 50:2063-2066.

Baum, B. R., and Lefkovitch, L. P. 1972a. A model for cultivar classification and identification with reference to oats (*Avena*). I. Establishment of the groupings by means of taximetric methods. Can. J. Bot. 50:121-130.

Baum, B. R., and Lefkovitch, L. P. 1972b. A model for cultivar classification and identification with reference to oats (*Avena*). II. A probabilistic definition of cultivar classes in oats and their Bayesian identification. Can. J. Bot. 50:131-138.

Baum, B. R., Rajhathy, T., Fleischmann, G., Martens, J. W., and Thomas, H. 1972. Wild oat gene pool. (A collection maintained by Can. Dep. Agric.) Canada Avena (CAV). Res. Branch Publ. 1475.

Baum, B. R. 1973a. Material for an international oat register. Inf. Can. Cat. No. A52-4772. Ottawa. 266 pp.

Baum, B. R. 1973b. The genus *Danthoniastrum,* about its circumscription, past and present status, and some taxonomic principles. Österr. Bot. Z. 122:51-57.

Baum, B. R. 1973c. Extrapolation of the predomesticated hexaploid cultivated oats. Evolution 27:518-623.

Baum, B. R., and Lefkovitch, L. P. 1973. A numerical taxonomic study of phylogenetic and phenetic relationships in some cultivated oats, using known pedigrees, Syst. Zool. 22:118-131.

Baum, B. R. 1974. Article 36 and numerical classification. Taxon 23:652-653.

Baum, B. R. 1975. Classification of the oat species (*Avena,* Poaceae) using various taximetric methods and an information theoretic model. Can. J. Bot. 52:2241-2262.

Baum, B. R., and Hadland, V. E. 1975. A scanning electron microscopic study of epicuticular waxes of glumes in *Avena magna, A. murphyi* and *A. sterilis.* Can. J. Bot. 51:2381-2383.

Baum, B. R., and Brach, E. J. 1975. Identification of oat cultivars by means of fluorescence spectrography—a pilot study aimed at automatic identification of cultivars. Can. J. Bot. 53:305-309.

Baum, B. R., Rajhathy, T., Martens, J. W., and Thomas, H. 1975. Wild oat gene pool. (A collection maintained by Can. Dep. Agric.) Canada Avena (CAV). Res. Branch Publ. No. 1475 (2nd edition). 7 + 100 pp.

Baum B. R., and Thompson, B. K. 1976. Classification of Canadian oat cultivars by quantifying the size-shape of their 'seeds': a step toward automatic identification. Can. J. Bot. 54:1472–1480.

Besser, W. S. J. G. 1826–1827. Page 526 *in* Schultes, J. A., and Schultes, J. H. Additamentum I. ad Mantissam III. Classis III. Stuttgart, Cotta. 717 pp.

Berlin, B., Breedlove, D. E., and Raven, P. H. 1974. Principles of tzeltal plant classification. An introduction to the botanical ethnography of a Mayan-speaking people of Highland Chiapas. Academic Press, New York. 660 pp.

Bonnett, O. T. 1961. The oat plant: its histology and development. Bull. 672. Univ. of Ill., Agric. Exp. Stn. Urbana, Ill. 112 pp.

Bor, N. L. 1960, The Grasses of Burma, Ceylon, India and Pakistan. Pergamon Press, New York. 767 p.

Bose, S. 1972. Comparison of wheat varieties on the basis of their relative phenolic contents. Hereditas 72:159-168.

Brach, E., and Baum, B. R. 1975. Identification of cultivars by fluorescence spectroscopy. Appl. Spectroscopy. 29:326-333.

Bretschneider, E. 1881. Botanicon Sinicum, Notes on Chinese botany from native and Western sources. J. North-China Branch R. As. Soc. N. S. 16:18-229.

Brouwer, W. 1971. Der Hafer. Pages 386-485 *in* Brouwer, W., Handbuch des Speziellen Pflanzenbaues. Paul Parrey, Berlin.

Cabezas, M., Navarro-Andres, F., and Cabezas, J. A. 1972. Estudio comparativo de las proteinas de cariopsides de algunos cereales por electroforesis de disco. Rev. Esp. Fisiol. 28:231-234.

Camin, J. H., and Sokal, R. R. 1965. A method for deducing branching sequences in phylogeny. Evolution 19:311-326.

Cannon, W. A. 1900. A morphological study of the flower and embryo of the wild oat, *Avena fatua* L. Proc. Calif. Acad. Sci. Third Ser. Bot. 1:329-364.

Chase, A., and Niles, C. D. 1962. Index to Grass Species. Hall, Boston. 3 vols.

Chu, Y. E., Chou, T. S., Li, Y. S., Shih, C. Y., and Woo, S. C. 1972. Identification of bamboo clones, *Dendrocalamus latiflorus* in Taiwan. Bot. Bull. Acad. Sin. (Taipei). 13:11-18.

Coffman, F. A. 1946. Origin of cultivated oats. J. Amer. Soc. Agron. 38:983-1002.

Coffman, F. A. 1964. Inheritance of morphologic characters in *Avena*. Techn. Bull. No. 1308. ARS. USDA. Washington. 101 p.

Coffman, F. A., ed. 1961. Oats and oat improvement. Madison, Wisconsin, Amer. Soc. Agron. 650 pp.

Coffman, F. A., and MacKey, J. 1959. Hafer. Pages 427-531 *in* Roemer, T., and Rudorf, W., Handbuch der Pflanzenzüchtung. 2nd ed. Paul

Parrey, Berlin.
Cosson, M. E. 1854. Classification des espèces du genre *Avena* du groupe de l'*Avena sativa* (Avena sect. Avenatypus), et considérations sur la composition et la structure de l'épillet dans la famille des graminées. Bull. Soc. Bot. Fr. 1:11-17.
Craig, I. L., Murray, B. E., and Rajhathy, T. 1972. Leaf esterase isozymes in *Avena* and their relationship to the genomes. Can. J. Genet. Cytol. 14:581-589.
Craig, I. L., Murray, B. E., and Rajhathy, T. 1974. *Avena canariensis*: morphological and electrophoretic polymorphism and relationship to the *A. magna–A. murphyi* complex and *A. sterilis*. Can. J. Genet. Cytol. 16:677-689.
De Candolle, A. 1882. Origine des plantes cultivées. Paris. viii + 377 pp.
Dixon, W. J., ed. 1970. Biomedical computer programs. Sch. of Med. U.C.L.A. p. 214
Ellis, R. P. 1971. The identification of wheat varieties by the electrophoresis of grain proteins. J. Natl. Inst. Agric. Bot. 12:223-235.
Estabrook, G. F. 1967. An information theory model for character analysis. Taxon 16:86-97.
Estabrook, G. F. 1972. Cladistic methodology: a discussion of the theoretical basis for the induction of evolutionary history. Annu. Rev. Ecol. Syst. 3:427-456.
Etheridge, W. C. 1916. A classification of the varieties of cultivated oats. N.Y. Agric. Exp. Stn. Ithaca. Mem. 10.
Faegri, K., and Iversen, J. 1964. Textbook of pollen analysis. 2nd ed. Blackwell, Oxford. 237 pp.
Farris, J. S. 1969. A successive approximations approach to character weighting. Syst. Zool. 18:374-385.
Farris, J. S. 1970. Methods for computing Wagner trees. Syst. Zool. 19:83-92.
Faust, M., and Shear, C. B. 1972. Fine structure of the fruit surface of Apple cultivars. Hortic. Sci. 7:44.
Gervais, C. 1973. Contribution à l'étude cytologique et taxonomique des avoines vivaces. Mem. Soc. Helv. Sci. Nat. 88:1-166.
Gilbert, E. S. 1968. On discrimination using qualitative variables. J. Am. Stat. Assoc. 63:1399-1412.
Good, R. 1974. The geography of the flowering plants. 4th ed. Longman, London.
Gower, J. C. 1966. Some distance properties of latent root and vector methods used in multivariate analysis. Biometrika 53:325-338.
Gower, J. C. 1967. Multivariate analysis and multidimensional geometry. The Statistician 17:13-28.
Gower, J. C. 1971. A general coefficient of similarity and some of its properties. Biometrics 27:857-871.
Gower, J. C., and Ross, G. J. S. 1969. Minimum spanning trees and single linkage cluster analysis. Appl. Stat. 18:54-64.
Grisebach, A. 1853. Pages 1842-1853 *in* Ledebour, K. F. von. Flora rossica, Stuttgart, Schweizerbart. Vol. 4.
Hall, D. M. 1967. The ultrastructure of wax deposits on plant leaf surfaces. II. Cuticular pores and wax formation. J. Ultrastruct. Res. 17:34-44.
Hallam, N. D., and Chambers, T. C. 1970. The leaf waxes of the genus *Eucalyptus* L'Heritier. Aust. J. Bot. 18:335-386.

Haussknecht, C. 1885. Über die Abstammung des Saathabers. Mitteil. Geogr. Ges. Thür. Jena 3:231-242.

Hawkes, J. G., and Hjerting, J. P. 1969. The Potatoes of Argentina, Brazil, Paraguay, and Uruguay. A biosystematic study. Clarendon Press, Oxford. 525 pp. 150 plates.

Hawksworth, F. G., Estabrook, G. F., and Rogers, D. J. 1968. Application of information theory model for character analysis in the genus *Arceuthobium* (Viscaceae). Taxon 17:605-619.

Hennig, W. 1950. Grundzüge einer theorie des phylogenetischen Systematik. Dtsch. Zentr., Berlin. 379 pp.

Hennig, W. 1966. Phylogenetic systematics. Urbana, Univ. Ill.

Henrard, J. Th. 1940. Notes on the nomenclature of some grasses. Blumea 3:425-431.

Holden, J. H. W. 1966. Species relationships in Avenae. Chromosoma 20:75-124.

Holub, J. 1958. Bemerkungen zur taxonomie der Gattung *Helictotrichon* Bess. Pages 101-133 *in* J. Klasterksy et coll., eds. Philipp Maximilian Opiz und seine Bedeutung für die Pflanzentaxonomie. Praha. Tschech. Akad. Wissenschaft.

Holub, J. 1962. Ein Beitrag zur Abgrenzung der Gattungen in der Tribus *Aveneae*: Die Gattung *Avenochloa* Holub. Acta Horti Bot. Pragensis 1962:75-86.

Hubbard, C. E. 1936. The species of *Helictotrichon* in Tropical Africa. Kew Bull. 1936:330-335.

Hubbard, C. E. 1937. *Helictotrichon* Bess. ex Schult. and *Avena* L. Pages 103-123 *in* A. W. Hill, ed. Flora of tropical Africa 10(1). Reeve and Co.

Huskins, C. L. 1946. Fatuoid, speltoid and related mutations of oats and wheat. Bot. Rev. 12:457-514.

Husnot, T. 1897. Graminées. Descriptions, Figures et Usages des Graminées Spontanées et Cultivées de France, Belgique, Iles Britanniques, Suisse. Cahan. 92 p. 33 pl.

Jain, S. K., and Marshall, D. R. 1967. Population studies in predominantly self-pollinating species. X. Variation in natural populations of *Avena fatua* and *A. barbata*. Am. Nat. 101:19-33.

Jain, S. K., and Singh, R. S. In press. Population biology of *Avena*. VI. Allozyme variation in relation to the genome analysis in the genus *Avena* Am. J. Bot.

Jardine, N., and Sibson, R. 1971. Mathematical Taxonomy. London, J. Wiley and Sons Ltd. App. 5:238-239.

Jessen, K., and Helbaek, H. 1944. Cereals in Great Britain and Ireland in Prehistoric and Early Historic Times. Kgl. Dan. Vidensk. Selsk. Biol. Skr. 3(2):1-68.

Jones, E. T. 1930. Morphological and genetical studies of fatuoid and other aberrant grain types in *Avena*. J. Genet. 23:1-68.

Kendrick, W. B., and Proctor, J. R. 1964. Computer taxonomy in the fungi imperfecti. Can. J. Bot. 42:65-88.

Ladizinski, G., and Zohary, D. 1971. Notes on species delimitation, species relationships and polyploidy in *Avena*. Euphytica 20:380-395.

Lance, G. N., and Williams, W. T. 1967. A general theory of classificatory sorting strategies. I. Hierarchical systems. Comput. J. 9:373-380.

Lanjouw, J., and Stafleu, F. A. 1964. Index Herbariorum. Part I. The Herbaria of the World. Utrecht. 251 pp.

Legendre, P., and Rogers, D. J. 1972. Characters and clustering in taxonomy: a synthesis of two taximetric methods. Taxon 21:567-606.
Linnaeus, C. 1753. Species Plantarum. A facsimile of the first edition. Ray Society, London. 1957, 1959.
Linnaeus, C. 1754. Genera Plantarum, 5th ed. Facsimile. With an introduction by W. T. Stearn. Ray Society, New York. 1960.
Maire, R. 1953. Flore de l'Afrique du Nord. Lechevalier, Paris. 2:264-312.
Maling, D. H. 1973. Coordinate Systems and Map Projections. Philip and Son, London.
Malzew, A. I. 1929. Novaya systema sect. Eu Avena Griseb. Bull. Appl. Bot. Genet. Plant Breed. 20:127-149.
Malzew, A. I. 1930. Wild and cultivated oats Section Eu Avena Griseb. Bull. Appl. Bot. Genet. Plant Breed. Suppl. 38th. 522 pp.
Marshall, D. R., and Allard, R. A. 1970. Isozyme polymorphisms in natural populations of *Avena fatua* and *A. barbata*. Heredity 29:373-382.
Martin, J. T., and Juniper, B. E. 1970*a*. The cuticles of plants. Edward Arnold. London. 347 pp.
Martin, J. T., and Juniper, B. E. 1970*b*. The cuticles of plants. Edward Arnold, London. p. 265.
McKee, G. W. 1973. Chemical and biochemical techniques for varietal identification. Seed Sci. Technol. 1:181-199.
Metcalfe, C. R. 1960. Anatomy of the Monocotyledons. I. Gramineae. Oxford.
Murray, B. E., Craig, I. L., and Rajhathy, T. 1970. A protein electrophoretic study of three amphiploids and eight species in *Avena*. Can. J. Genet. Cytol. 12:651-655.
Nakao, S. 1950. On the Mongolian naked oats, with special reference to their origins. Sci. Rep. of the Fac. of Agric. Naniwa Univ. No 1:7-24.
Nevski, S. A. 1934. Herbarium Florae Asiae Mediae ab Universitate Asiae Mediae editum. Fasc. XXI, N° 5036. *Avena ludoviciana* Dur.
Nilsson-Ehle, H. 1907. Om hafresorters constans. Sver. Utsadesfören. Tidskr. 1907:227-239.
Nilsson-Ehle, H. 1911. Über Falle spontanen Wegfallens eines Hemmungs faktors beim Hafer. Z. Indukt. Abstammungs.-Vererbungsl. 5:1-37.
Nilsson-Ehle, H. 1914. Uber einen als Hemmungsfaktor der Begrannung auftretenden Farbenfaktor beim Hafer. Z. Indukt. Abstammungs.-Vererbungsl. 12:36-55.
Nilsson-Ehle, H. 1921. Fortgesetzte Untersuchungen uber Fatuoidmutationen beim Hafer. Hereditas 2:401-409.
O'Mara, J. G. 1961. Cytogenetics. Vol. 8 *in* F. A. Coffman, ed. Oats and oat improvement. Am. Soc. Agron. Madison.
Page, C. N. 1974. Morphology and affinities of *Pinus canariensis*. Notes R. Bot. Gard. Edinb. 33:317-323.
Pankhurst, R. J. 1970. A computer program for generating diagnostic keys. Comput. J. 12:145-151.
Paunero, E. 1959. *Helictotrichon* Bess., *in* Las Aveneas espanolas. IV. An. Jard. Bot. Madrid. 17:257-290.
Pilger, R. 1954. Das System der Gramineae. Bot. Jahrb. 76:281-384.
Plarre, W. 1970. Hafer. Pages 134-157 *in* Hoffman, W., Mudra, A., and Plarre, W., Lehrbuch der Züchtung landwirtschaftlicher Kulturpflanzen. Paul Parrey, Berlin. Vol. 2.

Potztal, E. 1951. Anatomisch-systematische Untersuchungen an den Gattungen *Arrhenatherum* und *Helictotrichon*. Bot. Jahrb. 75:321-332.
Poushinsky, G., and Sheldrake, P. A space saving cluster listing method. Statistical Res. Service CDA Ottawa. Mimeographed ms.
Prim, R. C. 1957. Shortest connection matrix network and some generalizations. Bell Syst. Tech. J. 36:1389-1401.
Rajhathy, T., and Sadasivaiah, R. S. 1969. The cytogenetic status of *Avena magna*. Can. J. Genet. Cytol. 11:77-85.
Rajhathy, T., and Thomas, H. 1974. Cytogenetics of oats (*Avena* L.). Misc. Publ. No. 2, Genet. Soc. of Can. 90 pp.
Rao, C. R. 1952. Advanced statistical methods in biometric research. Wiley, New York.
Raven, P. H., Berlin B., and Breedlove, D. E. 1971. The origins of taxonomy. Science 174:1210-1213.
Reichenbach, H. G. L. 1830. Flora germanica excursiora. Knobloch, Leipzig. 878 pp. 1830-1832.
Richter, H. E. 1840. Codex Botanicus Linnaeanus. Wigand, Leipzig. 1101 pp.
Rikli, M. 1943–1948. Das pflanzenkleid der Mittlemeerlander. Huber, Bern. 3 vol.
Rottink, B. A., and Hanover, J. W. 1972. Identification of blue spruce cultivars by analysis of cortical oleoresin monoterpenes. Phytochemistry 11:3255-3257.
Rozhevits, R. Yu. 1934. Gramineae, in Flora of the USSR. Vol. 2. Leningrad. English translation, Israel Progr. Sci. Transl. Jerusalem, 1963.
Sadasivaiah, R. S., and Rajhathy, T. 1968. Genome relationships in tetraploid *Avena*. Can. J. Genet. Cytol. 10:655-669.
St.-Yves, A. 1931. Contribution à l'étude des Avena sect. *Avenastrum* (Eurasie et région Méditerranéenne) Candollea 4:353-504.
Sass, J. E. 1958 Botanical microtechnique. 3rd ed. Ames, Iowa. p. 102.
Schweickderdt, H. G. 1937. A revision of the South African species of *Helictotrichon* Bess. ex Schultes. Bothalia 3:185-203.
Seal, H. 1964. Multivariate statistical analysis for biologists. Methuen, London.
Singh, R. S., and Jain, S. K. 1971. Population biology of *Avena* II. Isozyme polymorphisms in populations of the Mediterranean region and central California. Theor. Appl. Genet. 41:79-84.
Singh, R. S., Jain, S. K., and Qualset, C. O. 1973. Protein electrophoresis as an aid to oat variety identification. Euphytica 22:98-105.
Sneath, P. H. A., and Sokal, R. R. 1973. Numerical taxonomy. The principles and practice of numerical classification. W. H. Freeman, San Francisco. 573 pp.
Stafleu, F. A., Bonner, C. E. B., McVaugh, R., Meikle, R. D., Rollins, R. C., Ross, R., Schopf, J. M., Schulze, G. M., De Vilmorin, R., and Voss, E. 1972. International code of botanical nomenclature. Oosthack Utrecht. 426 pp.
Stanton, T. R. 1955. Oat identification and classification. U.S. Dep. Agric. Tech. Bull. 1100.
Stapf, O. 1899. Gramineae. *In* W. T. Thiselton-Dyer, ed. Flora capensis 7(3):472-480. Ashford.
Takahoshi, O., Tani, T., and Naito, N. 1972. Scanning electron microscopy of crown rust appressorium produced on oat leaf surface. Kagawa Daigaku Nogakubu Gakuzyutu Hokoku 24:42-47 (In Japanese with

English summary).

Thellung, A. 1912. Über die Abstammung, den systematischen Wert und die Kulturgeschichte der Saathafer-Arten (Avenae sativae Cosson): Beiträge zu einer naturlichen systematik von *Avena* sect. Euavena. Vierteljahresschr. Naturforsch. Ges. Zür. 56:293-350.

Thellung, A. 1929. Die Übergangsformen vom Wildhafertypus (Avenae Agrestes) zum Saathafertypus (Avenae sativae) Rec. Trav. Bot. Neerland. 25:416-466.

Tournefort, J. 1700. Institutiones rei herbariae. Paris. 697 pp.

Tribe, I. S., Gaunt, J. K., and Parry, D. W. 1968. Cuticular lipids in the Gramineae. Biochem. J. 109:8-9.

Tulloch, A. P. 1973. Composition of leaf surface waxes of *Triticum* species, variation with age and tissue. Phytochemistry 12:2225-2232.

Tulloch, A. P., and Hoffman, L. L. 1973. Leaf wax of oats. Lipids 8:617-622.

Villaret-Von-Rochow, M. 1971. *Avena Ludoviciana* Dur. in the Late Neolithic of Switzerland, a Contribution to the origin of the oat (*Avena sativa* L.). Ber. Deut. Bot. Ges. 84(5):243-248.

Walker, J. T. 1969. Selection and quantitative characters in field crops. Biol. Rev. (Camb.) 44:207-243.

Werth, E. 1944. Der Hafer, Eine urnordische Getreideart. (Zur Geographie und Geschichte der Kulturpflanzen und Haustiere. XXV). Pflanzenzuecht. 26 (1/2):92-102.

Wettstein-Knowles, P. von. 1972. Genetic control of beta-diketone and hydroxy-beta-diketone synthesis in epicuticular waxes of Barley. Planta (Berl.) 106:113-130.

Wettstein-Knowles, P. von. 1974. Ultrastructure and origin of epicuticular wax tubes. J. Ultrastruct. Res. 46:483-498.

Wilson, E. O. 1974. Ecology, evolution, and population biology. Readings from Scientific American. Freeman, San Francisco. 315 pp.

Zohary, M. 1973. Geobotanical foundations of the Middle East. Fischer, Stuttgart. 739 pp.